The Subnuclear Series • Volume 53

Proceedings of the International School of Subnuclear Physics

THE FUTURE OF OUR PHYSICS INCLUDING NEW FRONTIERS

THE SUBNUCLEAR SERIES*

Series Editor: ANTONINO ZICHICHI, *European Physical Society, Geneva, Switzerland*

21. 1983 HOW FAR ARE WE FROM THE GAUGE FORCES?
22. 1984 QUARKS, LEPTONS, AND THEIR CONSTITUENTS
23. 1985 OLD AND NEW FORCES OF NATURE
24. 1986 THE SUPERWORLD I
25. 1987 THE SUPERWORLD II
26. 1988 THE SUPERWORLD III
27. 1989 THE CHALLENGING QUESTIONS
28. 1990 PHYSICS UP TO 200 TeV
29. 1991 PHYSICS AT THE HIGHEST ENERGY AND LUMINOSITY: To Understand the Origin of Mass
30. 1992 FROM SUPERSTRINGS TO THE REAL SUPERWORLD
31. 1993 FROM SUPERSYMMETRY TO THE ORIGIN OF SPACE–TIME
32. 1994 FROM SUPERSTRING TO PRESENT-DAY PHYSICS
33. 1995 VACUUM AND VACUA: The Physics of Nothing
34. 1996 EFFECTIVE THEORIES AND FUNDAMENTAL INTERACTIONS
35. 1997 HIGHLIGHTS OF SUBNUCLEAR PHYSICS: 50 Years Later
36. 1998 FROM THE PLANCK LENGTH TO THE HUBBLE RADIUS
37. 1999 BASICS AND HIGHLIGHTS IN FUNDAMENTAL PHYSICS
38. 2000 THEORY AND EXPERIMENT HEADING FOR NEW PHYSICS
39. 2001 NEW FIELDS AND STRINGS IN SUBNUCLEAR PHYSICS
40. 2002 FROM QUARKS AND GLUONS TO QUANTUM GRAVITY
41. 2003 FROM QUARKS TO BLACK HOLES: PROGRESS IN UNDERSTANDING THE LOGIC OF NATURE
42. 2004 HOW AND WHERE TO GO BEYOND THE STANDARD MODEL
43. 2005 TOWARDS NEW MILESTONES IN OUR QUEST TO GO BEYOND THE STANDARD MODEL
44. 2006 THE LOGIC OF NATURE, COMPLEXITY AND NEW PHYSICS: FROM QUARK-GLUON PLASMA TO SUPERSTRINGS, QUANTUM GRAVITY AND BEYOND
45. 2007 SEARCH FOR THE "TOTALLY UNEXPECTED" IN THE LHC ERA
46. 2008 PREDICTED AND TOTALLY UNEXPECTED IN THE ENERGY FRONTIER OPENED BY LHC
47. 2009 THE MOST UNEXPECTED AT LHC AND THE STATUS OF HIGH ENERGY FRONTIER
48. 2010 WHAT IS KNOWN AND UNEXPECTED AT LHC
49. 2011 SEARCHING FOR THE UNEXPECTED AT LHC AND THE STATUS OF OUR KNOWLEDGE
50. 2012 WHAT WE WOULD LIKE LHC TO GIVE US
51. 2013 REFLECTIONS ON THE NEXT STEP FOR LHC
52. 2014 STATUS OF THEORETICAL UNDERSTANDING AND OF EXPERIMENTAL POWER FOR LHC PHYSICS AND BEYOND
53. 2015 THE FUTURE OF OUR PHYSICS INCLUDING NEW FRONTIERS

Volume 1 was published by W. A. Benjamin, Inc., New York; 2–8 and 11–12 by Academic Press, New York and London; 9–10 by Editrice Compositori, Bologna; 13–29 by Plenum Press, New York and London; 30–53 by World Scientific, Singapore.

*For the complete list of titles in this series, please go to

http://www.worldscientific.com/series/ss

The Subnuclear Series • Volume 53

Proceedings of the International School of Subnuclear Physics

THE FUTURE OF OUR PHYSICS INCLUDING NEW FRONTIERS

Edited by

Antonino Zichichi

European Physical Society
Geneva, Switzerland

World Scientific

NEW JERSEY • LONDON • SINGAPORE • BEIJING • SHANGHAI • HONG KONG • TAIPEI • CHENNAI • TOKYO

Published by

World Scientific Publishing Co. Pte. Ltd.
5 Toh Tuck Link, Singapore 596224
USA office: 27 Warren Street, Suite 401-402, Hackensack, NJ 07601
UK office: 57 Shelton Street, Covent Garden, London WC2H 9HE

British Library Cataloguing-in-Publication Data
A catalogue record for this book is available from the British Library.

The Subnuclear Series — Vol. 53
THE FUTURE OF OUR PHYSICS INCLUDING NEW FRONTIERS
Proceedings of the 53rd Course of the International School of Subnuclear Physics

ISBN 978-981-3208-28-5

Desk Editor: Rhaimie Wahap

Printed in Singapore

PREFACE

In June/July 2015, a group of 171 physicists from 29 countries met in Erice to participate in the 53$^{\rm nd}$ Course of the International School of Subnuclear Physics. The countries represented by the participants were: Azerbaijan, Belgium, Brazil, China, Cuba, Cyprus, Czech, Denmark, France, Georgia, Germany, Hungary, India, Italy, Japan, Kazakhstan, Mexico, the Netherlands, Norway, Portugal, Russia, Slovak, South Africa, Spain, Switzerland, Turkey, Ukraine, United Kingdom and the United States of America.

The School was sponsored by the Academies of Sciences of Estonia, Georgia, Lithuania, Russia and Ukraine; the Chinese Academy of Sciences; the Commission of the European Communities; the European Physical Society; the Italian National Institute for Nuclear Physics; the Italian Ministry of Education, University and Scientific Research; the Sicilian Regional Government; the Weizmann Institute of Science; the World Federation of Scientists and the World Laboratory.

The main focus of this year's Course was the future of our Physics, including the new frontiers in other fields. In fact, during the last half century the School has been involved in all crucial steps of our Physics. Few examples: SU(3) flavour and SU(6) (with SU(2)-spin coupled with SU(3)-flavour) dismantled by the "No-Go-Theorem", the battle between S-Matrix and Field Theory, the Universality of the weak forces (started with the ε-parameter and the non existence of the "flavour changing-neutral currents" solved by the existence of "charm"), the experimental search for the 3$^{\rm rd}$ lepton in the early sixties before the discovery of the CP breaking, the birth of the Electroweak Unification and SSB (Spontaneous Symmetry Breaking) mechanism, the discovery of the negative sign of the β-function and of asymptotic freedom, the triumph of non Abelian field theories (QCD and QFD) with all consequences (including Instantons), the discovery of Supersymmetry (many years – and not few days – after the "No-Go-Theorem"). Now we would like LHC to give us the first sign for the existence of the Superworld. Finally we should not forget that, since complexity exists at the fundamental level, a totally unexpected discovery should be given to us by LHC or by some other Laboratory engaged at the frontier of our knowledge.

As usual, the lectures dealt with the latest and most significant achievements in both theoretical and experimental Physics. Some of them have been devoted to celebrate the genius of Ettore Majorana and the influence of his discoveries not only in Particle Physics, but also in other fields of Modern Physics such as Solid State Physics and Quantum Information Theory. Moreover, three groups of problems were open for competition: the first group was "Mixings", the second was "Elementary and composite states" and the third one was "Symmetry breakings".

An original feature of the School, introduced in 1996, is a series of special sessions devoted to "New Talents", with the aim of encouraging and promoting young physicists to achieve recognition at an international level. This is a problem in Experimental Physics, where collaborations count several hundreds of participants and it is almost impossible for young fellows to be known. Even if with much less emphasis, the problem exists also in Theoretical Physics. So it was decided to offer 47 young fellows the possibility to give an open presentation of the results of their studies, followed by a discussion. Among them 32 students were awarded a diploma and 2 were given a New Talent Award.

During the organization and the running of the Course, I enjoyed the collaboration of my colleague and friend, Gerardus 't Hooft, who shared with me the Directorship of the School. I would like to thank him, together with the group of invited scientists and all the people who have contributed to the success of the Course.

I hope the reader will enjoy the book as much as the students attending the lectures and the discussion sessions, which, as every year, have been the focal point of the School activity. Thanks to the work of the Scientific Secretaries, the discussions have been reproduced as faithfully as possible. At various stages of my work I have enjoyed the collaboration of many friends whose contributions have been extremely important for the School and are highly appreciated. I thank them most warmly. A final acknowledgement to all those in Erice, Bologna and Geneva who have helped me in so many occasions and to whom I feel very indebted.

Antonino Zichichi

Geneva, October 2015

CONTENTS

HIGHLIGHTS

DISCUSSIONS

CLOSING LECTURE

CLOSING CEREMONY

«ETTORE MAJORANA» FOUNDATION AND CENTRE FOR SCIENTIFIC CULTURE
TO PAY A PERMANENT TRIBUTE TO ARCHIMEDES AND GALILEO GALILEI, FOUNDERS OF MODERN SCIENCE AND TO ENRICO FERMI, THE "ITALIAN NAVIGATOR", FATHER OF THE WEAK FORCES

INTERNATIONAL SCHOOL OF SUBNUCLEAR PHYSICS
THE FUTURE OF OUR PHYSICS INCLUDING NEW FRONTIERS
53rd Course – ERICE-SICILY: 24 JUNE – 3 JULY 2015

OPENING SESSION

THE FUTURE OF OUR PHYSICS

ANTONINO ZICHICHI
University of Bologna and INFN, Italy
CERN, Geneva, Switzerland
Enrico Fermi Centre, Rome, Italy
Pontifical Academy of Sciences, Vatican City
World Federation of Scientists, Beijing, Geneva, Moscow, New York
Ettore Majorana Foundation and Centre for Scientific Culture, Erice, Italy

Thursday, 25 June 2015
Paul A.M. Dirac Lecture Hall – Patrick M.S. Blackett Institute

THE FUTURE OF OUR PHYSICS

INDEX

I PHYSICS PROBLEMS

EFFECTIVE ENERGY
- → Leading Effect
- → Quantum Number Flow
- → The High $P_\perp$ Myth

PHASE TRANSITIONS → EEE
- → LHC
- → AMS

↓ **Quark Gluon Plasma QGP**

↓ **Quark Gluon Coloured World** *

QGCW
n° of states much larger than all known

EFFECTIVE ENERGY
- → Leading Effect
- → Quantum Number Flow
- → The High $P_\perp$ Myth

The "**Effective Energy**"

EVIDENCE OF THE SAME MULTIPARTICLE PRODUCTION MECHANISM IN p-p COLLISIONS AS IN e^+e^- ANNIHILATION

M. Basile, G. Cara Romeo, L. Cifarelli, A. Contin, G. D'Alì, P. Di Cesare, B. Esposito, P. Giusti, T. Massam, F. Palmonari, G. Sartorelli, G. Valenti and A. Zichichi

Physics Letters 92B, 367 (1980).

"*The agreement between the momentum distributions obtained in e^+e^- annihilation and in pp collisions suggests that the mechanism for transforming energy into particles in these two processes, so far considered very different, must be the same*".

The physics of strong interactions was characterized by two classes of phenomena, one of "**static**" nature, the other of "**dynamic**" nature.

Both were affected by proliferation in the most fundamental component of this physics: its elementary particles.

The proliferation in the "**static**" **sector** of the strong interaction was the huge number of mesons and baryons [1].

This multitude of states was reduced by an order of magnitude through the octets and decuplets of Gell-Mann and Ne'eman $SU(3)_f$ [2].

The proliferation in the "dynamic" sector was the multitude of final states produced by pairs of interacting particles, in strong, electromagnetic and weak processes:

Strong	EM	Weak
π p	γ p	ν p
K p	e p	
p p	μ p	
p n	e^+e^-	
$\bar{p}$ p		

It is the introduction of the Effective Energy which allowed one to put all the different final states on the same basis.

This basis is the quantities measured in the multihadronic final states:

i) the average charged multiplicity; $< n_{ch} >$;

ii) the fractional energy distribution; ds / dx_i ;

iii) the transverse momentum distribution ds / dp_{t_i} ; etc. .... .

The results are the universality features measured in all multihadronic final states, no matter what is the pair of interacting particles in the initial state.

The universality features are a QCD **non-perturbative effect**.

Systematic study of the Leading Effect in
Strong EM Weak INTERACTIONS

THE "LEADING"-BARYON EFFECT IN STRONG, WEAK, AND ELECTROMAGNETIC INTERACTIONS

M. Basile, G. Cara Romeo, L. Cifarelli, A. Contin, G. D'Alì, P. Di Cesare, B. Esposito, P. Giusti, T. Massam, R. Nania, F. Palmonari, V. Rossi, G. Sartorelli, M. Spinetti, G. Susinno, G. Valenti, L. Votano and A. Zichichi.

Lettere al Nuovo Cimento 32, 321 (1981).

"This supports the idea that the "leading" phenomenon is generated by the quantum number «flow» from the initial to the final state".

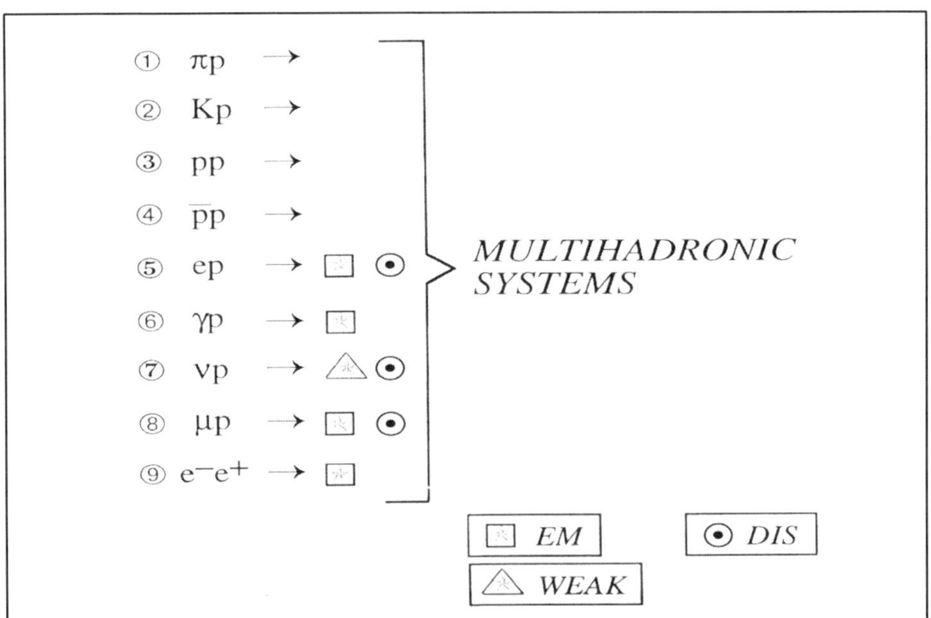

Flow of Quantum Numbers in (pp) **Strong Interactions**

THE «LEADING»-BARYON EFFECT IN STRONG, WEAK, ETC. 323

In fig. 1 we have plotted the values of L derived from the experimental x distributions of the different hadrons in the reactions (4*a*) to (4*g*). The final states are ordered according to the number of quarks propagating from the initial state (the proton) to the final state.

The results of fig. 1 show that for different hadrons the value of L is the same, provided the number of propagating quarks is the same

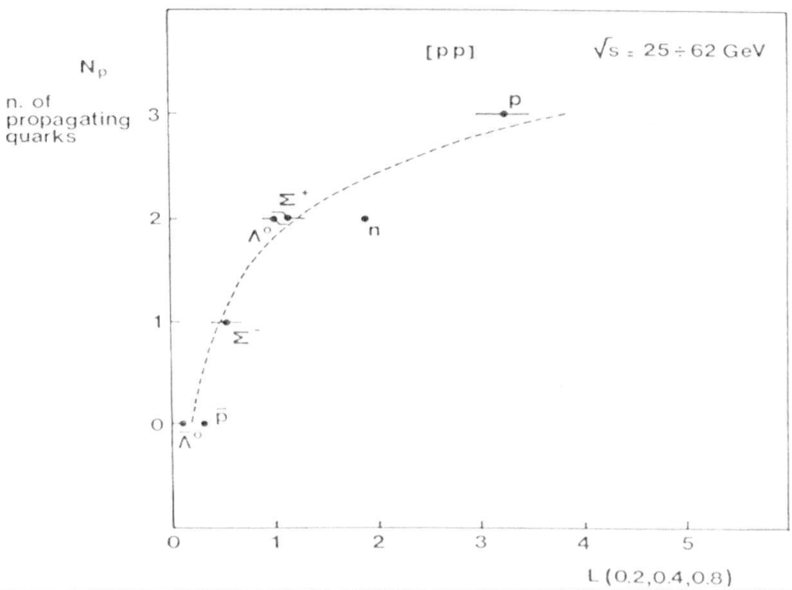

p → p (maximum Leading Effect)
p → n (less Leading Effect)

Figure 1: Figure from Reference 3 where the "leading" quantity L (0.2, 0.4, 0.8) derived for different types of baryons produced in (pp) collisions at CERN ISR is shown. The centre-of-mass energy ranges from 25 to 62 GeV. The hadrons are ordered according to the number of propagating quarks. The dotted curve superimposed is obtained by using a parametrization of the single-particle inclusive cross-section, $F(x) = (1 - x)^{\alpha}$, as described in section 3.

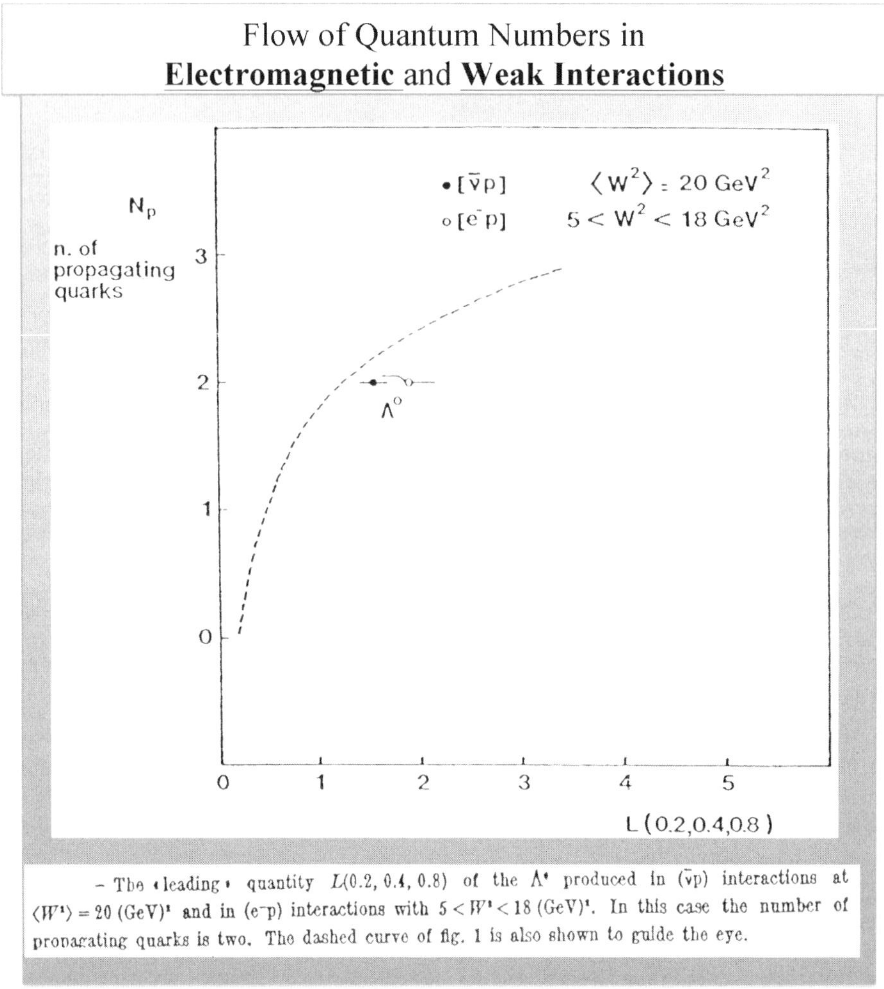

Figure 2: Reproduction of a Figure from Reference 3 where the "quantum number flow" from the initial to the final state is observed in an electromagnetic process ($e^- p \rightarrow \Lambda^0 x$) and in a weak process ($\bar{\nu} p \rightarrow \Lambda^0 x$).

It should be noticed, as discussed in ref. ([1]), that for Λ^0 production the energy dependence of L shows the same features as those observed for proton production. In fact, the three values of L obtained for the Λ^0 at ISR, at Fermilab, and at Cornell energies, suggest the following trend: the higher is the available energy, the lower is the value of L.

All the above results thus point out that it does not matter whether the hadron interacts strongly, weakly, or electromagnetically: its « leading » effect is always present.

5. *Conclusions.* – In baryon-baryon interactions, the « leading »-baryon effect shows up very clearly in the x range $(0.2 \div 0.8)$. This « leading » effect is maximum when the final-state hadron is the same as the initial-state hadron. However, the « leading » effect is present even when the initial-state quantum numbers differ from those of the final state (for instance, when a proton becomes a Λ^0). As the difference between the initial- and the final-state quark composition increases, the « leading » effect decreases. This supports the idea that the « leading » phenomenon is generated by the quantum number « flow » from the initial to the final state.

The « leading » baryon effect appears both in baryon-baryon and in lepton-baryon interactions. This means that a definite similarity must exist between processes in which a hadron is present in the initial state, no matter if the interaction is strong, weak or electromagnetic.

M. BASILE, et al.
14 Novembre 1981
Lettere al Nuovo Cimento
Serie 2, Vol. 32, pag. 321-326

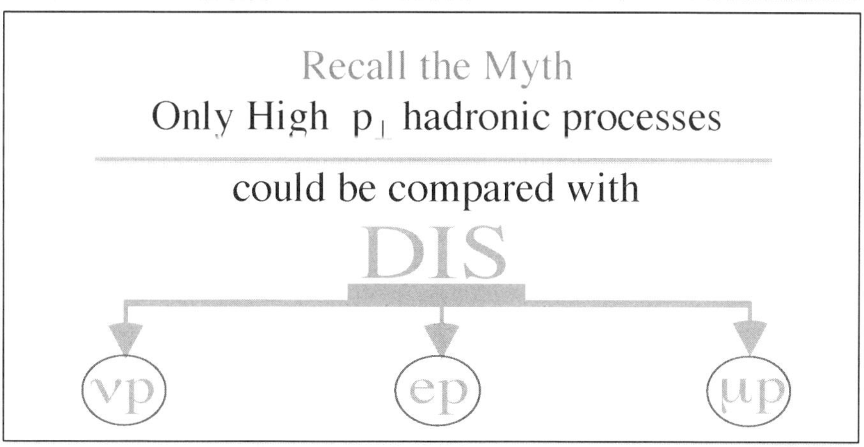

Figure 3A: A synthesis of the high transverse momentum myth.

The introduction of the Effective Energy has produced the result that multihadronic final states produced in high $p_\perp$ processes are analogous to those produced in low $p_\perp$ processes, provided the Effective Energies are the same.

THE END OF A MYTH: HIGH-P_T PHYSICS

"So far, the main picture of hadronic physics has been based on a distinction between high-p_T and low-p_T phenomena.

In the framework of parton model, high-p_T processes were the only candidates to establish a link between

- *purely hadronic processes*
- *(e^+e^-) annihilations*
- *(DIS) processes.*

The advent of QCD has emphasized in a dramatic way the privileged role of high-p_T physics due to the fact that, thanks to asymptotic freedom, QCD calculations via perturbative methods can be attempted at high-p_T and results successfully compared with experimental data [1]. The conclusion was: we can forget about everything else and limit ourselves to high-p_T physics.

Being theoretically off limits, low-p_T phenomena, which represent the overwhelming majority of hadronic processes (more than 99% of physics is here), have been up to now neglected. By subtracting the leading proton effects in order to derive the effective energy available for particle production and by using the correct variables, the BCF collaboration has performed a systematic study of the final states produced in low-p_T (pp) interactions at the ISR and has compared the results with those obtained in the processes listed below:

Process		*Data Sources*
(e^+e^-)		*SLAC, DORIS, PETRA*
(DIS)		*SPS/EMC*
(pp)	*Transverse physics*	*ISR (AFS)*
($\bar{p}p$)	*Transverse physics*	*SPS Collider (UA1)*
(e^+e^-)		*PETRA/TASSO (leading subtraction)*

The results of this study [2-18] show that, once a common basis for comparison is found by the use of the correct variables, remarkable analogies are observed in processes so far considered basically different like

- *low-p_T (pp) interactions*
- *(e^+e^-) annihilations*
- *(DIS) processes*
- *high-p_T (pp) and ($\bar{p}p$) interactions*

This is how universality features emerge, and this is the basis to proceed for a meaningful comparison, i.e.:

first identify the correct variables to establish a common basis,

then proceed to a detailed comparison."*

* *The root of this new approach to the study of hadronic interactions goes back a long time to a proposal by the CERN-Bologna group: "Study of deep inelastic high momentum transfer hadronic collisions" PMI/com-69/35, 8 July 1969."*

Figure 3B: Reproduction of the conclusions of a review paper [4].

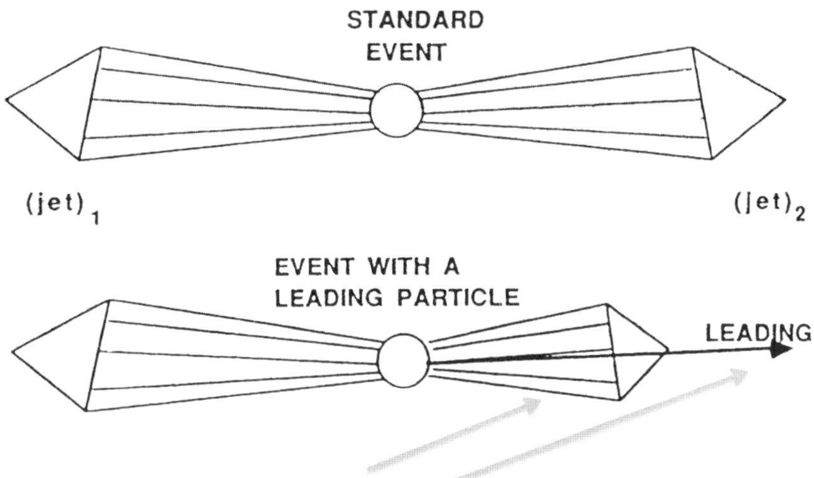

If the leading particle is not detected the Effective Energy can be deduced from all other particles detected.

PHASE TRANSITION

- If **smooth** and **uniform** nothing happens

PHASE TRANSITIONS

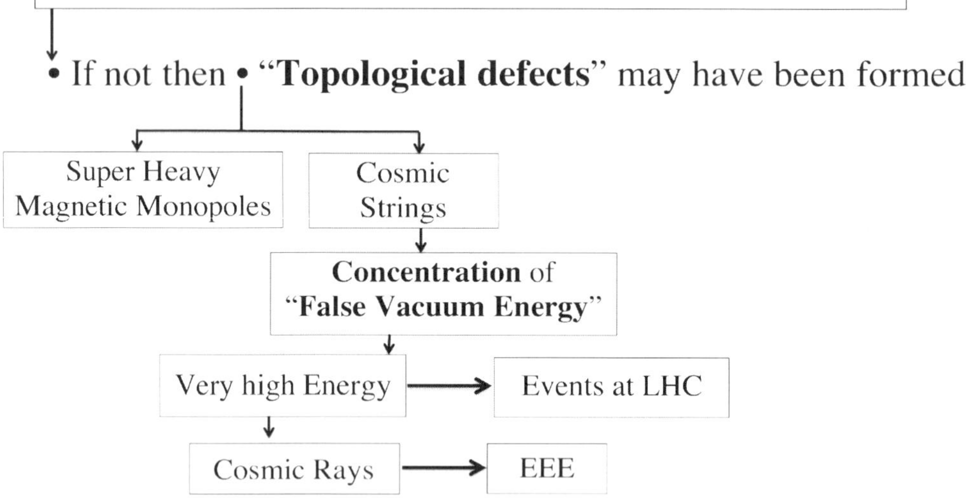

Time, Energy and Phase Transition

t(sec)	E(GeV)	Phase Transition
10^{-44}	10^{18}	Planck epoch ≡ Quantum Gravity ≡ Supergravity Superstring
10^{-35}	10^{16}	GUT
10^{-10}	10^{2}	Weak Symmetry Breaking ≡ Fermi epoch
10^{-5}	10^{-1}	Confinement Transition
$1\text{-}10^{2}$	$10^{-3}\text{-}10^{-2}$	Nucleo-Synthesis
10^{12}	10^{-9}	Recombination/Galaxy Formation
10^{17}	10^{-13}	Today

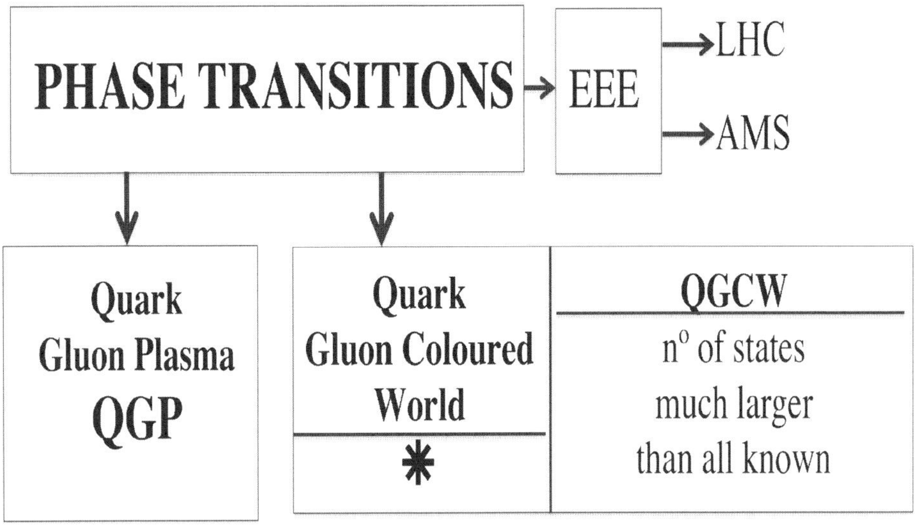

1st problem – In the QGCW there are all states allowed by the $SU(3)_c$ colour group. The number of possible states is by far more numerous than the number of colourless baryons and mesons, which have so far been built in all Labs, since the colourless condition is not needed. What are the consequences on the properties of the QGCW?

2nd problem – Light quarks versus heavy quarks. Are the **coloured quark masses the same** as the values we derive from the fact that baryons and mesons need to be in a **colourless state**? It could be that all six quark flavours are associated with nearly 'mass-less' states like those of the 1st family (u, d). In other words the reason why the 'top' quark appears to be so heavy ($\simeq 10^2$ GeV) could be due to the fact that it must satisfy some, **so far unknown**, condition related to the fact that the final state **must be QCD-'colourless'**. We know that confinement produces masses of the order of a GeV.

Therefore, according to our present understanding, **the QCD 'colourless' condition could not explain the heavy quark mass**, **but** since the origin of the quark masses is still not known, it cannot be excluded that in a QCD coloured world, the six quarks are **all nearly mass-less**. If this was the case, the masses we measure are heavier than the effective coloured quark masses. In this case all possible states generated by 'heavy' quarks would be produced in the **QGCW at much less temperature than** the one needed in our world made with baryons and mesons, **i.e. QCD colourless states**.

Here again we should try to see if with masses totally different from those expected, on the basis of what we know about colourless baryons and mesons, new **effects** could be detected due to the existence of all six flavours at relatively low temperature in the QGCW world.

3rd problem – We need to search for effects on the thermodynamic properties of the QGCW. Are these properties going to be along the 'extensivity' and / or 'non-extensivity' conditions? (Murray Gell-Mann and Constantino Tsallis).

4th problem – We need to **derive the equivalent** Stefan-Boltzmann Radiation Law for the QGCW. The relation between **energy density** at emission **U** and **Temperature T** of the source is

$$U = c \cdot T^4$$

in classical Thermodynamics.

In the QGCW the correspondence should be

$U \equiv p_{\perp}$ (transverse momentum)

$T \equiv$ average energy $\langle E \rangle$ in the CM system.

In the QGCW the production of 'heavy' flavours should be studied versus $\langle p_\perp \rangle$ and versus $\langle E \rangle$. The expectation is

$$\langle p_\perp \rangle \equiv C \cdot \langle E \rangle^4$$

and any deviation would be extremely important.

At extreme energies we do not know how many phase transitions can be involved. The higher the energy, the more complex is the interacting system of particles and more phase transitions can be involved.

There are **two sources** of unknown "phase transitions":

1) Those taking place at **CERN** (at LHC energy and its upgrading stages);

2) Those taking place in the **Universe**.

Both sources have in common a possible "signature":

Extreme Energy Events (EEE).

In fact the evolution of the Universe has gone through a series of phase transitions whose last step was at the Fermi Energy when

the $SU(2) \times U(1)$ generated QED and QFD.

This is the low energy level.

At the other extreme, there is E_{GUT} ($\simeq 10^{16}$ GeV) where the three gauge couplings ($\alpha_1\ \alpha_2\ \alpha_3$) converge, and E_{SU} ($\simeq 10^{18}$ GeV) the energy level where RQST (Relativistic Quantum String Theory) puts the origin of the gravitational force.

The **GAP** between E_{GUT} and E_{SU} could indeed be another source of phase transitions.

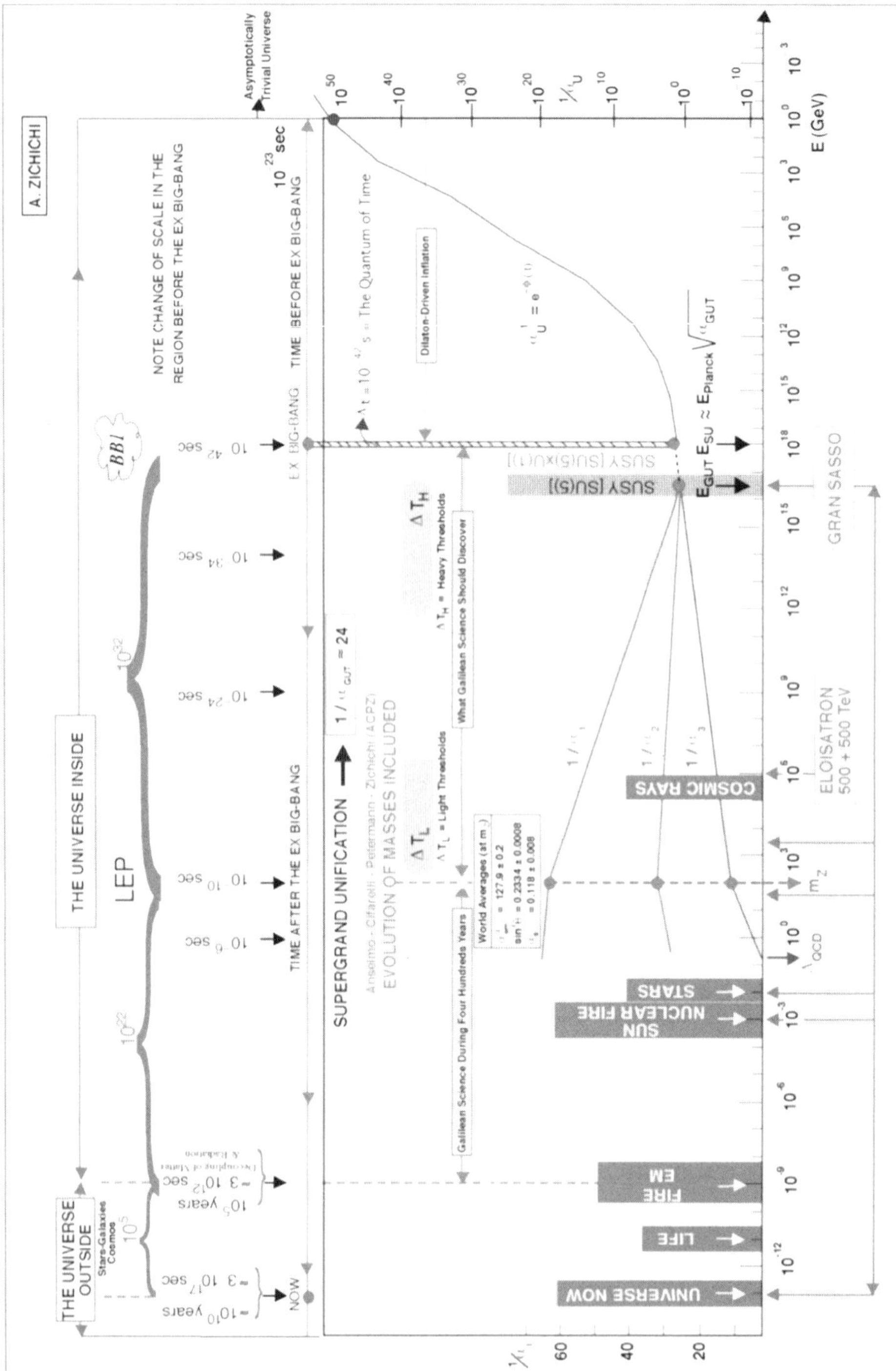
A. ZICHICHI
THE UNIVERSE INSIDE
THE UNIVERSE OUTSIDE
LEP
NOTE CHANGE OF SCALE IN THE REGION BEFORE THE EX BIG-BANG
TIME AFTER THE EX BIG-BANG
TIME BEFORE EX BIG-BANG
Asymptotically Trivial Universe
Dilaton-Driven Inflation
SUPERGRAND UNIFICATION
EVOLUTION OF MASSES INCLUDED
What Galilean Science Should Discover
Galilean Science During Four Hundreds Years
COSMIC RAYS
STARS
SUN NUCLEAR FIRE
FIRE EM
LIFE
UNIVERSE NOW
GRAN SASSO
ELOISATRON
500 + 500 TeV
E (GeV)

Figure 4

THE UNIFICATION OF ALL FUNDAMENTAL FORCES

The lines in Figure 4 result from calculations executed with a supercomputer using the following system of equations:

$$\mu \frac{d\alpha_i}{d\mu} = \frac{b_i}{2\pi} \alpha_i^2 + \sum_j \frac{b_{ij}}{8\pi^2} \alpha_i \alpha_j$$

This is a system of coupled non-linear differential equations where the existence of the Superworld is included. This system describes how the gauge couplings (α_1, α_2, α_3) vary with "μ", the basic parameter which depends on the energy of the elementary process, from the maximum level of Energy (Planck Scale) to the energy level of our world. **During more than ten years (from 1979 to 1991), no one had realized that the energy threshold for the existence of the Superworld was strongly dependent on the "running" of the masses**.

This is now called: **the EGM effect** (from the initials of Evolution of Gaugino Masses).

Figure 5

WHY WE NEED THE SUPERWORLD

There are fundamental reasons making the superworld a need.

1) The two energy scales, 10^{19} GeV (Planck) and 10^2 GeV (Fermi), must be kept separate.
2) The gravitational attraction of light must be prevented from being infinite. Otherwise we could see neither the light from Stars nor our light. The "gravitino" (supergravity) allows the gravitational attraction of light to be finite.
3) Gravitational attraction is powerful but it cannot be infinite. We would be stuck to the Sun. Space would not exist between Stars and Galaxies. Cosmic expansion would not exist. In order to have a finite gravitational attraction, theories are needed in which the Euclidean concept of point is abandoned. The point is replaced by a string. No more Point-like Theories but String Theories. These theories must be supersymmetric: the already quoted supersymmetry law ($F \equiv B$) must be valid in their mathematical structure. Otherwise "tachions" would appear. This is the origin of Relativistic Quantum String Theory (RQST).
4) If we want the unification of all fundamental phenomena – the synthesis of which is provided by three "gauge couplings", α_1 α_2 α_3 – running with the energy is needed. This "**running**" needs Supersymmetry.
5) An interesting detail: "no scale-supergravity" is an infrared solution of RQST. This might allow to understand the extremely small value of the Cosmological Constant.
6) Finally: why Three Columns and Three Forces? The answer to this question should come once we will be able to go from the compactification of the 43-dimensional superspace to our present world with (3+1) Space-Time dimensions.
7) **Note**: **A big problem**. Supersymmetry does not show up at our energy scale. Hence the problem arises to compute the energy above which the ($F \equiv B$) Law starts to act. Thanks to the **EGM effect**, this energy level is 700 times more accessible than thought so far.

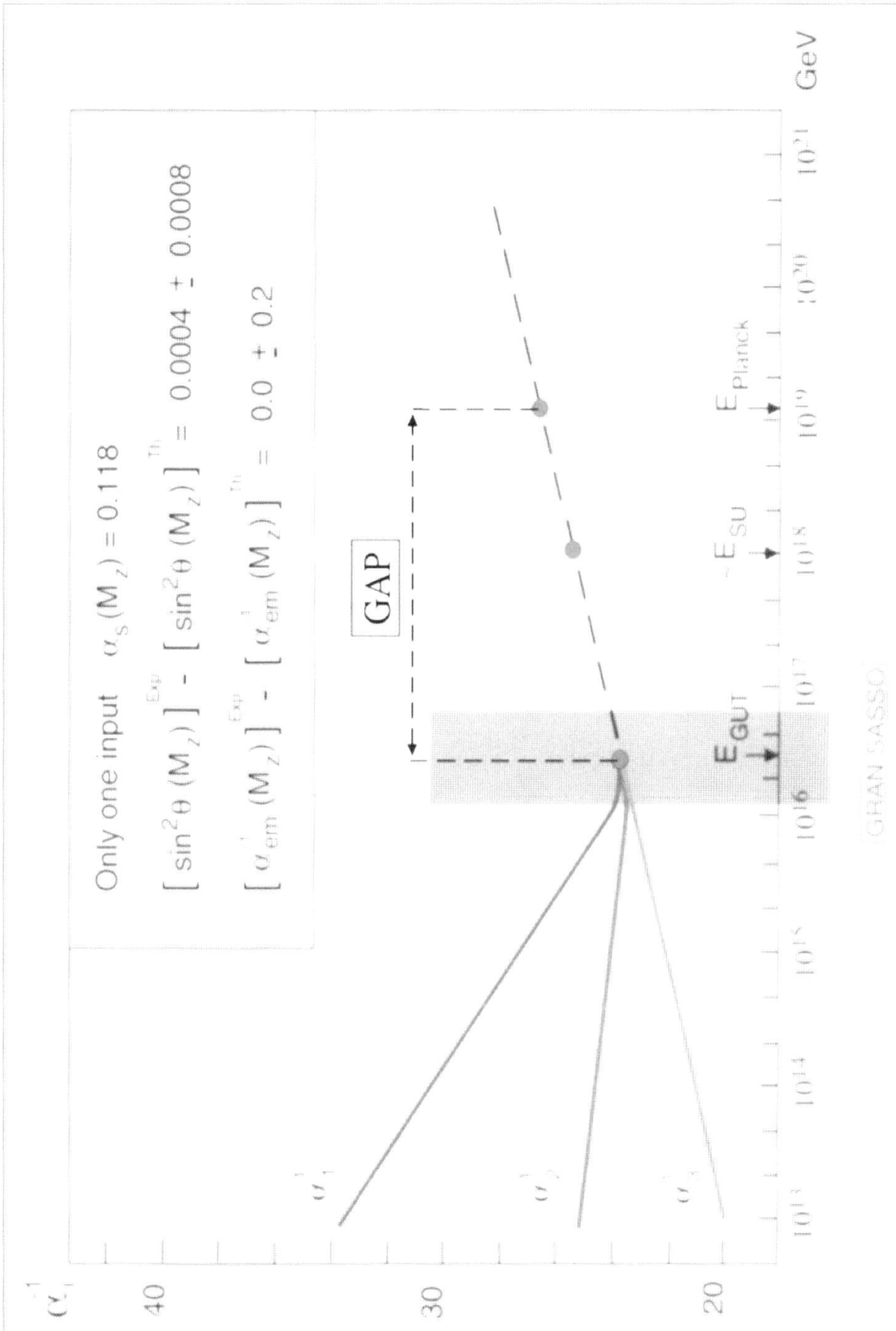
Only one input α_S (M_Z) = 0.118
[sin²θ (M_Z)]^Exp - [sin²θ (M_Z)]^Th = 0.0004 ± 0.0008
[α_em^-1 (M_Z)]^Exp - [α_em^-1 (M_Z)]^Th = 0.0 ± 0.2
GAP
E_GUT
E_SU
E_Planck
GRAN SASSO
GeV
40
30
20

Figure 6

If we could see the inner structure of the Black–Holes we would find that there are two types of Black–Holes. The primordial Black–Holes are made with matter whose charge is only the gravitational charge. The Black–Holes familiar to us are made with (p, n, e).

Conclusion.

All we could at present say on the correlation between the Subnuclear Universe and the one with Stars and Galaxies is to explain why: $N_{(p\,n\,e)} \simeq 10^{80}$ and $V(U) \simeq 98\%$; and to predict the existence of **two** types of Black–Holes: **Primordial Black–Holes** where matter has only the gravitational charge and **Standard Black–Holes** where matter is made with p, n, e.

On many occasions, during the activities of the International School of Cosmology and Gravitation, I have been discussing with friends and colleagues (including John Wheeler [5], Nathan Rosen [6] and Peter Bergmann [7]) how it happens that no one has been able so far to derive two basic values of our Universe:

① the number of protons, neutrons and electrons, $N_{(p\,n\,e)}$, which our Universe is made of, i.e.

① $N_{(p\,n\,e)} \simeq 10^{80}$;

and

② the volume of our Universe, V(U), which is empty, i.e.

② $V(U) \simeq 98\%$.

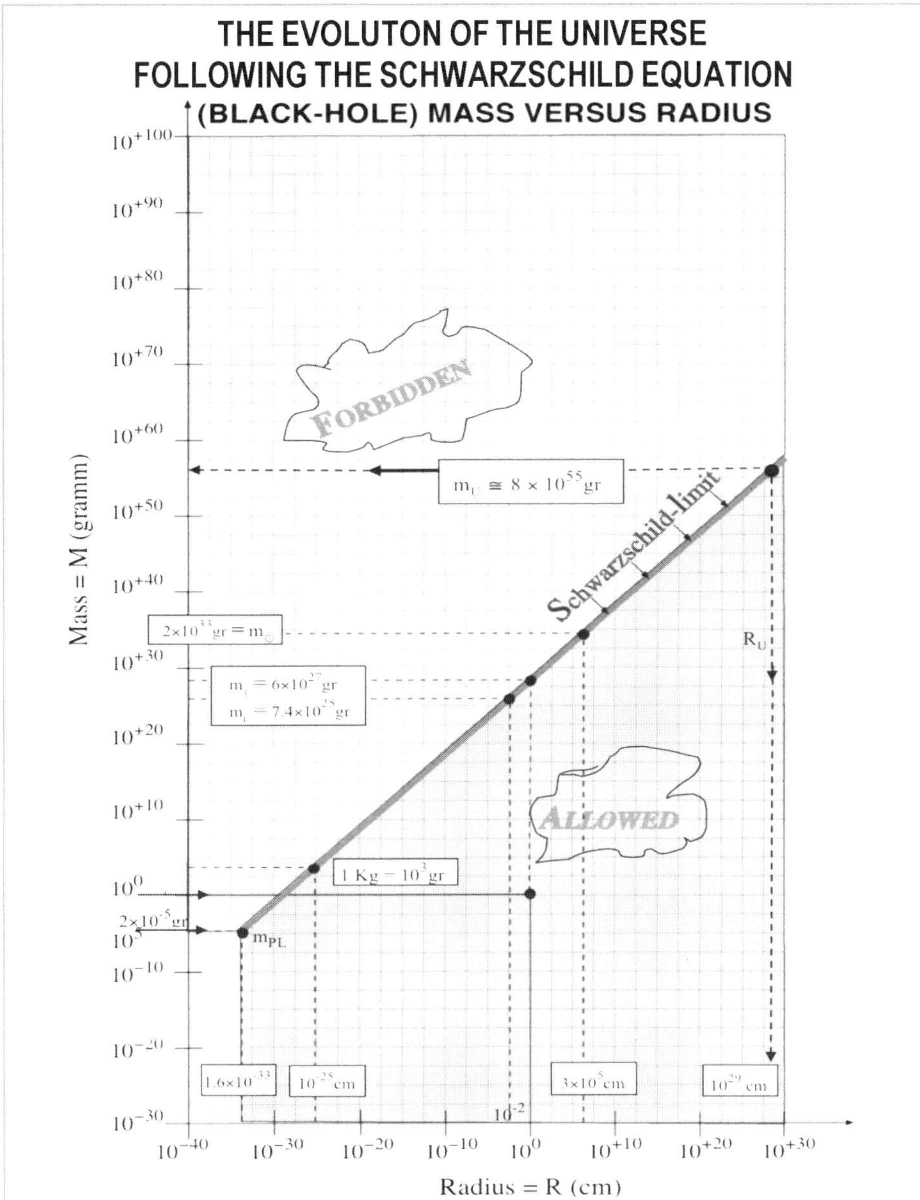

Figure 7

Conclusion: all effects discovered at $\simeq 10^2$ GeV $\Rightarrow$ need to be checked at $\mathbf{13 \times 10^3 \times GeV}$. **From** 10^2 GeV $\Rrightarrow$ $(13 \times 10^3 \times \text{GeV})$

SEARCH FOR EEE AT LHC.

II THE WHOLE OF OUR KNOWLEDGE

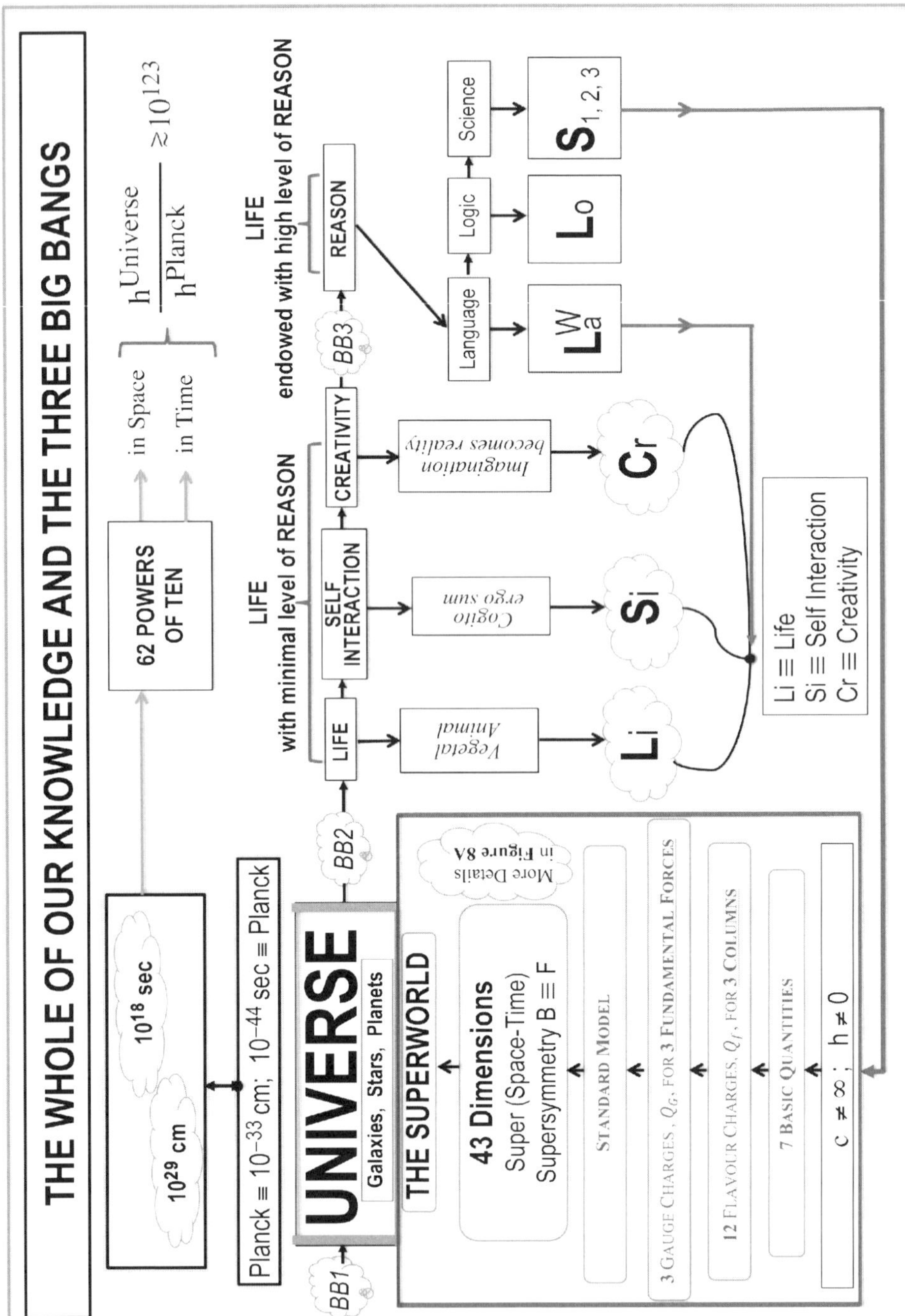

Figure 8

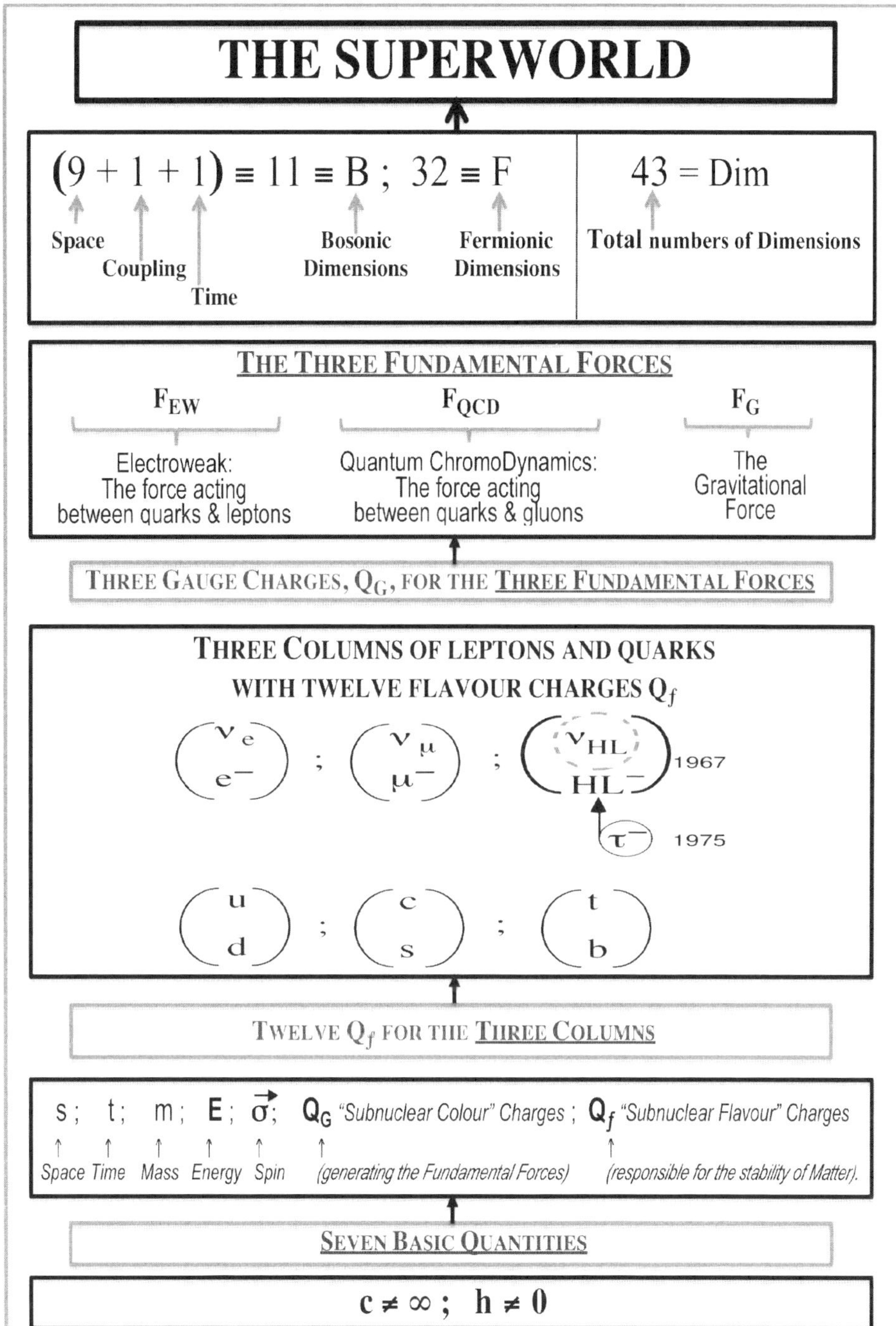
THE SUPERWORLD
(9 + 1 + 1) ≡ 11 ≡ B ; 32 ≡ F
Space
Coupling
Time
Bosonic Dimensions
Fermionic Dimensions
43 = Dim
Total numbers of Dimensions
THE THREE FUNDAMENTAL FORCES
F_{EW}
F_{QCD}
F_G
Electroweak: The force acting between quarks & leptons
Quantum ChromoDynamics: The force acting between quarks & gluons
The Gravitational Force
THREE GAUGE CHARGES, Q_G, FOR THE THREE FUNDAMENTAL FORCES
THREE COLUMNS OF LEPTONS AND QUARKS WITH TWELVE FLAVOUR CHARGES Q_f
ν_e ; ν_μ ; ν_{HL}
e^- ; μ^- ; HL^-
1967
τ^-
1975
u ; c ; t
d ; s ; b
TWELVE Q_f FOR THE THREE COLUMNS
s ; t ; m ; E ; $\vec{\sigma}$; Q_G "Subnuclear Colour" Charges ; Q_f "Subnuclear Flavour" Charges
Space Time Mass Energy Spin
(generating the Fundamental Forces)
(responsible for the stability of Matter).
SEVEN BASIC QUANTITIES
c ≠ ∞ ; h ≠ 0

Figure 8A

We physicists **cannot remain silent** when the great public shows a vivid interest for topics such as:

- Global warming
- The energy crisis
- The information security
- The environment
- The Intelligent Design
- The Evolution
- and other Problem coming from the "Whole of our Knowledge".

We have to convince the great public that the solution to all these **problems** requires **clarity** and **rigour** and that the best way to study these problems is through Science.

Since Physics is the "Queen of all Sciences", (**Enrico Fermi**), the solution of these problems needs physicists.

But the present trend is to study the topics mentioned above through the "new Science", the so called "Science of Complexity".

In fact the "Modern Culture" considers "Complexity" a source of new insights in **physics**, **biology**, **geology**, **cosmology**, **social sciences** and in **all intellectual activities** which look at the world through the lens of a standard analysis in terms of either "**Platonic Simplicity**" or "**Complexity**".

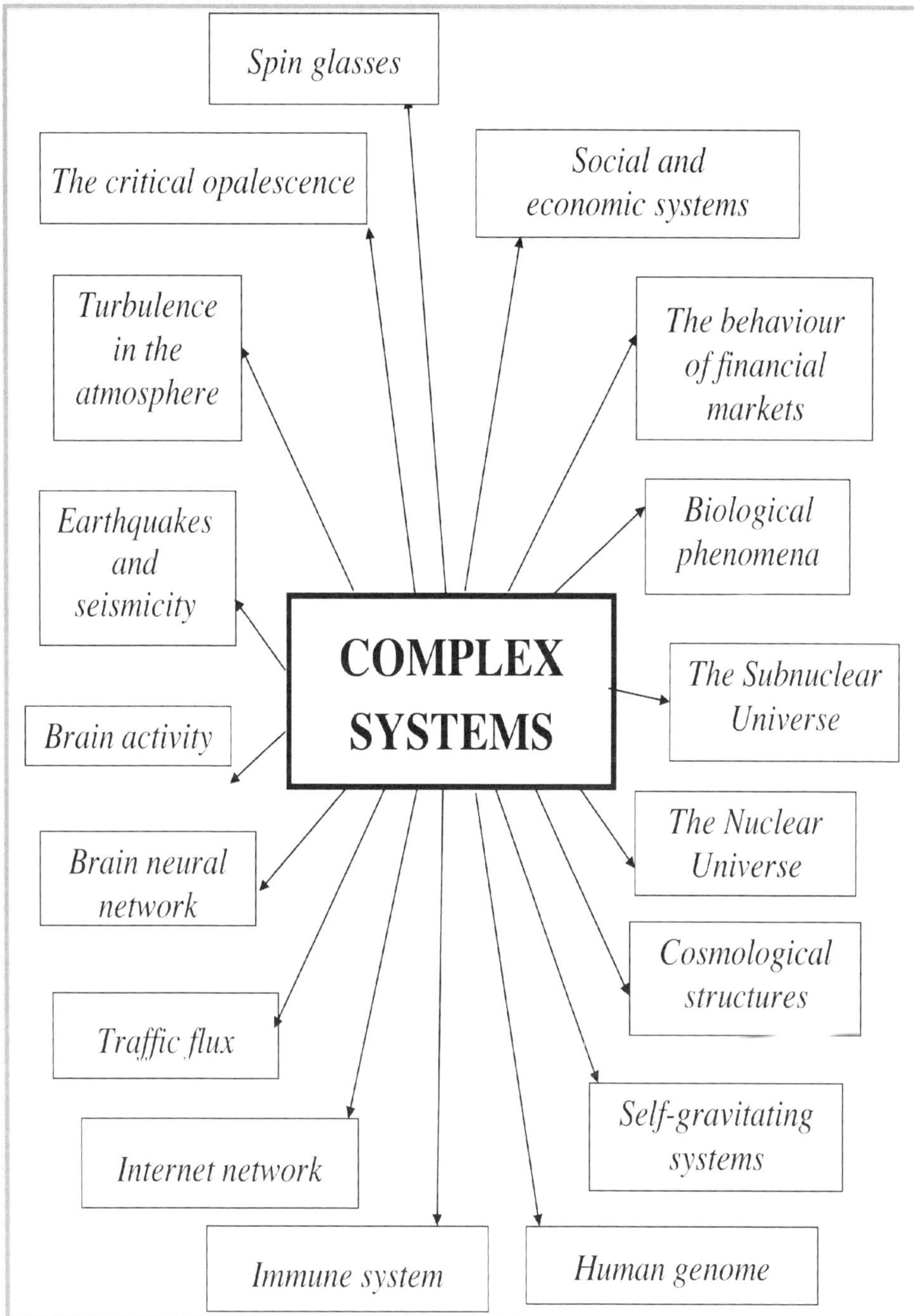
Spin glasses
The critical opalescence
Social and economic systems
Turbulence in the atmosphere
The behaviour of financial markets
Earthquakes and seismicity
Biological phenomena
COMPLEX SYSTEMS
The Subnuclear Universe
Brain activity
The Nuclear Universe
Brain neural network
Cosmological structures
Traffic flux
Self-gravitating systems
Internet network
Immune system
Human genome

Figure 9

The reason why we have to care about "Platonic Simplicity" is due to the emergence of this "new Science", which condemns "Reductionism" and promote "Holism".

But "**Complexity**" is ill-defined, as shown by the existence of very many definitions (at least seventy) of Complexity.

We will see that we have nothing to learn from this "new Science". In fact it is thanks to "Reductionism" that we are able to reach the following conclusion: despite the **seventy** definitions of "Complexity" there are **only two** experimentally observable effects which exist when "Complexity" is at work.

These two effects are the **UEEC** events and the **AFB** phenomena. Both are present in Physics.

Despite the seventy definitions of Complexity, the experimentally observable effects for the **existence** of **Complexity are only two**:

1) The Anderson-Feynman-Beethoven-type phenomena (**AFB**) i.e. phenomena whose laws and regularities ignore the existence of the Fundamental Laws of Nature from which they originate, and

2) The Sarajevo-type effects, i.e. Unexpected Events of quasi irrelevant magnitude which produce Enormous Consequences (**UEEC**).

The **SM&B** is the greatest synthesis of all times in the study of the fundamental phenomena governing the Universe in all its structures.

DETAILED INFORMATION ON THE STANDARD MODEL AND BEYOND

① RGEs $(\alpha_i\, (i \equiv 1, 2, 3);\ m_j\ (j \equiv q,\ l,\ G,\ H)) :\ f(k^2)$.

- GUT ($\alpha_{GUT} \cong 1/24$) & GAP ($10^{16} - 10^{18}$) GeV.
- SUSY (to stabilize $m_F/m_P \cong 10^{-17}$).
- RQST (to quantize Gravity).

② Gauge Principle (hidden and expanded dimensions).

– How a Fundamental Force is generated: SU(3); SU(2); U(1) and Gravity.

③ The Physics of Imaginary Masses: SSB.

– The Imaginary Mass in SU(2) × U(1) produces masses ($m_{W^\pm}$; m_{Z^0}; m_q; m_l), including $m_\gamma = 0$.

– The Imaginary Mass in SU(5)⇒SU(3)×SU(2)×U(1) or in any higher Symmetry Group (not containing U(1)) ⇒ SU(3) × SU(2) × U(1) produces Monopoles.

– The Imaginary Mass in $SU(3)_c$ generates Confinement.

④ Flavour Mixings & CP ≠ , T ≠ .

– No need for it but it is there.

⑤ Anomalies & Instantons.

– Basic Features of all Non-Abelian Forces.

Note:

q	≡	quark and squark;	m_F	≡	Fermi mass scale;
l	≡	lepton and slepton;	m_P	≡	Planck mass scale;
G	≡	Gauge boson and Gaugino;	k	≡	quadrimomentum;
H	≡	Higgs and Shiggs;	C	≡	Charge Conjugation;
RGEs	≡	Renormalization Group Equations;	P	≡	Parity;
GUT	≡	Grand Unified Theory;	T	≡	Time Reversal;
SUSY	≡	Supersymmetry;	≠	≡	Breakdown of Symmetry Operators.
RQST	≡	Relativistic Quantum String Theory;			
SSB	≡	Spontaneous Symmetry Breaking.			

The five basic steps in our understanding of Nature.

① The renormalization group equations (RGEs) imply that the gauge couplings (α_i) and the masses (m_j) all run with k^2. It is this running which allows GUT, suggests SUSY and produces the need for a non point-like description (RQST) of physics processes, thus opening the way to quantize gravity.

② All forces originate in the same way: the gauge principle.

③ Imaginary masses play a central role in describing nature.

④ The mass-eigenstates are mixed when the Fermi forces come in.

⑤ The Abelian force QED has lost its role of being the guide for all fundamental forces. The non-Abelian gauge forces dominate and have features which are not present in QED.

The basic achievements of the **SM&B** have been obtained via an impressive series of UEEC events; moreover the **SM&B** could not care less about the existence of **Platonic Simplicity**. **Let me repeat** the reason why we have to care about "Platonic Simplicity": the emergence of the "new science": the so called "Science of Complexity", to which "modern Culture" attributes an enormous importance: **new insights** in physics, biology, geology, cosmology, social sciences should come from Complexity.

EXAMPLES OF PLATONIC SIMPLICITY

PLATONIC SIMPLICITY. *THE PLATONIC GRAND UNIFICATION*

Let us now move towards the problem of Platonic Simplicity, taking as example the structure of a Grand Unification. The simplest way is to have one and only one basic fundamental particle, B.

This particle must obey the very simple symmetry law which puts fermions and bosons on the same basis (Figure 10).

This basic fundamental particle can therefore exist either as being a boson B_B or as being a fermion B_F.

Let us consider first

B_B

The Fundamental Forces

exist

because a

Basic Fundamental Boson

B_B

EXISTS

Figure 10

Figure 11 illustrates the simple sequence which generates all known forces of Nature.

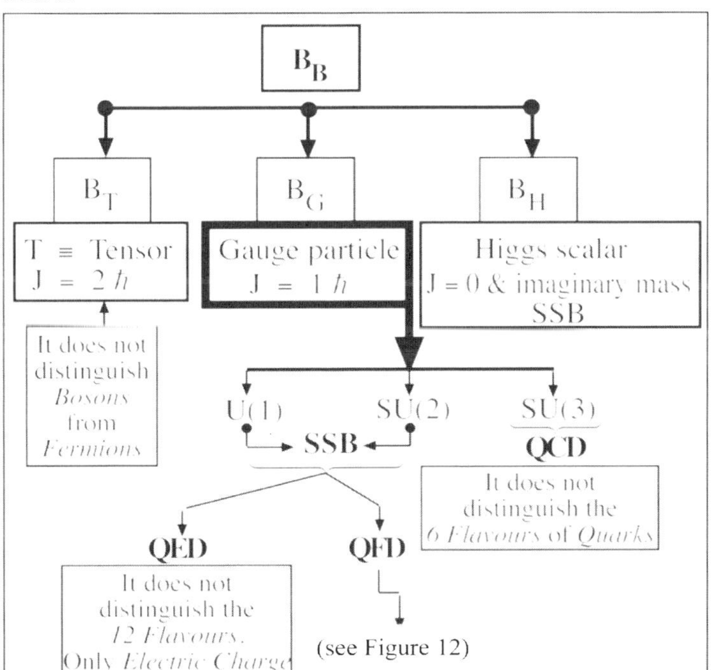

Figure 11

At the bottom of Figure 11 there is the force QFD, illustrated in Figure 12. The "Platonic" Simplicity suffers a further deviation.

QFD	
$(q^{u}_{m})_{1,2,3}$ $(q^{d}_{m})_{1,2,3}$	$(\ell^{u}_{m})_{1,2,3}$ $(\ell^{d}_{m})_{1,2,3}$
Experimentally proved	*Experimentally proved*
mixing & C ≠; P ≠; CP≠	**?**

Figure 12

In fact, we need to introduce many complications. The three quarks and the three leptons are "mixed" among them, not among "quarks" and "leptons. This mixing is indicated by the index m, while the indices "u" and "d" refer to the two types of flavours

(up-type) and

(down-type)

which are present in each of the three families:

1, 2, 3.

There is a further complication.

The two mixings for the "up" and the "down" flavours **must be different**.

In the case of the quark, this mixing is experimentally measured.

In the case of the leptons, the experimental results are with nearly half a century of delay, compared with the quark case.

Mixing and violation of Symmetry Laws for charge conjugation (C), parity (P), and the product of the two (CP) are well established in the quark case.

In the leptonic sector, only future experiments will tell us if the same Symmetry Laws are violated.

There is no known reason why all these details, **mixing of states** and **Symmetry Law violations**, are needed.

They have been experimentally discovered and show how many deviations from the **simple "Platonic" structure** are needed.

So far we have developed the sequence of **Platonic deviations from Simplicity**, starting from the basic fundamental boson B_B .

We now show in Figure 13 the **deviations needed** from the **Platonic Simplicity**, when we start from the basic fundamental fermion B_F .

It has to be with "**quark**" and "**lepton**" **flavours** and have two flavours in each class (called Family).

Total number of flavours 12: **6** for quarks
6 for leptons

Why so many? The answer will probably come from the Super Space with 43 Dimensions compactified into (3+1).

The quark sector interacts with two forces, QCD and QED, while the lepton sector interacts using only QED. The QFD force comes into play only after all the mixings come in.

No one knows why all these **Deviations** from the **Platonic Simplicity** are needed.

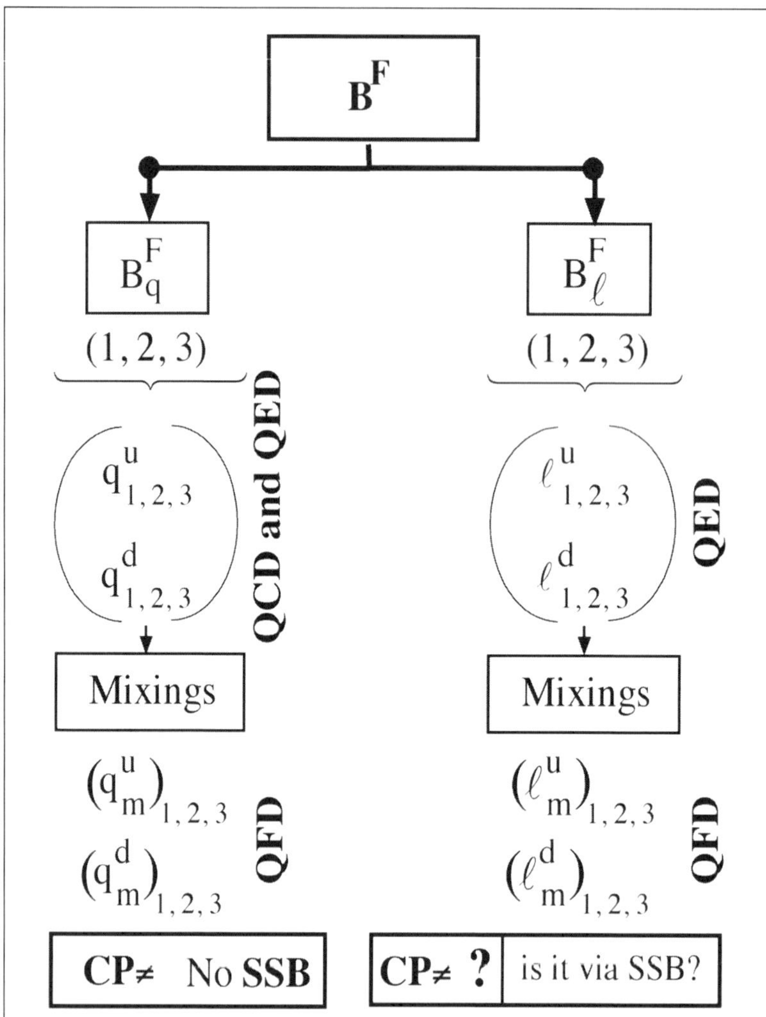

Figure 13

The bold symbols, QCD, QED in the column

$$B_q^F$$

indicate that the 6 quarks flavours interact via these two forces. In the lower part of the same column, the "mixing" indicates that the quark states are no longer "pure" states. They are "mixed"; only these mixed states

$$\left(q_m^u\right)_{1,\ 2,\ 3} \text{ and } \left(q_m^d\right)_{1,\ 2,\ 3}$$

interact via the QFD forces.

The column below

$$B_\ell^F$$

has the same structure, but the "mixings" are not the same as in the "quark" column. Furthermore, no one knows at present if the Symmetry CP is violated as it is in the quark case. This is why in the box CP ≠ there is a question mark. Another detail needs to be specified. In the quark case, the CP Symmetry breaking, CP ≠ , has been experimentally established not to be via the basic Standard Model mechanism SSB.

A further Deviation from Simplicity

In the leptonic case, we do not know if the CP Symmetry is violated. It could be it is. In this case it will be interesting to know if it follows the SSB mechanism. All these question marks are evidence of **further deviations** from the **simple Platonic descriptions** of natural phenomena.

In Figure 14 a synthesis of all deviations needed is illustrated.

THE PLATONIC GRAND UNIFICATION
AND THE DEVIATIONS NEEDED

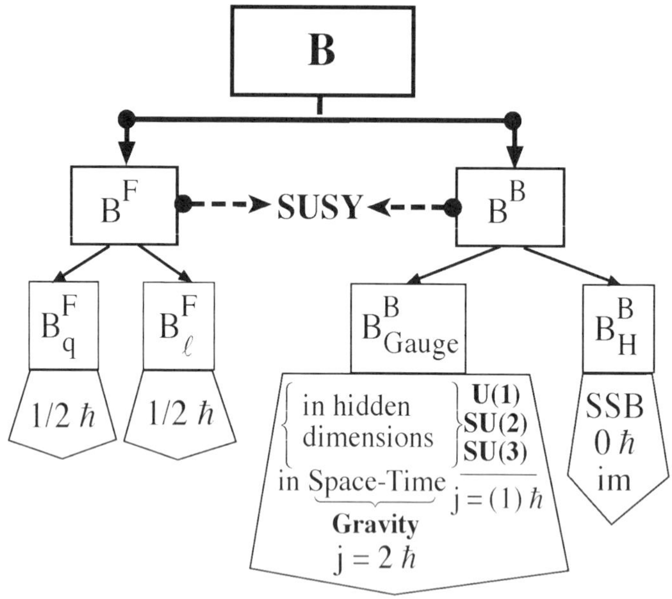

- The fundamental **forces** exist because a Basic Fundamental Boson **exists** ≡ B^B
- The fundamental **fermions** exist because a Basic Fundamental Fermion **exists** ≡ B^F
- The SSBs originate from the existence of a scalar with imaginary mass (im).

Figure 14

THE PLATONIC CONCEPT OF SUPERSYMMETRY

The Platonic Concept of Supersymmetry is schematically reported in the upper part of Figure 15, where the basic point for a Platonic Concept of Supersymmetry is given; i.e. the only fermions allowed to exist, would be the "gauginos" (with spin 1/2 ħ). To go from gauginos to fermions we need the 1st deviation from Platonic Supersymmetry. The 2nd deviation is needed in order to allow fermions to be quarks and leptons. The 3rd deviation is needed in order to have not one Family but three Families. The 4th deviation is needed in order to produce mixing. The 5th deviation is needed because we need at least two mixings, one for "up" and another one for "down" type quarks. All this is illustrated in Figure 15.

THE PLATONIC CONCEPT OF SUPERSYMMETRY

The **Gauge Principle** should generate a
Gauge Force → **Gauge Bosons**
If *NATURE* was platonically *SUPERSYMMETRIC*
Supersymmetry Transformation should generate **Gauginos**

1st **DEVIATION FROM PLATONIC SIMPLICITY**
OUR FERMIONS ARE NOT THE GAUGINOS

2nd **DEVIATION FROM PLATONIC SIMPLICITY**
THE FUNDAMENTAL FERMIONS ARE OF TWO DIFFERENT CLASSES: LEPTONS AND QUARKS

3rd **DEVIATION FROM PLATONIC SIMPLICITY**
THERE IS NOT ONLY ONE BUT THREE FAMILIES OF FUNDAMENTAL FERMIONS

4th **DEVIATION FROM PLATONIC SIMPLICITY**
THE FUNDAMENTAL FERMIONS BECOME MIXED WHEN THE WEAK FORCES ARE SWITCHED ON: MIXINGS EXIST

5th **DEVIATION FROM PLATONIC SIMPLICITY**
THERE ARE DIFFERENT MIXINGS

Figure 15

III THE FUTURE

THE FUTURE

MEMORY

On the occasion of the twenty-fifth anniversary of the Ettore Majorana Foundation and Centre for Scientific Culture (EMFCSC), in order to promote the values of scientific culture worldwide and following a proposal by the World Federation of Scientists (WFS), a special law was voted unanimously by the Sicilian Parliament to establish the

"Ettore Majorana Prize – Erice – Science for Peace".

The Prize has been awarded to fellows who played a leading role in promoting and implementing the goals outlined in the "Erice Statement" for a Science without secrets and without borders.

P.A.M. Dirac, P.L. Kapitza, A.D. Sakharov, E. Teller, V.F. Weisskopf, J.B.G. Dausset, S.D. Drell, M. Gell-Mann, H.W. Kendall, L.C. Pauling, A. Salam, C. Villi, R. Doll, J.C. Eccles, T.D. Lee, L. Montagnier, Qian Jaidong, J.S. Schwinger, U. Veronesi, G.M.C. Duby, R.L. Garwin, S.L. Glashow, D.C. Hodgkin, R.Z. Sagdeev, K.M.B. Siegbahn, Y.P. Velikhov, J. Karle, J.M.P. Lehn, A. Magnéli, N.F. Ramsey, H. Rieben, J.J. van Rood, C.S. Wu, R.L. Mössbauer, A. Müller, H. Kohl, M.S. Gorbachev, H.H. John Paul II, R. Clark, M. Cosandey, A. Peterman, R. Wilson, J. Alderdice, J.J. Friedman, M. Koshiba, S. Coleman, A.N. Chilingarov, P.C.W. Chu, L. Esaki, W.N. Lipscomb Jr., J. Szysko, M.-K. Wu, H.A. Hauptman, D.H. Hubel, R. Huber, B.I. Samuelsson, H. Sun, A.E. Yonath, G. 't Hooft, Y.T. Lee, W. Arber, S.C.C. Ting.

Last December 2013 there was a Ceremony illustrated in the following **Figure 16**.

INFN - EPS

Ceremony

European Physical Society – EPS Historic Site
The AdA Storage Ring at the INFN Frascati National Laboratories

AdA, ADONE, (J/ψ) AND THE 3rd LEPTON

Antonino Zichichi

INFN and University of Bologna, Italy
CERN, Geneva, Switzerland
World Federation of Scientists, Beijing, Geneva, Moscow, New York

INFN Frascati National Laboratories
Thursday, December 5th, 2013 – 11 a.m.

Figure 16

Let me show (**Figure 17**) one page of my Opening Lecture at the AdA-INFN-EPS unveiling Ceremony where it is proved the great value of Blackett teaching about Nature being smarter than all of us and about memory needed in order to never forget the consequences of neglecting this thanks to memory. But memory is needed not only in Physics but in our activities called Civilization.

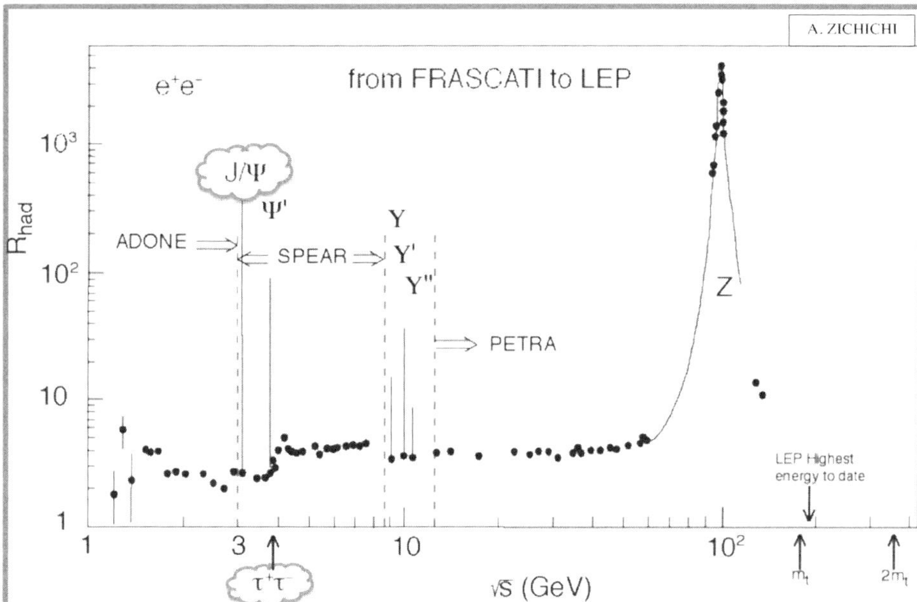

As show in the graph the (J/ψ) could have been discovered at ADONE. It was enough to increase the machine energy by 0.1 GeV as suggested by the CERN-Bologna group, but rejected by the responsible people.

Something similar did happen with PETRA whose lowest energy was not immediately after SPEAR but with the incredible gap where the Lederman's (Υ, Υ', Υ'') could have been discovered.

Let me go back to the origin of this new venture in elementary particle Physics based on the search for new Physics using non Bubble Chamber detectors. Needless to say that, in addition to the search for "narrow resonances", the search for the 3rd lepton (HL) was the reason for a stronger energy increase in ADONE, where the HL (now called τ) would have been discovered.

3

Figure 17

– 1963 –

PAPLEP

Proton **A**nti**P**roton into **Le**pton **P**airs

first search for
the 3rd lepton

and

$$\theta_{PS} \neq \theta_V .$$

The "pre-shower" technology implemented in the CERN experimental set-up for the study of the rare decay modes of the pseudoscalar and vector mesons.

THIS IS THE FIRST EXAMPLE of what is now "standard" in experimental subnuclear physics:
VERY LARGE ACCEPTANCE DETECTORS.
On the rails the "neutron missing mass spectrometer".

$\theta_{PS} \neq \theta_V$
INSTANTONS

In the Physics of Mesons the totally unexpected result was the difference existing between the two mesonic mixing angles, pseudoscalar θ_{PS} and vector θ_V. They should both be zero if $SU(3)_{uds}$ was a good Symmetry.

The experimental results gave $\theta_V = 51°$ and $\theta_{PS} = 10°$ **despite** $SU(3)_{uds}$. This is illustrated in Figure 18.

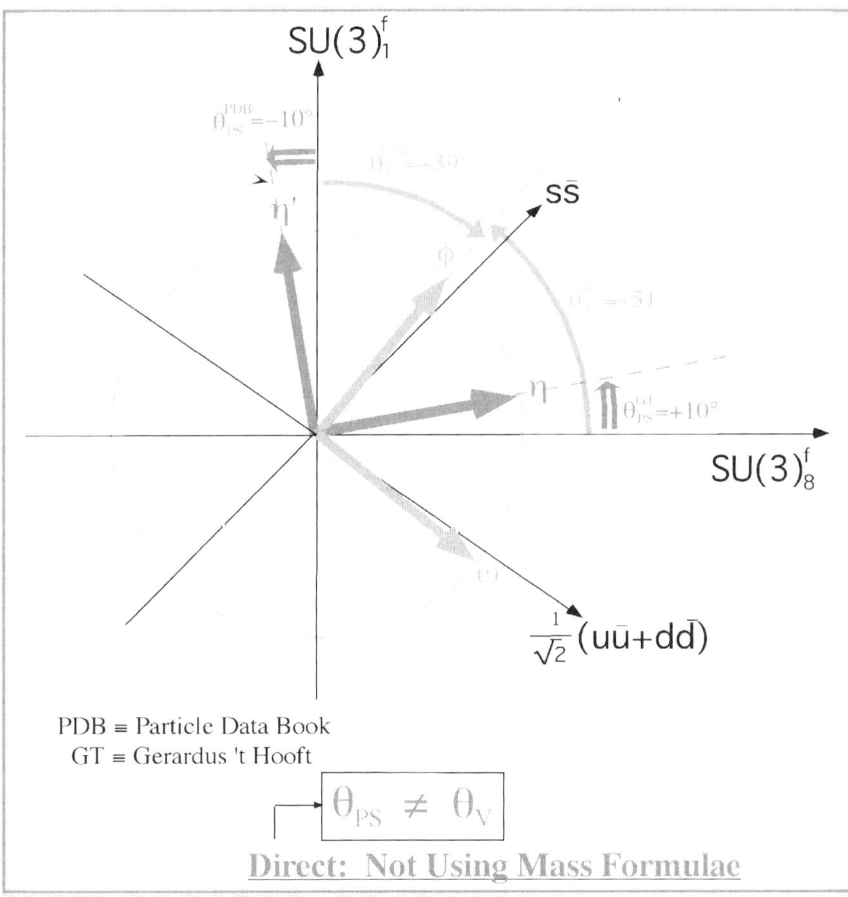

Figure 18

The Physics of Instantons in QCD is needed to explain θ_V and θ_{PS}, i.e. the mixing in the pseudoscalar and in the vector mesons. The existence of Instantons was not known. The Instantons came after the unexpected discovery that $\theta_{PS} \neq \theta_V \neq 0$.

$$\theta_{PS} \neq \theta_V$$

PRESHOWER $\pi/e = 5 \times 10^{-4}$

MUON PUNCH-THROUGH

To simultaneously detect $\mu^{\pm}e^{\mp}$ final states in $\bar{p}p$ annihilation.
The first experimental search for the **THIRD LEPTON** ($HL \equiv \tau$)
and the discovery of the Time-Like Structure of the Proton

$(HL^{\pm}) \equiv \tau^{\pm}$ & $F_p^{em}(q^2)_{\text{time-like}}$

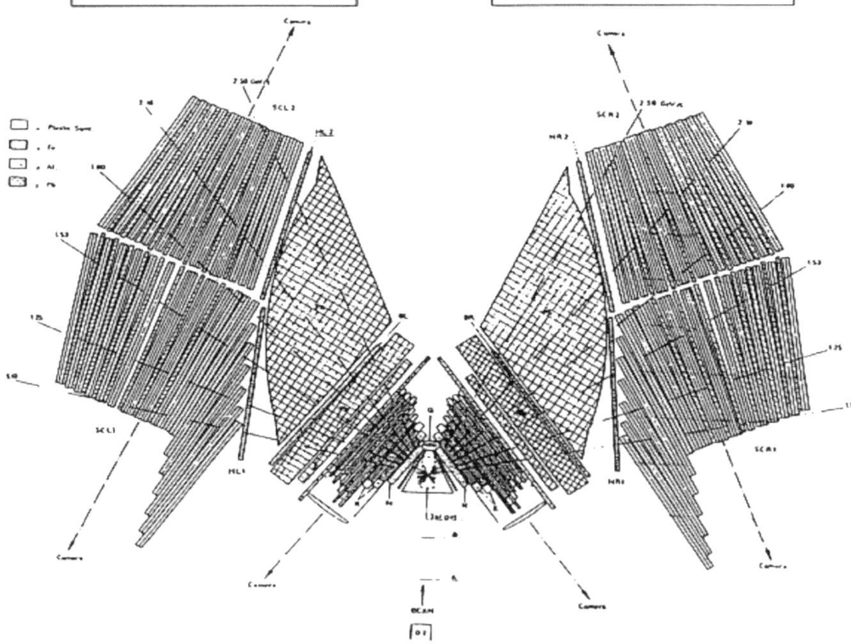

The Future. What about **Predictions? Recall**

UEEC

TOTALLY UNEXPECTED DISCOVERIES: A PERSONAL EXPERIENCE

SUPERSYNTHESIS

1 *Lorentz Invariance (Lorentz).*
2 *The quantum nature of the World (Planck).*
3 *Two worlds: "Time-like" and "Space-like" (Einstein).*
4 *The photon (Einstein).*
5 *Gauge Invariance (Weyl).*
6 *The Space-Time curvature (Einstein).*
7 *Bose-Einstein statistics (Bose-Einstein).*
8 *The structure of the atom (Bohr).*
9 *The wave nature of particles (de Broglie).*
10 *The wave function (Schrödinger) and its probabilistic interpretation (Born).*
11 *Nuclear Forces (Rutherford) and their "glue" (Yukawa).*
12 *The proof that Quantum Mechanics is self consistent (no contradictions) (Von Neumann).*
13 *The Weak Forces (Fermi).*
14 *The Exclusion Principle (Pauli).*
15 *The Uncertainty Principle (Heisenberg).*
16 *Fermi statistics (Fermi).*
17 *The antiparticles (Dirac).*
18 *The neutron (Chadwick).*
19 *Time Reversal Invariance (Wigner).*
20 *Other Invariance Laws (Wigner-Parity; Dirac, Weyl-charge conjugation; Pauli CPT).*
21 *The neutrino (Pauli-Fermi).*
22 *The Stars are "nuclear-fusion" candles (Fermi-Bethe).*
23 *Electronic computing (Von Neumann).*
24 ***The sequence of unexpected Fermi discoveries****: Fermi-coupling, Fermi-gas, Fermi-momentum, Fermi-temperature, Fermi-surface, Fermi-statistics, Fermi-transition, Fermi-length (plus the other two quoted above).*

A FEW EXAMPLES
WHERE I HAVE BEEN INVOLVED

① **The 3rd lepton**
despite *the abundance of neutrinos:* ν_e (ν_μ) *another* ***one*** (ν_{HL}).

② **Antimatter**
despite *S-matrix and C, P, CP breaking.*

③ **Nucleon Time-like** EM structure
despite *S-matrix.* **inst**

④ **No quarks** in violent (pp) collisions
despite *scaling.*

⑤ **Meson mixings**
$\theta_V \neq \theta_{PS}$: $(51^o) \neq (10^o) \neq 0$ ***despite*** $SU(3)_{uds}$. **inst**

⑥ **Effective energy:** *the QCD-light*
despite *QCD.*

⑦ **The running** of α_1 α_2 α_3 versus **energy** at ***a point*** E_{GU}.
(1979) **(1991)** ***despite*** *straight line convergence.*
EGM

Figura 19

1 Therefore **Totally Unexpected Effects** should **show up**.

2 **Effects**, which are impossible to be predicted on the basis of **present knowledge**.

3 Where these effects are most likely to be, no one knows.

4 **But**, with the advent of the Energy Level of **13 TeV** never reached so far with the LHC the ALICE experiment will study the properties of the **Quark-Gluon-Coloured-World (QGCW)**.

At this point Gerardus would ask me the question:

"How do we detect the totally unexpected effects?"

An example is illustrated in the Figure below where beams of known particles (p, n, γ, e, μ) bombard the QGCW and a special set of detectors measures the properties of the outcoming particles.

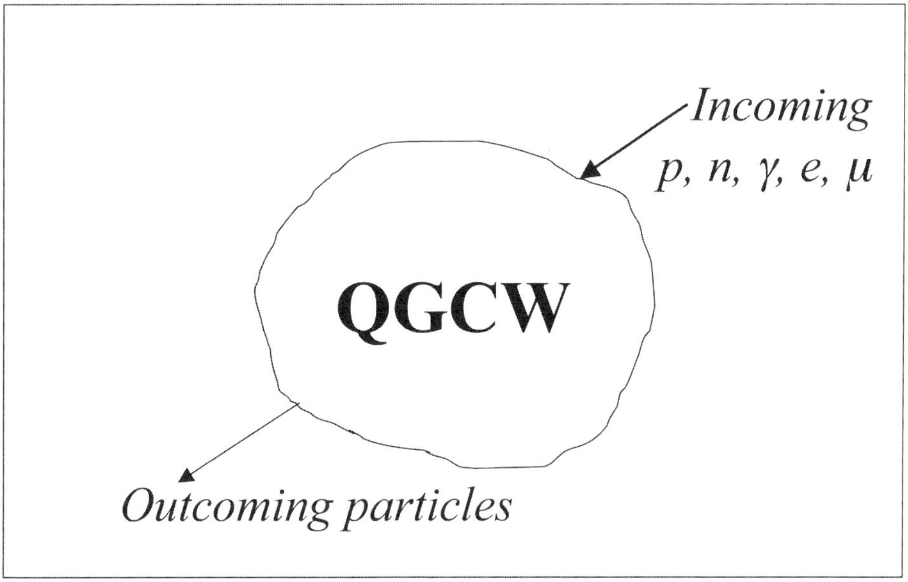

In the following years, much before the second 60th Gerardus Anniversary, we could celebrate another Gerardus Anniversary with the discovery of a totally unexpected effect.

It would be great if this happened using the ALICE apparatus at LHC.

This should indeed be the case if Nature follows the Logic of Complexity at the Fundamental level.

UEEC
"WHAT IF?"

	In History ≡ EWRL		In Science ≡ EBUS
I	What if Julius Caesar had been assassinated many years before?	*I*	What if Galileo Galilei had not discovered that $F = mg$?
II	What if Charles VII had not been able to win the 100 years war?	*II*	What if Newton had not discovered that $F = G\frac{m_1 \cdot m_2}{R_{12}^2}$?
III	What if America had been discovered a few centuries later?	*III*	What if Maxwell had not discovered the unification of electricity, magnetism and optical phenomena, which allowed him to conclude that light is a vibration of the EM field?
IV	What if Napoleon had not been born?	*IV*	What if Becquerel had not discovered radioactivity?
V	What if Louis XVI had been able to win against the 'Storming of the Bastille'?	*V*	What if Planck had not discovered that $h \neq 0$?
VI	What if the 1908 Tunguska Comet had fallen somewhere in Europe instead of Tunguska in Siberia?	*VI*	What if Lorentz had not discovered that space and time cannot both be real?
VII	What if the killer of the Austrian Archduke Franz Ferdinand had been arrested the day before the Sarajevo event?	*VII*	What if Einstein had not discovered the existence of time-like and space-like real worlds? Only in the time-like world, simultaneity does not change, with changing observer.
VIII	What if Lenin had been killed during his travelling through Germany?	*VIII*	What if Rutherford had not discovered the nucleus?
IX	What if Hitler had not been appointed Chancellor by the President of the Republic of Weimar Paul von Hindenburg?	*IX*	What if Hess had not discovered cosmic rays?
X	What if Pyotr Kapitza accepted to be the leader of the USSR H-bomb Project as wanted by Stalin?	*X*	What if Dirac had not discovered his equation, which opens new horizons, including the existence of the antiworld?
XI	What if Nazi Germany had defeated the Soviet Union?	*XI*	What if Fermi had not discovered weak forces?
XII	What if Karol Wojtyla had not been elected Pope, thus becoming John Paul II?	*XII*	What if Fermi and Dirac had not discovered the Fermi–Dirac statistics?
XIII	What if Gorbachev had not been defeated by Yeltsin?	*XIII*	What if Yukawa had not proposed the existence of a "meson" in order to have the nuclear glue?
XIV	What if the USSR had not collapsed?	*XIV*	What if the 'strange particles' had not been discovered in the Blackett Lab?

History is

"Evolution of the World in its Real Life" ≡ (**EWRL**)

Science is

"Evolution of our Basic Understanding of the laws governing the world in its Structure" ≡ (**EBUS**).

DARK MATTER

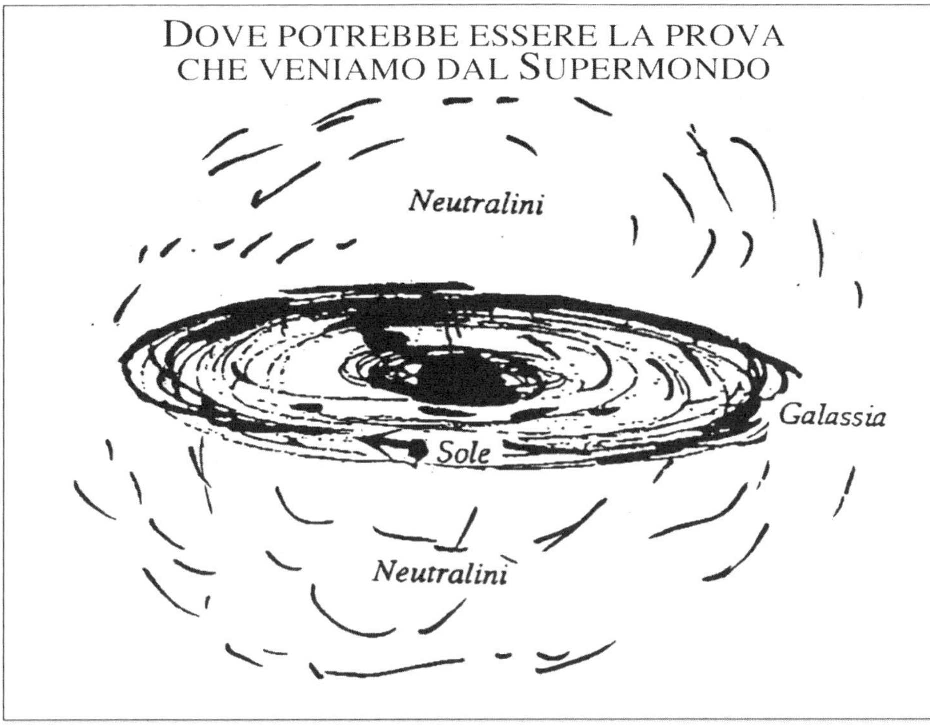

Figure 20

Question: What about UEEC in other fields? For example: condensed matter.

Tony Leggett, University of Illinois, Urbana - Champaign, USA, Nobel 2003 for "Superfluidity": "*It is relatively rare in Condensed-Matter Physics to predict discoveries, it is a field where you fall over them by accident*".

Our Community of **the Erice Subnuclear Physics School** has

Good Students **Great Fellows**

in their fields

GSGF

Recall: the 1st best Student **EUGENE Wigner. This year we have Francis Halzen.**

APPENDIX A
DIRAC Antiparticles & Antimatter

"Those who say that antihydrogen is antimatter should realize that we are not made of hydrogen and we drink water, not liquid hydrogen." These are words spoken by Paul Dirac to physicists gathered around him after his lecture "My life as a Physicist" at the Ettore Majorana Foundation and Centre for Scientific Culture in Erice in 1981 – 53 years after he had, with a single equation, opened new horizons to human knowledge.

CERN COURIER

Apr 29, 2009

Why antihydrogen and antimatter are different

As Paul Dirac realized, the existence of antihydrogen does not in itself prove the existence of antimatter. A look through the history of the subject, and in particular the role played by the CPT theorem, shows that ultimately it came down to experiment to prove the existence of antimatter through the discovery of the antideuteron at CERN in 1965.

Dirac - 1930

The existence of a single antiparticle,
the antielectron, guaranties the existence of
all antiparticles
antimatter
antistars
antigalaxies
because Nature obeys

C-INVARIANCE

1957: *discovery* of C $\neq$ and P $\neq$
by C.S. Wu et al. (Lee and Yang)
but CP ok

1964: *discovery* of CP $\neq$
by CCFT (Christenson, Cronin, Fitch and Turlay)

During the same Time

Triumph of the **S-Matrix** (G. Chew) and *crisis* of the **RQFT**

The existence of ANTIPARTICLES DOES NOT IMPLY the existence of ***ANTIMATTER***.

Dirac - 1965

Dirac at Erice

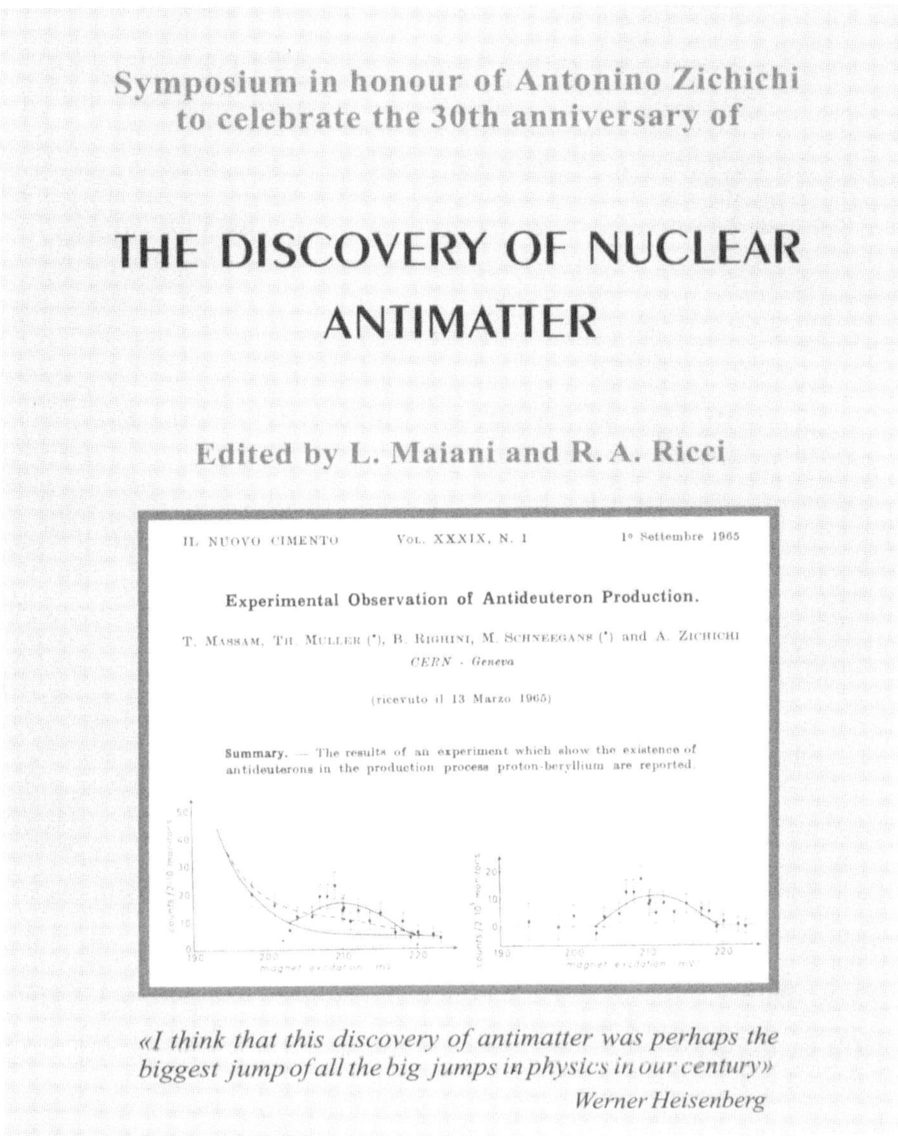

Symposium in honour of Antonino Zichichi
to celebrate the 30th anniversary of

THE DISCOVERY OF NUCLEAR ANTIMATTER

Edited by L. Maiani and R. A. Ricci

IL NUOVO CIMENTO Vol. XXXIX, N. 1 1° Settembre 1965

Experimental Observation of Antideuteron Production.

T. Massam, Th. Muller (*), B. Righini, M. Schneegans (*) and A. Zichichi
CERN - Geneva

(ricevuto il 13 Marzo 1965)

Summary. — The results of an experiment which show the existence of antideuterons in the production process proton-beryllium are reported.

«I think that this discovery of antimatter was perhaps the biggest jump of all the big jumps in physics in our century»
Werner Heisenberg

T.D. Lee - 1995

The CPT theorem rests on a foundation which has to be unsound, at least at the Planck length, and maybe at a much larger distance.

The symmetry of matter and antimatter must rest on experimental evidence.

Mass $\neq$ Matter

$$|\,m_i\rangle \;\equiv\; \textbf{Mass} \equiv \textbf{Antimass} \;\equiv\; |\,m_i\,\rangle$$

$i \equiv \mathbf{1}$ (Intrinsic); $i \equiv \mathbf{2}$ (Confinement); $i \equiv \mathbf{3}$ (Binding)

$$C\,|\,m_i\rangle \;=\; |\,m_i\,\rangle \qquad \boxed{*\,*\,*}$$

$$i \equiv 1, 2, 3$$

$$|\,m_i\,Q_j\,\rangle \;\equiv\; \textbf{Matter} \neq \textbf{Antimatter} \;\equiv\; |\,m_i\,\overline{Q}_j\,\rangle$$

$Q_j \equiv$ Flavour Charges

$$j \equiv (\,u \quad d \quad c \quad s \quad t \quad b\,) \;\equiv\; (1, 2, 3, 4, 5, 6)$$

$$(\,\nu_e \;\; e^- \;\; \nu_\mu \;\; \mu^- \;\; \nu_{HL} \;\; \underset{\tau^-}{\overset{HL^-}{\downarrow}}\,) \;\equiv\; (7, 8, 9, 10, 11, 12)$$

$$C\,|\,m_i\,Q_j\,\rangle \;=\; |\,m_i\,Q_j\,\rangle \qquad \boxed{*\,*\,*}$$

$$i \equiv 1, 2, 3\;;\;\; J \equiv 1, 2, 3, 4, 5, 6, 7, 8, 9, 10, 11, 12.$$

Figure 21

Many deviations from **simplicity are needed** to corroborate the **stability of matter: one charge** (the electric one) is not enough. Another unexpected discovery was the distinguish "colour" charge from "flavour" charge, with six colour charges and twelve flavour charges.

Six "Colour" charges (1 + 2 + 3) → 1 for QED, 2 for QFD, 3 for QCD.

Twelve "Flavour" charges (6 + 6) → 6 for quarks + 6 for leptons.

We can see in Figure 22 that there are **seven decades of developments**, started **from** the **antielectron** and **C-invariance**, **to** arrive at the experimental discovery of **Matter Antimatter Symmetry**. The detailed series of all these totally unexpected events are reported in the volume published on the occasion of the 30th anniversary of the discovery of antimatter [8].

THE INCREDIBLE STORY
TO DISENTANGLE THE ORIGIN OF THE STABILITY OF MATTER
FROM THE ANTIELECTRON TO ANTIMATTER
AND THE UNIFICATION OF ALL GAUGE FORCES

The validity of C invariance from 1927 to 1957.

After the discovery by Thomson in 1897 of the first example of an elementary particle, the Electron, it took the genius of Dirac to theoretically discover the Antielectron thirty years after Thomson.

1927 → Dirac equation [9]; the existence of the antielectron is, soon after, theoretically predicted. Only a few years were needed, after Dirac's theoretical discovery, to experimentally confirm (Anderson, Blackett and Occhialini [10]) the existence of the Dirac antielectron.

1930-1957→ **Discovery of the C operator** [(charge conjugation) H. Weyl and P.A.M. Dirac [11]]; discovery of the P Symmetry Operator [E.P. Wigner, G.C. Wick and A.S. Wightman [12, 13]]; discovery of the T operator (time reversal) [E.P. Wigner, J. Schwinger and J.S. Bell [14, 15, 16, 17]]; discovery of the CPT Symmetry Operator from RQFT (1955-57) [18].

1927-1957→ Validity of C invariance: e^+ [10]; $\bar{p}$ [19]; $\bar{n}$ [20]; $K_2^0 \to 3\pi$ [21] but see LOY [22].

The new era starts: C ≠ ; P ≠ ; CP ≠ (*).

1956 → Lee & Yang P ≠ ; C ≠ [23].

1957 → Before the experimental discovery of P ≠ & C ≠, Lee, Oehme, Yang (LOY) [22] point out that the existence of the second neutral K-meson, $K_2^0 \to 3\pi$, is proof neither of C invariance nor of CP invariance. Flavour antiflavour mixing does not imply CP invariance.

1957 → C.S. Wu et al. P ≠ ; C ≠ [24]; CP ok [25].

1964 → $K_2^0 \to 2\pi \equiv K_L$: CP ≠ [26].

1947-1967→ QED divergences & Landau poles.

1950-1970→ The crisis of RQFT & the triumph of S-matrix theory (i.e. the negation of RQFT).

1965 → Nuclear antimatter is (experimentally) discovered [27]. See also [8].

1968 → The discovery [28] at SLAC of Scaling (free quarks inside a nucleon at very high q^2) but in violent (pp) collisions no free quarks at the ISR are experimentally found [29]. Theorists consider Scaling as being evidence for RQFT not to be able to describe the Physics of Strong Interactions. The only exception is G. 't Hooft who discovers in 1971 that the β-function has negative sign for non-Abelian theories [30].

1971-1973→ β = – ; 't Hooft, Gross & Wilczek. The discovery of **non-Abelian** gauge theories. Asymptotic freedom in the interaction between quarks and gluons [30].

1974 → All gauge couplings $\alpha_1\ \alpha_2\ \alpha_3$ run with q^2 but they do not converge towards a unique point.

1979 → A.P. & A.Z. point out that the new degree of freedom due to SUSY allows the three couplings $\alpha_1\ \alpha_2\ \alpha_3$, **to converge towards a unique point** [31].

1980 → QCD has a "hidden" side: the multitude of final states for each pair of interacting particles: (e^+e^- ; $p\bar{p}$; πp; Kp; νp; pp; etc.)
The introduction of the Effective Energy allows to discover the Universality properties [32] in the multihadronic final states.

1992 → All gauge couplings converge towards a unique point at the gauge unification energy: $E_{GU} \cong 10^{16}$ GeV with $\alpha_{GU} \cong 1/24$ [33, 34].

1994 → The Gap [35] between E_{GU} & the String Unification Energy: $E_{SU} \cong E_{Planck}$.

1995 → **CPT loses its foundations at the Planck scale (T.D. Lee)** [36].

1995-1999→ **No CPT theorem from M-theory (B. Greene)** [37].

1995-2000→ A.Z. points out the need for new experiments to establish if matter-antimatter symmetry or asymmetry are at work.

(*) The symbol ≠ stands for "Symmetry Breakdown".

Figure 22

APPENDIX B
BLACKETT The discovery of the "Vacuum Polarization" (1932) [the 1st example of radiative effect: pre-the Lamb-shift (1947)]

Before the Lamb-shift

Blackett ≡experimental discovery of the "Vacuum Polarization"

The 1st Virtual Physics effect

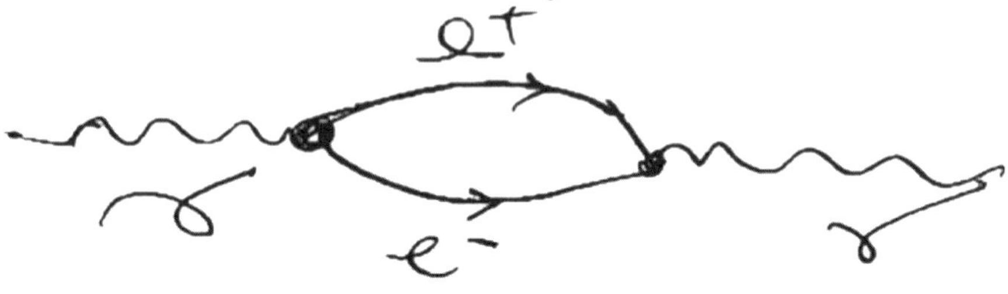

ANTONINO ZICHICHI

**MY TESTIMONY
ON LORD
PATRICK M.S. BLACKETT**

**A LESSON FOR THE FUTURE
OF
OUR SCIENCE**

The content of this book
has its basis in two lectures and two books.

The two lectures are:

"MY TESTIMONY ON LORD PATRICK M.S. BLACKETT"
delivered at the Clore Lecture Theatre – Imperial College, London, 30th April 2014 and the

"OPENING LECTURE"
at International School of Subnuclear Physics, Erice, 25 June 2014.

The two books:

"CREATIVITY IN SCIENCE"
[first edition (1996), World Scientific (1999), translated into Russian and published by YPCC, Moscow (2001)] and

"SUBNUCLEAR PHYSICS - THE FIRST FIFTY YEARS"
[a joint publication by the University and the Academy of Sciences of Bologna, Italy (1998), World Scientific, two editions (2000–2001)].

MY TESTIMONY ON LORD PATRICK M. S. BLACKETT

A LESSON FOR THE FUTURE OF OUR SCIENCE

ANTONINO ZICHICHI
University of Bologna and INFN, Italy
CERN, Geneva, Switzerland
Enrico Fermi Centre, Rome, Italy
Pontifical Academy of Sciences, Vatican City
World Federation of Scientists, Beijing, Geneva, Moscow, New York
Ettore Majorana Foundation and Centre for Scientific Culture, Erice, Italy

CONTENTS

APPENDIX C
THE NEW MANHATTAN PROJECT - CULTURAL HIROSHIMA- 72 PLANETARY EMERGENCIES

INTRODUCTORY SUMMARY

This book is a contribution to the THREE DAYS that the University of Bologna has dedicated to Science, Technology and Culture, so that they enter into the Heart of Everybody.

We are in the **Cultural Hiroshima**, as Fermi feared in the middle of the last Century, when (1945) we were in full Political Hiroshima.

The *Three Days* of Bologna has two objectives: to defeat the Cultural Hiroshima and to solve the problems of Planetary Emergencies, proposing the implementation of a project so that the great achievements of Science come soon in every day life.

The book has some parts in Italian and others in English, which is the universal language of Science.

The parts in Italian will be translated into all the other languages by our colleagues of WFS.

Figure 1 describes the Political Hiroshima (1945) and the Cultural Hiroshima (today).

Figure 2 illustrates the logic of the Project (The New Manhattan Project).

Figure 1

⊙ **POLITICAL HIROSHIMA (1945)**
The existence of Science is totally forgotten (for decisions at very high political level) as if it were not **thanks** to Science that the **new** war **Technologies**, million times more powerful, had been implemented.

⊙ **CULTURAL HIROSHIMA (NOW)**
The existence of Science is totally forgotten for political decisions (at any level) and in every day life as if it is not **thanks** to Science that the **new Technologies** have been invented: technologies which allow to win life-threatening illness, technologies that liberate from rough work and that allow humanity to have a standard of living never before achieved.

Figura 1

⊙ **HIROSHIMA POLITICA (1945)**
Ignorare la Scienza (nelle decisioni politiche di altissimo livello) e andare avanti come se non fosse **grazie** alla Scienza che sono state prodotte le **nuove Tecnologie** belliche, milioni di volte più potenti.

⊙ **HIROSHIMA CULTURALE (OGGI)**
Ignorare la Scienza nelle decisioni politiche (di qualsiasi livello) e nella vita di tutti i giorni come se non fosse **grazie** alla Scienza che sono state prodotte le **nuove Tecniche** che fanno superare malattie altrimenti letali, che liberano la forma di materia vivente cui apparteniamo da lavori spiacevoli e che permettono una qualità di vita mai prima raggiunta.

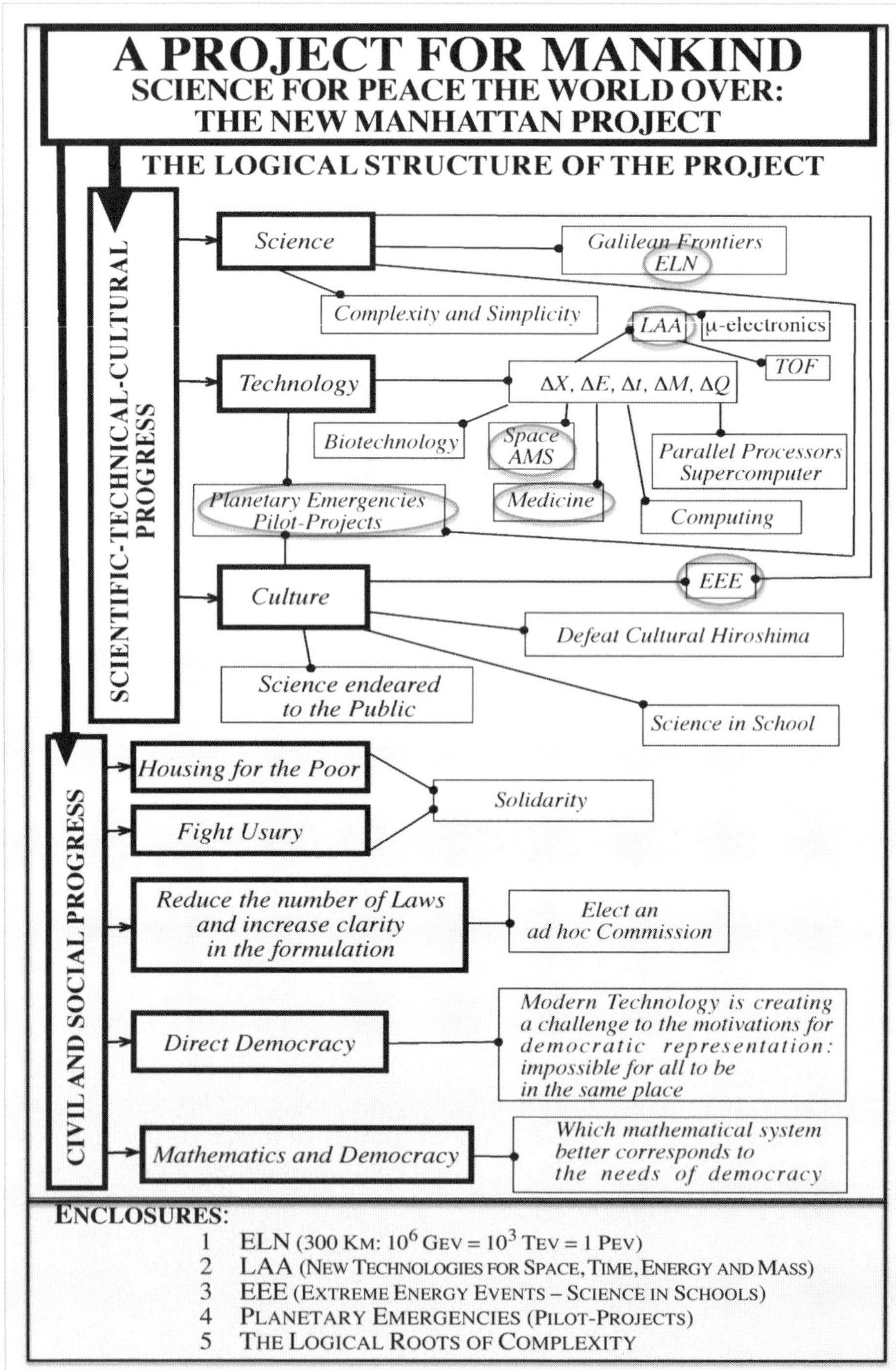
A PROJECT FOR MANKIND
SCIENCE FOR PEACE THE WORLD OVER:
THE NEW MANHATTAN PROJECT
THE LOGICAL STRUCTURE OF THE PROJECT
SCIENTIFIC-TECHNICAL-CULTURAL PROGRESS
Science
Galilean Frontiers
ELN
Complexity and Simplicity
LAA
μ-electronics
TOF
Technology
ΔX, ΔE, Δt, ΔM, ΔQ
Biotechnology
Space AMS
Parallel Processors Supercomputer
Planetary Emergencies Pilot-Projects
Medicine
Computing
Culture
EEE
Defeat Cultural Hiroshima
Science endeared to the Public
Science in School
CIVIL AND SOCIAL PROGRESS
Housing for the Poor
Fight Usury
Solidarity
Reduce the number of Laws and increase clarity in the formulation
Elect an ad hoc Commission
Direct Democracy
Modern Technology is creating a challenge to the motivations for democratic representation: impossible for all to be in the same place
Mathematics and Democracy
Which mathematical system better corresponds to the needs of democracy
ENCLOSURES:
1 ELN (300 Km: 10^6 GeV = 10^3 TeV = 1 PeV)
2 LAA (New Technologies for Space, Time, Energy and Mass)
3 EEE (Extreme Energy Events – Science in Schools)
4 Planetary Emergencies (Pilot-Projects)
5 The Logical Roots of Complexity

THE NEW MANHATTAN PROJECT

Science for Peace the World Over

TABLE OF CONTENTS

I L'Hiroshima Culturale potrebbe portarci a ripetere l'errore fatto con le due invenzioni mai capite dall'alba della Civiltà a Galilei

I-1 L'obiettivo della *Tre Giorni* di Bologna

La *Tre Giorni* che il Rettore Ivano Dionigi ha voluto dedicare alla Scienza alla Tecnica e alla Cultura, affinché entrino nel cuore del grande pubblico, ha come obbiettivo la lotta all'Hiroshima Culturale e la realizzazione del **Nuovo Progetto Manhattan** affinché le 72 Emergenze Planetarie possano essere affrontate e risolte.

La *Tre Giorni* di Bologna vuole portare al centro dell'attenzione mondiale l'Hiroshima Culturale proponendo la realizzazione di un "**Nuovo Progetto Manhattan**" affinché le grandi conquiste della Scienza producano invenzioni tecnologiche da usare nella vita di tutti i giorni. Non aspettando decine di migliaia di anni come avvenne con la **ruota** e il **fuoco**, né cento anni come avvenne con l'elettrone, ma facendo tesoro subito delle grandi scoperte scientifiche. Ecco perché è necessario che nessuno dimentichi la lezione che ci dà la Storia della nostra Civiltà. E cioè che se non fosse stato per Galilei avremmo potuto continuare a essere ancora oggi come i nostri antenati che, nel corso di decine di migliaia di anni, vissero con tecnologie totalmente equivalenti in quanto tutte erano basate sull'uso della **ruota** e del **fuoco** per un motivo molto semplice: nessuno era mai riuscito a capire **l'attrito** e la trasformazione della **massa in energia**.

Il **Nuovo Progetto Manhattan** ha le sue basi nei risultati ottenuti con 100 progetti-pilota realizzati da una comunità scientifica che coinvolge

150 Nazioni. I risultati ottenuti dimostrano che è possibile affrontare le Emergenze Planetarie e – se c'è volontà politica – superarle.

L'Università di Bologna ha il privilegio di essere, non solo la **più antica Università** del mondo occidentale, ma oggi il **fulcro** di un Progetto dalla cui realizzazione dipende il futuro che noi vogliamo assicurare ai nostri posteri, liberandoli dalle 72 Emergenze Planetarie.

Affinché il **Nuovo Progetto Manhattan**, possa diventare realtà è di vitale importanza che le grandi conquiste della Scienza e della Tecnica entrino nella vita di tutti i giorni, quindi nella Cultura del nostro tempo. Solo così può nascere la volontà politica in grado di liberare il futuro dell'Umanità dall'incubo dalle 72 Emergenze Planetarie. Una sintesi di queste Emergenze è riportata in questo volume. Per superarle è necessario capire le basi della posta in gioco.

Se **non c'è volontà politica** una scoperta scientifica impiega decine e decine di anni prima di entrare nella vita di tutti i giorni. La prova viene dalla scoperta del più piccolo "pezzettino di elettricità" cui venne dato il nome di "elettrone".

Viviamo l'era della tecnologia elettronica che – solo adesso – è entrata nella vita di tutti i giorni: telefonini, TV, radio, TAC, internet, e innumerevoli tecnologie mediche, non potrebbero esistere senza la scoperta dell'elettrone, fatta da J.J. Thomson oltre cent'anni fa (nel 1897).

Se ci fosse stato a quei tempi un Governo in grado di varare un **Progetto-tipo-Manhattan** per lo studio delle tecnologie che dovevano scaturire dalla scoperta dell'elettrone, oggi avremmo le invenzioni tecnologiche che i nostri posteri avranno fra cent'anni.

I-2 Cosa insegna il Progetto Manhattan

Il famoso **Progetto Manhattan** riuscì – in appena tre anni (1942-1945) – a trasformare una scoperta scientifica (la fissione nucleare) in "fuoco nucleare di pace" (reattori a fissione) e "fuoco nucleare di guerra" (Hiroshima e Nagasaki).

Il Progetto Manhattan insegna che – **se c'è volontà politica** – una scoperta scientifica può diventare tecnologia "pro" e "contro" la vita e la dignità umana nel giro di pochi anni. Attenzione però. È necessario impedire le invenzioni tecnologiche "contro".

Il motore del progresso è infatti la scoperta scientifica, che genera invenzioni tecnologiche: sta qui il problema.

Le invenzioni possono essere a scopi di pace ma anche a scopi di guerra. Sono infatti le invenzioni tecnologiche "pro" e mai "contro" che hanno portato all'attuale livello di vita: il più alto nella storia dell'Umanità.

Nel famoso incontro di Ginevra (1985), i due più potenti Capi di Stato (Ovest ed Est), Reagan e Gorbachev, dissero che il nemico n. 1 della pace nel mondo erano (e sono) i Laboratori segreti.

THE 72 PLANETARY EMERGENCIES IDENTIFIED BY THE ERICE SCIENTISTS AND CLASSIFIED INTO 15 BASIC SOURCES		NUMBER OF SOURCES
I	*WATER*	4
II	*SOIL*	3
III	*FOOD*	5
IV	*ENERGY*	5
V	*POLLUTION*	7
VI	*LIMITS OF DEVELOPMENT*	4
VII	*CLIMATIC CHANGES*	5
VIII	*GLOBAL MONITORING OF THE PLANET*	6
IX	*NEW MILITARY THREATS IN THE MULTIPOLAR WORLD*	3
X	*SCIENCE AND TECHNOLOGY FOR DEVELOPING COUNTRIES TO AVOID A NORTH-SOUTH ENVIRONMENTAL HOLOCAUST*	3
XI	*THE PROBLEM OF ORGAN SUBSTITUTION*	5
XII	*MEDICINE, INFECTIOUS AND OTHER DISEASES*	7
XIII	*CULTURAL POLLUTION*	8
XIV	*INFORMATION SECURITY AND COMMON DEFENSE AGAINST COSMIC OBJECTS*	2
XV	*THE HUGE MILITARY INVESTMENTS*	5
	Total	**72**

PLANETARY EMERGENCIES

(From left) Luigi Dadda, Pierre A. Piroué, Enrico Bignami, Yuval Ne'eman, Richard L. Garwin, John C. Eccles, Eugene P. Wigner, A.Z., Edward Teller, Paul A.M. Dirac, George Charpak. This is one of the first sessions devoted to the Erice Statement and to the future of our planet.

www.ilcigno.org

€

Finito di stampare
nel mese di giugno 2015
presso Tap Grafiche - Poggibonsi (SI)
per conto de
Il Cigno GG Edizioni
Piazza San Salvatore in Lauro, 15 00186 Roma
sito nel Complesso Monumentale di San Salvatore in Lauro,
un immobile dell'Ente morale Pio Sodalizio dei Piceni.

PIO SODALIZIO DEI PICENI

ALMA MATER STUDIORUM
UNIVERSITÀ DI BOLOGNA

RE UNI ON

IL PRIMO RADUNO MONDIALE DEI LAUREATI UNIVERSITÀ DI BOLOGNA

19-21 GIUGNO 2015

Partiti insieme per un lungo viaggio,
torniamo a casa per diverse vie.

Catullo

PROGRAMMA

11.00 **Palazzo Re Enzo | Sala di Re Enzo**

SCONFIGGERE L'HIROSHIMA CULTURALE: UN PO' DI SCIENZA NELLA VITA DI TUTTI I GIORNI

Perché le scoperte scientifiche impiegano decine di anni per diventare tecnologia? Un programma di informazione scientifica per convincere le organizzazioni mondiali a sostenere politicamente ed economicamente la ricerca scientifica e risolvere le emergenze planetarie.

- **Antonino Zichichi**, Fisico, intervistato da **Fabrizio Binacchi**, Direttore Rai Emilia Romagna

REFERENCES

[1] A record of the annual evolution in the multitude of baryonic and mesonic states can be found in the proceedings of the Erice Schools [1bis]. An example of the proliferation in the meson resonances is "*Meson Resonances and Related Electromagnetic Phenomena*", R.H. Dalitz and A. Zichichi (eds), Editrice Compositori, Proceedings of the EPS Conference, Bologna (1971).

[1bis] The progress of Subnuclear Physics as reported in the Erice School books whose titles follow.
1) *Strong, Electromagnetic and Weak Interactions* (1963); 2) *Symmetries in Elementary Particle Physics* (1964); 3) *Recent Developments in Particle Symmetries* (1965); 4) *Strong and Weak Interactions - Present Problems* (1966); 5) *Hadrons and their Interactions* (1967); 6) *Theory and Phenomenology in Particle Physics* (1968); 7) *Subnuclear Phenomena* (1969); 8) *Elementary Processes at High Energy* (1970); 9) *Properties of the Fundamental Interactions* (1971); 10) *Highlights in Particle Physics* (1972); 11) *Laws of Hadronic Matter* (1973); 12) *Lepton and Hadron Structure* (1974); 13) *New Phenomena in Subnuclear Physics* (1975); 14) *Understanding the Fundamental Constituents of Matter* (1976); 15) *The Whys of Subnuclear Physics* (1977); 16) *The New Aspects of Subnuclear Physics* (1978); 17) *Pointlike Structures Inside and Outside Hadrons* (1979); 18) *The High-Energy Limits* (1980); 19) *The Unity of Fundamental Interactions* (1981); 20) *Gauge Interactions: Theory and Experiments* (1982); 21) *How far are we from the Gauge Forces* (1983); 22) *Quarks, Leptons and their Constituents* (1984); 23) *Old and New Forces of Nature* (1985); 24) *The Super World I* (1986); 25) *The Super World II* (1987); 26) *The Super World III* (1988); 27) *The Challenging Questions* (1989); 28) *Physics up to 200 TeV* (1990); 29) *Physics at the Highest Energy and Luminosity: To Understand the Origin of Mass* (1991); 30) *From Superstrings to the Real Superworld* (1992); 31) *From Supersymmetry to the Origin of Space-Time* (1993); 32) *From Superstring to Present-day Physics* (1994); 33) *Vacuum and Vacua: the Physics of Nothing* (1995); 34) *Effective Theories and Fundamental Interactions* (1996); 35) *Highlights: 50 Years Later* (1997).
Vol. 1 was published by W.A. Benjamin, Inc., New York; 2-8 and 11-12 by Academic Press, New York and London; 9-10 by Editrice Compositori, Bologna; 13-29 by Plenum Press, New York and London; 30-36 by World Scientific.

[2] M. Gell-Mann, "*The Eightfold Way — A Theory of Strong-Interaction Symmetry*", California Institute Technology Synchrotron Lab. Report 20 (1961); "*Derivation of Strong Interactions from a Gauge Invariance*", Y. Ne'eman, *Nuclear Phys.* 26, 222 (1961); The experimental confirmation, in 1964, by Samios and collaborators [2bis] of the existence of the missing member of the baryonic decuplet, W^-, appeared to be, at the time, a triumph for the "eightfold way". The choice of the letter W, the last in the Greek alphabet, was due to the conviction that this particle was going to be the last

ever to be discovered. See also: M. Gell-Mann and Y. Ne'eman, "*The Eightfold Way*", W.A. Benjamin inc., New York and Amsterdam (1964).

[2bis] *Observation of a Hyperon with Strangeness Minus Three*
V.E. Barnes, P.L. Connolly, D.J. Crennell, B.B. Culwick, W.C. Delaney, W.B. Fowler, P.E. Hagerty, E.L. Hart, N. Horwitz, P.V.C. Hough, J.E. Jensen, J.K. Kopp, K.W. Lai, J. Leitner, J.L. Lloyd, G.W. London, T.W. Morris, Y. Oren, R.B. Palmer, A.G. Prodell, D. Radojictc, D.C. Rahm, C.R. Richardson, N.P. Samios, J.R. Sanford, R.P. Shutt, J.R. Smith, D.L. Stonehill, R.C. Strand, A.M. Thorndike, M.S. Webster, W.J. Willis and S.S. Yamamoto, *Phys. Rev. Lett.* 12, 204 (1964).

[3] *The "Leading"-Baryon Effect in Strong, Weak, and Electromagnetic Interactions*
M. Basile, G. Cara Romeo, L. Cifarelli, A. Contin, G. D'Alì, P. Di Cesare, B. Esposito, P. Giusti, T. Massam, R. Nania, F. Palmonari, V. Rossi, G. Sartorelli, M. Spinetti, G. Susinno, G. Valenti, L. Votano and A. Zichichi, *Lettere al Nuovo Cimento* 32, 321 (1981).

[4] For a complete set of references concerning this topic see "*The Creation of Quantum ChromoDynamics and the Effective Energy*" V.N. Gribov, G. 't Hooft, G. Veneziano and V.F. Weisskopf; L.N. Lipatov (ed), Academy of Sciences and University of Bologna, INFN, SIF, published by World Scientific (1998).

[5] INTERNATIONAL SCHOOL OF COSMOLOGY AND GRAVITATION
J. Wheeler, 1972–*High Energy Astrophysics and its Relation to Elementary Particle Physics*; and 1992–*String Quantum Gravity and Physics at the Planck Energy Scale.*

[6] INTERNATIONAL SCHOOL OF COSMOLOGY AND GRAVITATION
N. Rosen, 1977–*Theories of Gravitation.*

[7] INTERNATIONAL SCHOOL OF COSMOLOGY AND GRAVITATION
P.G. Bergmann, 1979–*Spin, Torsion, Rotation and Supergravity*; 1982–*Unified Field Theories of More Than 4 Dimensions Including Exact Solutions*; 1985–*Topological Properties and Global Structure of Space-Time*; 1987–*Gravitation Measurements, Fundamental Metrology and Constants*; 1990–*Symposium on the Problem of the Cosmological Constant in Honor of Peter Gabriel Bergmann's 75th Birthday*;
P.G. Bergmann and Zheniju Zhang, 1991–*Black Hole Physics*;
P.G. Bergmann, V. De Sabbata and T.-H. Ho, 1993–*Cosmology and Particle Physics*;
P.G. Bergmann, V. De Sabbata and H.-J. Treder, 1995–*Quantum Gravity*;
P.G. Bergmann, G. 't Hooft and G. Veneziano, 1998–*From the Planck Length to the Hubble Radius.*

[8] *The Discovery of Nuclear Antimatter*
Proceedings of the *Symposium to celebrate the 30th anniversary of the Discovery of Nuclear Antimatter*, L. Maiani and R.A. Ricci (eds), Conference Proceedings 53, 1, Italian Physical Society, Bologna, Italy (1995).

[9] *The Quantum Theory of the Electron*
P.A.M. Dirac, *Proc. Roy. Soc.* (London) A117, 610 (1928);
The Quantum Theory of the Electron, Part II
P.A.M. Dirac, *Proc. Roy. Soc.* (London) A118, 351 (1928).

[10] *The Positive Electron*
C.D. Anderson, *Phys. Rev.* 43, 491 (1933);
Some Photographs of the Tracks of Penetrating Radiation
P.M.S. Blackett and G.P.S. Occhialini, *Proc. Roy. Soc.* A139, 699 (1933).

[11] *Gruppentheorie und Quantenmechanik*
H. Weyl, 2nd ed., 234 (1931).

[12] *Unitary Representations of the Inhomogeneous Lorentz Group*
E.P. Wigner, *Ann. Math.*, 40, 149 (1939).

[13] *Intrinsic Parity of Elementary Particles*
G.C. Wick, E.P. Wigner, and A.S. Wightman, *Phys. Rev.* 88, 101 (1952).

[14] *Über die Operation der Zeitumkehr in der Quanten-mechanik*
E.P. Wigner, *Gött. Nach.* 546-559 (1931). Here for the first time an anti-unitary symmetry appears.

[15] E.P. Wigner, *Ann. Math.* 40, 149 (1939).

[16] J. Schwinger, *Phys. Rev.* 82, 914 (1951).

[17] *Time Reversal in Field Theory*
J.S. Bell, *Proc. Roy. Soc.* (London) A231, 479-495 (1955).

[18] To the best of my knowledge, the CPT theorem was first proved by W. Pauli in his article '*Exclusion Principle, Lorentz Group and Reflection of Space-Time and Charge*', in 'Niels Bohr and the Development of Physics' [Pergamon Press, London, page 30 (1955)], which in turn is an extension of the work of J. Schwinger [*Phys. Rev.* 82, 914 (1951); '*The Theory of Quantized Fields. II.*', *Phys. Rev.* 91, 713 (1953); '*The Theory of Quantized Fields. III.*', *Phys. Rev.* 91, 728 (1953); '*The Theory of Quantized Fields. VI.*', *Phys. Rev.* 94, 1362 (1954)] and G. Lüders, '*On the Equivalence of Invariance under Time Reversal and under Particle-Anti-particle Conjugation for Relativistic Field Theories*' [*Dansk. Mat. Fys. Medd.* 28, 5 (1954)], which referred to an unpublished remark by B. Zumino. The final contribution to the CPT theorem was given by R. Jost, in '*Eine Bemerkung zum CPT Theorem*' [*Helv. Phys. Acta* 30, 409 (1957)], who showed that a weaker condition, called 'weak local commutativity' was sufficient for the validity of the CPT theorem.

[19] *Observation of Antiprotons*
O. Chamberlain, E. Segrè, C. Wiegand, and T. Ypsilantis, *Physical Review* 100, 947 (1955).

[20] *Anti-Neutrons Produced from Anti-Protons in Charge Exchange Collisions*
B. Cork, G.R. Lambertson, O. Piccioni, W.A. Wenzel, *Physical Review* 104, 1193 (1957).

[21] *Observation of Long-Lived Neutral V Particles*
K. Lande, E.T. Booth, J. Impeduglia, L.M. Lederman, and W. Chinowski, *Physical Review* 103, 1901 (1956).

[22] *Remarks on Possible Noninvariance under Time Reversal and Charge Conjugation*
T.D. Lee, R. Oehme, and C.N. Yang, *Physical Review* 106, 340 (1957).

[23] *Question of Parity Conservation in Weak Interactions*
T.D. Lee and C.N. Yang, *Phys. Rev.* 104, 254 (1956).

[24] *Experimental Test of Parity Conservation in Beta Decay*
C.S. Wu, E. Ambler, R.W. Hayward, D.D. Hoppes, *Phys. Rev.* 105, 1413 (1957);

Observation of the Failure of Conservation of Parity and Charge Conjugation in Meson Decays: The Magnetic Moment of the Free Muon
R. Garwin, L. Lederman, and M. Weinrich, *Phys. Rev.* 105, 1415 (1957);

Nuclear Emulsion Evidence for Parity Non-Conservation in the Decay Chain $\pi^+\mu^+e^+$
J.I. Friedman and V.L. Telegdi, *Phys. Rev.* 105, 1681 (1957).

[25] *On the Conservation Laws for Weak Interactions*
L.D. Landau, *Zh. Éksp. Teor. Fiz.* 32, 405 (1957).

[26] *Evidence for the 2π Decay of the K_2^0 Meson*
J. Christenson, J.W. Cronin, V.L. Fitch, and R. Turlay, *Physical Review Letters* 113, 138 (1964).

[27] *Experimental Observation of Antideuteron Production*
T. Massam, Th. Muller, B. Righini, M. Schneegans, and A. Zichichi, *Nuovo Cimento* 39, 10 (1965).

[28] The first report on 'scaling' was presented by J.I. Friedman at the 14th International Conference on *High Energy Physics* in Vienna, 28 August-5 September 1968. The report was presented as paper n. 563 but not published in the Conference Proceedings. It was published as a SLAC preprint. The SLAC data on scaling were included in the Panofsky general report to the Conference where he says «... the apparent success of the parametrization of the cross-sections in the variable ν / q^2 in addition to the large cross-section

itself is at least indicative that point-like interactions are becoming involved». '*Low q^2 Electrodynamics, Elastic and Inelastic Electron (and Muon) Scattering*', W.K.H. Panofsky in Proceedings of 14th International Conference on *High Energy Physics* in Vienna 1968, J. Prentki and J. Steinberger (eds), page 23, published by CERN (1968). The following physicists participated in the inelastic electron scattering experiments: W.B. Atwood, E. Bloom, A. Bodek, M. Breidenbach, G. Buschhorn, R. Cottrell, D. Coward, H. DeStaebler, R. Ditzler, J. Drees, J. Elias, G. Hartmann, C. Jordan, M. Mestayer, G. Miller, L. Mo, H. Piel, J. Poucher, C. Prescott, M. Riordan, L. Rochester, D. Sherden, M. Sogard, S. Stein, D. Trines, and R. Verdier. For additional acknowledgements see J.I. Friedman, H.W. Kendall and R.E. Taylor, '*Deep Inelastic Scattering: Acknowledgements*', *Les Prix Nobel 1990*, (Almqvist and Wiksell, Stockholm/Uppsala 1991), also *Rev. Mod. Phys.* 63, 629 (1991). For a detailed reconstruction of the events see J.I. Friedman '*Deep Inelastic Scattering Evidence for the Reality of Quarks*' in '*History of Original Ideas and Basic Discoveries in Particle Physics*', H.B. Newman and T. Ypsilantis (eds), Plenum Press, New York and London, 725 (1994).

[29] *Quark Search at the ISR*
T. Massam and A. Zichichi, *CERN preprint*, June 1968;

Search for Fractionally Charged Particles Produced in Proton-Proton Collisions at the Highest ISR Energy
M. Basile, G. Cara Romeo, L. Cifarelli, P. Giusti, T. Massam, F. Palmonari, G. Valenti and A. Zichichi, *Nuovo Cimento* 40A, 41 (1977); and

A Search for quarks in the CERN SPS Neutrino Beam
M. Basile, G. Cara Romeo, L. Cifarelli, A. Contin, G. D'Alì, P. Giusti, T. Massam, F. Palmonari, G. Sartorelli, G. Valenti and A. Zichichi, *Nuovo Cimento* 45A, 281 (1978).

[30] *Subnuclear Physics - The First Fifty Years: Highlights from Erice to ELN*
A. Zichichi; O. Barnabei, P. Pupillo and F. Roversi Monaco (eds), Academy of Sciences and University of Bologna, published by World Scientific, Series in 20th Century Physics, Vol. 24 (2000).

[31] *New Developments in Elementary Particle Physics*
A. Zichichi, *Rivista del Nuovo Cimento* 2, n. 14, 1 (1979). The statement on page 2 of this paper, «*Unification of all forces needs first a Supersymmetry. This can be broken later, thus generating the sequence of the various forces of nature as we observe them*», was based on a work by A. Petermann and A. Zichichi in which the renormalization group running of the couplings using supersymmetry was studied with the result that the convergence of the three couplings improved. This work was not published, but perhaps known to a few. The statement quoted is the first instance in which it was pointed out that supersymmetry might play an important role in the convergence of the gauge couplings. In fact, the convergence of three straight

lines ($\alpha_1^{-1}\alpha_2^{-1}\alpha_3^{-1}$) with a change in slope is guaranteed by the Euclidean geometry, as long as the point where the slope changes is tuned appropriately. What is incorrect about the convergence of the couplings is that, with the initial conditions given by the LEP results, the change in slope needs to be at $M_{SUSY} \sim 1$ TeV as claimed by some authors not aware in 1991 of what was known in 1979 to A. Petermann and A. Zichichi.

[32] *The Creation of Quantum ChromoDynamics and the Effective Energy*
V.N. Gribov, G. 't Hooft, G. Veneziano and V.F. Weisskopf; L.N. Lipatov (ed), a joint publication by the University and the Academy of Sciences of Bologna, Italy (1998); World Scientific Series in 20th Century Physics, Vol. 25 (2000).

[33] *The Effective Experimental Constraints on M_{SUSY} and M_{GUT}*
F. Anselmo, L. Cifarelli, A. Petermann and A. Zichichi, *Nuovo Cimento* 104A, 1817 (1991).

[34] *The Simultaneous Evolution of Masses and Couplings: Consequences on Supersymmetry Spectra and Thresholds*
F. Anselmo, L. Cifarelli, A. Petermann and A. Zichichi, *Nuovo Cimento* 105A, 1179 (1992).

[35] *A Study of the Various Approaches to M_{GUT} and α_{GUT}*
F. Anselmo, L. Cifarelli and A. Zichichi, *Nuovo Cimento* 105A, 1335 (1992). This is the paper on the study of the Gap given as Reference [35bis].

[35bis] *Searching for the Superworld*
S. Ferrara and R.M. Mössbauer (eds), in honour of A. Zichichi on the occasion of the Sixth Centenary celebrations of the University of Turin, Italy. World Scientific Series in 20th Century Physics, Vol. 39 (2007).

[36] *Are Matter and Antimatter Symmetric?*
T.D. Lee, in Proceedings of the '*Symposium to celebrate the 30th anniversary of the Discovery of Nuclear Antimatter*', L. Maiani and R.A. Ricci (eds), Conference Proceedings 53, page 1, Italian Physical Society, Bologna, Italy (1995).

[37] *String Theory: the Basic Ideas*
B. Greene, Erice Lectures - Discussion 1999 in '*Basics and Highlights in Fundamental Physics*', A. Zichichi (ed), World Scientific (2001).

LECTURER: A. ZICHICHI

The Future of our Physics

Scientific Secretaries: P. Owen, L. Parisi

DISCUSSION

- *Campbell:*

What is the difference between the Quark Gluon Colour World (QGCW) and Quark Gluon Plasma (QGP)?

- *Zichichi:*

The QGCW is made at higher energy and should produce surprising new signatures which are not yet defined. It could be that the quark masses are all nearly zero, even the very high ones. It could be the zero-colour QCD state that makes quarks massive.

- *Campbell:*

Can you use jets from quarks or gluons to study the QGCW?

- *Zichichi:*

One can study the QGCW with an external probe (e,p,n) and hope for an anomaly. We need a big discovery, probably unpredicted.

- *Meyer:*

What were you referring to by saying 'the universe is 98% empty'?

- *Zichichi:*

If you take the Schwarzschild equation and you plot the mass radius relation, then it fits perfectly well, as I have shown. We would like to know why the universe has 10^80 protons, neutrons and electrons and is 90% empty. There is only one answer to this question, view the Schwarzschild equation.

- *Stopyra:*

Do you think there is a place for holistic theories like complexity in physics?

Zichichi:

Complexity is everywhere. It can be described in terms of energy and any other parameter that you can think of. I have provided you with examples of phenomena which are considered complex. I have collected at least seventy definitions of complexity, then I have selected seven definitions among the others and we have discussed them. The important point is, no matter what the definition of complexity is, the experimental evidences of the existence of complexity are two: the Anderson-Feynman-Beethoven-type phenomena and the UEEC events, i.e. Unexpected Events which produce Enormous Consequences like the Sarajevo-type effects. So, referring to the presentation, what is the relation between history and physics? History is the Evolution of the World in its Real Life (EWRL), science is the Evolution of our Basic Understanding of the laws governing the world in its Structure (EBUS). They have in common cultural Evolution and both of them are dominated by UEEC (on page 109 of the presentation, many examples are provided). The conclusion is that in both fields evolution is present and implies complexity.

- *Kovalenko:*

Are there similarities in particle production related to string fragmentation mechanism when we talk about number of charged particles and transverse momentum?

- *Zichichi:*

The point is that whatever you are looking at if the effective energy is the same the physics should be the same regardless of the final states you observe.

- *Bellantuono:*

Are there features and behaviours common to a general class of complex phenomena? At present is it possible to try to study complexity with a unified top down approach starting from such common features?

- *Zichichi:*

Yes, the general class is the Anderson-Feynman-Beethoven one. The phenomena belonging to such class have laws and regularities ignoring the existence of the fundamental laws from which they originate. The transition from QCD to nuclear physics is very important in this sense: The study of nuclear physics started long before the formulation of QCD.

- *'t Hooft:*

As for the QGCW, I think you have to be more precise in its definition. Regardless whether you have hadrons or quarks, confined or not, all physical states are locally gauge-invariant, hence locally colorless. A better definition of the different phases is whether there are flux tubes confining quarks or not or whether there are only finite-size colorless bubbles called hadrons, surrounded by a region that looks like the vacuum.

Zichichi:

I completely agree with such a definition.

Majorana Fermions in Particle Physics, Solid State and Quantum Information

L. Borsten[1] **and M. J. Duff**[2]

[1] *School of Theoretical Physics, Dublin Institute for Advanced Studies, 10 Burlington Road, Dublin 4, Ireland*

[2] *Theoretical Physics, Blackett Laboratory, Imperial College London, London SW7 2AZ, United Kingdom*

leron@stp.dias.ie
m.duff@imperial.ac.uk

ABSTRACT

This review is based on lectures given by M. J. Duff summarising the far reaching contributions of Ettore Majorana to fundamental physics, with special focus on Majorana fermions in all their guises. The theoretical discovery of the eponymous fermion in 1937 has since had profound implications for particle physics, solid state and quantum computation. The breadth of these disciplines is testimony to Majorana's genius, which continues to permeate physics today. These lectures offer a whistle-stop tour through some limited subset of the key ideas. In addition to touching on these various applications, we will draw out some fascinating relations connecting the normed division algebras $\mathbb{R}, \mathbb{C}, \mathbb{H}, \mathbb{O}$ to spinors, trialities, K-theory and the classification of stable topological states of symmetry-protected gapped free-fermion systems.

Contents

1 Introduction: Majorana's vision

In April 1937, but eight years into a prolific career tragically cut short just eleven months later, Ettore Majorana discovered that it is theoretically consistent for a fermion to be its own anti-particle [1]. Majorana fermions, as they have come to be known in a befitting tribute to their originator, were at first slow to find their raison d'être, but have since had a profound impact on an impressive array of seemingly disparate fields including particle physics, condensed matter and quantum computing. They are also crucial to the notion of supersymmetry and hence related approaches to unification and quantum gravity such as M-theory. This particular story is described in the sister contribution dedicated to Majorana [2] appearing in these same proceedings. Here instead we attempt to review Majorana fermions in particle physics, solid state and quantum information, guided chiefly by the excellent accounts given in [3–13]. Although all are tied tightly to Majorana's pioneering insights some examples rather stretch the original incarnation of 1937 and others even the very notion of a fermion. It would therefore serve us well to begin, more modestly, as Majorana did. Here we adopt west coast conventions.

When Dirac first formulated his equation for a relativistic electron,

$$(i\gamma^\mu\partial_\mu - m)\psi = 0, \qquad \{\gamma^\mu, \gamma^\nu\} = 2\eta^{\mu\nu}, \tag{1.1}$$

he hit upon a realisation of the four matrices γ^μ containing both real and imaginary elements, the implication being that ψ itself must be complex valued. What a triumph; this single observation anticipated the existence of antimatter. Electrons and positrons are simultaneously accommodated by ψ and ψ^* distinct. However, this success does not imply the mathematical or theoretical necessity of a complex ψ. Motivated in part by mathematical elegance Majorana questioned this apparent implication and in doing so discovered an alternative realisation in which the four γ^μ are all pure imaginary and hence ψ may be consistently taken as real valued. Pure imaginary γ and real ψ are referred to as a "really real" Majorana representation and constitute a special, but entirely equivalent, case of the more general reality condition used to define Majorana fermions,

$$\overline{\psi}^D = \overline{\psi}^M, \tag{1.2}$$

where $\overline{\psi}^D$ and $\overline{\psi}^M$ denote the Dirac and Majorana conjugates, respectively.

Projecting (1.1) with the usual chiral operator, $P_R = \frac{1}{2}(1+\gamma_5)$, in the really real Majorana representation we obtain

$$i\gamma^\mu\partial_\mu\psi_L - m\psi_L^* = 0, \tag{1.3}$$

where we have used $P_R\psi = P_L^*\psi = (P_L\psi)^*$. Hence, the global phase symmetry $\psi \to e^{i\theta}\psi$ enjoyed by the Dirac equation is broken by the mass term. Consequently, massive Majorana fermions cannot be coupled to a U(1) gauge potential and are uncharged. Note, for commuting variables the kinetic and mass terms of the action,

$$S[\psi] = -\frac{1}{2}\int \bar{\psi}(i\gamma^\mu\partial_\mu - m)\psi, \tag{1.4}$$

are symmetric and antisymmetric respectively. Hence the kinetic term is a total derivative while the mass term vanishes identically, rendering the action trivial. This led Majorana to invoke Grassmann valued fields. That these are now a pedestrian part of our mathematical description of fermions again speaks to Majorana's vision.

We conclude: the tenets of relativity and quantum theory allow, in principle, for neutral fermions that are their own anti-particles. This of course leaves their relevance to fundamental physics unaddressed. Majorana himself speculated that such fermions could represent neutrinos on the basis that they do not carry any electromagnetic charge, but at the time of writing neutrinos were largely hypothetical and little of their would-be properties was known. The picture of neutrinos that evolved over the subsequent years appeared to rule out Majorana's tentative hypothesis. In the standard model of particle physics they are described by massless Weyl spinors with distinct anti-particles. However, this framework has since been eroded by neutrino oscillations [14], which imply, at least without some new and rather radical ingredients, that neutrinos carry a small mass. The off-diagonal mass terms required for neutrino oscillations (or any mass terms for that matter) are incompatible with the hypothesis that neutrinos are Weyl fermions. Of course, this does not necessarily mean that neutrinos are Majorana as opposed to Dirac, but there are processes which could in principle distinguish the two possibilities, such as double beta decay. See [15] and the references therein.

Single beta decay is energetically forbidden in most nuclei with even atomic number Z and even neutron number N. However, a large class of even Z, even N nuclei do allow the second order process of double beta decay, accompanied with the emission of two anti-neutrinos conserving lepton number. This process has been observed in a number of isotopes and is consistent with the standard model. If, however, neutrinos are Majorana then there is another possible double beta decay process that does not produces any neutrinos and hence changes the lepton number by two. If neutrinos are their own anti-particles then beta decay is possible via either the emission or absorption equivilantly. Hence, double beta decay can be mediated by two neutrons exchanging a single neutrino, a process forbidden within standard model. Although there exist a number of alternative mechanisms, the observation of neutrino-less double beta decay would unambiguously imply neutrinos are Majorana due to a theorem of Schechter and Valle [16]. The decay rate of Majorana neutrino mediated neutrino-less double beta decay is quadratic in the effective Majorana neutrino mass, consistent with the observation that in the massless limit Dirac and Majorana neutrinos are in general physically indistinguishable. Although neutrino oscillation measurements cannot separate Dirac and Majorana neutrinos, they do provide some constraints on the effective Majorana neutrino mass. If the heavier masses observed in early measurements of atmospheric neutrino oscillations provide the dominant contribution, then the effective Majorana neutrino mass would be expected to fall somewhere between 15 and 50 MeV. The current generation of double beta decay experiments should probe the Majorana hypothesis down to an effective Majorana mass of around 100 MeV while the next generation is anticipated to cover the full range of 15-50 MeV in the coming decade. A nice summary of these experimental efforts is given in [9].

In addition to having some promising experimental prospects, the Majorana neutrino hypothesis is also phenomenologically attractive. In particular, the seesaw mechanism [17–21] simultaneously accounts for light left-handed and very heavy right-handed neutrinos, as required by their non-observation, while providing a consistent mechanism for leptogenesis. Could it be that Majorana left us with an important piece of the puzzle explaining the asymmetry between matter and anti-matter and hence our origins? Evidently, there are some provocative reasons to believe that Majorana's original plan for his fermion may yet be vindicated.

Regardless, it is often the case that particularly profound insights inspire developments that go well beyond the intended domain of application. Majorana fermions are no exception and in the following we will explore some these unanticipated pay-offs. This is a nice example of how curiosity driven pure science often pays its way many times over; ask good scientific questions and good answers, more broadly understood, will follow. We will begin in section 2 by reviewing spinors from a slightly more mathematical perspective, preparing the ground for some fascinating connections described in section 3.2. We will then give two examples of Majorana's idea appearing in solid state physics. First in section 3.1 we discuss emergent Majorana fermions as quasiparticle states in superconductors. Second, we explore the idea of Majorana zero modes (Mzm) in section 3.2. Although Mzm are, in the spirit of Majorana, their own antiparticles, they obey anyonic statistics and we therefore avoid using the term fermion. This feature makes Mzm an exciting prospect for fault-tolerant quantum computing as we discuss in section 3.2.2. We will then discuss the definition and classification of topological phases for gapped free-fermion systems more generally in section 3.2.3. This leads us to some fascinating relations connecting spinors, normed division algebras, K-theory and the classification of stable topological states of symmetry-protected gapped free-fermion systems [22–32]. In this manner we go full circle, starting from Majorana's work on spacetime spinors, then onto Mzm in condensed matter, their place in the classification of topological phases, and finally back to the theory of spinors via K-theory and division algebras. Octonions, the largest of the four normed division algebras, have appeared in fundamental physics in a variety of guises, see for example [5,33–44] and the references therein, primarily through their intriguing connections to spacetime geometry, supersymmetry and exceptional Lie groups. Here we see another, rather subtle, occurrence in the context of condensed matter systems. It is not yet clear, however, whether or not the octonions enter the physics in any meaningful way, let alone if this perspective sheds new light on the problem. See Appendix A for a lightening tour of the division algebras.

In light of the broad and interdisciplinary scope of such a review we have sought to be as elementary as possible. Our hope is that someone unfamiliar with the various topics covered will at least be able to take home some limited intuition without too much further reading. Given our space constraints we are, however, doomed to failure in this regard, for which we apologise in advance. For the same reason we regretfully will have to omit numerous important topics, but hopefully the references provided will help in this regard.

2 Spinors, trialities and algebras

We shall begin by briefly recalling the theory of spinors from the perspective of Clifford algebras as laid out in [4,45]. We shall then describe how these structures, together with the notion of triality [3], lead us to the four normed division algebras, $\mathbb{A} = \mathbb{R}, \mathbb{C}, \mathbb{H}, \mathbb{O}$ as described in [3,5]. For the bare essentials on the division algebras see Appendix A. With this relation in hand we shall outline the connections between normed division algebras, Bott periodicity and real K-theory with a view to their application to topological phases. The interconnected relations shared by Clifford algebras, Bott periodicity and K-theory constitute an elegant and well-known story described in a number of excellent treatises, such as [46,47]. The special role played by the octonions was emphasised by, in particular, Adams and Baez in [3,5], which we follow closely here.

For an d-dimensional pseudo-Euclidean vector space V with inner-product $\langle , \rangle$ of signature $s+t=d$ we define the Clifford algebra on V, denoted $\text{Cliff}(V)$, as the associative algebra freely generated by V modulo the Clifford relations,

$$xy + yx = -2\langle x, y\rangle \tag{2.1}$$

for all $x, y \in V$. Note, the opposite sign convention is taken in [45]. As a vector space (but not as an algebra)

$$\text{Cliff}(V) \cong \bigoplus_{k=0}^{d} \wedge^k V, \tag{2.2}$$

and so $\dim \text{Cliff}(V) = 2^d$. If $V = \mathbb{R}^{s,t}$ with the canonical inner-product we will write $\text{Cliff}(s,t)$.

The representation theory is underpinned by the fundamental lemma of Clifford algebras: Suppose $\mathcal{A}$ is a unital associative algebra. Then any linear map $\phi : V \to \mathcal{A}$ such that,

$$\phi(x)\phi(x) = -\langle x, x\rangle \mathbb{1}, \quad \forall x \in V, \tag{2.3}$$

admits a unique extension to an algebra homomorphism, $\hat{\phi} : \text{Cliff}(V) \to \mathcal{A}$.

For a unital algebra $\mathcal{A}$ let $\mathcal{A}[n]$ denote the set of $n \times n$ matrices with entries in $\mathcal{A}$. All Clifford algebras $\text{Cliff}(s,t)$ are isomorphic to the (direct sum of) matrix algebras $\mathbb{A}[n]$ for $\mathbb{A} = \mathbb{R}, \mathbb{C}, \mathbb{H}$, as given in Table 1. Here we already see the Bott periodic (mod 8 repetition) features discovered by Cartan [48].

$$\text{Cliff}(s+8,t) \cong \text{Cliff}(s,t+8) \cong \text{Cliff}(s,t) \otimes \mathbb{R}[16], \quad \text{where} \quad \text{Cliff}(8,0) \cong \text{Cliff}(0,8) \cong \mathbb{R}[16]. \tag{2.4}$$

Note, starting from $s - t = 5$, say, we have the useful mnemonic:

$$\mathbb{C} \quad \mathbb{R} \quad \mathbb{R}\mathbb{R} \quad \mathbb{R} \quad \mathbb{C} \quad \mathbb{H} \quad \mathbb{H}\mathbb{H} \quad \mathbb{H} \quad \mathbb{C} \quad \mathbb{R} \quad \mathbb{R}\mathbb{R} \quad \mathbb{R} \quad \mathbb{C} \quad \mathbb{H} \quad \mathbb{H}\mathbb{H} \quad \mathbb{H}\ldots \tag{2.5}$$

If we complexify this pattern simplifies. Since all complex inner product spaces with the same dimension $d = s+t$ are isometric and the isometry lifts to a unique Clifford algebra automorphism there is a unique complex Clifford algebra for a given dimension, which is given by

$$\mathrm{Cliff}_{\mathbb{C}}(d) = \mathrm{Cliff}(s,t) \otimes_{\mathbb{R}} \mathbb{C}. \tag{2.6}$$

The order eight pattern of the $\mathrm{Cliff}(s,t)$ matrix representations given in Table 1 is reduced to an order two "Bott periodic" sequence:

$$\begin{array}{ccc} d \mod 2 & \mathrm{Cliff}_{\mathbb{C}}(d) & \mathrm{Pin}_{\mathbb{C}}(d) \text{ irreps} \\ 0 & \mathbb{C}[n] & \mathbb{C}^n \\ 1 & \mathbb{C}[n] \oplus \mathbb{C}[n] & \mathbb{C}_+[n] \quad \mathbb{C}_-[n] \end{array} \tag{2.7}$$

where n is fixed by the dimension of $\mathrm{Cliff}_{\mathbb{C}}(d)$, $n = 2^{d/2}$ for $d = 0 \mod 2$ and $n = 2^{(d-1)/2}$ for $d = 1 \mod 2$.

The isomorphism with matrix algebras implies that the non-trivial $\mathrm{Cliff}(s,t)$-modules are isomorphic to either $\mathbb{A}^n$ or $\mathbb{A}^n \oplus \mathbb{A}^n$, as given in Table 1, and $\mathbb{C}^n$ or $\mathbb{C}^n \oplus \mathbb{C}^n$ in the complexified case. These are referred to as pinor representations for reasons that will become clear shortly. In particular, we have

$s-t \mod 8$	$\mathrm{Cliff}(s,t) \cong \mathrm{Cliff}_0(s+1,t)$	$\mathrm{Pin}(s,t)$ irreps P	$\mathrm{Spin}(s,t)$ irreps S
0	$\mathbb{R}[n]$	$\mathbb{R}^n$	$\mathbb{R}^n_+ \quad \mathbb{R}^n_-$
1	$\mathbb{C}[n]$	$\mathbb{C}^n$	$\mathbb{R}^n$
2	$\mathbb{H}[n]$	$\mathbb{H}^n$	$\mathbb{C}^n$
3	$\mathbb{H}[n] \oplus \mathbb{H}[n]$	$\mathbb{H}^n_+ \quad \mathbb{H}^n_-$	$\mathbb{H}^n$
4	$\mathbb{H}[n]$	$\mathbb{H}^n$	$\mathbb{H}^n_+ \quad \mathbb{H}^n_-$
5	$\mathbb{C}[n]$	$\mathbb{C}^n$	$\mathbb{H}^n$
6	$\mathbb{R}[n]$	$\mathbb{R}^n$	$\mathbb{C}^n$
7	$\mathbb{R}[n] \oplus \mathbb{R}[n]$	$\mathbb{R}^n_+ \quad \mathbb{R}^n_-$	$\mathbb{R}^n$

Table 1: The Clifford algebras and minimal pin and spin representations. The parameters n are determined by the real dimension of the Clifford algebra: $n^2 \dim \mathbb{A} = 2^d$ for $s-t \mod 8 = 0,1,2,4,5,6$ and $2n^2 \dim \mathbb{A} = 2^d$ for $s-t \mod 8 = 3,7$.

$$\mathrm{Cliff}(0,0) \cong \mathbb{R}[1], \qquad \mathrm{Cliff}(1,0) \cong \mathbb{C}[1], \qquad \mathrm{Cliff}(2,0) \cong \mathbb{H}[1]. \tag{2.8}$$

and their corresponding pinor representations, $\mathbb{R}, \mathbb{C}, \mathbb{H}$. Although $\mathbb{O}$, being non-associative (see Appendix A), is not itself a Clifford algebra, it does form a representation of $\mathrm{Cliff}(\mathrm{Im}\mathbb{O}) \cong \mathrm{Cliff}(7,0)$ [3]. More generally, for a normed division algebra $\mathbb{A}$ consider the linear map,

$$\begin{array}{cccc} \phi: & \mathrm{Im}\mathbb{A} & \to & L(\mathrm{Im}\mathbb{A}), \\ & a & \mapsto & L_a \end{array} \tag{2.9}$$

where we have defined the left multiplication operator $L_a(x) := ax, \forall a, x \in \mathbb{A}$. Since $\mathrm{Im}\mathbb{A}$ can be regarded as the tangent space to the unit sphere in $\mathbb{A}$ and L_a is norm-preserving for $||a|| = 1$, the set of maps in $L(\mathrm{Im}\mathbb{A})$ are skew-symmetric and satisfy,

$$L_a L_a = -\langle a, a \rangle. \tag{2.10}$$

By the fundamental lemma of Clifford algebras we therefore have a faithful (as $\mathrm{Im}\mathbb{A}$ is orthogonal to the span of the identity in $\mathbb{A}$) representation of $\mathrm{Cliff}(\mathrm{Im}\mathbb{A})$ on $\mathbb{A}$.

Let us now turn to the defintion of the pin and spin groups, $\mathrm{Pin}(s,t)$ and $\mathrm{Spin}(s,t)$, which sit inside $\mathrm{Cliff}(s,t)$. The involution on V given by $\tilde{v} = -v, v \in V$ induces an automorphism on $\mathrm{Cliff}(V)$. Let the even part of $\mathrm{Cliff}(V)$ be defined by,

$$\mathrm{Cliff}_0(V) := \{x \in \mathrm{Cliff}(V) \,|\, \tilde{x} = x\}. \tag{2.11}$$

For $V \cong \mathbb{R}^{s,t}$ we then have the canonical Clifford algebra isomorphisms [4]:

$$\begin{aligned} \mathrm{Cliff}(s-1,t) &\cong \mathrm{Cliff}(t-1,s), \\ \mathrm{Cliff}(s,t) &\cong \mathrm{Cliff}_0(s+1,t), \\ \mathrm{Cliff}(s,t) &\cong \mathrm{Cliff}_0(s,t+1), \\ \mathrm{Cliff}_0(s,t) &\cong \mathrm{Cliff}_0(t,s). \end{aligned} \tag{2.12}$$

The multiplicative group $\mathrm{Cliff}^*(s,t)$ in $\mathrm{Cliff}(s,t)$ is given by,

$$\mathrm{Cliff}^*(s,t) := \{x \in \mathrm{Cliff}(s,t) \,|\, x = v_1 \cdots v_r, \quad \langle v_i, v_i \rangle \neq 0, v_i \in V\}. \tag{2.13}$$

The double cover $\mathrm{Pin}(s,t)$ of $\mathrm{O}(s,t)$ is given by the pin subgroup in $\mathrm{Cliff}^*(s,t)$ generated by unit vectors,

$$\mathrm{Pin}(s,t) := \{x \in \mathrm{Cliff}^*(s,t) \,|\, x = v_1 \cdots v_r, \quad |\langle v_i, v_i \rangle| = 1, v_i \in V\}. \tag{2.14}$$

Similarly, the spin group $\mathrm{Spin}(s,t)$ double covers $\mathrm{SO}(s,t)$ and is given by the subgroup in $\mathrm{Cliff}^*(s,t)$ generated by an even number of unit vectors,

$$\mathrm{Spin}(s,t) := \mathrm{Pin}(s,t) \cap \mathrm{Cliff}_0(s,t). \tag{2.15}$$

From these definitions it follows that the irreducible representations of $\mathrm{Cliff}(s,t)$ and $\mathrm{Cliff}_0(s,t)$ restrict to irreducible representations of $\mathrm{Pin}(s,t)$ and $\mathrm{Spin}(s,t)$, respectively, as given in Table 1. Including the conjugate representations for the complex cases completes the classification of all pinor and spinor representations in all signatures as real, complex or quaternionic vector spaces.

The standard nomenclature used in high energy physics typically applies the term "spinor", with various qualifiers (Dirac, Majorana, Weyl, symplectic etc), to cover all cases. See for example the discussion in [49]. In particular, "Majorana spinor" as used in section 1 corresponds to the pinor representation of $\mathrm{Cliff}(1,3)$, which according to Table 1 is isomorphic to $\mathbb{R}^4$, as we would expect. To reconnect to the more familiar path to Majorana spinors, note that the complex pinor[1] representation $P_\mathbb{C} \cong \mathbb{C}^4$ of the complexified Clifford algebra $\mathrm{Cliff}_\mathbb{C}(4) \cong \mathrm{Cliff}(1,3) \otimes_\mathbb{R} \mathbb{C}$ admits an invariant real structure: a linear map $\varphi : P_\mathbb{C} \to P_\mathbb{C}$ such that $\varphi(\alpha\psi) = \alpha^*\varphi(\psi)$ and $\varphi^2 = 1$. The elements in $P_\mathbb{C}$ invariant under φ are said to be Majorana. This is equivalent to the usual definition that the Dirac and Majorana conjugates are identical on Majorana spinors. This follows from the fact that given an invariant Hermitian form $\langle -, - \rangle : P_\mathbb{C} \times P_\mathbb{C} \to \mathbb{C}$ (which always exists) and an invariant real structure on $P_\mathbb{C}$ one can construct an invariant non-degenerate complex symmetric bilinear form $B : P_\mathbb{C} \times P_\mathbb{C} \to \mathbb{C}$,

$$B(\psi, \chi) := \langle \varphi(\psi), \chi \rangle. \tag{2.16}$$

On Majorana spinors $\varphi(\psi) = \psi$ we therefore have $B(\psi, \chi) := \langle \psi, \chi \rangle$ for all χ. In a basis-dependent form

$$\psi^T B \chi = \psi^\dagger A \chi, \qquad \forall \chi, \tag{2.17}$$

[1]Since the usual irreducible conjugate Weyl spinors $S, \overline{S}$ of $\mathrm{Spin}(1,3)$ (which are equivalent as $\mathbb{R}$-representations under the intertwiner given by conjugation on $P_\mathbb{C}$) can be obtained from $P_\mathbb{C}$ by projecting on the positive and negative eigenspaces of a unit volume element, $S \cong \{x \in P_\mathbb{C} \,|\, \lambda x = ix\}$ and $\overline{S} \cong \{x \in P_\mathbb{C} \,|\, \lambda x = (-)ix\}$, we can regard $P_\mathbb{C}$ as their direct sum. The Majorana condition then relates the Weyl spinors in the familiar manner.

where B, A here are symmetric and hermitian matrices, respectively. Hence

$$\psi^T B = \psi^\dagger A, \tag{2.18}$$

where the left-hand side defines the Majorana conjugate, the matrix B is the charge conjugation matrix typically denoted C, while the right-hand side defines the Dirac conjugate.

We have already seen that the four normed division algebras, $\mathbb{R}, \mathbb{C}, \mathbb{H}$ and $\mathbb{O}$, form representations of the Clifford algebras Cliff(Im$\mathbb{R}$), Cliff(Im$\mathbb{C}$), Cliff(Im$\mathbb{H}$) and Cliff(Im$\mathbb{O}$). We would now like to make the connection between $\mathbb{R}, \mathbb{C}, \mathbb{H}, \mathbb{O}$ and spinors even more concrete, using the notion of triality [3, 5]. Here we focus on the Euclidean case $(s, t) = (d, 0)$. The group Pin(d) has, besides the pinor representations P, a vector representation $V \cong \mathbb{R}^d$. The obvious inclusion $V \hookrightarrow$ Cliff(d) allows us to restrict the Clifford algebra action on P to an intertwiner of Pin(d) representations.

$$\begin{aligned} \tilde{m}: \quad & V \times P^{\pm} \quad \rightarrow \quad P^{\mp} \quad d = 3, 7 \bmod 8 \\ \tilde{m}: \quad & V \times P \quad \rightarrow \quad P \quad \text{otherwise} \end{aligned} \tag{2.19}$$

Decomposing the pinor representations under Spin(d) we obtain

$$\begin{aligned} \tilde{m}: \quad & V \times S^{\pm} \quad \rightarrow \quad S^{\mp} \quad d = 0, 4 \bmod 8 \\ \tilde{m}: \quad & V \times S \quad \rightarrow \quad S \quad \text{otherwise} \end{aligned} \tag{2.20}$$

All such spinor spaces admit an inner product (see for example [4]) and so can be dualised to give trilinear maps

$$\begin{aligned} \tilde{t}: \quad & V \times S^{\pm} \times S^{\mp} \quad \rightarrow \quad \mathbb{R} \quad d = 0, 4 \bmod 8 \\ \tilde{t}: \quad & V \times S \times S \quad \rightarrow \quad \mathbb{R} \quad \text{otherwise} \end{aligned} \tag{2.21}$$

which we could refer to as "pre-trialities", for reasons that we will now make clear.

Given three real vector spaces a triality is a non-degenerate trilinear map,

$$t : V_1 \times V_2 \times V_3 \rightarrow \mathbb{R}. \tag{2.22}$$

We can dualise to produce a bilinear map

$$m : V_1 \times V_2 \rightarrow V_3^*, \tag{2.23}$$

which can be interpreted as left (right) multiplication by elements in V_1 (V_2). The non-degeneracy of t then implies a set of isomorphisms $V_1 \cong V_2 \cong V_3 \cong V$ such that $m : V \times V \rightarrow V$ defines a division algebra; m really does stand for multiplication. Conversely, all division algebras define a triality. In particular, a normed division algebra gives a normed triality satisfying,

$$|t(x, y, z)| \leq \|x\| \, \|y\| \, \|z\|, \tag{2.24}$$

where for all x, y there is a z such that the bound is saturated. Conversely, any normed triality gives a normed division algebra.

We now can appreciate the key observation: the pre-trilaities defined on the vector and spinor representations in (2.21) will yield bona fide normed trialties if and only if $\dim V = \dim S^{(\pm)} = d$. Consulting Table 1, we see that this happens precisely for $d = 1, 2, 4, 8$:

$$\begin{aligned} t_1: \quad & V_1 \times S_1 \times S_1 \quad \rightarrow \quad \mathbb{R} \quad \Leftrightarrow \quad \mathbb{R} \\ t_2: \quad & V_2 \times S_2 \times S_2 \quad \rightarrow \quad \mathbb{R} \quad \Leftrightarrow \quad \mathbb{C} \\ t_4: \quad & V_4 \times S_4^+ \times S_4^- \quad \rightarrow \quad \mathbb{R} \quad \Leftrightarrow \quad \mathbb{H} \\ t_8: \quad & V_8 \times S_8^+ \times S_8^- \quad \rightarrow \quad \mathbb{R} \quad \Leftrightarrow \quad \mathbb{O} \end{aligned} \tag{2.25}$$

In particular, the octonions provide a representation for the three 8-dimensional representations of Spin(8).

We have now seen two ways in which the the normed division algebras are connected to Clifford algebras and their representations. The number eight appears to be omnipresent and one might wonder if there is a deeper connection between the Bott periodic patterns displayed by Clifford algebras and the appearance of the normed division algebras, in particular the 8-dimensional octonions. Indeed there is and in more ways than one. Of particular interest to us is the relation mediated by real K-theory.

Before we come to that, let us take a look at Bott periodicity through the division algebraic lens [5]. We have a relationship to $(\dim \mathbb{A})$-dimensional spheres via the division algebraic projective lines treated as smooth manifolds,

$$\mathbb{AP}^1 \cong S^{\dim \mathbb{A}}. \tag{2.26}$$

One has to take extra care for $\mathbb{OP}^1$ when defining "a line through the origin", see for example the treatment of the Cayley plane given in [4], but otherwise it is conceptually entirely equivalent to the familiar Riemann sphere $\mathbb{CP}^1$. Now, consider the maps $f_\mathbb{A}$, as in [5], sending norm-one elements $a \in \mathbb{A}$ to the orthogonal linear operators on $\mathbb{A}$ given by right-multiplication $R_a(x) = xa$,

$$\begin{aligned} f_\mathbb{A} : \quad S^{\dim \mathbb{A}-1} &\rightarrow \mathrm{O}(\dim \mathbb{A}), \\ a &\mapsto R_a \end{aligned} \tag{2.27}$$

Any real vector bundle over $S^{\dim \mathbb{A}}$ with $(\dim \mathbb{A})$-dimensional fibres can be form by gluing the trivial bundles on the northern and southern hemispheres along the equator using a map $f : S^{\dim \mathbb{A}-1} \rightarrow \mathrm{O}(\dim \mathbb{A})$. See for example [47]. For $f = f_\mathbb{A}$ we obtain the canonical line bundles $L_\mathbb{A}$ [5]; the fibre at a point in $\mathbb{AP}^1$ is the corresponding line through the origin, forming a copy of $\mathbb{A}$. The homotopy classes of the maps $f_\mathbb{A}$, denoted $[f_\mathbb{A}]$, generate the non-trivial homotopy groups of the topological group $\mathrm{O}(\infty) = \mathrm{inj}\lim \mathrm{O}(n)$ (or just O).

$$\begin{array}{rcrcl} [f_\mathbb{R}] & \rightarrow & \pi_0[\mathrm{O}(\infty)] & \cong & \mathbb{Z}_2, \\ [f_\mathbb{C}] & \rightarrow & \pi_1[\mathrm{O}(\infty)] & \cong & \mathbb{Z}_2, \\ & & \pi_2[\mathrm{O}(\infty)] & \cong & \varnothing, \\ [f_\mathbb{H}] & \rightarrow & \pi_3[\mathrm{O}(\infty)] & \cong & \mathbb{Z}, \\ & & \pi_4[\mathrm{O}(\infty)] & \cong & \varnothing, \\ & & \pi_5[\mathrm{O}(\infty)] & \cong & \varnothing, \\ & & \pi_6[\mathrm{O}(\infty)] & \cong & \varnothing, \\ [f_\mathbb{O}] & \rightarrow & \pi_7[\mathrm{O}(\infty)] & \cong & \mathbb{Z}, \end{array} \tag{2.28}$$

as computed by Bott [50], who also established the periodic behaviour bearing his name,

$$\pi_i[\mathrm{O}(\infty)] \cong \pi_{i+8}[\mathrm{O}(\infty)]. \tag{2.29}$$

Note, the non-trivial cases occur for $i = \dim \mathbb{A} - 1 \bmod 8$. Of course, this follows from the fact that $f_\mathbb{A}$ defines a map from $S^{\dim \mathbb{A}-1}$ to $\mathrm{O}(\dim \mathbb{A}) \hookrightarrow \mathrm{O}(\infty)$ [5].

In 1961 Atiyah and Hirzebruch showed [51] that Bott periodicity (2.29) is related to algebraic K-theory. In particular, the sequence of discrete groups (2.32) is replicated by the reduced real K-theory of n-spheres for $n = 1, 2 \ldots 8$:

$$\widetilde{KO}(S^n) \cong \pi_{n-1}[\mathrm{O}(\infty)]. \tag{2.30}$$

Here, the nontrivial cases correspond to $S^{\dim \mathbb{A}}$ and the normed division algebras are again, from a certain perspective, responsible. To describe this observation, let us briefly recall the basic notions behind K-theory. K-theory is a deep set of ideas that we cannot possibly do justice to here; hopefully our very limited sketch will be enough to develop some heuristic intuition. The interested reader is encouraged to consult as a starting point the excellent introductions [52][2] and [47].

Real K-theory probes the structure of a topological space X by constructing a ring from the set of isomorphism classes, $\mathrm{Vect}(X)$, of real vector bundles over X. Note, $\mathrm{Vect}(X)$ is already a semi-ring (i.e. a ring with the axiom of additive inverses relaxed) with addition and multiplication given by the direct sum and tensor product, respectively, of representative bundles. This semi-ring can be completed to produce a ring[3]. A familiar example of this procedure is given by the construction of the integers $\mathbb{Z}$ from the natural numbers $\mathbb{N}$. More generally, the ring completion of a semi-ring R_0 is a pair (R, τ), where R is a ring and τ a semi-ring homomorphism $\tau : R_0 \to R$ such that for any semi-ring homomorphism $f : R_0 \to S$, S a ring, there exists a unique homomorphism $g : R \to S$ satisfying $g \circ \tau = f$. One construction of (R, τ) is given by $R = R_0 \times R_0/\sim$, where

$$(a, b) \sim (a', b') \quad \Leftrightarrow \quad \exists c \in R_0 \quad \text{s.t.} \quad a + b' + c = a' + b + c, \tag{2.31}$$

and $\tau(a) = [(a, 0)]$. The additive identity is $[(0, 0)]$ and a quick check confirms that the inverse of $[(a, b)]$ is $[(b, a)]$. Applying this construction to $\mathrm{Vect}(X)$ we obtain a ring denoted $KO(X)$, which is referred to as the real K-theory of X. For a pointed space X, the homomorphism rk_0 sending an isomorphism class in $\mathrm{Vect}(X)$ to the dimension of a representative fibre at the basepoint of X extends, by construction, to a homomorphism $rk : KO(X) \to \mathbb{Z}$ such that $rk \circ \tau = rk_0$. The reduced real K-theory of X that we alluded to in (2.30) is then defined as the kernel of rk and denoted $\widetilde{KO}(X)$. A brief remark on notion: it is often convenient to denote the K-theory of complex and real vector bundles by $K_\mathbb{C}$ and $K_\mathbb{R}$ (instead of K and KO), respectively, reserving the bare K for the unspecified case.

An n-sphere S^n with the north pole marked constitutes a pointed space and the reduced real K-theory for n mod 8 is given by

$$\begin{array}{rcrcl}
[L_\mathbb{R}] & \to & \widetilde{KO}(S^1) & \cong & \mathbb{Z}_2, \\
[L_\mathbb{C}] & \to & \widetilde{KO}(S^2) & \cong & \mathbb{Z}_2, \\
 & & \widetilde{KO}(S^3) & \cong & \varnothing, \\
[L_\mathbb{H}] & \to & \widetilde{KO}(S^4) & \cong & \mathbb{Z}, \\
 & & \widetilde{KO}(S^5) & \cong & \varnothing, \\
 & & \widetilde{KO}(S^6) & \cong & \varnothing, \\
 & & \widetilde{KO}(S^7) & \cong & \varnothing, \\
[L_\mathbb{O}] & \to & \widetilde{KO}(S^8) & \cong & \mathbb{Z},
\end{array} \tag{2.32}$$

where we have indicated that the non-trivial cases $\widetilde{KO}(S^{\dim \mathbb{A}})$ are generated by the isomorphism class of the canonical $\mathbb{A}$-line bundle [5]. Hence $\widetilde{KO}(S^n) \cong \pi_{n-1}[\mathrm{O}(\infty)]$. We can moreover construct a $\mathbb{Z}$-graded ring,

$$\widetilde{KO} = \bigoplus_n \widetilde{KO}(S^n), \tag{2.33}$$

with multiplication induced by smash products of vector bundles. Multiplication by $[L_\mathbb{O}]$ gives an isomorphism $\widetilde{KO}(S^n) \cong \widetilde{KO}(S^{n+8})$. In this sense it is the octonions that generate Bott periodicity!

[2] This focusses on complex K-theory, but the key concepts largely transfer to the real case.

[3] In general, the multiplicative composition of the semi-ring is not required and we can consider instead an Abelian monoid, an Abelian group with the requirement of the existence of inverses relaxed. Then we have instead the weaker, in that we do not consider any multiplicative structure, group completion with the same conditions as the ring completion otherwise. The set of isomorphism classes of objects in an additive category $\mathcal{C}$ forms an Abelian monoid. The group completion in this case is referred to as the Grothendieck group, denoted $K(\mathcal{C})$, and constitutes the fundamental example of this notion. See [47] for a proper account of these constructions.

Of course, the shared Bott periodic features of Clifford algebras, spinors, division algebras and real K-theory are no coincidence. In particular, Atiyah, Bott and Shapiro made a detailed instigation [45] into Clifford algebras, spinors and real K-theory. First of all, let M_k denote the Abelian group freely generated by the irreducible $\mathbb{Z}_2$-graded Cliff(k)-modules. The isomorphism $\hat{\phi}$: Cliff$(k+l) \to$ Cliff$(k) \hat{\otimes}$ Cliff(l) given by the linear extension of

$$\phi(e_i) = \begin{cases} e_i \otimes \mathbb{1} & i \leq k \\ \mathbb{1} \otimes e_i & k < i \end{cases} \tag{2.34}$$

induces a pairing of Cliff(k) and Cliff(l) modules giving

$$M_\star = \bigoplus_k M_k \tag{2.35}$$

a $\mathbb{Z}$-graded ring structure. Multiplication by the class of an irreducible module in M_8 gives an isomorphism $M_k \cong M_{k+8}$. Define A_k as the cokernal of the homomorphism $M_{k+1} \to M_k$ induced by the inclusion Cliff$(k) \hookrightarrow$ Cliff$(k+1)$. Then $A_\star = \bigoplus A_k$ is an ideal in $M_\star$, with an inherited ring structure, and we have an isomorphism with the reduced real K-theory.

$$A_k \cong \widetilde{KO}(S^k) \tag{2.36}$$

where the Bott periodicity of A_k essentially follows from that of Cliff(k) via $M_k \cong M_{k+8}$. When we discuss the classification of topological phases, we will return to K-theory. To set the scene we must first return to the subject of Majorana fermions.

3 Majorana in condensed matter

While the jury is still out on whether or not Majorana fermions will be discovered in the context of fundamental particle physics, condensed matter systems allow us to elaborate on what Nature naively offers, presenting new opportunities to realise physically Majorana's insight. The fundamental constituents of condensed matter systems are invariably electrons, which carry charge and have distinct antiparticles. Clearly, if Majorana fermions are to be found here they must arise as emergent excitations.

Superconductors, in particular, represent a promising setting for such emergent behaviour since the gauge symmetry associated to charge conservation is broken. Indeed, the prospect of observing Majorana quasiparticles in midgap excitations of a chiral p-wave superconductor has a reasonably long history [53–59], with early indications appearing in the particle physics literature more than 30 years ago [60]. Much of the focus has been on the idea of Majorana zero-modes (Mzm), also dubbed Majorinos [11], appearing in topologically non-trivial phases of matter. Such quanta are Majorana in the sense that they are their own antiparticles, but the term "fermion" is intentionally avoided here because they generically obey non-Abelian anyonic exchange statistics. Their topological character and non-Abelian statistics make Mzm particularly exciting from the perspective of quantum computing, motivating substantial experimental and theoretical efforts. We will come to this subject in section 3.2, but first we will briefly review the emergence of Majorana fermions, in the original sense of the term, in ordinary (non-topological) superconductors. Although less enticing with respect to quantum computing, these examples are closest in spirit to Majorana's original analysis and there are even proposals to observe them in the laboratory that directly mirror experiments in particle physics [61].

3.1 Majorana fermions in superconductors

In spite of the apparent contradiction Majorana fermions (note the very intentional use of the term fermion here) in ordinary superconductors are actually rather ubiquitous. Although this has been generally appreciated for some time now, it was treated carefully and emphasised only rather recently in [62].

In [11] Wilczek presented a very simple toy model that captures the key features while avoiding the complication of actually having to deal with superconductors and spinors. To help set the scene, we will repeat this example here. First, recall that the mass term of the Majorana Lagrangian (1.4) in terms of the chiral projection ψ_L is given by unusual form,

$$\mathcal{L}_m \sim \psi_L^* \gamma_0 \psi_L^* + \psi_L \gamma_0 \psi_L. \tag{3.1}$$

Again, we emphasise that this Majorana mass term breaks the U(1) symmetry $\psi_{L/R} \to e^{i\theta}\psi_{L/R}$ of the Dirac Lagrangian and is lepton number violating. Now consider a model with two complex scalar fields, ϕ and φ, with a global U(1) symmetry, $\phi \to e^{i\theta}\phi$ and $\varphi \to e^{i2\theta}\varphi$. The key point is that the expectation value of φ can break the U(1) symmetry such that a Majorana mass term is generated. Let the mass and cubic interaction terms, which must initially be compatible with the U(1) symmetry, be given by

$$\mathcal{L} = -m^2\phi^*\phi - \frac{\lambda}{2}\left(\varphi^*\phi^2 + c.c.\right). \tag{3.2}$$

If $\langle\varphi\rangle = 0$ then the global U(1) is preserved and the ϕ quanta are a particle-antiparticle pair of equal mass m and opposite charge. If, however, $\langle\varphi\rangle = \Delta \in \mathbb{R}$ the global U(1) is broken and we have the very suggestive scalar analog of a Majorana mass term,

$$\mathcal{L} \to -(m^2 + \lambda\Delta)(\mathrm{Re}\phi)^2 - (m^2 - \lambda\Delta)(\mathrm{Im}\phi)^2. \tag{3.3}$$

The quanta of the *real* fields Reϕ and Imϕ are their own antiparticles with definite and distinct masses, $\sqrt{m^2 \pm \lambda\Delta}$. Note, if the symmetry breaking is small it is possible that the other interactions (not considered above) are more close to being diagonal in the ϕ, ϕ^* basis. Then for time scales less than the ϕ, ϕ^* oscillation time $\sim m/\lambda\Delta >> 1$ the Majorana nature of the system will be hidden for practical purposes. This serves to highlight the point that even if particles are their own antiparticles they can to arbitrary accuracy behave as if they are not [11], a fact we have in mind when discussing neutrinos.

With this illustrative example in hand we now turn to the emergence of Majorana fermions in superconductors. Our starting point is the Bogoliubov-de Gennes (BdG) formalism for superconductivity with spatial inhomogeneity. See for example [63]. The mean-field BdG Hamiltonian is given by,

$$\mathcal{H} = \int d^3x \left(\psi^\dagger h\psi + \frac{i}{2}(\Delta\psi^\dagger\sigma_y\psi^\dagger + h.c.) - \frac{1}{V}|\Delta|^2\right), \tag{3.4}$$

Here $\psi_s^\dagger(\mathbf{x}), s = \uparrow, \downarrow$ creates an electron of spin s at spatial point $\mathbf{x}$. The first term $\psi^\dagger h\psi$ gives the kinetic and single-electron potential energy. The remaining terms are derived from the Bogoliubov mean-field decoupling of the standard attractive ($V > 0$) interaction term,

$$V\psi_\uparrow^\dagger\psi_\downarrow^\dagger\psi_\uparrow\psi_\downarrow \quad\longrightarrow\quad V\left(\langle\psi_\uparrow^\dagger\psi_\downarrow^\dagger\rangle\psi_\uparrow\psi_\downarrow + \psi_\uparrow^\dagger\psi_\downarrow^\dagger\langle\psi_\uparrow\psi_\downarrow\rangle - \langle\psi_\uparrow^\dagger\psi_\downarrow^\dagger\rangle\langle\psi_\uparrow\psi_\downarrow\rangle\right), \tag{3.5}$$

where we have defined the superconducting order parameter,

$$\Delta(\mathbf{x}) = V\langle\psi_\uparrow\psi_\downarrow\rangle, \tag{3.6}$$

and the expectation values are taken with respect to (3.4). Note that the resulting interaction given by the middle term of (3.4) for constant Δ is very much akin to the conventional number-violating Majorana mass term, as in (3.1), with Majorana mass $|\Delta|$. This is number violating ($\delta L = 2$) in the sense that a pair of electrons can be absorbed/produced by the superconducting condensate treated in the mean-field approximation. These processes have observable consequences, produced by pair-wise annihilation of Bogoliubov quasiparticles, that parallel those of Majorana fermions in particle physics [61].

We can recast (3.4) in the standard BdG form by introducing the Nambu spinor,

$$\Psi = \begin{pmatrix}\psi_s \\ \psi_s^\dagger\end{pmatrix} \tag{3.7}$$

such that

$$\mathcal{H} = \int d^3x \left(\Psi^\dagger H \Psi - \frac{1}{V}|\Delta|^2 \right), \tag{3.8}$$

where

$$H = \begin{pmatrix} h & \Delta \\ \Delta^* & -\sigma_y h^* \sigma_y \end{pmatrix}. \tag{3.9}$$

This form brings us even closer to the original Majorana analysis, since the Nambu spinor is constrained to satisfy the reality condition,

$$C\Psi^* = \Psi, \quad \text{where} \quad C = \sigma_y \otimes \sigma_y. \tag{3.10}$$

Moreover, expressed in terms of the eigenfuctions of the stationary BdG equation,

$$H\Psi_n = E_n \Psi_n, \tag{3.11}$$

used to solve (3.8) with the self-consistency condition (3.6), the BdG Hamiltonian is exactly of the Majorana Hamiltonian form,

$$\mathcal{H} = \sum_{E_n > 0} E_n a_n^\dagger a_n + \text{const}, \quad \text{where} \quad a_n = \int d^d x \Psi_n^\dagger \Psi. \tag{3.12}$$

As a result the quasiparticles of the BdG formalism have all the key features of Majorana fermions: they are neutral fermions that are their own antiparticles with number-violating Majorana mass given by the superconducting order Δ.

As emphasised in [62] these Majorana features are a generic consequence of superconductivity and fermionic statistics. Fermionic statistics alone, without additional assumptions regarding any other symmetries of the system, very broadly implies that the BdG Hamiltonian allows real-valued solutions for the Nambu spinor that can be quantised as Majorana fields. They proceed by introducing a generic set of fermionic degrees of freedom, Ψ_a, where $a = (\mathbf{x}, \alpha, s, i)$ is a composite index labelling position $\mathbf{x}$, possible flavours α, any half-integer spin s and Nambu grading $i = \pm$, i.e. $\Psi_{+(-)} = \psi^{(\dagger)}$. The Hamiltonian density describing superconductivity for such a system can then be written as,

$$\Psi^\dagger H_{\text{gen}} \Psi, \quad \text{where} \quad H_{\text{gen}} = \begin{pmatrix} h & \Delta \\ \Delta^\dagger & -h^* \end{pmatrix} \tag{3.13}$$

and $h = h^\dagger, \Delta^T = -\Delta$. For consistency, the operators Ψ must satisfy the constraint

$$\Psi^\dagger = C\Psi, \quad \text{where} \quad C = \begin{pmatrix} 0 & \mathbb{1} \\ \mathbb{1} & 0 \end{pmatrix} \tag{3.14}$$

and the anticommutation relations are given by,

$$\{\Psi_a, \Psi_b\} = C_{ab}, \qquad \{\Psi_a, \Psi_b^\dagger\} = \delta_{ab}. \tag{3.15}$$

The fermionic nature of Ψ implies that any BdG-type H_{gen} has a conjugation symmetry of the form,

$$-H_{\text{gen}}^* = C^* H_{\text{gen}} C, \tag{3.16}$$

where $C^* = C^{-1}$. The point highlighted by [62] is that there is a unitarily related basis in which H_{gen} is pure imaginary. Consider a unitary V under which

$$H_{\text{gen}} \to \tilde{H}_{\text{gen}} = V H_{\text{gen}} V^\dagger. \tag{3.17}$$

From (3.16) we have,

$$\tilde{C} = VCV^T, \tag{3.18}$$

which respects $\tilde{C}^* = \tilde{C}^{-1}$. It was shown in [62] that for generic H_{gen} there exists a V such that $C = \mathbb{1}$. Then from (3.16) we have $\tilde{H}^*_{\text{gen}} = -\tilde{H}_{\text{gen}}$ so that Schrödinger's equation,

$$\partial_t \tilde{\Psi} = i\tilde{H}_{\text{gen}}\tilde{\Psi} \tag{3.19}$$

admits real solutions, just as in Majorana's original analysis. Since the BdG formalism, which builds on the standard Bardeen, Cooper and Schrieffer model, is experimentally very successful, we are led to infer that Majorana fermions in this sense are common features of superconductors. Majorana zero-modes associated to localised states in topologically non-trivial phases are much rarer however. Let us now turn to this topic.

3.2 Majorana zero-modes

An important theme of current research in condensed matter is the notion of topological phases. These are not distinguished in the familiar manner by symmetries, but instead by topological invariants. They now come in a variety of guises, several of which have be experimentally realised [64–72]. Typically topological insulators or superconductors are fermion systems that have a gap in the bulk, but admit topologically protected gapless boundary states.

The theoretical and experimental existence of such states has dramatic implications, both in terms of fundamental physics and potential applications: they represent a genuinely new feature of quantum theory while offering an alternative path to fault-tolerant quantum computation. The susceptibility of qubit systems to decoherence, leading to phase and spin-flip errors, remains the principal obstacle to realising a scalable quantum computer. One approach is to use fault-tolerant error-correcting protocols, algorithmically redundant portions of code that serve specifically to detect and correct decoherence errors [73, 74]. If the error occurrence rate can be kept under a threshold of around 10^{-4} [75, 76], then fault-tolerant error-correction would allow for large scale reliable quantum computation. Topological quantum computing [77] is another approach to fault tolerance, complementary to the detect and correct paradigm. The idea is to use physical features of the system, specifically topological obstructions, to protect against errors appearing in the first place. Majorana zero-modes arising in topologically non-trivial superconducting phases are a striking example of how these ideas may be realised in practise.

In section 3.2.1 we will discuss the appearance of Mzm in a paradigmatic example, the Majorana (or Kitaev) chain [78]. We then briefly consider there application to topological quantum computing in section 3.2.2. To conclude, we will address the definition and classification of topological phases for gapped free-fermion systems more carefully and generally in section 3.2.3.

3.2.1 The Majorana chain

The simplest model admitting Mzm is the 1-dimensional Majorana (or Kitaev) chain [78]. A length N Majorana chain consists of N spinless fermions on a 1-dimensional lattice. The spinless condition initially gave the impression that this example would be limited to an unrealistic, albeit illuminating, toy model. However, there are now several plausible physical realisations, as we shall briefly review later.

Before turning to Kitaev's model, let us discuss some generic features and expectations. The basic idea is that each fermionic site $i = 1, \ldots N$ is described by the usual fermionic annihilation and creation operators, $a_i, a_i^\dagger$, with the usual anti-commutation relations, $\{a_i^\dagger, a_j\} = \delta_{ij}$, $\{a_i^{(\dagger)}, a_j^{(\dagger)}\} = 0$. Here, $a_i^\dagger$ and a_i can be thought of as creating a "spinless electron" or hole, respectively, at site i. These can be formally arranged into a set of so-called Majorana operators,

$$\gamma_{2i-1} = a_i^\dagger + a_i, \qquad \gamma_{2i} = i(a_i^\dagger - a_i) \tag{3.20}$$

which satisfy the Clifford and Majorana like relations

$$\{\gamma_\alpha, \gamma_\beta\} = 2\delta_{\alpha\beta}, \qquad \gamma_\alpha^\dagger = \gamma_\alpha, \qquad \alpha = 1, \dots 2N. \tag{3.21}$$

These modes are their own antiparticles in the sense that $\gamma^\dagger = \gamma$, but note that $\gamma^2 = 1$ and therefore is not fermionic in conventional terms. Generically this is of course just a formal manipulation and we should not think of the γ modes as representing quasiparticles. It is the $a_i = \gamma_{2i-1} + i\gamma_{2i}$ operators that typically have a well-defined occupation number; the Majorana modes pair up into ordinary "Dirac" modes with distinct antiparticles, $a^\dagger \neq a$.

However, Kitaev demonstrated [78] that there is a natural family Hamiltonians that give rise to Majorana modes as the effective low-energy degrees of freedom. The key observation is that, in special circumstances, two single Majorana modes can be localised at the boundaries of the chain, rendering them effectively free and un-paired [78]. For such systems the Majorana operators are not merely a formal tool, but are actually required to capture the correct physics.

Kitaev considered the Hamiltonian,

$$\mathcal{H} = \sum_i \left[-t(a_i^\dagger a_{i+1} + h.c.) - \mu(a_i^\dagger a_i - \frac{1}{2}) + (\Delta a_i a_{i+1} + h.c.) \right], \tag{3.22}$$

where t is the hopping amplitude, μ is the chemical potential and $\Delta = e^{i\theta}|\Delta|$ the induced superconductiing gap. For $\Delta \neq 0$ this represents a superconducting quantum wire of spinless fermions.

Ignoring the phase $\theta = 0$ (via the obvious gauge transformation) for simplicity we have

$$\mathcal{H} = \frac{i}{2} \sum_i \left[\mu \gamma_{2i}\gamma_{2i-1} + (\Delta + t)\gamma_{2i}\gamma_{2i+1} + (\Delta - t)\gamma_{2i-1}\gamma_{2i+2} \right] \tag{3.23}$$

in terms of the Majorana operators. Kitaev showed that this Hamiltonian admits two phases divided by the lines $2|t| = |\mu|$:

(*a*) $2|t| < |\mu|$, topologically trivial with all pairs of Majorana modes belonging to a common site i bound into an ordinary "Dirac" fermion.

(*b*) $2|t| > |\mu|, \Delta \neq 0$, topologically non-trivial with un-paired Majorana zero-modes appearing at the ends of the chain.

To understand these two phases it is instructive to consider a simple limiting example from each class.

For (*a*) let $t = \Delta = 0, \mu < 0$ so that

$$\mathcal{H} = \frac{i}{2}(-\mu) \sum_i \gamma_{2i-1}\gamma_{2i} = -\mu \sum_i \left(a_i^\dagger a_i - \frac{1}{2} \right). \tag{3.24}$$

We immediately see that the Majorana operators $\gamma_{2i-1}, \gamma_{2i}$ associated to a given site i are paired together and the ground state has all sites unoccupied. The system is obviously gapped since it costs energy μ to excite a quasiparticle.

For (*b*) let $t = \Delta > 0, \mu = 0$ so that

$$\mathcal{H} = it \sum_i^{N-1} \gamma_{2i}\gamma_{2i+1} = 2t \sum_i^{N-1} \left(c_i^\dagger c_i - \frac{1}{2} \right), \tag{3.25}$$

where

$$c_i = \frac{1}{2}(\gamma_{2i} + i\gamma_{2i+1}), \quad c_i^\dagger = \frac{1}{2}(\gamma_{2i} - i\gamma_{2i+1}). \tag{3.26}$$

We see that in the bulk Majorana operators $\gamma_{2i}, \gamma_{2i+1}$ associated to adjacent but *different* sites, i and $i+1$, are paired together. The extremal operators $\gamma_1 = \gamma_L$ and $\gamma_{2N} = \gamma_R$ are Majorana zero-modes;

they do not appear in the Hamiltonian at all and are unpaired. From the second form of (3.25) it is apperent that the ground states satisfies $c_i|\Omega_0\rangle = 0$ for all $i < N$. Since $\gamma_{L,R}$ obviously commute with the Hamiltonian there is a two-fold degeneracy of the ground state. In particular, the non-local "Dirac" fermion

$$f = \frac{1}{2}(\gamma_L + i\gamma_R), \tag{3.27}$$

costs zero energy and if $f|\Omega_0\rangle = 0$ then $|\Omega_1\rangle = f^\dagger|\Omega_0\rangle$ is necessarily a second orthogonal ground state. We can write,

$$-i\gamma_L\gamma_R|\Omega_0\rangle = |\Omega_0\rangle, \qquad -i\gamma_L\gamma_R|\Omega_1\rangle = -|\Omega_1\rangle \tag{3.28}$$

from which we see that $|\Omega_0\rangle$ is even and $|\Omega_1\rangle$ is odd under the fermionic parity operator

$$P = \prod_i (-i\gamma_{2i-1}\gamma_{2i}). \tag{3.29}$$

As for $t = \Delta = 0, \mu < 0$ the bulk has a gapped spectrum since it cost t to produce an excitation. In fact, the bulk properties of the two cases are identical. However, the boundary conditions remain distinct; only for $t = \Delta > 0, \mu = 0$ can unpaired Majorana zero-modes exists. In this sense $t = \Delta > 0, \mu = 0$ belongs to a topologically non-trivial phase. We will make this statement more precise in section 3.2.3.

In summary, these two cases represent two distinct phases of the Majorana chain that have the same bulk features, but differing boundary modes. Only in the latter are there unpaired zero energy Majorana modes. The properties of these two phases extend to the full parameter spaces defined in (a) and (b) above. This relies on the fact the bulk spectrum is gapped in both phases. For a finite length chain this suppresses exponentially in N the amplitude for the fermionic pair described by $\gamma_{L,R}$ to tunnel across the chain. To study the bulk energy spectrum for generic values of the parameters we can impose periodic boundary conditions. The Hamiltonian in momentum space is then given by

$$\mathcal{H} = \sum_k \left[(-2t\cos(k) - \mu)a_k^\dagger a_k + \Delta(i\sin(k)a_k a_{-k} + h.c.)\right] \tag{3.30}$$

and the energy is given by

$$E(k) = \pm\sqrt{(2t\cos(k) + \mu)^2 + (\Delta\sin(k))^2} \tag{3.31}$$

so that for $\Delta \neq 0$ the spectrum is fully gapped unless $2t = \pm\mu$. The key condition for the two phases to extend to the full parameter spaces (a) and (b) is the absence and existence, respectively, of zero-energy boundary modes linear in the Majorana operators. That is, solutions to the equation

$$[\mathcal{H}, \sum_{\alpha=1}^{2N} c_\alpha\gamma_\alpha] = 0. \tag{3.32}$$

A very clear treatment of this problem is given in [11], so we will not repeat it here. Instead, let us summarise the key attributes of the solution. As one would expect, it splits into two cases. For the parameter range specified by (a) the required boundary conditions (for an open chain) cannot be satisfied and hence there are no zero-mode solutions. If, on the other hand, $2t > |\mu|, \Delta \neq 0$ then all boundary conditions are satisfied (up to corrections exponentially suppressed in N) and we have two boundary zero-modes b_L and b_R localised at $i = 1$ and $i = N$, respectively. If $2t > -|\mu|, \Delta \neq 0$ then b_L and b_R are simply interchanged and we conclude that boundary Majorana zero-modes exists for the whole of the (b) parameter range.

For $N \to \infty$ this result is exact. For finite N there is a weak interaction between b_L and b_R that can accounted for with an effective Hamiltonian,

$$\mathcal{H}_{\text{int}} \propto e^{-N/l_0} b_L b_R. \tag{3.33}$$

Here l_0^{-1} is the characteristic length scale given by the smallest of $|\ln|\mu_+||$ and $|\ln|\mu_-||$, where

$$\mu_\pm = \frac{-\mu \pm \sqrt{\mu^2 - 4(\Delta^2 - t^2)}}{2(\Delta + t)}. \tag{3.34}$$

Note, for (*b*) we have $|\mu_\pm| < 1$. The interaction term drives a re-combination of b_L and b_R back into a conventional doublet of creation and annihilation operators. However, for a long wire the operators b_L, b_R can be effectively regarded as separated and local up to exponentially suppressed corrections. The degeneracy of the grounds states is lifted by a separation of e^{-N/l_0} and up to exponentially suppressed corrects the approximate Majorana zero-modes satisfy the characteristic equation $b_{L,R}^2 = 1$. Note, we have neglected the treatment of (weak) interaction terms beyond the mean-field approximation and the possible effect they may have on the existence of Majorana modes, although they are expected to persist. For a recent discussion of these issues see [79, 80] and the references therein.

Note, the above analysis of Mzm rested on the assumption that the chain is made up of spinless fermions. Were it otherwise, the degeneracy of every eigenstate would be doubled and, in particular, the boundaries would support two Mzm re-assembled into a single ordinary fermion. That is not to say the Kitaev chain cannot be realised using electron systems, only that the their spin degree of freedom must be essentially frozen out. There are now a number of rather ingenious and realistic proposals that address this issue, while meeting the other two key requirements of superconductivity and a bulk gap. These include the boundary of 2-dimensional topological insulators [81–84] or nanowires made from a 3-dimensional topological insulator [85], both with proximity induced superconductivity.

The experimentally most successful approach to date is the use of semiconducting quantum wires. The pioneering proposals of [86, 87] use a semiconducting wire placed on the surface of a block of conventional 3-dimensional s-wave superconductor with an external magnetic field applied. If the wire has an appreciable spin-orbit coupling then the magnetic field can be used to prefer a particular spin direction, effectively rendering the constituent electrons spinless. Remarkably, the proximity effects of the s-wave superconductor, together with the spin-orbit coupling, can effectively induce p-wave superconductivity in the wire as required by the Cooper-pairing of spinless fermions. For a review of this phenomenon see [7]. The freedom to use readily available s-wave superconductors is obviously of great technological significance.

The first experimental evidence supporting the existence of Mzm was observed in the breakthrough work of the Delft group [88]. They placed a single InSb crystal wire, which has good proximity induced superconductivity, on a substrate. One portion of the wire is placed in contact with a superconducting metal and another disconnected portion is placed in contact with a normal metal. The small middle section of the wire not in contact with either is depleted of electrons to create an approximately insulating bridge connecting the two portions. Theoretically, with an appropriate external B field applied we would expect the superconducting portion to enter a topological phase with Mzm localised at the ends. A voltage bias is applied across the bridge and the differential tunnelling conductance is measured. To a very good approximation the conductance is proportional to the density of states at the end of the superconducting portion of the wire adjacent to the bridge. For a small external magnetic field, less than about 90 mT, the observed conductance suggests a superconducting gap at around 260 μeV and nothing special happening at zero bias, consistent with a wire in the expected topologically trivial phase. As the B field is increased a peak in the conductance is seen at zero bias suggesting a transition to a topological phase and the emergence of Mzm qualitatively consistent with the theoretical prediction. This result has since been reproduced by a number of groups [89–92], amounting to a compelling case for the existence of Mzm. It should be noted, however, that a similar zero-voltage bias conductance peak can be induced by disorder and multiple bands even when in a topologically trivial phase [93], so perhaps some caution is still required.

3.2.2 Topological quantum computing

As we have emphasised several times Mzm are not fermions in the conventional sense. Instead they are *anyons*, a term coined by Wilzcek [94]; identical particles with non-trivial quantum exchange statistics that is neither bosonic nor fermionic. The possibility of anyonic statistics was first demonstrated in [95,96]. Note, in three space dimensions quantum statistics is necessarily either bosonic or fermionic. Adiabatically exchanging a pair of particles twice is equivalent to adiabatically sending one particle all the way around the other, which in three space dimensions is topologically equivalent to having done nothing and thus must act on the wavefunction as the identity. The exchange operator squares to the identity implying one of the two standard possibilities: boson versus fermion. In two space dimensions, however, this is no longer true [95,96]; a path enclosing a particle cannot be smoothly deformed to a point without being cut by the particle. When a pair of *Abelian* anyons are interchanged the wavefunction picks up an complex phase factor and, as such, they can be thought of as interpolating between bosons and fermions. Abelian ayons are now recognised as a ubiquitous feature of fractional quantum hall states [97,98].

More recently, it has been realised that there are ayonic system for which particle exchange induces a unitary change of state going beyond a mere phase factor. In this case the ordering of exchanges is important and they are accordingly referred to as *non-Abelian* anyons. Quasiparticles with non-Abelian aynionic statistics first arose in conformal field theory [99] and Chern-Simons theory [100]. Later, Moore and Read [101] showed that fractional quantum Hall states could support non-Abelian statistics. There is now significant theoretical and experimental evidence for non-Abelian statistics in quantum Hall states with filling factor 5/2 [102–104]. A key step with respect to our present story was the discovery [105] that these Moore-Read states share universal features, including non-Abelian anyonic statistics, with those of topological 2-dimensional spinless $p + ip$ superconductors. In groundbreaking work, Fu and Kane [106] argued that the required topological phase in two dimensions would emerge at the interface of a topological insulator and a conventional s-wave superconductor. Mzm that commute with the Hamiltonian are localised at the centres of Abrikosov vortices[4] that form in the topological superconducting interface and are subject to non-Abelian exchange statistics. Experimental realisations of the Fu-Kane model are in progress [107–110] and there have been some early indications of Mzm reported in [111].

Such systems, realising non-Abelian Mzm in two dimensions, represent a compelling framework for topological quantum computing, as first proposed by Kitaev [77,78,112,113]. Unfortunately, we cannot even scratch the surface here. Reference [114] is recommended for the interested reader. Consider a distribution of $2N$ Mzm $\gamma_i, i = 1, 2, \ldots, 2N$ (localised by vortices) arranged in a 2-dimensional plane and protected by a gap. As in the Kitaev chain, the Mzm are paired into N non-local fermion operators $f_i = (\gamma_{2i-1} + i\gamma_{2i})/2$, which generate 2^N orthogonal zero-energy states that can be either occupied or empty, $n_i = 0, 1$. The subspace of ground states $\mathcal{H}_0$ is thus 2^N-dimensional and spanned by,

$$|n_1 n_2, \ldots n_N\rangle, \qquad n_i \in \{0, 1\}, \tag{3.35}$$

where the n_i are the eigenvalues of the corresponding number operators,

$$n_i = f_i^\dagger f_i = \frac{1}{2}\left(1 + i\gamma_{2i-1}\gamma_{2i}\right). \tag{3.36}$$

The ground state manifold of $2N$ Mzm thus renders us a basis for N topological qubits [77,115]. The inherent non-locality of the f_i operators dramatically suppresses environmentally induced decoherence, since it requires coordinated "measurements" at spatially separated locations.

If the temperature is kept sufficiently below the bulk gap then an initial qubit state will remain confined to the ground state manifold under the adiabitic manipulation of the vortices and, hence, the state space explored is protected against local perturbations. However, because of the non-Abelian statistics the adiabatic exchange of two vortices will generically induce a unitary transformation within the qubit state

[4] For our purposes it is sufficient to regard a vortex as a point at which the superconducting order Δ goes to zero and around which its phase picks up a 2π shift.

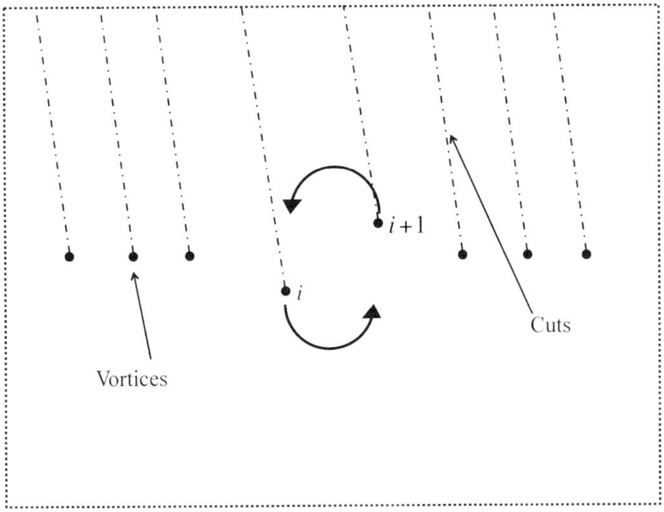

Figure 1: Exchanging vortices.

space. Not only do the topological features of the system protect against decoherence errors, a large class of unitary operations acting on the qubit state can be implemented by simple sequences of adiabatic swaps!

Let us explore this last idea a little further. The non-Abelian nature of Mzm in the Fu-Kane model (and its ilk) essentially follows from geometric phase and explicit monodromy considerations, see for example [56, 101, 116, 117]. Ivanov [118] developed a heuristically useful picture that reduces the essence of the phenomenon to a set of rules for its effect on the Mzm operators γ_i as the vortices are interchanged. Consider a set of vortices in the plane as in Figure 1. We let the vortices adiabatically move around before returning to the same set of positions, but possibly re-ordered. To encode the phases picked up Ivanov attached a cut extending from each vortex to a common boundary. Each time a vortex passes through a cut the corresponding Mzm operator acquires a phase shift of $\pm\pi$ and so picks up a sign. Roughly speaking this is related to the phase of the superconducting order parameter Δ, which is regarded as single-valued away from the cuts and jumps by 2π across the cuts. From (3.4) a jump of 2π corresponds to a phase shift of π for the Mzm operators. From this rule we see that if two adjacent vortices i and $i+1$, as in Figure 1, are interchanged in a counterclockwise fashion (without interfering with any other vortices) we obtain the following transformation:

$$T_i : \begin{cases} \gamma_i & \mapsto \gamma_{i+1} \\ \gamma_{i+1} & \mapsto -\gamma_i \\ \gamma_j & \mapsto \gamma_j, \quad \forall j \neq i, i+1 \end{cases} \tag{3.37}$$

Here we are using T_i to denote the operation induced by counterclockwise exchanging the vortex at postition i with vortex at position $i+1$. The operation induced by the vortex at position i encircling the vortex at position $i+1$ is therefore given by

$$T_i^2 : \begin{cases} \gamma_i & \mapsto -\gamma_i \\ \gamma_{i+1} & \mapsto -\gamma_{i+1} \\ \gamma_j & \mapsto \gamma_j, \quad \forall j \neq i, i+1 \end{cases} \tag{3.38}$$

consistent with the observation that each vortex must cross the others cut once. Note, it is also simple to check that these rules imply $T_i^4 = \mathbb{1}$ and the relations:

$$\begin{aligned} T_i T_j &= T_j T_i, \quad |i-j| > 1 \\ T_i T_j T_i &= T_j T_i T_j, \quad |i-j| = 1. \end{aligned} \tag{3.39}$$

These are the relations obeyed by the braid group,

$$B_n = \langle\, T_1, \ldots, T_{n-1} \mid T_i T_j = T_j T_i \quad |i-j| > 1 \quad \text{and} \quad T_i T_j T_i = T_j T_i T_j \quad |i-j| = 1 \,\rangle. \tag{3.40}$$

Indeed, if we adopt a vortex history or world-line point of view, sequences of oriented exchanges are obviously mapped one-to-one to braids. The braid group is generated by the elementary exchanges T_i modulo the relations (3.39), where composition is given by sequential application of the braid operations. The action of the braid group generators on the Mzm operators induces a projective representation $\tau : \mathcal{H}_0 \to \mathcal{H}_0$ acting on the Hilbert space of ground states. Since the basis (3.35) can be constructed using the the Mzm operators, the generating elements of B_{2N} take a simple explicit form in terms of Mzm operators [118],

$$\tau(T_i) = \exp\left(\frac{\pi}{4}\gamma_{i+1}\gamma_i\right) = \frac{1}{\sqrt{2}}(1 + \gamma_{i+1}\gamma_i). \tag{3.41}$$

This is essentially the Jones representation of the braid group [119], which is reducible for an even number of strands. Indeed, since the generating operators or even in Mzm operators they preserve the fermion number mod 2 and the Hilbert space has two 2^{N-1}-dimensional invariant subspaces. This same representation was found previously by Nayak and Wilczek [115] using conformal field theory. As an example consider the action on the 2-qubit (four Mzm) basis vectors of T_2^{-1} (*clockwise* exchange of vortex two and three),

$$\tau(T_2^{-1})|n_1 n_2\rangle = \exp\left(-\frac{\pi}{4}\gamma_3\gamma_2\right)|n_1 n_2\rangle = \frac{\pi}{4}\left(|n_1 n_2\rangle + i(-1)^{n_1}\sigma_x \otimes \sigma_x |n_1 n_2\rangle\right). \tag{3.42}$$

Given the importance of two dimensions here it would seem naively that the 1-dimensional Kitaev chain is not of use; two quasiparticles necessarily have to pass through each other when exchanged in one dimension, thus overlapping and breaking the ground state degeneracy. Indeed, Kitaev originally envisaged the chain as a form of topologically stable quantum memory for this reason. However, this obstacle is circumvented by fabricating 2-dimensional, or even 3-dimensional, networks of quantum wires [120–122]. Through an ingenious mechanism the Mzm at junctions of the network can be adiabatically swapped using locally tuneable voltage gates that shift the topological domains along the wires [120, 122]. The simplest example is the three-point move at a T-junction, which allows the two possible exchanges of two Mzm. Since the argument for non-Abelian statistics in topological 2-dimensional spinless $p + ip$ superconductors relied on the exchange of vortices binding the Mzm it is not immediately obvious that the quantum wire networks should exhibit the same behaviour. However, it was shown in [120–122] that the effective p-wave pairing induced in the wires by the s-wave superconductor contacts leads to precisely the behaviour described by Ivanov for Mzm bound to vortices, so that the networks have the same topological quantum computing potential. A nice pedagogical discussion of the related algebraic structures that emerge is given in [11].

Finally, we should note that the braiding operations of 2-dimensional spinless $p + ip$ superconductors or quantum wire networks alone are not sufficient for universal quantum computing, despite the fact they can implement a significant subset of the required unitary transformations. Fortunately the required additional operations can be implemented non-topologically without paying too high a price [123, 124]. Alternatively, one can search for generalisations of the quantum wire networks for which the braiding moves yield quantum operations rich enough to generate all the gates required for universal quantum computation [125].

3.2.3 The periodic table of topological phases

We previously noted that the Majorana chain has two phases, one of which is topologically non-trivial. We will now make this statement more precise following [22, 78, 126]. In particular, it will be shown how the two phases are distinguished by a topological invariant. We will then place this example within the general classification of gapped free-fermion systems and their phases [22–24, 30, 127].

For the Majorana chain with open boundary conditions the two phases have the same bulk symmetries and physics, but differ in their edge states. This situation is clearly not captured by the standard Landau paradigm, raising the question of whether or not the phases can be distinguished by studying the bulk physics. Indeed they can using a topological invariant that depends only on the properties of the bulk Hamiltonian. More generally, two gapped phases are said to be *topologically equivalent* if there is a continuous path in the phase diagram connecting them without closing the gap at any point. Phases that are topologically equivalent to a collection of independent atoms are said to be *topologically trivial*, otherwise they are *topologically non-trivial*. These phases are distinguished not by symmetries, but by *topological invariants* that are insensitive to smooth deformations of the underlying topological space. These words are, well, just words and there are a number of approaches to making them more mathematically precise, building largely on the pioneering classification schemes of [22–24, 127]. Later we will comment on this challenge, but first let us return to the treatment of the Majorana chain.

Assuming an energy gap, for each 1-dimensional Hamiltonian $\mathcal{H}$ Kitaev argued on physical grounds for the existence of a $\mathbb{Z}_2$ valued topological invariant, the Majorana number $\mathcal{M}(\mathcal{H}) \in \{-1, 1\}$, that take fermionic parity into account. Consider two parallel weakly interacting open Majorana chains. The un-paired Mzm localised on the boundary of each chain may pair-up breaking the ground state degeneracy so that if $\mathcal{M}(\mathcal{H}) = -1$ indicates the existence of Mzm then the Majorana number should satisfy $\mathcal{M}(\mathcal{H} \oplus \mathcal{H}) = \mathcal{M}(\mathcal{H})\mathcal{M}(\mathcal{H}')$ for non-interacting chains. For a fixed Hamiltonian $\mathcal{H}$, Kitaev showed that for two closed chains of lengths N_1 and N_2 the Majorana number can be expressed as

$$\mathcal{M}(\mathcal{H}) = \frac{P(\mathcal{H}(N_1 + N_2))}{P(\mathcal{H}(N_1))P(\mathcal{H}(N_2))}. \tag{3.43}$$

Here $\mathcal{H}(N)$ denotes the Hamiltonian on a closed chain of length N and $P(\mathcal{H}(N))$ is the fermionic parity (3.29) of the unique ground state. This identity follows essentially from considering the two possible ways the two chains can be closed, either into two separate chains of length N_1 and N_2 or a single chain of length $N_1 + N_2$.

Generically, the Majorana number will be difficult to calculate, but for a gapped Kitaev chain it can computed quite simply. Using the Majorana operators a completely generic (i.e. arbitrary, but quadratic) Hamiltonian, subsuming the specific case of (3.22), for such a system with N sites can be written,

$$\mathcal{H} = \frac{i}{4} \sum_{\alpha,\beta=1}^{2N} \gamma_\alpha A_{\alpha\beta} \gamma_\beta, \tag{3.44}$$

where A is real and antisymmetric. We also assume A is non-degenerate. Consequently there is a $S \in \mathrm{O}(2N)$ such that

$$\tilde{A} = SAS^T = \bigoplus_{i=1}^{N} \lambda_i \varepsilon, \qquad \varepsilon = \begin{pmatrix} 0 & 1 \\ -1 & 0 \end{pmatrix}. \tag{3.45}$$

is block-diagonal and $\lambda_i > 0, \forall i$. Here, $\pm i\lambda_i$ are the pure imaginary eigenvalues of A. In this basis the ground state has even parity since it is annihilated by $\tilde{a}_i = (\tilde{\gamma}_{2i-1} + i\tilde{\gamma}_{2i})/2$, which implies

$$\prod_i (-i\tilde{\gamma}_{2i-1}\tilde{\gamma}_{2i})|0\rangle = \prod_i (\tilde{\gamma}_{2i-1}\tilde{\gamma}_{2i-1})|0\rangle = |0\rangle. \tag{3.46}$$

For $\det S = 1$ or $\det S = -1$ the transformation (3.45) is parity preserving or parity reversing, respectively. Hence, on transforming back we obtain,

$$P(\mathcal{H}) = \operatorname{sgn}\det(S) = \operatorname{sgn}\,\mathrm{Pf}(A), \tag{3.47}$$

where we have used $\mathrm{Pf}(SAS^T) = \mathrm{Pf}(\tilde{A}) > 0$ and the property of the Pfaffian, $\mathrm{Pf}(SAS^T) = \mathrm{Pf}(A)\det(S)$. Note, for an even length chain we therefore have $\mathcal{M}(\mathcal{H}) = \operatorname{sgn}\,\mathrm{Pf}(A)$. Since we are dealing with closed chains we have periodicity N and more generally we can consider any periodicity $L|N$. Writing

$$\mathcal{H} = \frac{i}{4}\sum_{x,y=1}^{L}\sum_{a,b=1}^{2N/L}\gamma_{xa}A_{ab}(y-x)\gamma_{yb}, \tag{3.48}$$

where $y-x$ is taken mod L, we can use the Fourier transform

$$\bar{A}_{ab}(k) = \sum_{z} e^{ikz}A_{ab}(z), \quad k = \frac{2\pi n}{L} \mod 2\pi \quad \text{for} \quad n \in \mathbb{Z}_N \tag{3.49}$$

to write the Pfaffian as

$$\mathrm{Pf}(A) = \prod_{k=-k}\mathrm{Pf}\,\bar{A}(k)\prod_{k\neq -k}\det\bar{A}(k). \tag{3.50}$$

The assumption that the system is gapped implies that $\det\bar{A}(k)$ is positive [78] and hence, recalling k is defined mod 2π,

$$\mathcal{M}(\mathcal{H}) = \operatorname{sgn}\,\mathrm{Pf}(A) = \operatorname{sgn}\left[\mathrm{Pf}\,\bar{A}(0)\,\mathrm{Pf}\,\bar{A}(\pi)\right]. \tag{3.51}$$

For the Majorana chain the momentum space Hamiltonian was given in (3.30). Arranging the Majorana operators into doublets

$$\Gamma_i = (\gamma_{2i-1}, \gamma_{2i})^T \tag{3.52}$$

momentum space Hamiltonian (3.30) in the Majorana basis is given by

$$\mathcal{H} = i\sum_{k}\Gamma_k^T\left[-(2t\cos(k)-\mu)i\sigma_y + \Delta\sin(k)\sigma_x\right]\Gamma_{-k} \tag{3.53}$$

so that

$$\bar{A}(0) = -(2t-\mu)i\sigma_y, \quad \bar{A}(\pi) = -(-2t-\mu)i\sigma_y \tag{3.54}$$

and

$$\operatorname{sgn}\,\mathrm{Pf}(A) = -\operatorname{sgn}(2t-\mu)(2t+\mu) = \begin{cases} 1 & 2|t| < |\mu| \\ -1 & 2|t| > |\mu| \end{cases} \tag{3.55}$$

in agreement with the analysis of the phases given section 3.2.1.

The geometric significance of the Majorana number as given in (3.51) is not immediately apparent. However, there is also a $\mathbb{Z}_2$ valued Chern-Simons invariant defined on the Majorana chain [128] and it was shown in [126] that the two forms are equivalent. Using $\det(S) = \operatorname{sgn}\,\mathrm{Pf}(A)$ we can alternatively write

$$\mathcal{M}(\mathcal{H}) = \det\bar{S}(0)\det\bar{S}(\pi), \tag{3.56}$$

where the Fourier transform $\bar{S}(k)$ is unitary and satisfies $\bar{S}(k)^* = \bar{S}(-k)$ from the reality of S. Letting $\det\bar{S}(k) = \exp(i\phi_k)$ we then have

$$\mathcal{M}(\mathcal{H}) = (-1)^{\Delta\phi/\pi}, \tag{3.57}$$

where $\Delta\phi = \phi_0 - \phi_\pi$ is quantised in units of π, since the reality constraint implies $\phi_k^* = -\phi_{-k} \mod 2\pi$. In the large N limit the phase difference can be expressed in terms of the Zak-Berry phase,

$$\Delta\phi = \int_{-\pi}^{\pi}\mathrm{Tr}\,\mathcal{A}(k)dk \tag{3.58}$$

where $\mathcal{A}(k)$ is the non-Abelian Zak-Berry connection on the set of occupied Bloch states. Hence, the Majorana number is identified as a quantised Chern-Simons invariant in one dimension $\mathcal{M}(H) = \exp[i2\pi \mathrm{CS}_1(\mathcal{A})]$, where

$$\mathrm{CS}_1(\mathcal{A}) = \frac{i}{2\pi}\int \mathrm{Tr}\,\mathcal{A}. \tag{3.59}$$

This is but one of many topological invariants appearing in the classification of topological phases. See [13] for a detailed and highly accessible review.

Having seen an isolated example let us now turn to the classification of all gapped free fermion phases as specified by their dimensions, symmetries and topological properties. In [22] Altland and Zirnbauer classified disordered fermion systems into ten symmetry classes that correspond to the classical symmetric spaces of compact type. These same ten classes apply to the Hamiltonians of free fermions systems (irrespective of whether or not they are gapped). The classification of topological phases for gapped free fermion systems was pioneered by Schnyder, Ryu, Furusaki, and Ludwig in [22] and reviewed in [129]. In particular, for a given dimension they showed that topologically non-trivial phases exist in five of the ten symmetry classes. Of these they established that three are furnished with a $\mathbb{Z}$-valued invariant, while the remaining two admit a $\mathbb{Z}_2$-valued invariant. Pulling these observations together, Kitaev developed a compelling and influential picture of this classification system, now known as the "Periodic Table for topological insulators and superconductors" [23]. He introduced a set mathematical organising principles that revealed the Bott periodic features of the classification. In particular, he uncovered the important role played by Clifford algebras and K-theory in understanding the classification of the topological phases. These ideas have since been developed in number of directions. A systematic and elucidating account of Kitaev's perspective was presented in [25] and the role of Clifford algebras was emphasised and completed in [130]. A unified framework on a more mathematically firm footing was developed by Freed and Moore in a clear and comprehensive work [26], which, in particular, treated the symmetries carefully and introduced twisted K-theory [131] as key tool. Many of the mathematical ambiguities (as well certain physical puzzles) where clarified in [32], again in more rigorous mathematical terms that rely primarily on *unreduced* K-theory. We will comment further on some of these and other perspectives once we have introduced more fully the key ideas put forward in original classification schemes following [22, 23, 25].

Our starting point is the "ten-fold way" of Altland and Zirnbauer [127], generalising Dyson's three symmetry classes for many-body systems [132]. Consider a generic matrix H, indices going over position, spin, flavour,..., that can be used to build a arbitrary quadratic fermionic Hamiltonian. The possible matrices H are separated into ten classes under the presence or absence of three discrete transformations $\mathcal{C}$, $\mathcal{T}$ and $\mathcal{P} = \mathcal{C}\cdot\mathcal{T}$ [22].

$$\begin{aligned} \mathcal{C}H\mathcal{C}^{-1} &= -H \qquad \mathcal{C}^2 = \pm\mathbb{1}, \\ \mathcal{T}H\mathcal{T}^{-1} &= H \qquad \mathcal{T}^2 = \pm\mathbb{1}. \end{aligned} \tag{3.60}$$

The particle-hole symmetry (PHS) $\mathcal{C}$ and time-reversal symmetry (TRS) $\mathcal{T}$ are antiunitary operations, while sublattice symmetry (SLS) $\mathcal{P}$ is unitary. In a given basis their action on H can be represented by complex matrices satisfying:

$$\begin{aligned} CH^*C^{-1} &= H \qquad C^*C = \pm\mathbb{1}, \\ TH^*T^{-1} &= H \qquad T^*T = \pm\mathbb{1}, \\ PHP^{-1} &= -H \qquad P = P^\dagger, P^2 = \mathbb{1}. \end{aligned} \tag{3.61}$$

For example, recall the BdG Hamiltonian describing superconducting order (3.4),

$$\mathcal{H} = \Psi^\dagger H \Psi = \frac{1}{2}\begin{pmatrix}\psi^\dagger & \psi\end{pmatrix}\begin{pmatrix} h & \Delta \\ \Delta^\dagger & -h^T\end{pmatrix}\begin{pmatrix}\psi \\ \psi^\dagger\end{pmatrix}, \qquad h^\dagger = h, \quad \Delta^T = -\Delta. \tag{3.62}$$

By construction the Nambu spinor Ψ satisfies the constraint $\Psi^\dagger = (C\Psi)^T$ where $C = \sigma_x$ squares to $+\mathbb{1}$ and

$$CH^*C^{-1} = -H \tag{3.63}$$

so that by definition the BdG system respects the PHS with $C^*C = +\mathbb{1}$, in this case better referred to as the particle-hole *constraint*. Recall, we can rewrite the BdG Hamiltonian in a Majorana basis,

$$\mathcal{H} = \Gamma^T(iX)\Gamma, \qquad X^* = X, \quad X^T = -X, \tag{3.64}$$

so that X is an element of $\mathfrak{so}(4n)$.

Examining the various possibilities leads to ten symmetry classes as shown in Table 2. There are three possibilities for PHS: no symmetry, symmetry with $\mathcal{C}^2 = \mathbb{1}$, and symmetry with $\mathcal{C}^2 = -\mathbb{1}$. Similarly, there are three possibilities for TRS, giving a total of nine. Only for the case of no symmetry under both PHS and TRS is the SLS action not fixed. Hence there are ten classes in total. These are in one-to-one correspondence with the $10 = 2 + 8$ classical families of symmetric spaces of compact type as classified by Cartan, see Table 2. Again this is easily understood. Roughly speaking the discrete symmetries can be used to construct an involution, $\theta : \mathfrak{g} \to \mathfrak{g}, \theta^2 = \mathbb{1}$, so that,

$$\mathfrak{g} = \mathfrak{h} + \mathfrak{p}, \quad [\mathfrak{h}, \mathfrak{h}] \subset \mathfrak{h}, \quad [\mathfrak{h}, \mathfrak{p}] \subset \mathfrak{p}, \quad [\mathfrak{p}, \mathfrak{p}] \subset \mathfrak{h} \tag{3.65}$$

where $\mathfrak{h}$ ($\mathfrak{p}$) is the positive (negative) eigenspace of θ. For each symmetry class, specified by the presence/absence of a PHS/TRS/SLS, there is an associated Lie algebra $\mathfrak{g}$ and involution θ such that iH will lie in the corresponding negative eigenspace $\mathfrak{p}$. For example, in the case of no symmetries H is a $n \times n$ Hermitian matrix so that $iH \in \mathfrak{u}(1) \oplus \mathfrak{su}(n)$ and $\exp[iH]$ can be regarded as an element of the type II symmetric space $\mathrm{U}(n) \times \mathrm{U}(n)/\mathrm{U}(n) \cong \mathrm{U}(n)$ of class A (up to Abelian factors). The adjoint action of the SLS operation squares to identity so that it defines an involution. If H respects the SLS, that is anti-commutes with $\mathcal{P}$, then iH must lie in the negative eigenspace $\mathfrak{p} \cong \mathbb{C}^k \otimes \mathbb{C}^m$, where $\mathfrak{u}(n) = \mathfrak{u}(k) \oplus \mathfrak{u}(m) + \mathfrak{p}$ and $n = k + m$. This condition is preserved by the adjoint action of $\mathrm{U}(k) \times \mathrm{U}(m)$ so that $\exp[iH]$ is an element of the type I symmetric space of class $AIII$ $\mathrm{U}(n)/\mathrm{U}(k) \times \mathrm{U}(m)$. This pair of symmetric spaces, or equivalently symmetry classes, repeats with the order two Bott periodicity of complexified Clifford algebras (2.7).

The correspondence between the remaining symmetric spaces and symmetry classes of H can be made systematic [25] such that they appear in particular order that brings out their mod eight Bott periodic features first noted in [23]. This approach turns around that of Altland-Zirnbauer by starting with the symmetric spaces, extracting the corresponding Hamiltonian families and only then identifying their symmetry class. Consider $\mathrm{O}(16r)$ in the inductive $r \to \infty$ limit and its vector representation $V \cong \mathbb{R}^{16r}$. Now introduce a set of orthogonal anti-commuting complex structures J_i acting on V,

$$J_i J_j + J_j J_i = -2\delta_{ij}\mathbb{1}, \qquad i, j = 1, 2, \ldots m, \tag{3.66}$$

constituting a reducible representation of $\mathrm{Cliff}(m, 0)$. The subgroup in $\mathrm{O}(16r)$ commuting with J_1, being a complex structure, is $\mathrm{U}(8n)$. Now J_2 anti-commutes with the complex structure J_1 and so is anti-linear on $\mathbb{C}^{8r}$ and thus defines a quaternionic structure. The subgroup in $\mathrm{U}(8n)$ commuting with J_2 is therefore $\mathrm{Sp}(4r) \cong \mathrm{U}(4r, \mathbb{H})$. Introducing one-by-one further J_i generates a sequence of subgroups repeating modulo eight [25]:

$$\begin{array}{ccccccccccccccc} \mathbb{R} & & \mathbb{C} & & \mathbb{H} & & \mathbb{H}\mathbb{H} & & \mathbb{H} & & \mathbb{C} & & \mathbb{R} & & \mathbb{R}\mathbb{R} \\ \mathrm{O}(16r) & \subset & \mathrm{U}(8r) & \subset & \mathrm{Sp}(4r) & \subset & \mathrm{Sp}(2r) \times \mathrm{Sp}(2r) & \subset & \mathrm{Sp}(2r) & \subset & \mathrm{U}(2r) & \subset & \mathrm{O}(2r) & \subset & \mathrm{O}(r) \times \mathrm{O}(r) \end{array} \tag{3.67}$$

where we have indicated their corresponding real, complex and quaternionic nature,

$$\begin{aligned} \mathfrak{so}(n) &\cong \{x \in \mathbb{R}[n] \,|\, x^\dagger = -x\}, \\ \mathfrak{u}(n) &\cong \{x \in \mathbb{C}[n] \,|\, x^\dagger = -x\}, \\ \mathfrak{sp}(n) &\cong \{x \in \mathbb{H}[n] \,|\, x^\dagger = -x\}, \end{aligned} \tag{3.68}$$

which reflects the Bott periodic mnemonic (2.5) for the Clifford matrix algebras. Let G be a given group in this sequence and K its adjacent subgroup, then G/K is a homogeneous symmetric space. If G_k is the

Class	q	Coset space	PHS	TRS	SLS	0	1	2	3	4	5	6	7
A	0	$\mathrm{U}(n)\times\mathrm{U}(n)/\mathrm{U}(n)$	0	0	0	$\mathbb{Z}$	0	$\mathbb{Z}$	0	$\mathbb{Z}$	0	$\mathbb{Z}$	0
$AIII$	1	$\mathrm{U}(n)/\mathrm{U}(k)\times\mathrm{U}(m)$	0	0	+	0	$\mathbb{Z}$	0	$\mathbb{Z}$	0	$\mathbb{Z}$	0	$\mathbb{Z}$
D	1	$\mathrm{O}(n)\times\mathrm{O}(n)/\mathrm{O}(n)$	+	0	0	$\mathbb{Z}_2$	$\mathbb{Z}_2$	$\mathbb{Z}$	0	0	0	$\mathbb{Z}$	0
$DIII$	2	$\mathrm{O}(2n)/\mathrm{U}(n)$	+	−	+	0	$\mathbb{Z}_2$	$\mathbb{Z}_2$	$\mathbb{Z}$	0	0	0	$\mathbb{Z}$
AII	3	$\mathrm{U}(2n)/\mathrm{Sp}(n)$	0	−	0	$\mathbb{Z}$	0	$\mathbb{Z}_2$	$\mathbb{Z}_2$	$\mathbb{Z}$	0	0	0
CII	4	$\mathrm{Sp}(n)/\mathrm{Sp}(k)\times\mathrm{Sp}(m)$	−	−	+	0	$\mathbb{Z}$	0	$\mathbb{Z}_2$	$\mathbb{Z}_2$	$\mathbb{Z}$	0	0
C	5	$\mathrm{Sp}(n)\times\mathrm{Sp}(n)/\mathrm{Sp}(n)$	−	0	0	0	0	$\mathbb{Z}$	0	$\mathbb{Z}_2$	$\mathbb{Z}_2$	$\mathbb{Z}$	0
CI	6	$\mathrm{Sp}(n)/\mathrm{U}(n)$	−	+	+	0	0	0	$\mathbb{Z}$	0	$\mathbb{Z}_2$	$\mathbb{Z}_2$	$\mathbb{Z}$
AI	7	$\mathrm{U}(n)/\mathrm{O}(n)$	0	+	0	$\mathbb{Z}$	0	0	0	$\mathbb{Z}$	0	$\mathbb{Z}_2$	$\mathbb{Z}_2$
BDI	0	$\mathrm{O}(n)/\mathrm{O}(k)\times\mathrm{O}(m)$	+	+	+	$\mathbb{Z}_2$	$\mathbb{Z}$	0	0	0	$\mathbb{Z}$	0	$\mathbb{Z}_2$

Table 2: Classification table for topological insulators and superconductors [22, 23, 129]. The first column list the Cartan labels of the symmetric spaces corresponding to the Hamiltonian symmetry classes as determined by the PHS, TRS and SLS columns (here 0 denotes no symmetry and $\pm$ denotes a symmetry that squares to $\pm\mathbb{1}$). The second column lists the corresponding coset spaces, where $n = k + m$. Note, on taking the inductive limit $n \to \infty$ we obtain the classifying spaces R_q given in (3.71) and (3.70). The remaining columns contain the classification of phases for topological insulators and superconductors with translation invariance in free fermion systems in spatial dimensions 0 through 7. Here a 0 means there is no non-trivial topological phase, while $\mathbb{Z}_2$ and $\mathbb{Z}$ denote the sets of distinct phases. The real cases are given by the reduced real K-theory groups described in section 2, $\widetilde{KO}(S^{q+1-d}) \cong \pi_0(R_{q+1-d})$. Note, we have the same Bott periodic pattern in both symmetry class and spatial dimension. We see that the Majorana chain belonging to the D class has a $\mathbb{Z}_2$ set of phases, one trivial and the other admitting Mzm on the boundaries.

subgroup defined by commuting with all J_i up to $i = k$ then the adjoint action of J_{k+1} on its Lie algebra $\mathfrak{g}_k$ is an involution and $\mathfrak{g}_{k+1}$ (the Lie algebra of $G_{k+1} \equiv K_k$) is its positive eigenspace. Note, for fixed $J_1, \ldots, J_k$ the symmetric space G_k/G_{k+1} parametrises the possible choices of J_{k+1}, since, essentially, the stability group of J_{k+1} as an element of the adjoint representation of G_k is G_{k+1}. This process gives us an ordered list of classical symmetric spaces repeating with order eight, labelled by $q \in \mathbb{F}_8$ in Table 2. We note that the exceptional symmetric spaces, which are naturally associated with octonionic geometries [5], are not naively applicable since being of fixed size they do not admit a thermodynamic limit. It is perhaps conceivable, although rather unlikely, that a large many-body system could have repeating subsystems constrained by exceptional symmetries, a trace of which might survive the thermodynamic limit.

Given an element in $\mathfrak{p} \cong T(G/K)$ for a specific G/K we can extract a Hamiltonian and determine its symmetries [25]. We have already seen that for the PHS respecting BdG Hamiltonian with $\mathcal{C}^2 = \mathbb{1}$ and no further symmetries that $iX = H$ where $X \in \mathfrak{so}(4n)$ corresponding to the type II symmetric space in class D, $\mathrm{O}(4n)\times\mathrm{O}(4n)/\mathrm{O}(4n) \cong \mathrm{O}(4n)$. Conversely, consider an arbitrary element $X \in \mathfrak{so}(4n)$ and map $i \mapsto -i\sigma_y \otimes \mathbb{1}$ so that we can identify a Hamiltonian $H = -i\sigma_y \otimes X$. Then $\mathcal{C} = \sigma_z \otimes \mathbb{1}$ defines a real structure anti-commuting with H and hence a PHS with $\mathcal{C}^2 = \mathbb{1}$, placing H in symmetry class

D. Lets move one step along the symmetric spaces to $\mathrm{O}(4n)/\mathrm{U}(2n)$. By construction the elements of $\mathfrak{p} \cong T(\mathrm{O}(4n)/\mathrm{U}(2n)) \cong \wedge^2\mathbb{C}^{2n}$ are real anti-symmetric matrices $X \in \wedge^2\mathbb{R}^{4n}$ that anti-commute with a single complex structure $JX = -XJ$. As before let $H = -i\sigma_y \otimes X$ so that $\mathcal{C} = \sigma_z \otimes \mathbb{1}$ is again a PHS. Now $\mathcal{C}(\mathbb{1} \otimes J) = \sigma_z \otimes J$ is anti-linear (anti-commutes with $-i\sigma_y \otimes \mathbb{1}$), commutes with H and hence defines a TRS such that $\mathcal{T}^2 = -\mathbb{1}$. The product $\mathcal{C} \cdot \mathcal{T}$ is given by $\mathbb{1} \otimes J$, which is linear (commutes with $-i\sigma_y \otimes \mathbb{1}$), anti-commutes with H and and hence defines a SLS. We conclude that H indeed respects the PHS, TRS and SLS specified by Table 2 for class *DIII*. One can continue in this manner for all eight classes [25]. This construction kept things symmetric so that one can continue indefinitely for large enough r and so expose the periodic features, but in general for the relevant cases we can have $G(n)/K(k) \times K(m)$ with $n = k + m$. Note, with the ordering dictated by (3.67) the presence of SLS symmetry is alternating while the PHS follows the same Bott periodic pattern as the symmetric spaces,

$$\begin{matrix} \mathbb{RR} & \mathbb{R} & \mathbb{C} & \mathbb{H} & \mathbb{HH} & \mathbb{H} & \mathbb{C} & \mathbb{R} & \mathbb{RR} & \mathbb{R} & \mathbb{C} & \mathbb{H} & \mathbb{HH} & \mathbb{H}\cdots \\ + & + & 0 & - & - & - & 0 & + & + & + & 0 & - & - & -\cdots \end{matrix} \tag{3.69}$$

as does TRS, but shifted back two places. Here we are denoting the absence of the PHS symmetry by 0 and its presence by $\pm$ for $\mathcal{C}^2 = \pm\mathbb{1}$.

For the classification of topological phases we are then interested in the partition of a given space $\mathcal{M}$ of admissible H determined by the symmetry class and gap condition, which is presumed to form a topological space, into its path-connected components $\pi_0(\mathcal{M})$. Heuristically this corresponds to the equivalence of two Hamiltonians if they can be identified by adiabatic perturbations that do not close the gap or violate any of the symmetries at any point. If we consider families of Hamiltonians $H(x)$ parametrised by some topological space $x \in X$ (for example a Brillouin zone) then a vector bundle picture merges, typically by associating an isolated family of m negative energy bands with a rank m complex vector bundle, see for example [27]. We are then interested in homotopy equivalent vector bundles over X, as given by homotopy classes of maps from X to some classifying space[5]. Of course, we have to include the action of the transformations for a given symmetry class, on both the base space and the fibres, furnishing these vector bundles with additional structure. For example, the TRS of class *AI* implies the use of "real" vector bundles developed in [133]. In a series of papers [27,28,31] the appropriate categories of vector bundles with extra structure where constructed for classes *AI*, *AII*, *AIII* (referred to as 'real', 'quaternionic', and 'chiral', respectively), and the classification of topological phases for a broad class of compact base spaces was completed. All ten classes where treated in [29,30] over either spheres or tori equipped with an involution and fibres consisting of complex vector spaces subject to appropriate constraints reflecting the symmetry class. Here too we see the Bott periodic features discovery by Kitaev in his original classification scheme [23], which we now discuss.

Seeking a manageable classification that captures the essential physics Kitaev appealed to the notion of *stable equivalence*, which gives a courser partitioning than homotopy alone and naturally leads us to a K-theoretic perspective. The physical picture is that we are allowed to augment our system of interest with additional degrees of freedom that are typically trivial with respect to the relevant physics and then look for equivalence in this larger space. For example, we may include additional flat bands for an insulator by augmenting with inner atomic shells. Two initially inequivalent phases may become equivalent once we allow for argumentation, while preserving the features we would like to capture in the classification. Kitaev also adopted (in general) a *relative* perspective, again naturally in keeping with his K-theoretic approach. Here one phase is considered relative to another; in essence we classify topological obstructions between Hamiltonians as opposed to the Hamiltonians themselves. The relative approach was also recently advocated in [134], motivated in part by its naturalness, but also because the "absolute" picture requires a canonical notion of the trivial or zero phase, which is not always available leading to certain ambiguities. Indeed, for the class *AIII* the equivalence under isomorphism and homotopy in the

[5]We could have alternatively considered isomorphism classes, then the K-theory groups $K(X)$ would arise naturally, but homotopy seems physically natural and appears to be the correct choice especially when the vector bundles are furnished with extra structure coming from the symmetries classes [31].

category of chiral vector bundles do not coincide and the homotopy classification depends on a choice of reference isomorphism so that only phases relative to each other have an absolute meaning [31].

With these considerations in mind Kitaev defined two admissible matrices $A, B \in \mathcal{M}$ to be equivalent $A \sim B$ if and only if $A \oplus C$ and $B \oplus C$ belong to the same element of $\pi_0(\mathcal{M})$ for some matrix C of arbitrary size. Note, $\mathcal{M}$ here places no restriction on the size of admissible matrices. Second, for a pair of same size matrices (A, B) he defined the difference class $d(A, B)$ as the equivalence class with representative (A, B) of same size pairs, where two pairs (A, B) and (A', B') are equivalent if and only if $A \oplus B' \sim A' \oplus B$. Note, this definitions allows for A and A' (and so B and B') to have different sizes. These definitions capture the notion of augmentation and relativity, respectively. Because we allow for arbitrary size augmentations we might anticipate that the classification for families of Hamiltonians will depend on $[X, R]$, the set of homotopy classes of maps from the parameter space X to some classifying space R. The action of the symmetries makes this rather subtle, as we shall see. But first let us get a handle on the set of classifying spaces that we expect to be involved.

We have already seen that the symmetry requirements imply that the admissible A (from here on in we will consider iH as opposed to H) belong to $\mathfrak{p} \cong T(G/K)$ for G/K belonging to one of the ten classes of symmetric spaces in Table 2. Since we are considering gapped systems Kitaev also required that the admissable A be full rank. As we are interested in topological phases the magnitude of the eigenvalues is irrelevant; we can always use a homotopy equivalent spectrally flattened matrix satisying $A^2 = -\mathbb{1}$ with eigenvalues $\pm i$. Let $\mathfrak{p}_l = T(G_l/G_{l+1})$, then for all $A \in \mathfrak{p}_l, A^2 = -\mathbb{1}$ the matrix $J_{l+1}A$ anticommutes with $J_i, i = 1 \ldots l+1$ and is therefore a candidate J_{l+2}. Hence, the set of possible $A \in \mathfrak{p}_l, A^2 = -\mathbb{1}$ is given by the symmetric space one step along the Bott periodic sequence G_{l+1}/G_{l+2}. The space of admissible $A \in \mathfrak{p}_k = T(G_l/G_{l+1})$ of a fixed size is G_{l+1}/G_{l+2}, where for the cases $G(n)/K(k) \times K(m)$ we have to take the union of the possible spits $n = k + m$. Hence, allowing for augmentation we expect the sequence of classifying spaces given by the inductive limits of the G/K given Table 2. For the real order eight case we have $R_q, q \in \mathbb{F}_8$ where,

$$\begin{array}{cccccccc} R_1 & R_2 & R_3 & R_4 & R_5 & R_6 & R_7 & \mathbb{Z} \times R_0 \\ \mathrm{O} & \mathrm{O}/\mathrm{U} & \mathrm{U}/\mathrm{Sp} & \mathbb{Z} \times \mathrm{BSp} & \mathrm{Sp} & \mathrm{Sp}/\mathrm{U} & \mathrm{U}/\mathrm{O} & \mathrm{BO} \end{array} \tag{3.70}$$

For the complex order two case we have $C_q, q \in \mathbb{F}_2$ where

$$\begin{array}{cc} C_1 & C_0 \\ \mathrm{U} & \mathbb{Z} \times \mathrm{BU} \end{array} \tag{3.71}$$

For example, if X is a d-sphere with no further structure and we label the symmetry classes (symmetric spaces) by the same q, as in Table 2, then for class q we would have $[S^d, R_{q+1}] \cong \pi_d(R_{q+1}) \cong \pi_0(q+d+1)$ [25]. In general, however, X will carry extra structure such as involution.

Kitaev approached the problem via Clifford algebras and K-theory [23]. As we have seen each set of symmetries corresponding to a given class can be mapped to a Clifford algebra $\mathrm{Cliff}(p, q)$ for $q = 0$. Moreover, the spectrally flattened $A^2 = -\mathbb{1}$ can be used to extend the Clifford algebra to $\mathrm{Cliff}(p+1, q)$ and we need to classify the possible extensions in the inductive limit. The key object is the K-theory group $K^{p,q}(X)$. Loosely speaking this is the Grothendieck group of the restriction functor, induced by the inclusion $\mathrm{Cliff}(p, q) \subset \mathrm{Cliff}(p, q+1)$, from the category of $\mathrm{Cliff}(p, q+1)$-modules to the category of $\mathrm{Cliff}(p, q)$-modules over a base space X, which we will assume compact. See [47] chapter II section 2.13 and chapter III section 4. In particular, for compact X we have the isomorphisms,

$$K_{\mathbb{R}}^{p,q}(X) \cong [X, R_{q-p}], \tag{3.72}$$

where $q - p$ is taken module 8, and

$$K_{\mathbb{C}}^{p,q}(X) \cong [X, C_{q-p}], \tag{3.73}$$

where $q-p$ is taken module 2 [47]. Note, for the problem at hand we can (and Kitaev did) use the Clifford period relations,

$$\begin{aligned} \mathrm{Cliff}(0,d+2) &\cong \mathrm{Cliff}(d,0) \otimes \mathrm{Cliff}(0,2) \\ \mathrm{Cliff}(d+2,0) &\cong \mathrm{Cliff}(0,d) \otimes \mathrm{Cliff}(2,0) \\ \mathrm{Cliff}(s,t) &\cong \mathrm{Cliff}(s,t) \otimes \mathrm{Cliff}(1,1) \end{aligned} \tag{3.74}$$

to send $p \to 0$ and $q \to p+2$ up to Morita equivalence.

For translationally invariant systems in an arbitrary spatial dimension d we can consider the family of $A(k)$ for $k \in \tilde{\mathbb{R}}^d$, where the tilde indicates standard Euclidean space equipped with an involution $\tau(k) = -k$. The K-theory of 'real' vector bundles over a topological space X with an involution $\tau : X \to X, \tau^2 = \mathbb{1}$ was developed by Atiyah [133], where it was denoted $KR(X)$. See also [27] for a rigorous treatment of 'real' vector bundles applied to class AI. To obtain a reasonable classification Kitaev considered fixed $|k| \to \infty$ asymptotics for $A(k)$ so that we can regard infinity as the boundary $\tilde{S}^{d-1}$ of a large ball $\tilde{B}^d$. Then for real symmetry class q in d spatial dimensions the classification is given by the relative[6] K-theory group [23],

$$K_{\mathbb{R}}^{0,q+1}(\tilde{B}^d, \tilde{S}^{d-1}) \cong \pi_0(R_{q+1-d}) \cong \widetilde{KO}(S^{q+1-d}), \tag{3.75}$$

which yields Table 2 as in [23, 129]. The $d = 1$ Majorana chain from section 3.2.1 belongs to class D ($q = 1$) and is thus predicted correctly to have two phases $\pi_0(R_1) \cong \mathbb{Z}_2$, one of which is topologically non-trivial, or alternatively a single $\mathbb{Z}_2$-valued invariant separating the phases. To give another example, topological superconductors in $d = 1$ and class $DIII$ (as considered, for example, in the context of B phase superfluid ^{3}He [22]) have phases labeled by a $\mathbb{Z}$-valued topological invariant, the winding number. For a comprehensive review of Table 2 along with various examples realising its entries see [13]. We should mention that there are a number of subtleties hidden by Table 2. For example, the use of stable equivalence has led to the prediction of no topological phases in $d = 3$ for class A, yet if one restricts to two bands only then non-trivial phases appear [30], although they are unstable under the addition of trivial bands. Finally, let us conclude by noting that the pattern appearing in Table 2 can be reinterpreted in terms of the division algebras by passing through Clifford algebras as in section 2. On the surface at least this appears to be nothing more than curiosity; whether or not the division algebras could play a more direct role is left for future consideration.

4 Conclusions

We have touched on Majorana's impact in particle physics, condensed matter and quantum computation. In section 1 we reviewed Majorana's original vision, which may yet prove crucial to our understanding of neutrino physics. We then moved on to the role of Majorana quasiparticles as emergent states appearing in condensed matter systems with superconducting order in section 3. This included both Majorana fermions and their more exotic cousins, Majorana zero-modes exhibiting non-Abelian ayononic statistics. We indicated how attributes of the latter lend themselves to fault-tolerant topological quantum computation. To close we examined the classification of gapped free fermion phases in condensed matter, which tied together our understanding of many-body systems and the theory of spacetime spinors appearing in fundamental particle physics via Clifford algebras and K-theory.

The impact of topological phases in quantum computing represents a marvellous and rather rare example of fundamental theoretical advances progressing hand-in-hand with their technological applications [6]. One should also keep in mind, however, that this synthesis has its germ in numerous fundamental results

[6] A relative K-theory group $K(X, Y)$, for Y closed in X, is the Grothendieck group of the functor induced by the restriction of bundles on X to those on Y. Note, we have an isomorphism with the standard reduced K-theory, $K(X, Y) \cong \tilde{K}(X/Y)$, which is sometimes taken as the definition. $K^{p,q}(X, Y)$ is the appropriate corresponding generalisation of $K^{p,q}(X)$. See [47] chapter II section 2 and chapter III section 5.

of curiosity driven science, stretching from Majorana to todays cutting edge research in condensed matter. The pioneers of these varied fields could have scarcely anticipated their current technological applications, let alone have been motivated by them. This brings us back to the sentiment expressed in the closing remarks of section 1. Good science seeking to expand our basic understanding needs no other motivation; rewards of a broader nature surely will follow. The intrinsic quality of this form of technological progress is in the very fact that it cannot be anticipated in the present.

Our survey of Majorana fermions has a glaring omission: the crucial role they play in supersymmetry, supergravity, string theory and M-theory. This was left to [2]. However, we cannot resist concluding with some remarks that connect to aspects of string/M-theory. A crucial feature of topological phases that we have not properly touched on, although Mzm in the open Kitaev chain are an explicit example, is the bulk-boundary correspondence. See [13, 22] and the references therein. Topologically trivial and non-trivial phases must be separated by a quantum phases transition. The implication is that if we put together two d-dimensional systems, one in a trivial phase and the other in a non-trivial phase, then the $(d-1)$-dimensional interface must support gapless states, typically Dirac or Majorana zero-modes. These states are topologically protected against perturbations as long as the bulk gap and symmetries are preserved.

This bulk-boundary picture leads us to another perspective: topological phases can be characterised by anomalies [135, 136]. A theory well-defined on a d-dimensional spatial manifold without boundary, is not necessarily well-defined when a boundary is introduced. To construct a sensible theory some additional degrees of freedom must be introduced on the boundary. Particularly relevant is the case when such degrees of freedom are gapless. The boundary theory must have an anomaly in order to cancel that of the bulk theory; this is the idea of anomaly inflow [137]. The standard example in condensed matter is the integer quantum Hall effect in two spatial dimensions. The spacetime theory has a Chern-Simons coupling of the gauge field, which is not gauge invariant when there is a boundary. This deficiency is compensated by the existence of chiral modes on the boundary. These same ideas have been be applied to a significant proportion of the fermionic topological phases considered in Table 2 [138]. In particular, the analysis of "global anomalies" in [136] revealed that phases of a 3-dimensional topological superconductor in class $DIII$, where the boundary theory consists of gapless Majorana fermions once again, are reduced from $\mathbb{Z}$ to $\mathbb{Z}_{16}$, as required by the presence of interactions [139]. That it is 16 and not 8 is rather subtle and follows from an ambiguity in the sign of the boundary partition function.

In general, the "anomaly framework" makes use of a relativistic formulation, which is possible because the boundary fermions have an emergent relativistic symmetry. On the one hand, reliance on a relativistic framework might be regarded as unnatural for condensed matter systems, but on the other it offers an intriguing perspective on genuinely relativistic systems. For example, one could ask for a condensed matter analog of the M2-M5 brane system in M-theory as was done in [140]. The partial answer given there was a topological superconductor with its boundary split into two regions, one of which has as usual gapless modes, while the other has symmetry respecting gapped modes.

The sharing of ideas between condensed matter and string/M-theory is not a one-way street, either. A key idea underpinning M-theory is that of U-duality, a set of quantum dualities that interchange dual descriptions of the same physics. The best understood class of such operations go by the name of T-duality, which for example interchanges type IIA and type IIB string theories. It turns out that T-duality admits a direct analog for topological insulators [141] that simplifies the bulk-boundary correspondence!

Acknowledgments

MJD is grateful to Prof Zichichi for his hospitality. The work of LB was supported by a Schrödinger Fellowship. The work of MJD is supported by the STFC under rolling grant ST/G000743/1.

A Division algebras

An algebra $\mathbb{A}$ defined over $\mathbb{R}$ with identity element e_0, is said to be *composition* if it has a non-degenerate quadratic form[7] $\mathbf{n} : \mathbb{A} \to \mathbb{R}$ such that,

$$\mathbf{n}(ab) = \mathbf{n}(a)\mathbf{n}(b), \quad \forall \; a, b \in \mathbb{A}, \tag{A.1}$$

where we denote the multiplicative product of the algebra by juxtaposition. Regarding $\mathbb{R} \subset \mathbb{A}$ as the scalar multiples of the identity $\mathbb{R}e_0$ we may decompose $\mathbb{A}$ into its "real" and "imaginary" parts $\mathbb{A} = \mathbb{R} \oplus \mathbb{A}'$, where $\mathbb{A}' \subset \mathbb{A}$ is the subspace orthogonal to $\mathbb{R}$. An arbitrary element $a \in \mathbb{A}$ may be written $a = \text{Re}(a) + \text{Im}(a)$. Here $\text{Re}(a) \in \mathbb{R}e_0$, $\text{Im}(a) \in \mathbb{A}'$ and

$$\text{Re}(a) = \frac{1}{2}(a + \bar{a}), \qquad \text{Im}(a) = \frac{1}{2}(a - \bar{a}), \tag{A.2}$$

where we have defined conjugation using the bilinear form,

$$\bar{a} := \langle a, e_0 \rangle e_0 - a, \qquad \langle a, b \rangle := \mathbf{n}(a+b) - \mathbf{n}(a) - \mathbf{n}(b). \tag{A.3}$$

A composition algebra $\mathbb{A}$ is said to be *division* if it contains no zero divisors,

$$ab = 0 \quad \Rightarrow \quad a = 0 \quad \text{or} \quad b = 0,$$

in which case $\mathbf{n}$ is positive semi-definite and $\mathbb{A}$ is referred to as a normed division algebra. Hurwitz's celebrated theorem states that there are exactly four normed division algebras [142]: the reals, complexes, quaternions and octonions, denoted respectively by $\mathbb{R}, \mathbb{C}, \mathbb{H}$ and $\mathbb{O}$. They may be constructed via the Cayley-Dickson doubling procedure, $\mathbb{A}' = \mathbb{A} \oplus \mathbb{A}$ with multiplication in $\mathbb{A}'$ defined by

$$(a, b)(c, d) = (ac - d\bar{b}, \bar{a}d + cb). \tag{A.4}$$

With each doubling a property is lost as summarised here:

$\mathbb{A}$	*Construction*	*Dim*	*Division*	*Associative*	*Commutative*	*Ordered*
$\mathbb{R}$	$\mathbb{R}$	1	*yes*	*yes*	*yes*	*yes*
$\mathbb{C}$	$\mathbb{R} \oplus \mathbb{R}$	2	*yes*	*yes*	*yes*	*no*
$\mathbb{H}$	$\mathbb{C} \oplus \mathbb{C}$	4	*yes*	*yes*	*no*	*no*
$\mathbb{O}$	$\mathbb{H} \oplus \mathbb{H}$	8	*yes*	*no*	*no*	*no*
$\mathbb{S}$	$\mathbb{O} \oplus \mathbb{O}$	16	*no*	*no*	*no*	*no*

On doubling the octonions, $\mathbb{S} \cong \mathbb{O} \oplus \mathbb{O}$, the division property fails and we will not consider such cases here. Note that, while the octonions are not associative they are alternative:

$$[a, b, c] := (ab)c - a(bc) \tag{A.5}$$

is an alternating function under the interchange of its arguments. This property is crucial for supersymmetry.

An element $a \in \mathbb{O}$ may be written $a = a^a e_a$, where $a = 0, \ldots, 7$, $a^a \in \mathbb{R}$ and $\{e_a\}$ is a basis with one real e_0 and seven $e_i, i = 1, \ldots, 7$, imaginary elements. The octonionic multiplication rule is,

$$e_a e_b = (\delta_{a0}\delta_{bc} + \delta_{0b}\delta_{ac} - \delta_{ab}\delta_{0c} + C_{abc})\, e_c, \tag{A.6}$$

where C_{abc} is totally antisymmetric and $C_{0bc} = 0$. The non-zero C_{ijk} are given by the Fano plane. See Figure 2.

[7] A *quadratic norm* on a vector space V over a field $\mathbb{R}$ is a map $\mathbf{n} : V \to \mathbb{R}$ such that: (1) $\mathbf{n}(\lambda a) = \lambda^2 \mathbf{n}(a), \lambda \in \mathbb{R}, a \in V$ and (2) $\langle a, b \rangle := \mathbf{n}(a+b) - \mathbf{n}(a) - \mathbf{n}(b)$ is bilinear.

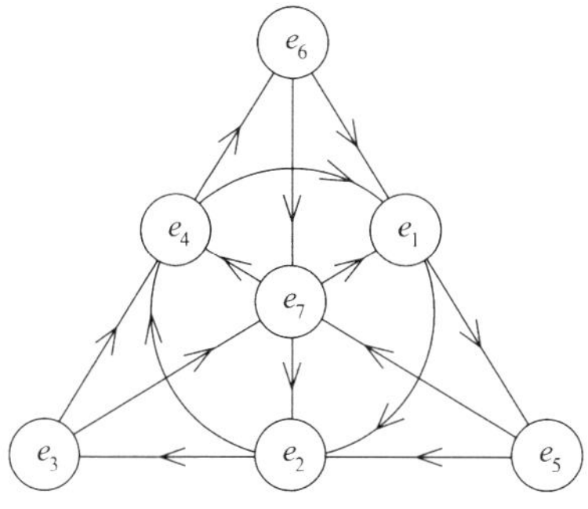

Figure 2: The Fano plane. The structure constants are determined by the Fano plane, $C_{ijk} = 1$ if ijk lies on a line and is ordered according as its orientation. Each oriented line follows the rules of quaternionic multiplication. For example, $e_2e_3 = e_5$ and cyclic permutations; odd permutations go against the direction of the arrows on the Fano plane and we pick up a minus sign, e.g. $e_3e_2 = -e_5$.

References

[1] E. Majorana, "Theory of the Symmetry of Electrons and Positrons," *Nuovo Cim.* **14** (1937) 171–184.

[2] S. Ferrara, A. Marrani, and A. Yeranyan, "Majorana Fermions, Supersymmetry Breaking, and Born-Infeld Theory," in *53rd International School of Subnuclear Physics: The Future of our Physics Including New Frontiers (ISSP 2015) Erice, Italy, June 24-July 3, 2015*. 2015. `arXiv:1510.01658 [hep-th]`. `http://inspirehep.net/record/1396357/files/arXiv:1510.01658.pdf`.

[3] J. F. Adams, Z. Mahmud, and M. Mimura, *Lectures on exceptional Lie groups*. University of Chicago Press, 1996.

[4] F. R. Harvey, *Spinors and calibrations*. academic press New York, 1990.

[5] J. C. Baez, "The octonions," *Bull. Amer. Math. Soc.* **39** (2001) 145–205, `arXiv:math/0105155`.

[6] C. Nayak, S. H. Simon, A. Stern, M. Freedman, and S. Das Sarma, "Non-Abelian anyons and topological quantum computation," *Rev. Mod. Phys.* **80** (2008) 1083–1159.

[7] J. Alicea, "New directions in the pursuit of majorana fermions in solid state systems," *Reports on Progress in Physics* **75** (2012) no. 7, 076501.

[8] C. Beenakker, "Search for majorana fermions in superconductors," *Annu. Rev. Condens. Matter Phys.* **4** (2013) no. 1, 113–136.

[9] S. R. Elliott and M. Franz, "Colloquium: Majorana Fermions in nuclear, particle and solid-state physics," *Rev. Mod. Phys.* **87** (2015) 137, `arXiv:1403.4976 [cond-mat.supr-con]`.

[10] E. Akhmedov, "Majorana neutrinos and other Majorana particles:Theory and experiment," 2014. `arXiv:1412.3320 [hep-ph]`. `http://inspirehep.net/record/1333680/files/arXiv:1412.3320.pdf`.

[11] F. Wilczek, "Majorana and Condensed Matter Physics," `arXiv:1404.0637 [cond-mat.supr-con]`.

[12] C. W. J. Beenakker, "Random-matrix theory of majorana fermions and topological superconductors," *Rev. Mod. Phys.* **87** (Sep, 2015) 1037–1066. http://link.aps.org/doi/10.1103/RevModPhys.87.1037.

[13] C.-K. Chiu, J. C. Teo, A. P. Schnyder, and S. Ryu, "Classification of topological quantum matter with symmetries," *arXiv preprint arXiv:1505.03535* (2015) .

[14] **Super-Kamiokande** Collaboration, Y. Fukuda *et al.*, "Evidence for oscillation of atmospheric neutrinos," *Phys. Rev. Lett.* **81** (1998) 1562–1567, arXiv:hep-ex/9807003 [hep-ex].

[15] F. T. Avignone, III, S. R. Elliott, and J. Engel, "Double Beta Decay, Majorana Neutrinos, and Neutrino Mass," *Rev. Mod. Phys.* **80** (2008) 481–516, arXiv:0708.1033 [nucl-ex].

[16] J. Schechter and J. W. F. Valle, "Neutrinoless Double beta Decay in SU(2) x U(1) Theories," *Phys. Rev.* **D25** (1982) 2951.

[17] P. Minkowski, "$\mu \to e\gamma$ at a Rate of One Out of 10^9 Muon Decays?," *Phys. Lett.* **B67** (1977) 421–428.

[18] R. N. Mohapatra and G. Senjanovic, "Neutrino Mass and Spontaneous Parity Violation," *Phys. Rev. Lett.* **44** (1980) 912.

[19] T. Yanagida, "HORIZONTAL SYMMETRY AND MASSES OF NEUTRINOS," *Conf. Proc.* **C7902131** (1979) 95–99.

[20] S. L. Glashow, "The Future of Elementary Particle Physics," *NATO Sci. Ser. B* **61** (1980) 687.

[21] M. Gell-Mann, P. Ramond, and R. Slansky, "Complex Spinors and Unified Theories," *Conf. Proc.* **C790927** (1979) 315–321, arXiv:1306.4669 [hep-th].

[22] A. P. Schnyder, S. Ryu, A. Furusaki, and A. W. W. Ludwig, "Classification of topological insulators and superconductors in three spatial dimensions," *Phys. Rev. B* **78** (Nov, 2008) 195125. http://link.aps.org/doi/10.1103/PhysRevB.78.195125.

[23] A. Kitaev, "Periodic table for topological insulators and superconductors," *AIP Conf. Proc.* **1134** (2009) 22–30, arXiv:0901.2686 [cond-mat.mes-hall]. [,22(2009)].

[24] S. Ryu, A. P. Schnyder, A. Furusaki, and A. W. W. Ludwig, "Topological insulators and superconductors: tenfold way and dimensional hierarchy," *New Journal of Physics* **12** (2010) no. 6, 065010. http://stacks.iop.org/1367-2630/12/i=6/a=065010.

[25] M. Stone, C.-K. Chiu, and A. Roy, "Symmetries, dimensions and topological insulators: the mechanism behind the face of the bott clock," *Journal of Physics A: Mathematical and Theoretical* **44** (2011) no. 4, 045001. http://stacks.iop.org/1751-8121/44/i=4/a=045001.

[26] D. S. Freed and G. W. Moore, "Twisted equivariant matter," *Annales Henri Poincare* **14** (2013) 1927–2023, arXiv:1208.5055 [hep-th].

[27] G. De Nittis and K. Gomi, "Classification of real bloch-bundles: topological quantum systems of type ai," *Journal of Geometry and Physics* **86** (2014) 303–338.

[28] G. De Nittis and K. Gomi, "Classification of quaternionic" bloch-bundles," *Communications in Mathematical Physics* **339** (2015) no. 1, 1–55.

[29] R. Kennedy and C. Guggenheim, "Homotopy theory of strong and weak topological insulators," *Phys. Rev. B* **91** (Jun, 2015) 245148. http://link.aps.org/doi/10.1103/PhysRevB.91.245148.

[30] R. Kennedy and M. R. Zirnbauer, "Bott Periodicity for $\mathbb{Z}_2$ Symmetric Ground States of Gapped Free-Fermion Systems," *Commun. Math. Phys.* **342** (2016) no. 3, 909–963, arXiv:1409.2537 [math-ph].

[31] G. De Nittis and K. Gomi, "Chiral vector bundles: A geometric model for class aiii topological quantum systems," *arXiv preprint arXiv:1504.04863* (2015) .

[32] G. C. Thiang, "On the K-theoretic classification of topological phases of matter," *Annales Henri Poincare* **17** (2016) no. 4, 757–794, arXiv:1406.7366 [math-ph].

[33] M. Gunaydin, "Exceptional Realizations of Lorentz Group: Supersymmetries and Leptons," *Nuovo Cim.* **A29** (1975) 467.

[34] F. Gursey, "Octonionic Structures in Particle Physics," *Lect.Notes Phys.* **94** (1979) 508–521.

[35] B. Julia, "Group disintegrations," in *Superspace and Supergravity*, S. Hawking and M. Rocek, eds., vol. C8006162 of *Nuffield Gravity Workshop*, pp. 331–350. Cambridge University Press, 1980.

[36] T. Kugo and P. K. Townsend, "Supersymmetry and the division algebras," *Nucl. Phys.* **B221** (1983) 357.

[37] M. Günaydin, G. Sierra, and P. K. Townsend, "Exceptional supergravity theories and the magic square," *Phys. Lett.* **B133** (1983) 72.

[38] A. Sudbery, "Division algebras, (pseudo)orthogonal groups, and spinors," *J. Phys.* **A17** (1984) no. 5, 939–955.

[39] L. Borsten, D. Dahanayake, M. J. Duff, H. Ebrahim, and W. Rubens, "Black Holes, Qubits and Octonions," *Phys. Rep.* **471** (2009) no. 3–4, 113–219, arXiv:0809.4685 [hep-th].

[40] J. C. Baez and J. Huerta, "Division algebras and supersymmetry i," in *Superstrings, Geometry, Topology, and C*-Algebras, eds. R. Doran, G. Friedman and J. Rosenberg, Proc. Symp. Pure Math*, vol. 81, pp. 65–80. 2009. arXiv:0909.0551 [hep-th].

[41] L. Borsten, M. J. Duff, L. J. Hughes, and S. Nagy, "A magic square from Yang-Mills squared," *Phys.Rev.Lett.* **112** (2014) 131601, arXiv:1301.4176 [hep-th].

[42] A. Anastasiou, L. Borsten, M. J. Duff, L. J. Hughes, and S. Nagy, "Super Yang-Mills, division algebras and triality," *JHEP* **1408** (2014) 080, arXiv:1309.0546 [hep-th].

[43] A. Anastasiou, L. Borsten, M. J. Duff, L. J. Hughes, and S. Nagy, "An octonionic formulation of the M-theory algebra," *JHEP* **1411** (2014) 022, arXiv:1402.4649 [hep-th].

[44] L. Borsten and M. J. Duff, "Gravity as the square of YangMills?," *Phys. Scripta* **90** (2015) 108012, arXiv:1602.08267 [hep-th].

[45] M. Atiyah, R. Bott, and A. Shapiro, "Clifford modules," *Topology* **3, Supplement 1** (1964) 3 – 38. http://www.sciencedirect.com/science/article/pii/0040938364900035.

[46] D. Husemöller, *Fibre Bundles*, vol. 20 of *Graduate texts in mathematics*. Springer-Verlag New York, 1993.

[47] M. Karoubi, *K-theory: An introduction*, vol. 226. Springer Science & Business Media, 2008.

[48] E. Cartan, "Nombres complexes," in *Encyclopédie des sciences mathématiques*, J. Molk, ed., vol. 1, pp. 329–468. 1908.

[49] J. M. Figueroa-OFarrill, "Majorana spinors," *Lecture notes, University of Edinburgh* (2006) .

[50] R. Bott, "The stable homotopy of the classical groups," *Proceedings of the National Academy of Sciences of the United States of America* **43** (1957) no. 10, 933–935. http://www.jstor.org/stable/89403.

[51] M. Atiyah and F. Hirzebruch, "Vector bundles and homogeneous spaces," *Proc. Sympos. Pure Math.* **III** (1961) 7–38.

[52] M. F. Atiyah and D. Anderson, *K-theory*, vol. 2. WA Benjamin New York, 1967.

[53] N. Kopnin and M. Salomaa, "Mutual friction in superfluid he 3: Effects of bound states in the vortex core," *Physical Review B* **44** (1991) no. 17, 9667.

[54] G. E. Volovik, "Fermion zero modes on vortices in chiral superconductors," *JETP Lett.* **70** (1999) 609–614, arXiv:cond-mat/9909426 [cond-mat]. [Pisma Zh. Eksp. Teor. Fiz.70,601(1999)].

[55] T. Senthil and M. P. Fisher, "Quasiparticle localization in superconductors with spin-orbit scattering," *Physical Review B* **61** (2000) no. 14, 9690.

[56] N. Read and D. Green, "Paired states of fermions in two-dimensions with breaking of parity and time reversal symmetries, and the fractional quantum Hall effect," *Phys. Rev.* **B61** (2000) 10267, arXiv:cond-mat/9906453 [cond-mat].

[57] A. Y. Kitaev, "Unpaired majorana fermions in quantum wires," *Physics-Uspekhi* **44** (2001) no. 10S, 131.

[58] O. Motrunich, K. Damle, and D. A. Huse, "Griffiths effects and quantum critical points in dirty superconductors without spin-rotation invariance: One-dimensional examples," *Phys. Rev. B* **63** (May, 2001) 224204. http://link.aps.org/doi/10.1103/PhysRevB.63.224204.

[59] S. D. Sarma, C. Nayak, and S. Tewari, "Proposal to stabilize and detect half-quantum vortices in strontium ruthenate thin films: Non-abelian braiding statistics of vortices in a p x+ i p y superconductor," *Physical Review B* **73** (2006) no. 22, 220502.

[60] R. Jackiw and P. Rossi, "Zero modes of the vortex-fermion system," *Nuclear Physics B* **190** (1981) no. 4, 681–691.

[61] C. W. J. Beenakker, "Annihilation of colliding bogoliubov quasiparticles reveals their majorana nature," *Phys. Rev. Lett.* **112** (Feb, 2014) 070604. http://link.aps.org/doi/10.1103/PhysRevLett.112.070604.

[62] C. Chamon, R. Jackiw, Y. Nishida, S. Y. Pi, and L. Santos, "Quantizing Majorana Fermions in a Superconductor," *Phys. Rev.* **B81** (2010) 224515, arXiv:1001.2760 [cond-mat.str-el].

[63] P. G. De Gennes, *Superconductivity of metals and alloys (advanced book classics)*. Addison-Wesley Publ. Company Inc Redwood City, 1999.

[64] C. L. Kane and E. J. Mele, "Z_2 topological order and the quantum spin hall effect," *Phys. Rev. Lett.* **95** (Sep, 2005) 146802. http://link.aps.org/doi/10.1103/PhysRevLett.95.146802.

[65] B. A. Bernevig and S.-C. Zhang, "Quantum spin hall effect," *Phys. Rev. Lett.* **96** (Mar, 2006) 106802. http://link.aps.org/doi/10.1103/PhysRevLett.96.106802.

[66] J. E. Moore and L. Balents, "Topological invariants of time-reversal-invariant band structures," *Phys. Rev. B* **75** (Mar, 2007) 121306. http://link.aps.org/doi/10.1103/PhysRevB.75.121306.

[67] M. König, S. Wiedmann, C. Brüne, A. Roth, H. Buhmann, L. W. Molenkamp, X.-L. Qi, and S.-C. Zhang, "Quantum spin hall insulator state in hgte quantum wells," *Science* **318** (2007) no. 5851, 766–770. http://science.sciencemag.org/content/318/5851/766.

[68] L. Fu, C. L. Kane, and E. J. Mele, "Topological insulators in three dimensions," *Phys. Rev. Lett.* **98** (Mar, 2007) 106803. http://link.aps.org/doi/10.1103/PhysRevLett.98.106803.

[69] D. Hsieh, D. Qian, L. Wray, Y. Xia, Y. S. Hor, R. J. Cava, and M. Z. Hasan, "A topological dirac insulator in a quantum spin hall phase," *Nature* **452** (04, 2008) 970–974. http://dx.doi.org/10.1038/nature06843.

[70] R. Roy, "Z_2 classification of quantum spin hall systems: An approach using time-reversal invariance," *Phys. Rev. B* **79** (May, 2009) 195321. http://link.aps.org/doi/10.1103/PhysRevB.79.195321.

[71] D. Hsieh, Y. Xia, L. Wray, D. Qian, A. Pal, J. H. Dil, J. Osterwalder, F. Meier, G. Bihlmayer, C. L. Kane, Y. S. Hor, R. J. Cava, and M. Z. Hasan, "Observation of unconventional quantum spin textures in topological insulators," *Science* **323** (2009) no. 5916, 919–922. http://science.sciencemag.org/content/323/5916/919.

[72] Y. L. Chen, J. G. Analytis, J.-H. Chu, Z. K. Liu, S.-K. Mo, X. L. Qi, H. J. Zhang, D. H. Lu, X. Dai, Z. Fang, S. C. Zhang, I. R. Fisher, Z. Hussain, and Z.-X. Shen, "Experimental realization of a three-dimensional topological insulator, bi2te3," *Science* **325** (2009) no. 5937, 178–181. http://science.sciencemag.org/content/325/5937/178.

[73] D. Aharonov and M. Ben-Or, "Fault-tolerant quantum computation with constant error," in *Proceedings of the twenty-ninth annual ACM symposium on Theory of computing*, pp. 176–188, ACM. 1997.

[74] E. Knill, R. Laflamme, and W. Zurek, "Threshold accuracy for quantum computation," *arXiv preprint quant-ph/9610011* (1996) .

[75] C. Zalka, "Threshold estimate for fault tolerant quantum computing," arXiv:quant-ph/9612028 [quant-ph].

[76] A. M. Steane, "Active stabilization, quantum computation, and quantum state synthesis," *Phys. Rev. Lett.* **78** (Mar, 1997) 2252–2255. http://link.aps.org/doi/10.1103/PhysRevLett.78.2252.

[77] A. Yu. Kitaev, "Fault tolerant quantum computation by anyons," *Annals Phys.* **303** (2003) 2–30, arXiv:quant-ph/9707021 [quant-ph].

[78] A. Y. Kitaev, "Unpaired majorana fermions in quantum wires," *Physics-Uspekhi* **44** (2001) no. 10S, 131.

[79] G. Kells, "Many-body majorana operators and the equivalence of parity sectors," *Phys. Rev. B* **92** (Aug, 2015) 081401. http://link.aps.org/doi/10.1103/PhysRevB.92.081401.

[80] G. Kells, "Multiparticle content of majorana zero modes in the interacting p-wave wire," *Phys. Rev. B* **92** (Oct, 2015) 155434. http://link.aps.org/doi/10.1103/PhysRevB.92.155434.

[81] L. Fu and C. L. Kane, "Josephson current and noise at a superconductor/quantum-spin-hall-insulator/superconductor junction," *Phys. Rev. B* **79** (Apr, 2009) 161408. http://link.aps.org/doi/10.1103/PhysRevB.79.161408.

[82] M. Z. Hasan and C. L. Kane, "*Colloquium* : Topological insulators," *Rev. Mod. Phys.* **82** (Nov, 2010) 3045–3067. http://link.aps.org/doi/10.1103/RevModPhys.82.3045.

[83] X.-L. Qi and S.-C. Zhang, "Topological insulators and superconductors," *Rev. Mod. Phys.* **83** (Oct, 2011) 1057–1110. http://link.aps.org/doi/10.1103/RevModPhys.83.1057.

[84] M. Franz and L. Molenkamp, *Topological Insulators*, vol. 6. Elsevier, 2013.

[85] A. Cook and M. Franz, "Majorana fermions in a topological-insulator nanowire proximity-coupled to an s-wave superconductor," *Phys. Rev. B* **84** (Nov, 2011) 201105. http://link.aps.org/doi/10.1103/PhysRevB.84.201105.

[86] R. M. Lutchyn, J. D. Sau, and S. Das Sarma, "Majorana fermions and a topological phase transition in semiconductor-superconductor heterostructures," *Phys. Rev. Lett.* **105** (Aug, 2010) 077001. http://link.aps.org/doi/10.1103/PhysRevLett.105.077001.

[87] Y. Oreg, G. Refael, and F. von Oppen, "Helical liquids and majorana bound states in quantum wires," *Phys. Rev. Lett.* **105** (Oct, 2010) 177002. http://link.aps.org/doi/10.1103/PhysRevLett.105.177002.

[88] V. Mourik, K. Zuo, S. M. Frolov, S. R. Plissard, E. P. A. M. Bakkers, and L. P. Kouwenhoven, "Signatures of majorana fermions in hybrid superconductor-semiconductor nanowire devices," *Science* **336** (2012) no. 6084, 1003–1007. http://science.sciencemag.org/content/336/6084/1003.

[89] H. O. H. Churchill, V. Fatemi, K. Grove-Rasmussen, M. T. Deng, P. Caroff, H. Q. Xu, and C. M. Marcus, "Superconductor-nanowire devices from tunneling to the multichannel regime: Zero-bias oscillations and magnetoconductance crossover," *Phys. Rev. B* **87** (Jun, 2013) 241401. http://link.aps.org/doi/10.1103/PhysRevB.87.241401.

[90] A. Das, Y. Ronen, Y. Most, Y. Oreg, M. Heiblum, and H. Shtrikman, "Zero-bias peaks and splitting in an al-inas nanowire topological superconductor as a signature of majorana fermions," *Nature Physics* **8** (2012) no. 12, 887–895.

[91] M. T. Deng, C. L. Yu, G. Y. Huang, M. Larsson, P. Caroff, and H. Q. Xu, "Anomalous zero-bias conductance peak in a nbinsb nanowirenb hybrid device," *Nano Letters* **12** (2012) no. 12, 6414–6419. http://dx.doi.org/10.1021/nl303758w. PMID: 23181691.

[92] A. D. K. Finck, D. J. Van Harlingen, P. K. Mohseni, K. Jung, and X. Li, "Anomalous modulation of a zero-bias peak in a hybrid nanowire-superconductor device," *Phys. Rev. Lett.* **110** (Mar, 2013) 126406. http://link.aps.org/doi/10.1103/PhysRevLett.110.126406.

[93] J. Liu, A. C. Potter, K. T. Law, and P. A. Lee, "Zero-bias peaks in the tunneling conductance of spin-orbit-coupled superconducting wires with and without majorana end-states," *Phys. Rev. Lett.* **109** (Dec, 2012) 267002. http://link.aps.org/doi/10.1103/PhysRevLett.109.267002.

[94] F. Wilczek, "Quantum mechanics of fractional-spin particles," *Phys. Rev. Lett.* **49** (Oct, 1982) 957–959. http://link.aps.org/doi/10.1103/PhysRevLett.49.957.

[95] J. M. Leinaas and J. Myrheim, "On the theory of identical particles," *Il Nuovo Cimento B (1971-1996)* **37** (2007) no. 1, 1–23. http://dx.doi.org/10.1007/BF02727953.

[96] F. Wilczek, "Magnetic flux, angular momentum, and statistics," *Phys. Rev. Lett.* **48** (Apr, 1982) 1144–1146. http://link.aps.org/doi/10.1103/PhysRevLett.48.1144.

[97] B. I. Halperin, "Statistics of quasiparticles and the hierarchy of fractional quantized hall states,"*Phys. Rev. Lett.* **52** (Apr, 1984) 1583–1586.
http://link.aps.org/doi/10.1103/PhysRevLett.52.1583.

[98] D. Arovas, J. R. Schrieffer, and F. Wilczek, "Fractional statistics and the quantum hall effect,"*Phys. Rev. Lett.* **53** (Aug, 1984) 722–723.
http://link.aps.org/doi/10.1103/PhysRevLett.53.722.

[99] G. Moore and N. Seiberg, "Polynomial equations for rational conformal field theories," *Physics Letters B* **212** (1988) no. 4, 451 – 460.
http://www.sciencedirect.com/science/article/pii/0370269388917960.

[100] E. Witten, "Quantum field theory and the jones polynomial," *Comm. Math. Phys.* **121** (1989) no. 3, 351–399. http://projecteuclid.org/euclid.cmp/1104178138.

[101] G. Moore and N. Read, "Nonabelions in the fractional quantum hall effect," *Nuclear Physics B* **360** (1991) no. 23, 362 – 396.
http://www.sciencedirect.com/science/article/pii/055032139190407O.

[102] M. Dolev, M. Heiblum, V. Umansky, A. Stern, and D. Mahalu, "Observation of a quarter of an electron charge at the ν= 5/2 quantum hall state," *Nature* **452** (2008) no. 7189, 829–834.

[103] I. P. Radu, J. B. Miller, C. M. Marcus, M. A. Kastner, L. N. Pfeiffer, and K. W. West, "Quasi-particle properties from tunneling in the v = 5/2 fractional quantum hall state," *Science* **320** (2008) no. 5878, 899–902. http://science.sciencemag.org/content/320/5878/899.

[104] R. L. Willett, L. N. Pfeiffer, and K. W. West, "Measurement of filling factor 5/2 quasiparticle interference with observation of charge e/4 and e/2 period oscillations," *Proceedings of the National Academy of Sciences* **106** (2009) no. 22, 8853–8858.
http://www.pnas.org/content/106/22/8853.abstract.

[105] N. Read and D. Green, "Paired states of fermions in two dimensions with breaking of parity and time-reversal symmetries and the fractional quantum hall effect,"*Phys. Rev. B* **61** (Apr, 2000) 10267–10297. http://link.aps.org/doi/10.1103/PhysRevB.61.10267.

[106] L. Fu and C. L. Kane, "Superconducting proximity effect and majorana fermions at the surface of a topological insulator,"*Phys. Rev. Lett.* **100** (Mar, 2008) 096407.
http://link.aps.org/doi/10.1103/PhysRevLett.100.096407.

[107] J. R. Williams, A. J. Bestwick, P. Gallagher, S. S. Hong, Y. Cui, A. S. Bleich, J. G. Analytis, I. R. Fisher, and D. Goldhaber-Gordon, "Unconventional josephson effect in hybrid superconductor-topological insulator devices,"*Phys. Rev. Lett.* **109** (Jul, 2012) 056803.
http://link.aps.org/doi/10.1103/PhysRevLett.109.056803.

[108] S. Cho, B. Dellabetta, A. Yang, J. Schneeloch, Z. Xu, T. Valla, G. Gu, M. J. Gilbert, and N. Mason, "Symmetry protected josephson supercurrents in three-dimensional topological insulators," *Nature communications* **4** (2013) 1689.

[109] P. Zareapour, A. Hayat, S. Y. F. Zhao, M. Kreshchuk, A. Jain, D. C. Kwok, N. Lee, S.-W. Cheong, Z. Xu, A. Yang, *et al.*, "Proximity-induced high-temperature superconductivity in the topological insulators bi2se3 and bi2te3," *Nature communications* **3** (2012) 1056.

[110] E. Wang, H. Ding, A. V. Fedorov, W. Yao, Z. Li, Y.-F. Lv, K. Zhao, L.-G. Zhang, Z. Xu, J. Schneeloch, *et al.*, "Fully gapped topological surface states in bi2se3 films induced by a d-wave high-temperature superconductor," *Nature physics* **9** (2013) no. 10, 621–625.

[111] J.-P. Xu, C. Liu, M.-X. Wang, J. Ge, Z.-L. Liu, X. Yang, Y. Chen, Y. Liu, Z.-A. Xu, C.-L. Gao, D. Qian, F.-C. Zhang, and J.-F. Jia, "Artificial topological superconductor by the proximity effect," *Phys. Rev. Lett.* **112** (May, 2014) 217001. http://link.aps.org/doi/10.1103/PhysRevLett.112.217001.

[112] M. H. Freedman, "P/np, and the quantum field computer," *Proceedings of the National Academy of Sciences* **95** (1998) no. 1, 98–101.

[113] A. Kitaev, "Anyons in an exactly solved model and beyond," *Annals of Physics* **321** (2006) no. 1, 2–111.

[114] C. Nayak, S. H. Simon, A. Stern, M. Freedman, and S. Das Sarma, "Non-abelian anyons and topological quantum computation," *Rev. Mod. Phys.* **80** (Sep, 2008) 1083–1159. http://link.aps.org/doi/10.1103/RevModPhys.80.1083.

[115] C. Nayak and F. Wilczek, "2n-quasihole states realize 2 n- 1-dimensional spinor braiding statistics in paired quantum hall states," *Nuclear Physics B* **479** (1996) no. 3, 529–553.

[116] V. Gurarie and C. Nayak, "A plasma analogy and berry matrices for non-abelian quantum hall states," *Nuclear Physics B* **506** (1997) no. 3, 685–694.

[117] A. Stern, F. von Oppen, and E. Mariani, "Geometric phases and quantum entanglement as building blocks for non-abelian quasiparticle statistics," *Physical Review B* **70** (2004) no. 20, 205338.

[118] D. A. Ivanov, "Non-abelian statistics of half-quantum vortices in p-wave superconductors," *Phys. Rev. Lett.* **86** (Jan, 2001) 268–271. http://link.aps.org/doi/10.1103/PhysRevLett.86.268.

[119] V. F. R. Jones, "Hecke algebra representations of braid groups and link polynomials," *Annals of Mathematics* **126** (1987) no. 2, 335–388. http://www.jstor.org/stable/1971403.

[120] J. Alicea, Y. Oreg, G. Refael, F. von Oppen, and M. P. Fisher, "Non-abelian statistics and topological quantum information processing in 1d wire networks," *Nature Physics* **7** (2011) no. 5, 412–417.

[121] J. D. Sau, D. J. Clarke, and S. Tewari, "Controlling non-abelian statistics of majorana fermions in semiconductor nanowires," *Phys. Rev. B* **84** (Sep, 2011) 094505. http://link.aps.org/doi/10.1103/PhysRevB.84.094505.

[122] B. I. Halperin, Y. Oreg, A. Stern, G. Refael, J. Alicea, and F. von Oppen, "Adiabatic manipulations of majorana fermions in a three-dimensional network of quantum wires," *Physical Review B* **85** (2012) no. 14, 144501.

[123] S. Bravyi and A. Kitaev, "Universal quantum computation with ideal clifford gates and noisy ancillas," *Phys. Rev. A* **71** (Feb, 2005) 022316. http://link.aps.org/doi/10.1103/PhysRevA.71.022316.

[124] S. Bravyi, "Universal quantum computation with the $\nu = 52$ fractional quantum hall state," *Phys. Rev. A* **73** (Apr, 2006) 042313. http://link.aps.org/doi/10.1103/PhysRevA.73.042313.

[125] M. H. Freedman, M. Larsen, and Z. Wang, "A modular functor which is universal for quantum computation," *Communications in Mathematical Physics* **227** (2002) no. 3, 605–622.

[126] J. C. Budich and E. Ardonne, "Equivalent topological invariants for one-dimensional majorana wires in symmetry class d," *Physical Review B* **88** (2013) no. 7, 075419.

[127] A. Altland and M. R. Zirnbauer, "Nonstandard symmetry classes in mesoscopic normal-superconducting hybrid structures," *Physical Review B* **55** (1997) no. 2, 1142.

[128] Y. Hatsugai, "Quantized berry phases as a local order parameter of a quantum liquid," *Journal of the Physical Society of Japan* **75** (2006) no. 12, 123601.

[129] A. P. Schnyder, S. Ryu, A. Furusaki, and A. W. Ludwig, "Classification of topological insulators and superconductors," in *American Institute of Physics Conference Series*, vol. 1134, pp. 10–21. 2009.

[130] G. Abramovici and P. Kalugin, "Clifford modules and symmetries of topological insulators," *International Journal of Geometric Methods in Modern Physics* **9** (2012) no. 03, 1250023.

[131] M. Atiyah and G. Segal, "Twisted K-theory," arXiv:math/0407054 [math-kt].

[132] F. J. Dyson, "Statistical theory of the energy levels of complex systems. i," *Journal of Mathematical Physics* **3** (1962) no. 1, 140–156.

[133] M. ATIYAH, "K-theory and reality," *The Quarterly Journal of Mathematics* **17** (1966) no. 1, 367–386.

[134] G. C. Thiang, "Topological phases: isomorphism, homotopy and K-theory," *Int. J. Geom. Meth. Mod. Phys.* **12** (2015) no. 09, 1550098, arXiv:1412.4191 [math-ph].

[135] X. Chen, Z.-C. Gu, Z.-X. Liu, and X.-G. Wen, "Symmetry protected topological orders and the group cohomology of their symmetry group," *Phys. Rev.* **B87** (2013) no. 15, 155114, arXiv:1106.4772 [cond-mat.str-el].

[136] E. Witten, "Fermion Path Integrals And Topological Phases," arXiv:1508.04715 [cond-mat.mes-hall].

[137] C. G. Callan, Jr. and J. A. Harvey, "Anomalies and Fermion Zero Modes on Strings and Domain Walls," *Nucl. Phys.* **B250** (1985) 427.

[138] S. Ryu, J. E. Moore, and A. W. W. Ludwig, "Electromagnetic and gravitational responses and anomalies in topological insulators and superconductors," *Phys. Rev. B* **85** (Jan, 2012) 045104. http://link.aps.org/doi/10.1103/PhysRevB.85.045104.

[139] L. Fidkowski, X. Chen, and A. Vishwanath, "Non-abelian topological order on the surface of a 3d topological superconductor from an exactly solved model," *Phys. Rev. X* **3** (Nov, 2013) 041016. http://link.aps.org/doi/10.1103/PhysRevX.3.041016.

[140] E. Witten, "Lecture at strings 2015 icts-titr,". https://strings2015.icts.res.in/talkDetails.php?sId=70&tId=14.

[141] K. C. Hannabuss, V. Mathai, and G. C. Thiang, "T-duality simplifies bulk-boundary correspondence: the general case," arXiv:1603.00116 [hep-th].

[142] A. Hurwitz, "Uber die komposition der quadratishen formen von beliebig vielen variabeln," *Nachr. Ges. Wiss. Gottingen* (1898) 309–316.

LECTURER: M. J. DUFF

Majorana Fermions in Particle Physics, Solid State and Quantum Information Theory

Scientific Secretaries: A. Anastasiou, A. Marrani

DISCUSSION

- *Sopenko*:

In your lecture you mentioned a condensed matter analogue of an M2 brane ending on an M5 brane as a 3+1 dimensional superconductor. Could you please explain this point?

- *Duff*:

This morning I presented a slide that Witten showed in his talk at STRINGS 2015 in Bangalore on Monday, which I downloaded from the internet. Witten was talking about Majorana fermions in 3 space dimensions, and remarked that they are governed by the same equations occurring in the theory of topological superconductors. Remarkably, high-energy physicists and condensed matter physicists are asking, for different reasons, whether this system is consistent, and it seems that they both reached the conclusion that in general it is not, unless the number of Majorana fermions is a multiple of 16.

- *Zichichi:*

Why 16?

- *Duff*:

I am not an expert in this field, but I think that the answer concerns the index associated with Majorana fermions in 3 dimensions: this index has to acquire values multiple of 16 in order for the partition function to be well defined. I do not know whether 16 can be explained from some deep philosophical point of view, but it is quite amazing that condensed matter theorists reached the same conclusion from a different perspective. Now, coming back to Witten's slide, the relevant question is "How does this fit with M-theory?" M-theory is an 11-dimensional theory which unifies all the five 10-dimensional string theories, and also incorporates branes as well as strings. It is the best candidate for a unified theory of gravity and quantum mechanics. It has not yet proved its worth in particle physics, but we live in hope. Concerning the brane content of M-Theory, there exists a 2-brane, named M2-brane (an object which has 2 spatial and 1 time dimensions), and its magnetic dual, a 5-brane named M5-brane. Indeed, just as a charged particle couples to a vector field and a string couples to a rank-2antisymmetric tensor field, a 2-brane couples to a rank-3 antisymmetric tensorfield, whose corresponding field strength is a 4-form. In 11 dimensions a 4-form is dual to a 7-form, whose associated potential is a 6-form, which thus is related to a 5-brane. So, the M5-brane is like the magnetic monopole in 11 dimensions, just as the M2-brane generalizes the electric monopole to higher dimensions. Just as in 10-dimensional string theory open strings can end on D-branes, in 11-dimensions M2-branes can end on M5-branes. This means that M2-brane generally do not live in a closed manifold, but rather in a manifold whose boundary is the M5-brane itself. In his very recent talk, Witten remarked that a very similar situation takes place in 4-dimensional condensed matter, in which a superconductor may have a boundary split into two parts, and he speculated on the intriguing possibility that such a condensed matter configuration and the corresponding setup in M-theory share the same mathematical description.

- *Zichichi:*

When you say that half of the boundary is in contact with some other material, what do you precisely mean?

- *Duff:*

I mean that the superconductor may not be composed of a single substance, but rather by various substances each one in contact with others through a boundary.

- *Sopenko:*

What's the relevance of Majorana particles to superconductors?

- *Duff:*

As I explained in my lecture of this morning, let me stress again that actually Majorana zero modes – and not Majorana fermions - are relevant to superconductors. In condensed matter systems, one can have electrons, holes, Cooper pairs, bound states... but not a Majorana particle.

- *Sokolov:*

I would like you to elucidate the notion of measure of entanglement. This morning, you mentioned the 2-tangle τ_{AB}, calculated as a matrix determinant. Then, you considered the non-locality measure in the CHSH, and related it to another measure of entanglement. What is the difference between these measures?

- *Duff:*

For simplicity's sake, let me consider ordinary (and not super) qubits. The simplest case in which entanglement may arise is given by two qubits: entanglement takes place for non-separable states, corresponding to a non-vanishing 2-tangle (i.e. non-zero determinant of the 2x2 matrix associated to the two qubits state). Therefore, the 2-tangle is a measure of the degree of entanglement of the state, and upon proper normalization it lies between 0 (no entanglement) and 1 (maximal entanglement). Let me now consider the CHSH game. The rules are quite complicated, and I will not go through them; let me just mention that there is a referee along with Alice and Bob, and Alice and Bob are competing not against each other but against the referee. The referee sends each of them a qubit, and in turn they send back a qubit; but they can decide beforehand some strategy between themselves of how they will respond. If they share an entangled state, they can score better than they could otherwise. Tsirelson proved that there is a maximum winning probability, called theTsirelson bound, which cannot be exceeded in ordinary quantum mechanics. Quantum information theorists have invented what they dub "black boxes"; these are some as yet undiscovered theories - which may or may not exist – generally performing better than ordinary quantum mechanics. In the list of fundamental properties of "black boxes", one must of course insist that causality cannot be violated, and thus nothing can travel faster than light, thus implying the bound 1. Consequently, there is a region lying in between the hypothetical perfection of "black boxes" and the real world represented by usual quantum mechanics with Tsirelson's bound. The question we asked ourselves is whether superqubits could probe this region by performing better that ordinary qubits in relation to the CHSH game, and we surprisingly found that super qubits can exceed Tsirelson's bound, thus yielding more non-locality than in ordinary quantum mechanics.

- *Bolov:*

As far as I understood, you mentioned that degenerate states of some superconductors may be associated with Majorana fermions. Do you actually mean that the degenerate states of the quasi-particle holes may be associated with Majorana fermions?

- *Duff:*

From my partial reading of literature (of which I am not expert), I can say that there are at least two different stories, depending on whether 1-dimensional or higher dimensional systems are considered. For instance, Witten was talking about real Majorana fermions, which obey Fermi statistics and are of the kind that Majorana himself had in mind. But, as far as I know, they have no application to quantum computing (I may be wrong on that, but that's the outcome of my analysis of literature). On the other hand, 1-dimensional systems which occur in quantum computing are characterized by quantities which are referred to as 'Majorana fermions', but which are not really fermions at all, for the simple fact that Fermi statistics does not exist in 1 dimension. In fact, they obey anyonic statistics, which are intermediate between Bose and Fermi ones. Consequently, such 1-dimensional analogues of Majorana fermions neither commute nor anti-commute; however, for technical reasons they also have to be non-Abelian anyons. A better description of such entities would be "Majorana zero-modes'; their Majorana nature resides in the fact that they are their own anti-particles.

- *Saluki:*

My question is about the measure of entanglement. In the case of 2 particles, the measure of the entanglement is the determinant, while for 3 particles this is given by the hyper determinant. If the latter vanishes, does this still imply that the state is separable?

- *Duff:*

No. Remember that 2-qubit states can be entangled or not, depending on the vanishing or not of the 2-tangle. For 3 qubits, the story is more complicated but also more interesting: Alice, Bob and Charlie can form a state |a>|b>|c> which is separable, but also a state |ab>|c>, in which Alice is entangled with Bob but not with Charlie (and there are 3 of those). Moreover, one can also have a state |abc>, which is genuinely 3-way entangled. It turns out that there are 2 ways of having 3-way entanglement, one corresponding to the so-called W (e.g., |001> + |100> + |100>) and another one provided by the so-called GHZ (e.g., |000> + |111>). The hyper determinant is non-vanishing only for GHZ. Thus, the W state is entangled but the corresponding hyper determinant vanishes. Due to the symmetries of the system, one can actually prove that 3-qubit entanglement splits into these various classes (orbits), depending on whether there is bi-partite entanglement or one of the 3-way entanglements.

- *Saluki:*

So the hyper determinant is not really sufficient to determine entanglement ?

- *Duff:*

Yes, exactly. This is due to the fact that there is more than one invariant/covariant of the SLOCC operations one has to take into account. By the way (although this has nothing to do with my talk), I would like to point out that in recent years we found some analogies between these entanglement measures and black hole entropies. Indeed, there are lots of interesting coincidences where the same mathematics that appears in quantum information theory also appears in black hole physics. So, we asked ourselves the question: ''How many ways are there to entangle 4 qubits?" The relevant mathematical formalism had already been developed by high-energy theoretical physicists studying black holes. We just had to translate those results into the right QI language. It turned out that there are 31 different ways of entangling the black holes, thus implying that there exist 31 in equivalent ways of entangling 4 qubits. It is remarkable to observe that this was a case in which string theory was used to make a prediction which in principle could be tested in the laboratory.

- *Minkowski:*
In one of your slides, you considered an action in 3 dimensions. What's their signature ?

- *Duff:*
I think that Witten intended 3 space dimensions.

- *Minkowski:*
I suppose you were assuming the existence of a Majorana fermion mode for that action, and you were also understanding that this action is real, positive, etc. Can it be compared to the normal 4-dimensional field theory in the quadratic approximation?

- *Duff:*
I guess you are asking me whether Witten's action integrated over time yields the action of a time independent 4-dimensional Majorana spinor. I am not sure but I think the answer is yes.

- *Minkowski:*
Ok, but all in all can you solve this 3-dimensional system, like it is known to be feasible, e.g. in 4 dimensions ?

- *Duff:*
Yes, I do not see why not.

- *Chen:*
I would like you to briefly reconsider the CHSH game. I would like to understand the reason of the enhancement of the maximal correlation in the case of superqubits. Indeed, as far as I understand, probability quantum enhancements are generally related to interference. As a consequence, the upper limit in ordinary quantum mechanics should be ultimately connected to probability conservation laws. Therefore, I would appreciate if you could spend some more words on the maximum probability enhancement taking place in super qubit systems.

- *Duff:*
A crucial point for the quantum mechanics based on super qubits is the extraction of real probabilities from Grassmann numbers. Even though a well defined procedure is not yet established, in literature there exist at least 2 approaches to such a problem. The first approach is based on De Witt's book on super-Hilbert spaces, and it does not exceed the ordinary quantum mechanics upper probability bound. Consequently, this case is consistent, but not so interesting. The second approach based on (a suitable generalization of) Rodgers' norm is more interesting, because Tsirelson's bound can actually be exceeded, however at the price of having negative probabilities, Along the years, various theorists tried to make sense of Quantum Mechanics, but their role and use in this context is still under investigation.

- *Chen:*
The formula describing a 2-superqubit state you have been showing in your slides looks like a

superposition of an ordinary qubit and of a superqubit. Is this true? Can different components interfere with each other?

- *Duff:*

One can check that the 2-supequbits state is entangled (i.e., non-separable) when the superdeterminant of the corresponding 3x3 matrix (extending the purely bosonic 2-matrix pertaining to 2 qubits) is non-zero. Thus, at the level of the measure of entanglement, superqubits provide a consistent and reasonable generalization of the 2 qubits setup. In this framework, the main problem resides in giving a physical interpretation/meaning to the Grassmanian amplitudes occurring in superqubits quantum mechanics.

Majorana Fermions, Supersymmetry Breaking, and Born-Infeld Theory

S. Ferrara [a,b,c], A. Marrani [d,e], and A. Yeranyan [b,d]

[a] *Th-Ph Department, CERN*
CH - 1211 Geneva 23, SWITZERLAND

[b] *INFN - Laboratori Nazionali di Frascati*
Via Enrico Fermi 40, I-00044 Frascati, Italy

[c] *Department of Physics and Astronomy,*
U.C.L.A., Los Angeles CA 90095-1547, USA

[d] *Centro Studi e Ricerche "Enrico Fermi",*
Via Panisperna 89A, I-00184, Roma, Italy

[e] *Dipartimento di Fisica e Astronomia "Galileo Galilei", and INFN,*
Università di Padova, Via Marzolo 8, I-35131 Padova, Italy

sergio.ferrara@cern.ch
alessio.marrani@pd.infn.it
ayeran@lnf.infn.it

Abstract

This review is devoted to highlight some aspects of the relevance of Majorana fermions in rigid supersymmetry breaking in four spacetime dimensions. After introducing some basic facts on spinors, and on their symmetries and reality properties, we consider Goldstino actions describing partial breaking of rigid supersymmetry, then focussing on Born-Infeld non-linear theory, its duality symmetry, and its supersymmetric extensions, also including multi-field generalizations exhibiting doubly self-duality.

Contents

1 Introduction

The present review is aimed to contribute to the Proceedings of the 53rd Course of the International School of Subnuclear Physics (held in Erice in Summer 2015), which also celebrated the Triumph of Ettore Majorana, the great disappeared Sicilian Physicist whose ideas and invention have permeated both theoretical and experimental physics.

Section 2 is devoted to Majorana spinors, their definition in higher dimensions and their role in Supersymmetry. Example of such particles are the gravitino and the Goldstino in $D = 4$, since the related anticommuting supersymmetry parameters are Majorana spinors. The same holds true in $D = 3, 9, 11$ while in $D = 10$ (as in $D = 2$) the notion of Majorana-Weyl spinor occurs. As an application, the role of Majorana and Dirac neutrino masses in the see-saw mechanism of GUT's is explained.

In Section 3 some basic features of spontaneously broken rigid supersymmetry and partial supersymmetry breaking are described. The Volkov-Akulov and the Born-Infeld Lagrangians naturally emerge in this framework, and they both contribute to the description of partially broken rigid $\mathcal{N} = 2$ supersymmetry.

The electric-magnetic self duality enjoyed by the Born-Infeld theory is then extended in Section 4, in order to describe more general theories, following the work of Gaillard and Zumino [1]. These generalizations involve the Schrödinger formulation of Born-Infeld-type theories, forms of higher degree as well as geometrical mass terms in higher dimensions.

2 Majorana Fermions and Supersymmetry

Supersymmetry deals with basic spinorial entities called *Majorana spinors.* This notion originates from a fundamental observation of the Italian (Sicilian) physicist Ettore Majorana, who noticed that the four-dimensional Dirac equation admits "real" solutions if the γ-matrices are suitably chosen [2].

This is the so-called Majorana representation, in which the γ-matrices are real. Therefore the Dirac equation

$$\left(\gamma^{\mu}\partial_{\mu}+m\right)\psi=0 \tag{1}$$

admits manifestly real solutions.

In the Majorana representation, the γ^{i} are symmetric, while γ^{0} and $\gamma^{5}=\gamma^{0}\gamma^{1}\gamma^{2}\gamma^{3}$ are antisymmetric, and they satisfy the Clifford algebra relations

$$\left\{\gamma^{\mu},\gamma^{\nu}\right\}=2\eta^{\mu\nu},\qquad \eta^{\mu\nu}=diag\left(-,+,+,+\right). \tag{2}$$

How about *chiral* spinors? They cannot be real, since they are eigenstates of γ^{5}, corresponding to the eigenvalues $\pm i$ $\left(\left(\gamma^{5}\right)^{2}=-1\right)$:

$$\gamma^{5}\psi_{\pm}=\pm i\psi_{\pm}, \tag{3}$$

thus implying that[1]

$$\psi_{\pm} \; : \; =\frac{1}{2}\left(1\mp i\gamma^{5}\right)\psi; \tag{4}$$

$$\psi_{+}+\psi_{-} \; = \; \psi; \quad \psi_{\pm}^{*}=\psi_{\mp}. \tag{5}$$

Spinors which are eigenstates of γ^{5} are called *Weyl spinors.* From the previous properties, it follows that Weyl spinors have two components instead of four, and they are complex in $D=4$.

Symmetries of γ-matrices depend on the dimension $D=s+t$ of spacetime, and the reality properties of spinors on its signature $\rho:=|s-t|$ (*modulo* 8 : Bott periodicity). There are eight cases :

$$\text{complex spinors} \; : \; \rho=2,6; \tag{6}$$

$$\text{real spinors} \; : \; \rho=0,1,7; \tag{7}$$

$$\text{pseudoreal spinors} \; : \; \rho=3,4,5, \tag{8}$$

[1]In the present Section, the upperscript "$*$" denotes complex conjugation.

where

$$\psi \text{ real} \Leftrightarrow \psi^* = \psi; \tag{9}$$

$$\psi \text{ pseudoreal} \Leftrightarrow \left(\psi_\alpha^A\right)^* = \Omega^{AB}\mathbb{C}_{\alpha\beta}\psi_\beta^B, \tag{10}$$

with Ω and $\mathbb{C}$ respectively denoting the anti-involutive invariant structures of the automorphism group of the supersymmetry algebra and of the spinor space ($\Omega^2 = -1$, $\mathbb{C}^2 = -1$).

For ordinary D-dimensional Minkowski spacetime $(s,t) = (D-1,1)$, and thus $\rho = D-2$. In this case, it holds that

$$D \text{ even } (mod\ 8) \quad : \quad \left\{ \begin{array}{l} \text{complex spinors} : D = 4,8; \\ \text{real spinors} : D = 2,10; \\ \text{pseudoreal spinors} : D = 6, \end{array} \right. \quad \text{applying to Weyl spinors } \psi_\pm. \tag{11}$$

$$D \text{ odd } (mod\ 8) \quad : \quad \left\{ \begin{array}{l} \text{real spinors} : D = 3,9,11; \\ \text{pseudoreal spinors} : D = 5,7. \end{array} \right. \tag{12}$$

For instance, in $D = 10$ the Majorana representation is valid for both ψ_+ and ψ_-, since $\left(\gamma^{11}\right)^2 = 1$; thus, ψ_+ and ψ_- in $D = 10$-dimensional ordinary Minkowski spacetime are named *Majorana-Weyl spinors*.

Since we have been mentioning them a few lines above, it is here worth recalling also the anticommutator sectors of supersymmetry algebras (without central extensions) in D-dimensional Minkowski spacetime:

$$\left\{\mathcal{Q}_\alpha^A, \mathcal{Q}_\beta^B\right\} = \Omega^{AB}\left(\gamma^\mu\right)_{\alpha\beta} P_\mu \text{ in } D = 5,6,7; \tag{13}$$

$$\left\{\mathcal{Q}_\alpha^A, \bar{\mathcal{Q}}_{\dot{\alpha}B}\right\} = \delta_B^A\left(\gamma^\mu\right)_{\alpha\dot{\alpha}} P_\mu \text{ in } D = 4,8; \tag{14}$$

$$\left\{\mathcal{Q}_\alpha^A, \mathcal{Q}_\beta^B\right\} = \delta^{AB}\left(\gamma^\mu\right)_{\alpha\beta} P_\mu \text{ in } D = 3,9,10,11, \tag{15}$$

where P_μ is the momentum operator. The Latin capital indices run over the number $\mathcal{N}$ of independent spinor charges : $A, B = 1, ..., \mathcal{N}$; for D even, the automorphism group of the supersymmetry algebra (named $\mathcal{R}$-symmetry group) reads

$$SO\left(\mathcal{N}_+\right) \times SO\left(\mathcal{N}_-\right) \text{ in } D = 2\ mod\ 8\ \left(\mathcal{N} = \mathcal{N}_+ + \mathcal{N}_-\right); \tag{16}$$

$$Usp\left(2\mathcal{N}_+\right) \times USp\left(2\mathcal{N}_-\right) \text{ in } D = 6\ mod\ 8\ \left(\mathcal{N} = 2\mathcal{N}_+ + 2\mathcal{N}_-\right); \tag{17}$$

$$U\left(\mathcal{N}\right) \text{ in } D = 4,8\ mod\ 8. \tag{18}$$

In $D = 11$, if the massless spectrum is bound to have 2 as the highest spin, then $\mathcal{N}_{\max} = 1$, and the corresponding M-theory superalgebra with central extensions reads as follows :

$$\left\{\mathcal{Q}_\alpha, \mathcal{Q}_\beta\right\} = \underset{\text{point particle}}{\left(\gamma^\mu\right)_{\alpha\beta} P_\mu} + \underset{\text{M2-brane}}{\left(\gamma^{\mu\nu}\right)_{\alpha\beta} Z_{\mu\nu}} + \underset{\text{M5-brane}}{\left(\gamma^{\mu\nu\rho\sigma\delta}\right)_{\alpha\beta} Z_{\mu\nu\rho\sigma\delta}}. \tag{19}$$

Concerning the counting of (*off-shell*) independent components of spinors, it goes as follows :

$$D \text{ even} \quad : \quad \text{each } \psi_\pm \text{ has } 2^{\frac{D-2}{2}} \text{ components (which double for complex or pseudoreal ones)} \tag{20}$$

$$D \text{ odd} \quad : \quad \text{each } \psi \text{ has } 2^{\frac{D-1}{2}} \text{ components (which double for pseudoreal ones).} \tag{21}$$

Consequently, (pseudo)real spinor components of dimension 8, 16 and 32 are possible for $D \leqslant 6$, $D \leqslant 10$ and $D \leqslant 11$, respectively.

The dimension D *mod* 8 determines the symmetry of the tensor products of spinor representations (denoted by ψ), which are defined by the *morphism* map

$$A : \psi \times \psi \to \Lambda^k, \tag{22}$$

where Λ^k denotes the space of k-forms. The resulting symmetry of A, depending on D and k, is presented in Table 1.

D (mod 8)	k even		k odd	
	morphism	symmetry	morphism	symmetry
0	$\psi_\pm \otimes \psi_\pm \to \Lambda^k$	$(-1)^{k(k-1)/2}$	$\psi_\pm \otimes \psi_\mp \to \Lambda^k$	
1	$\psi \otimes \psi \to \Lambda^k$	$(-1)^{k(k-1)/2}$	$\psi \otimes \psi \to \Lambda^k$	$(-1)^{k(k-1)/2}$
2	$\psi_\pm \otimes \psi_\mp \to \Lambda^k$		$\psi_\pm \otimes \psi_\pm \to \Lambda^k$	$(-1)^{k(k-1)/2}$
3	$\psi \otimes \psi \to \Lambda^k$	$-(-1)^{k(k-1)/2}$	$\psi \otimes \psi \to \Lambda^k$	$(-1)^{k(k-1)/2}$
4	$\psi_\pm \otimes \psi_\pm \to \Lambda^k$	$-(-1)^{k(k-1)/2}$	$\psi_\pm \otimes \psi_\mp \to \Lambda^k$	
5	$\psi \otimes \psi \to \Lambda^k$	$-(-1)^{k(k-1)/2}$	$\psi \otimes \psi \to \Lambda^k$	$-(-1)^{k(k-1)/2}$
6	$\psi_\pm \otimes \psi_\mp \to \Lambda^k$		$\psi_\pm \otimes \psi_\pm \to \Lambda^k$	$-(-1)^{k(k-1)/2}$
7	$\psi \otimes \psi \to \Lambda^k$	$(-1)^{k(k-1)/2}$	$\psi \otimes \psi \to \Lambda^k$	$-(-1)^{k(k-1)/2}$

Table 1: Properties of the morphism map A [3]

For a comprehensive treatment of spinors and Clifford algebras for arbitrary (s, t)-signatures of spacetime, see *e.g.* [3].

It should be recalled that the massless Dirac equation ((1) with $m = 0$) has an extra symmetry, which in the Majorana representation takes the form

$$\psi' = e^{\alpha\gamma^5}\psi; \tag{23}$$

this is indeed a $U(1)$ rotation, since γ^5 is anti-involutive : $(\gamma^5)^2 = -1$.

However, this symmetry is broken by the mass term for a Majorana fermion (called *Majorana mass*). Defining a complex Dirac spinor ψ_D as

$$\psi_D \quad : \quad = \psi_1 + i\psi_2; \tag{24}$$

$$\psi_D^* \quad = \quad \psi_1 - i\psi_2, \tag{25}$$

where ψ_1 and ψ_2 are Majorana spinors, one can them have a $U(1)$ symmetry which rotates ψ_1 into ψ_2.

Then, one can define

$$\psi_L \;:=\; \frac{1}{2}\left(1-i\gamma^5\right)\psi_D; \quad \psi_R := \psi_L^* = \frac{1}{2}\left(1+i\gamma^5\right)\psi_D^*; \tag{26}$$

$$\chi_L \;:=\; \frac{1}{2}\left(1-i\gamma^5\right)\psi_D^*; \quad \chi_R := \chi_L^* = \frac{1}{2}\left(1+i\gamma^5\right)\psi_D. \tag{27}$$

Therefore, ψ_L and χ_L (and ψ_R and χ_R) have opposite $U(1)$ phases, whereas ψ_L and χ_R (and ψ_R and χ_L) have identical $U(1)$ phases. The $U(1)$-invariant Dirac equation in the chiral (Weyl) notation becomes

$$\gamma^\mu \partial_\mu \psi_L + m\chi_R = 0 \;\; (+h.c.)\,; \quad \gamma^\mu \partial_\mu \chi_L + m\psi_R = 0 \;\; (+h.c.)\,. \tag{28}$$

Thus, the Dirac mass term, in this notation, is $m\psi_L\chi_L + h.c.$, which is of course $U(1)$-invariant.

In principle, if one gives up the $U(1)$ symmetry, *a Dirac fermion can have three types of mass terms*, namely :

$$m\psi_L\chi_L, \qquad M\psi_L\psi_L, \qquad N\chi_L\chi_L, \tag{29}$$

where m is named *Dirac mass*, and M and N *Majorana masses*. As a result, the matrix

$$\mathcal{M} := \begin{pmatrix} M & m \\ m & N \end{pmatrix} \tag{30}$$

has two eigenvalues :

$$m_{1,2} := \frac{1}{2}\left[M + N \pm \sqrt{(M-N)^2 + 4m^2}\right]. \tag{31}$$

In particular, for $N = 0$ and $M \gg m$, the eigenvalues become

$$m_{1,2} \cong \left(M, -\frac{m^2}{M}\right). \tag{32}$$

This result is at the basis of the *see-saw mechanism* [4, 5].

Concerning the superpartner of the graviton, this is a spin-$\frac{3}{2}$ Rarita-Schwinger field, called *gravitino*. Under the Lorentz group $SO(1, D-1)$ it is a vector-spinor $\psi_{\mu\alpha}$. If it is massless (namely, in the case of unbroken supersymmetry), it lies in an irreducible, γ-traceless representation of the little group $SO(D-2)$; so, assuming real (semi)spinors, the gravitino has a number of (*on-shell*) independent components given by

$$D \text{ odd}, (\psi \text{ real}) \;:\; \#\left(\psi_{\mu\alpha}\right) = (D-2)\,2^{\frac{D-3}{2}} - 2^{\frac{D-3}{2}} = (D-3)\,2^{\frac{D-3}{2}}; \tag{33}$$

$$D \text{ even}, (\psi_\pm \text{ real}) \;:\; \#\left(\psi_{\mu\alpha,\pm}\right) = (D-2)\,2^{\frac{D-4}{2}} - 2^{\frac{D-4}{2}} = (D-3)\,2^{\frac{D-4}{2}}, \tag{34}$$

which for example yields

$$D \;=\; 11 : \#\left(\psi_{\mu\alpha}\right) = 8\cdot 2^4 = 128; \tag{35}$$

$$D \;=\; 10 : \#\left(\psi_{\mu\alpha,\pm}\right) = 7\cdot 2^3 = 56. \tag{36}$$

The simplest Lagrangian density of supergravity couples Einstein gravity with the Rarita-Schwinger density, yielding

$$\mathcal{L}_{\rm sugra} = R - \bar{\psi}_\mu \gamma^{\mu\nu\rho} D_\nu \psi_\rho. \tag{37}$$

Assuming 2 as the highest spin in the massless spectrum (pertaining to the graviton), after [6] it is known that Poincaré supersymmetry coupled to gravity is possible for $D \leqslant 11$, while conformal supersymmetry is possible for $D \leqslant 6$; this result also stems from the classification of superalgebras done by Kac [7].

3 Spontaneous Symmetry Breaking of Rigid Supersymmetry : Goldstino Actions

As holding for any action invariant under continuous symmetries, in theories invariant under one or more supersymmetry transformations there are conserved *Majorana vector-spinor Noether currents* [8] (in van der Waerden notation):

$$\partial^\mu \; J^A_{\mu\,\alpha}(x) \;=\; 0 \quad (A = 1, .., \mathcal{N}). \tag{38}$$

The corresponding charges

$$\mathcal{Q}^A_\alpha \;=\; \int d^3\mathbf{x} \;\, J^A_{0\,\alpha}(x) \tag{39}$$

generate the ($\mathcal{N}$-extended) supersymmetry algebra, whose non-trivial part in $D = 4$ (which we consider throughout this Section) is given by (14).

In $\mathcal{N} = 1$ spontaneous breaking, the order parameter μ enters a term linear in the supercurrent [9, 10]

$$J_{\mu\,\dot{\alpha}} \;=\; \mu \;\, (\gamma_\mu \mathcal{G})_{\dot{\alpha}} + \;\dots\;, \tag{40}$$

where $\mathcal{G}$ is a Majorana field, the *Goldstino; note that* μ has dimension 2 in natural units.

The supersymmetry current algebra implies that

$$\int d^3\mathbf{y}\, \{J_{0\,\dot{\alpha}}(y), J_{\mu\alpha}(x)\} = (\sigma^\nu)_{\alpha\dot{\alpha}}\, T_{\nu\mu}(x) + \text{derivatives}, \tag{41}$$

so that

$$\langle 0|\left\{\bar{\mathcal{Q}}_{\dot{\alpha}}\,, J_{\mu\alpha}(x)\right\}|0\rangle \;=\; (\sigma^\nu)_{\alpha\,\dot{\alpha}} \;\langle 0|T_{\nu\mu}|0\rangle \;=\; (\sigma_\mu)_{\alpha\,\dot{\alpha}} \;\mu^2 \;. \tag{42}$$

By generalizing this for several supercharges, one can conclude that (in absence of central extensions, as well as in the rigid case - no gravity coupled -, as we assume throughout this Section), either all supersymmetries are unbroken or they are all broken at the same scale f :

$$J_{\mu\,\dot{\alpha} A} \;=\; \mu \;\, (\gamma_\mu \mathcal{G}_A)_{\dot{\alpha}} + \;\dots\;. \tag{43}$$

One can evade this result by modifying the current algebra via terms not proportional to δ_A^B, namely by adding a contribution proportional to a constant matrix C_A^B in the adjoint of $SU(\mathcal{N})$ (*i.e.*, traceless and Hermitian):

$$\langle 0|\left\{\bar{\mathcal{Q}}_{\dot\alpha}\,,J_{\mu\alpha}(x)\right\}|0\rangle \;=\; (\sigma^\nu)_{\alpha\dot\alpha}\;\langle 0|T_{\nu\mu}|0\rangle\;+(\sigma_\mu)_{\alpha\dot\alpha}\,C_A^B. \tag{44}$$

As it holds in general for Goldstone particles, the Goldstino $\mathcal{G}$ is a massless fermion, with low–energy self–interactions described by a Lagrangian invariant under a non–linear realization of supersymmetry, and depending on both $\mathcal{N}$ and C_A^B, because it can be shown that

$$\delta_A\chi^i\;\delta^B\overline{\chi}_i\;=\;V\;\delta_A^B\;+\;C_A^B\quad\left(\delta\chi^i\;=\;\delta_A\chi^i\,\epsilon^A\right)\,. \tag{45}$$

If k supersymmetries are unbroken, the Hermitian matrix $V\mathbf{1}+\mathbf{C}$ has rank $\mathcal{N}-k$ in the vacuum, and the $\mathcal{N}-1$ possible scales of partial supersymmetry breaking are the classified by the $\mathcal{N}-1$ Casimirs of $SU(\mathcal{N})$. Amusingly, the resulting characteristic equation can be solved in algebraically closed form up to the quartic order, corresponding in this setting to $\mathcal{N}=4$, the maximal value of $\mathcal{N}$ for rigid supersymmetry.

For $\mathcal{N}=2$, there are various examples realizing a spontaneous supersymmetry breaking [11, 12, 13, 14, 15], and the previous analysis in terms of $SU(2)$ invariants was performed in [16]. The rigid limit of $\mathcal{N}=2$ supergravity theory reproducing partial supersymmetry breaking was considered in [17]. The unique Casimir of $SU(2)$ is the squared norm of a 3–vector ξ^x constructed via electric and magnetic Fayet–Iliopoulos (FI) terms

$$\xi^x\;\;:=(Q_y\wedge Q_z)\,\epsilon^{xyz}; \tag{46}$$

$$Q_y\wedge Q_z\;\;:=m_y^\Lambda\;e_{z\Lambda}\;-\;m_z^\Lambda\;e_{y\Lambda}\;, \tag{47}$$

where $Q_x=(m_x^\Lambda,e_{x\,\Lambda})$, and $\Lambda=1,\ldots,n$, with n denoting the number of $\mathcal{N}=2$ vector multiplets. The wedge product $\wedge$ is with respect to the symplectic structure inherited from the rigid special Kähler geometry endowing the vector multiplets' scalar manifold, and the matrix C_A^B acquires the explicit form

$$\delta_A\chi^i\;\delta^B\overline{\chi}_i\;=\;V\;\delta_A^B\;+\;(\sigma_x)_A^{\;B}\;\xi^x\;=\;\begin{pmatrix} V-\xi^3 & \xi^1+i\,\xi^2\\ \xi^1-i\,\xi^2 & V+\xi^3\end{pmatrix}, \tag{48}$$

with eigenvalues

$$\lambda_\pm=V\mp\sqrt{\xi^x\xi^x}. \tag{49}$$

Thus, when $V=\sqrt{\xi^x\xi^x}$, $\lambda_+=0$ and $\mathcal{N}=2$ rigid supersymmetry is broken down to $\mathcal{N}=1$, with order parameter

$$\mu=\left(\xi^x\xi^x\right)^{1/4}. \tag{50}$$

The last term in the r.h.s. of Eq. (44) corresponds to a "vector" central charge $Z^A_{\mu|B}$ [18, 19]; the only field not being inert is the gauge field of $\mathcal{N} = 2$ vector multiplets.

For a general $\mathcal{N}$-extended rigid supersymmetry broken down to $\mathcal{N} = 0$ (corresponding, from the previous reasoning, to $C^B_A = 0$), the low–energy effective Lagrangian for N Goldstino fields is an obvious extension of the Volkov–Akulov (VA) action [20] derived in the $\mathcal{N} = 1$ case, and in general reads

$$\mathcal{L}_{VA}\left(\mu, \mathcal{G}^A_\alpha\right) = \mu^2 \left[1 - \sqrt{-\det\left(\eta_{\mu\nu} + \frac{i}{\mu^2}\left(\overline{\mathcal{G}}_A \gamma_\mu \partial_\nu \mathcal{G}^A - h.c.\right)\right)}\right] . \tag{51}$$

It should be stressed that this Lagrangian contains a finite number of terms, because of the *nilpotency* of the Goldstino fields $\mathcal{G}^A$. Furthermore, $\mathcal{L}_{VA}$ has a $U(\mathcal{N})$ $\mathcal{R}$–symmetry, being invariant under the following non–linear supersymmetry transformations :

$$\delta\ \mathcal{G}^A_\alpha(x) = \mu\ \epsilon^A_\alpha + \frac{i}{\mu}\left(\overline{\mathcal{G}}_B \gamma^\mu \epsilon^B - \overline{\epsilon}_B \gamma^\mu \mathcal{G}^B\right) \partial_\mu\ \mathcal{G}^A_\alpha . \tag{52}$$

On the other hand, the Goldstino action for $\mathcal{N} = 2$ rigid supersymmetry partially broken down to $\mathcal{N} = 1$ is described by the supersymmetric Born–Infeld (SBI) theory : this is given by a *non-linear* Lagrangian $\mathcal{L}_{SBI}\left(\mathcal{G}_\alpha, F_{\mu\nu}\right)$ for an $\mathcal{N} = 1$ vector multiplet whose chiral superfield strength $W_\alpha = \overline{D}^2 D_\alpha V$ contains the goldstino $\mathcal{G}$ at $\theta = 0$ and the self–dual Maxwell field strength at the next order in θ, with the following two properties :

$$\mathcal{L}_{SBI}\left(\mathcal{G}_\alpha, F_{\mu\nu} = 0\right) \longrightarrow \mathcal{L}_{VA}\ ; \tag{53}$$

$$\mathcal{L}_{SBI}\left(\mathcal{G}_\alpha = 0, F_{\mu\nu}\right) = \mu^2 \left[1 - \sqrt{-\,det\left(\eta_{\mu\nu} + \frac{1}{\mu} F_{\mu\nu}\right)}\right] = \mathcal{L}_{BI}, \tag{54}$$

where $\mathcal{L}_{BI}$ is given by (94) further below.

$\mathcal{L}_{SBI}$ exhibits two types of super–invariances: a manifest $\mathcal{N} = 1$ supersymmetry, linearly realized in $\mathcal{N} = 1$ superspace, and a second supersymmetry which is non–manifest and non–linearly realized. In terms of microscopic parameters, the scale of the broken supersymmetry is the $SU(2)$- and symplectic- invariant quantity (50); see Sec. 4.4 further below.

It is here worth pointing out that the Goldstino action for a partial breaking $\mathcal{N} \to \mathcal{N} - k$ of rigid supersymmetry is only known for $\mathcal{N} = 2$, corresponding to the non–linear limit of a quadratic action of $\mathcal{N} = 2$ vector multiplets endowed with FI terms. The simplest case is given by a single $N = 2$ vector multiplet, which under the spontaneous breaking $\mathcal{N} = 2 \to 1$ reduces to an $\mathcal{N} = 1$ vector multiplet W_α and to an $\mathcal{N} = 1$ chiral multiplet X. The presence of a non-vanishing matrix C^B_A allows the chiral multiplet X to acquire a non-zero mass m_X :

$$\underbrace{\left(1, 2\left(\frac{1}{2}\right), 2\,(0)\right)}_{\mathcal{N}=2\text{ vector mult.}} \overset{\mathcal{N}=2\to 1}{\longrightarrow} \underbrace{\left(1, \frac{1}{2}\right)}_{\mathcal{N}=1\text{ vector mult., } m_V=0} + \underbrace{\left(\frac{1}{2}, 2\,(0)\right)}_{\mathcal{N}=1\text{ chiral mult., } m_X\neq 0} . \tag{55}$$

For m_X large enough, X can be integrated out, giving rise to a non–linear theory for the fields $\mathcal{G}_\alpha$ and $F_{\mu\nu}$ of the $\mathcal{N} = 1$ vector multiplet, described by the aforementioned $\mathcal{L}_{SBI}(\mathcal{G}_\alpha, F_{\mu\nu})$ [21, 23].

4 Electric-Magnetic Duality, Born–Infeld and Generalizations

Electric-magnetic (e.m.) duality is one of the most fascinating symmetries of non-linear (and thus interacting) theories of gauge fields.

4.1 Maxwell Theory

Free Maxwell theory is the prototype of e.m. duality invariant theory. In the vacuum, the Eqs. of motion read ($F_{\mu\nu} := \partial_\mu A_\nu - \partial_\nu A_\mu$)

$$\partial_\mu F^{\mu\nu} = 0, \quad \partial_\mu \star F^{\mu\nu} = 0 \tag{56}$$

where

$$\star F_{\mu\nu} := \frac{1}{2}\epsilon_{\mu\nu\rho\sigma}F^{\rho\sigma} , \tag{57}$$

and they are invariant under $SO(2)$ rotations:

$$\begin{pmatrix} F \\ \star F \end{pmatrix}' = \begin{pmatrix} \cos\alpha & -\sin\alpha \\ \sin\alpha & \cos\alpha \end{pmatrix} \begin{pmatrix} F \\ \star F \end{pmatrix}. \tag{58}$$

This latter transformation, in terms of electric and magnetic fields $F_{\mu\nu} = \left(\vec{E}, \vec{B}\right)$ reads

$$\begin{pmatrix} \vec{E} \\ \vec{B} \end{pmatrix}' = \begin{pmatrix} \cos\alpha & -\sin\alpha \\ \sin\alpha & \cos\alpha \end{pmatrix} \begin{pmatrix} \vec{E} \\ \vec{B} \end{pmatrix}. \tag{59}$$

Also in this notation it is immediate to see that Maxwell Equations in absence of sources ($F_{oi} = E_i$, $F_{ij} = \epsilon_{ijk}B_k$)

$$\vec{\nabla} \cdot \vec{E} = 0, \quad \partial_t \vec{E} = \vec{\nabla} \times \vec{B}, \tag{60}$$

$$\vec{\nabla} \cdot \vec{B} = 0, \quad \partial_t \vec{B} = -\vec{\nabla} \times \vec{E} \tag{61}$$

are invariant.

Note that the Hamiltonian

$$\mathcal{H} = \frac{1}{2}\left(\left|\vec{E}\right|^2 + \left|\vec{B}\right|^2\right) \tag{62}$$

is e.m. duality invariant, while the Lagrangian

$$\mathcal{L} = \frac{1}{2}\left(\left|\vec{E}\right|^2 - \left|\vec{B}\right|^2\right) \tag{63}$$

is not. It should also be pointed out that the *e.m. duality symmetry is not an internal symmetry, since it rotates tensors with pseudo-tensors* (as such, it can be regarded as a sort of "bosonic chiral transformation").

Electric-magnetic dualities are transformations among two-form field strengths and their *duals*. These transformations extend to $(p+2)$-form field strengths and their duals, which, in D space-time dimensions, are $(D-p-2)$-forms (for $D=4$ and $p=0$, one obtains that $p+2=D-p-2$). In order to generalize the e.m. duality to p-forms, one demands that $D/2=p+2$, so that one has $(p+1)$-form gauge fields coupled to sources, which are p-extended objects: $p=0$ (pointlike), $p=1$ (string), $p=2$ (membrane), $p=3$ (three-brane), *etc.*:

$$F_{\mu_1 \cdots \mu_{p+2}} = \partial_{[\mu_1} A_{\mu_2 \cdots \mu_{p+2}]}. \tag{64}$$

When $D/2=p+2$ we have that $D-p-2=p+2$, and p-dimensional extended objects can source both electric and magnetic fields. We thus have *dyons*, extended objects that carry both an electric and a magnetic charge. In all other cases, electric and magnetic objects extends in different space dimensions, $(p, D-p-4)$ respectively, so dyons *cannot* exist.

The Dirac-Schwinger-Zwanziger quantization condition reads as follows:

- for $D/2 \neq p+2$:

$$em' = 2\pi k, \quad k \in \mathbb{Z}, \tag{65}$$

- for $D/2 = p+2$

$$em' + e'm = 2k\pi, \quad k \in \mathbb{Z}\ (p \text{ odd}); \tag{66}$$

$$em' - e'm = 2k\pi, \quad k \in \mathbb{Z}\ (p \text{ even}). \tag{67}$$

Note that the latter condition is not only invariant under $SO(2)$, but also under $Sp(2,\mathbb{R})$, isomorphic to $SL(2,\mathbb{R})$.

4.2 Non-Linear Electric–Magnetic Duality

The very idea of duality is to generalize the duality of free Maxwell theory to the case of interactions. Interactions introduce non-linearities, which can be due either to the Maxwell field itself, or to other matter fields (as in electrodynamics) which interact with the e.m. field.

Non-linear (pure) electromagnetism can be seen as an effective theory of Maxwell theory in a medium that has a nonlinear response to the electric and magnetic fields. These nonlinearities are captured by the relations expressing the *electric displacement* $\vec{D}$ and the *magnetic field* $\vec{H}$ in terms of the *magnetic induction* $\vec{B}$ and the *electric field* $\vec{E}$. The field equations (in absence of sources) are thus divided in a set of linear ones that are independent from the medium

$$\vec{\nabla}\cdot\vec{D} = 0, \quad \partial_t\vec{D} = \vec{\nabla}\times\vec{H}, \tag{68}$$

$$\vec{\nabla}\cdot\vec{B} = 0, \quad \partial_t\vec{B} = -\vec{\nabla}\times\vec{E}, \tag{69}$$

and a set describing the nonlinearities, i.e. the properties of the medium, hence these equations are called constitutive relations,

$$\vec{D} = \vec{D}\left(\vec{E},\vec{B}\right) = \vec{E} + ..., \tag{70}$$

$$\vec{H} = \vec{H}\left(\vec{E},\vec{B}\right) = \vec{B} + ..., \tag{71}$$

Maxwell theory is recovered when

$$\vec{D} = \vec{E}, \quad \vec{H} = \vec{B}\,. \tag{72}$$

The dots ... in Eq. (71) stand for some higher powers of the electric field $\vec{E}$ and of the magnetic induction $\vec{B}$, so that the nonlinear theories we consider are deformations of linear electromagnetism because (70) and (71) reduce to (72) for weak fields.

In a theory based on a Lagrangian density $\mathcal{L} = \mathcal{L}\left(\vec{E},\vec{B}\right)$, one has the Eqs.

$$\vec{D} = 2\frac{\delta\mathcal{L}\left(\vec{E},\vec{B}\right)}{\delta\vec{E}}, \quad \vec{H} = -2\frac{\delta\mathcal{L}\left(\vec{E},\vec{B}\right)}{\delta\vec{B}}. \tag{73}$$

Note that Eqs. (68) are invariant under the linear transformations

$$\begin{pmatrix}\vec{B}\\ -\vec{D}\end{pmatrix}' = \begin{pmatrix}A & B\\ C & D\end{pmatrix}\begin{pmatrix}\vec{B}\\ -\vec{D}\end{pmatrix}; \tag{74}$$

$$\begin{pmatrix}\vec{E}\\ \vec{H}\end{pmatrix}' = \begin{pmatrix}A & B\\ C & D\end{pmatrix}\begin{pmatrix}\vec{E}\\ \vec{H}\end{pmatrix}. \tag{75}$$

Hence the full set of equations of motion are invariant under the $\begin{pmatrix}A & B\\ C & D\end{pmatrix}$ duality rotations if the constitutive relations are compatible with these rotations, i.e. if the transformations $\vec{D}(\vec{E},\vec{B}) \to \vec{D}(\vec{E}',\vec{B}')$ and $\vec{H}(\vec{E},\vec{B}) \to \vec{H}(\vec{E}',\vec{B}')$ are the same as the rotations (74), (75).

It is convenient to rewrite the nonlinear equations and the duality rotations using a relativistic formalism. We set as usual $F_{\mu\nu} = \left(\vec{E}, \vec{B}\right)$ and $G_{\mu\nu} = \left(\vec{H}, -\vec{D}\right)$, so that in the vacuum:

$$\vec{D} = \vec{E}, \quad \vec{H} = \vec{B} \quad \Leftrightarrow \quad G_{\mu\nu} = \star F_{\mu\nu}. \tag{76}$$

The dynamical equations then read $dF = 0, dG = 0$ i.e.,

$$\partial_\mu \star F^{\mu\nu} = 0, \quad \partial_\mu \star G^{\mu\nu} = 0 \ , \tag{77}$$

while the constitutive relations become

$$G_{\mu\nu} = G_{\mu\nu}\left(F, \star F\right), \tag{78}$$

$$G_{\mu\nu} = \star F_{\mu\nu} + \ldots \ , \tag{79}$$

and, if we have a Lagrangian,

$$\star G^{\mu\nu} = 2\frac{\delta \mathcal{L}\left(F\right)}{\delta F_{\mu\nu}}. \tag{80}$$

In order to have compatibility with the equations of motion, this demands the following integrability conditions :

$$\frac{\delta \star G^{\mu\nu}}{\delta F^{\rho\sigma}} = \frac{\delta \star G_{\rho\sigma}}{\delta F_{\mu\nu}}. \tag{81}$$

The duality rotations correspond to transform

$$\begin{pmatrix} F \\ G \end{pmatrix} \to \begin{pmatrix} F \\ G \end{pmatrix}' = \begin{pmatrix} A & B \\ C & D \end{pmatrix} \begin{pmatrix} F \\ G \end{pmatrix}, \tag{82}$$

and are a symmetry of the theory if the transformation $G(F) \to G(F') = G(AF + BG)$ is the same as the rotation $G(F) \to CF + DG(F)$. This highly non-linear constraint restricts the possible forms of G (the constitutive relations), and also the linear transformations (82).

Gaillard and Zumino [1] have proven that the most general e.m. duality rotation for a non-linear theory, also depending on extra matter fields (such as fermions, scalars, and possibly in a curved gravitational background) is $Sp(2n, \mathbb{R})$, where n is the number of vector field strengths F^Λ $(\Lambda = 1, \ldots, n)$. This means that the matrix

$$\mathcal{S} := \begin{pmatrix} A & B \\ C & D \end{pmatrix} \tag{83}$$

instead of being a $GL(2n, \mathbb{R})$ matrix is a symplectic matrix, indeed it satisfies

$$\mathcal{S}^T \Omega \mathcal{S} = \Omega, \quad \Omega := \begin{pmatrix} 0 & -\mathbb{I} \\ \mathbb{I} & 0 \end{pmatrix}, \tag{84}$$

which equivalently reads

$$A^T D - C^T B = \mathbb{I}, \quad A^T C = C^T A, \quad B^T D = D^T B.$$

In absence of scalars, or if scalars are inert under duality rotations, the e.m. duality group is at most $U(n)$ (in particular, for $n = 1$, one retrieves Maxwell e.m. duality group: $U(1) \sim SO(2)$). The constraints on the n field strength F^Λ and G^Λ in this case read

$$F^\Lambda \star F^\Sigma + G^\Lambda \star G^\Sigma = 0; \tag{85}$$

$$F^\Lambda \star G^\Sigma - F^\Sigma \star G^\Lambda = 0. \tag{86}$$

For $n = 1$ (namely, $U(1)$), one gets

$$F \star F + G \star G = 0. \tag{87}$$

For general e.m. theories where also matter fields are present the constitutive relations will also depend on these fields, that we generically denote ζ^i, (and on their derivatives that we omit writing) so that we have $G_{\mu\nu} = G_{\mu\nu}(F, \zeta^i)$. Duality rotations now will include also a (nonlinear) transformation on the matter fields ζ^i.

As shown by Gaillard and Zumino [1] (and reviewed in [24]) in this general case, under an infinitesimal duality rotation parametrized by

$$s = \begin{pmatrix} a & b \\ c & d \end{pmatrix} \in \mathfrak{sp}(2n, \mathbb{R}), \text{ i.e. } a = -d^T, \; b = b^T, c = c^T \;, \tag{88}$$

the Lagrangian varies by

$$\delta\mathcal{L} = \frac{1}{4}(Fc \;\star F + Gb \;\star G)\,,$$

where $\delta L \equiv L(F + \delta F, \zeta^i + \delta\zeta^i) - L(F, \zeta^i)$ is the total infinitesimal variation of the Lagrangian under the duality rotations of the field strengths and of the matter fields. If we expand δF in terms of F and G using the infinitesimal version of the duality rotation (82) then we obtain

$$\delta_\zeta L := L(F, \zeta^i + \delta\zeta^i) - L(F, \chi^i) = \frac{1}{4}(Fc \;\star F - Gb \;\star G) + \frac{1}{2}Fd \;\star G. \tag{89}$$

If there are no matter fields the left hand side is zero, the duality rotation group reduces then to $U(n)$, and in this case the infinitesimal parameters satisfy $c = -b$ and $d = -d^T$. Then Eq. (89) reproduces just the previous Eqs. (85) and (86) for $U(n)$.

Note that for e.m. duality rotations the total variation $\delta\mathcal{L}$ never vanishes. If $c \neq 0, b = 0$ the infinitesimal variation $\delta\mathcal{L}$ is a total derivative because of the equations of motion $dF = 0$. If

$c = 0, b \neq 0$ it is a total derivative because of the equations of motion $dG = 0$. If both $b \neq 0$ and $c \neq 0$ we have to use both equations of motion $dF = 0$ and $dG = 0$. We therefore see that the Lagrangians of theories with duality rotation symmetry change by a term that is a total derivative only on shell of the equations of motion $dF = 0$ and $dG = 0$.

We also have that under a finite duality rotation the action changes by a term that is a total derivative only on shell of both $dF = 0$ and $dG = 0$ equations of motions. Indeed under a finite $U(1)$ rotation of angle α we have [25]

$$\mathcal{L}(F') - \mathcal{L}(F) = \frac{1}{8}\Big(sin(2\alpha)\,(F \star F - G \star G) - 4\,sin^2(\alpha)\,F \star G\Big)\,. \tag{90}$$

We notice that in the special case of Maxwell theory $G = \star F$ and therefore the infinitesimal variation $\mathcal{L}(F') - \mathcal{L}(F)$ is a total derivative just on shell of the relation $G = \star F$ and of the equation $dF = 0$ (*i.e.* without using the equation of motion $dG = 0$). In other words, if we solve the equation $dF = 0$ by introducing a gauge potential so that $F = dA$ and hence see the Lagrangian as dependent on A, then for infinitesimal angle α the variation (90) is a total derivative *off shell* of the equations of motion for the gauge potential A. This off shell infinitesimal symmetry of Maxwell theory (that acts non-locally on the gauge potential A) was studied in [26]; however it does not correspond to a finite *off shell* symmetry. As we see from the second term in (90) in order for $F \star G$ to be a total derivative we have to use also the equations of motion $dG = 0$, that in Maxwell theory simply read $d \star F = 0$.

Two important results are obtained from the previous formulæ for general nonlinear theories.

1. *Any* Lagrangian density can be written as $\frac{1}{4}F \star G$ plus a duality invariant Lagrangian $\mathcal{L}_{inv}$,

$$\mathcal{L} = \frac{1}{4}F \star G + \mathcal{L}_{inv}\,, \quad \delta\mathcal{L}_{inv} = 0. \tag{91}$$

   The term $\mathcal{L}_{inv}$ provides information about the constitutive relations (*i.e.*, the medium), in particular $\mathcal{L}_{inv} = 0$ vanishes in linear Maxwell theory without coupling to matter. On the other hand, for (quadratic) theories coupled to matter $\mathcal{L}_{inv}$ provides information about some matter couplings (such as Pauli-like terms). In supergravity it is in fact possible to construct pairs of antisymmetric Lorentz tensors $(H_{\mu\nu}(\zeta)\,, I_{\mu\nu}(\zeta))$ (typically bilinear in fermions) which transforms as the couple (F, G). In this case, one has

$$\mathcal{L}_{inv} = \frac{1}{4}(FI - GH) + \mathcal{L}_{inv}(\zeta)\,. \tag{92}$$

2. The energy-momentum tensor

$$\Theta^{\mu}_{\lambda} = -\partial_{\lambda}\zeta^{i}\frac{\partial\mathcal{L}}{\partial(\partial_{\mu}\zeta^{i})} + \delta^{\mu}_{\lambda}\mathcal{L} + \star G^{\Lambda|\mu\nu}F_{\Lambda|\nu\lambda} \tag{93}$$

is duality invariant if duality is a symmetry of the theory. In the absence of matter fields the trace of the energy momentum tensor is proportional to $\mathcal{L}_{inv}$: $\frac{1}{4}\Theta^{\mu}_{\mu} = \mathcal{L}_{inv}$.

4.3 Born-Infeld Theory

The Born-Infeld (BI) model [27]

$$\begin{aligned}\mathcal{L}_{BI} &= \mu^2\left(1-\sqrt{-\det\left(\eta_{\mu\nu}+\frac{F_{\mu\nu}}{\mu}\right)}\right) &(94)\\ &= \mu^2\left(1-\sqrt{1+\frac{1}{2\mu^2}F^2-\frac{1}{16\mu^4}\left(F\ \star F\right)^2}\right) &(95)\\ &= \mu^2\left(1-\sqrt{1+\frac{1}{\mu^2}\left(\left|\vec{B}\right|^2-\left|\vec{E}\right|^2\right)-\frac{1}{\mu^4}\left(\vec{E}\cdot\vec{B}\right)^2}\right) &(96)\end{aligned}$$

is the simplest non-linear e.m. duality invariant theory. It was introduced to remove the divergence of the electrons self-energy in classical electrodynamics. This is obtained by having an upper bound on the electric field; indeed, for $\vec{B} = 0$ $\mathcal{L}_{BI}$ yields the bound

$$\left|\vec{E}\right| \leqslant \mu\ . \tag{97}$$

In fact, $\mathcal{L}_{BI}$ has unique properties against instabilities created by the medium, and not many generalizations are known in the multi-field case (*i.e.*, in presence of more than one Maxwell field). Furthermore, by expanding $\mathcal{L}_{BI}$ in power series of the coupling constant $\frac{1}{\mu}$ one obtains

$$\star G_{\mu\nu}\left(F,\star F;\mu\right) = -F_{\mu\nu} + \ldots \tag{98}$$

and hence for $\mu \to \infty$, *i.e.*, for weak fields, one recovers Maxwell theory.

The equations of motion of BI theory are a particular realization of the constitutive relations

$$\star G_{\mu\nu} = \ \star G_{\mu\nu}\left(F,\star F;\mu\right), \tag{99}$$

(here μ is a dimensionful coupling constant typically present in nonlinear theories, which was previously omitted for sake of brevity). Moreover, they satisfy the e.m. duality conditions (87). By recalling (94) and the definition (80), the explicit dependence of G on F can be computed to read

$$G_{\mu\nu}\left(F,\star F;\mu\right) = \frac{\star F_{\mu\nu} + \frac{1}{4\mu^2}\left(F\ \star F\right)F_{\mu\nu}}{\sqrt{1+\frac{1}{2\mu^2}F^2-\frac{1}{16\mu^4}\left(F\ \star F\right)^2}}. \tag{100}$$

The Hamiltonian depends on the variables $\left(\vec{D}, \vec{B}\right)$, and is obtained by a Legendre transform with respect to $\vec{E}$, namely by defining

$$\vec{D} = \frac{\delta\mathcal{L}(E,B)}{\delta\vec{E}} \quad , \quad \mathcal{H} = \vec{D}\cdot\vec{E} - \mathcal{L} \, . \tag{101}$$

It is worth stressing that the Hamiltonian is perfectly regular for all values of these fields variables:

$$\mathcal{H}_{BI} = \mu^2 \left(\sqrt{1 + \frac{1}{\mu^2}\left(\left|\vec{D}\right|^2 + \left|\vec{B}\right|^2\right) + \frac{1}{\mu^4}\left|\vec{D}\wedge\vec{B}\right|^2} - 1 \right) . \tag{102}$$

In particular in the limits of weakly and of strongly valued fields:

$$\lim_{\mu\to\infty} \mathcal{H}_{BI} \;=\; \frac{1}{2}\left(\left|\vec{D}\right|^2 + \left|\vec{B}\right|^2\right) \; ; \tag{103}$$

$$\lim_{\mu\to 0} \mathcal{H}_{BI} \;=\; \left|\vec{D}\wedge\vec{B}\right| . \tag{104}$$

As yielded by the property (54), the BI Lagrangian $\mathcal{L}_{BI}$ is the bosonic part of the $\mathcal{N} = 1$ non-linear Goldstino Lagrangian for the aforementioned partial breaking $\mathcal{N} = 2 \to 1$ of rigid supersymmetry; in this context, the Maxwell field is the $\mathcal{N} = 1$ superpartner of the spin-$\frac{1}{2}$ Goldstino field [28, 29, 11]. For various recent developments on BI theory, supersymmetry, and e.m. duality, see *e.g.* [30, 31, 32, 33, 34].

4.4 Supersymmetry, Special Geometry, and Multi–Field Extension

The supersymmetric BI Lagrangian follows from an algebraic constraint in Superspace among various multiplets [35, 36].

This constraint enforces a non–linear relation between the two $\mathcal{N} = 1$ supermultiplets X and W_α that build the $\mathcal{N} = 2$ vector multiplet [11, 37, 38]:

$$W^2 \;+\; X\left(m_1 \;-\; \bar{D}^2\bar{X}\right) \;=\; 0 \, , \tag{105}$$

where m_1 corresponds to a particular choice of a magnetic FI term. The non–linear relation (105) determines X as a non–linear function of $D^2\, W^2$ and $\overline{D}^2\, \overline{W}^2$ and implies the nilpotency constraints

$$X^2 \;=\; 0 \, , \quad X\, W_\alpha \;=\; 0 \, . \tag{106}$$

The supersymmetric BI Lagrangian then acquires the form

$$\mathcal{L}_{SBI} \;=\; Im \int d^2\theta \, (e_1 + i e_2)\, X\left(m_1, D^2\, W^2, \overline{D}^2\, \overline{W}^2\right) \, , \tag{107}$$

where e_1 and e_2 are electric FI terms whose labels correspond to the first two directions in an $SU(2)$ triplet. For canonically normalized vectors, one can then recognize that the scale μ and the theta–angle ϑ can be expressed in terms of m_1, e_1 and e_2 as

$$\mu \;=\; \sqrt{m_1\, e_2}\;, \qquad \vartheta \;=\; \frac{e_1}{e_2}\;. \tag{108}$$

In the bosonic limit, the quadratic constraint becomes

$$F_+^2 + \mathcal{F}\left(m_1 - \bar{\mathcal{F}}\right) = 0, \tag{109}$$

where F_+ is the *self-dual curvature* of the Maxwell field :

$$F_{+|\mu\nu} := \frac{1}{2}\left(F_{\mu\nu} + i\ \star F_{\mu\nu}\right) \tag{110}$$

and $\mathcal{F}$ is an auxiliary complex scalar field such that

$$\begin{aligned}\mathcal{L}_{BI} &= e_1\ \mathrm{Im}\,(\mathcal{F}) + e_2\ \mathrm{Re}\,(\mathcal{F}) = \\ &= -\vartheta F_{\mu\nu} \star F^{\mu\nu} + \frac{\mu^2}{2}\left[1 \;-\; \sqrt{1 \;+\; \frac{4}{\mu^2}\ F_{\mu\nu}\,F^{\mu\nu} \;-\; \frac{4}{\mu^4}\ \left(F_{\mu\nu} \star F^{\mu\nu}\right)^2}\;\right]. \end{aligned}\tag{111}$$

In the multi-field case, the generalization of $\mathcal{L}_{SBI}$ rests on the constraints ($A = 1, ..., n$) [21, 22]

$$d_{ABC}\left[W^B, W^C + Y^B\left(m^C - \bar{D}^2 \bar{Y}^C\right)\right] \;=\; 0\;, \tag{112}$$

and the complete Lagrangian reads

$$\mathcal{L} \;=\; -\ Im \int d^2\theta\,\left[U_{AB}\,W^A\,W^B \;+\; \mathcal{W}(X) \;+\; \frac{1}{2}\ \bar{D}^2\,\left(X^A\,\bar{U}_A \;-\; \bar{X}^A, U_A\right)\right]\;. \tag{113}$$

The θ^2 component of eq. (112),

$$d_{ABC}\left[F_+^B \cdot F_+^C \;+\; \mathcal{F}^B\left(m^C \;-\; \bar{\mathcal{F}}^C\right)\right] \;=\; 0\;, \tag{114}$$

where F_+^A are self–dual field strength combinations and $\mathcal{F}^A$ is the auxiliary field, is the multi–field generalization of the BI constraint (109) that is induced by the Special Geometry. Moreover, it holds that

$$\Re\, F_+^A\ F_+^B \;= F^A\ F^B\ , \qquad \Im\, F_+^A\ F_+^B \;= F^A\ \star F^B\ . \tag{115}$$

d_{ABC} is a rank-3 completely symmetric tensor, encoding the coefficients of the cubic term of the holomorphic prepotential of the aforementioned rigid special Kähler geometry characterizing the $\mathcal{N} = 2$ vector multiplets' scalar manifold, whose classification is based on the singularity structure of cubic varieties [22, 23].

The real parts of Eqs. (114) are n quadratic equations that are generally coupled. Letting

$$H^A = \frac{m^A}{2} - \Re\mathcal{F}^A , \qquad R^{AB} = F^A F^B + \frac{m^A m^B}{4} - \Im\mathcal{F}^B \Im\mathcal{F}^C , \tag{116}$$

they take the form

$$d_{ABC}\left(H^B H^C - R^{BC}\right) = 0 . \tag{117}$$

On the other hand, the imaginary parts of Eqs. (114) are n linear equations for $\Im\mathcal{F}^A$:

$$d_{ABC}\left(F^B \star F^C + \Im\mathcal{F}^B m^C\right) = 0 . \tag{118}$$

The bosonic part of the multi-field Lagrangian (113) can be expressed in terms of the $\mathcal{F}^A$ and of the real magnetic changes m^A. It can also be written in terms of additional complex charges $e_A = e_{1A} + i\, e_{2A}$, as

$$\mathcal{L}_{\text{Bose}} = e_{2A}\left(\frac{m^A}{2} - H^A\right) + C_{AB}\left(H^A H^B - R^{AB}\right) + \epsilon_{1A}\Im\mathcal{F}^A . \tag{119}$$

The matrix C_{AB} is needed whenever the matrix $d_{AB} := d_{ABC}\, q^C$ is not positive definite. Moreover, by a change of symplectic basis one could also eliminate the real parts e_{1A} of the electric charges, which multiply combinations of the fields that are total derivatives. These Lagrangians combine, in general, a quadratic Maxwell–like term with additional higher–order contributions. For $n = 1$, or whenever the matrix d_{AB} is positive definite, one is not compelled to introduce the C_{AB} and the Lagrangian takes the simpler form (111). In all cases, however, the difficult step in the construction of the Lagrangians is the solution of the quadratic constraints, and in particular of the non–linear ones given in (117).

4.5 Schrödinger Formulation

The general study and classification of constitutive relations admitting e.m. duality symmetry can be attacked by adopting complex and chiral variables, indeed these variables simply transform with a phase under duality symmetry rotations. We therefore define[2]:

$$T := F - iG \Leftrightarrow \overline{T} := F + iG, \tag{120}$$

so that the e.m. duality condition (87) equivalently reads

$$\star\overline{T}T = 0. \tag{121}$$

[2]Here and below, the bar denotes complex conjugation.

The corresponding chiral combinations read

$$T^{\pm} := \frac{1}{2}\left(F^{\pm} - iG^{\pm}\right) \Leftrightarrow \overline{T^{\pm}} := \frac{1}{2}\left(F^{\mp} + iG^{\mp}\right), \tag{122}$$

where $F^{\pm} = \frac{1}{2}(F \pm i \star F)$, $G^{\pm} = \frac{1}{2}(G \pm i \star G)$. The equations of motion of linear electromagnetism in these variables read

$$T^{+} = 0 \Leftrightarrow (F - iG) + i\,(\star F - i\ \star G) = 0 \Leftrightarrow G = \ \star F. \tag{123}$$

The constitutive relations in these new variables, studied in [39], and independently and more generally in [40, 41], read as follows:

$$T^{+} = \frac{\partial I\left(T^{-}, \overline{T^{-}}\right)}{\partial \overline{T^{-}}}, \quad \overline{T^{+}} = \frac{\partial I\left(T^{-}, \overline{T^{-}}\right)}{\partial T^{-}}. \tag{124}$$

These are six relations if the arbitrary function I (the "Lagrangian" of these new variables) is real. Moreover, duality invariance is obtained if I is invariant under duality rotations. This is the case if

$$I\left(T^{-}, \overline{T^{-}}, g\right) = I(u),$$

where

$$u = 2g^2|(T^{-})^2| = g(|T^2| + |T\,{}^*T|)\ , \quad \text{i.e.,} \quad u^2 = 4g^4(T^{-})^2(\overline{T}^{-})^2\ . \tag{125}$$

We note that the function I is $U(1)$-duality-invariant because the Lorentz scalar $(T^{-})^2(\overline{T^{-}})^2$ is duality invariant, indeed under a $U(1)$ duality rotation of angle α, the fields T and $\overline{T}$ transform with opposite phases $e^{-i\alpha}$ and $e^{i\alpha}$.

By use of a Legendre transform [39, 42, 25] between the functions $\mathcal{L}(F)$ and $I(T^{-}, \overline{T^{-}})$, the constitutive relations (124) in the T-variables can be shown to be equivalent to the original constitutive relations (80).

To summarize, the freedom in writing a self-dual and non-linear theory of electromagnetism can be traced back to a real function I of a real (Lorentz-invariant and $U(1)$-duality invariant) variable $(T^{-})^2\left(\overline{T^{-}}\right)^2$, or its square root u. Arbitrary functions $I(u)$ are in one-to-one correspondence with Lagrangians $\mathcal{L}\,(F)$ such that the equations (124) are equivalent to the original constitutive relations and self duality condition (80) and (87).

Furthermore, it was shown by Schrödinger [43] (see also [44]) that the constitutive relations for the BI theory are

$$\star T_{\mu\nu} = -\frac{T^2}{\star TT}T_{\mu\nu} - \frac{g^2}{8}\ \star TT\,\overline{T}_{\mu\nu}. \tag{126}$$

In the general case the following constitutive relations *à la Schrödinger* define a nonlinear and self-dual theory [42]

$$\star T_{\mu\nu} = -\frac{T^2}{\star TT} T_{\mu\nu} - \frac{g^2}{8}\frac{f(u)}{u} \star TT\,\overline{T}_{\mu\nu}, \tag{127}$$

where we recall that $u = 2|(T^-)^2| = |T^2| + |T \star T|$, and where we require $lim_{u\to 0} f(u) = 0$, so to recover Maxwell theory for weak fields. For example the duality conditions (121) follow contracting (127) with $\star T^{\mu\nu}$. Contracting (127) with $T^{\mu\nu}$ and with $\overline{T}^{\mu\nu}$ it can also be shown that

$$\left(1 - t^2\right)^2 f(u) = 32t; \tag{128}$$

where

$$t = \frac{T\overline{T}}{|T^2| + |T \star T|}\ . \tag{129}$$

In the first Ref. of [41] the function $t(u)$ corresponding to the BI Lagrangian density was found through an iterative procedure order by order in u; the first coefficients of its power series expansion were observed to match those of a particular generalized hypergeometric function, yielding to the conclusion that

$$t(u) = \frac{u}{32}\,{}_3F_2\left(\frac{1}{2},\frac{3}{4},\frac{5}{4};\frac{4}{3},\frac{5}{3};-\frac{u^2}{3^3\cdot 2^2}\right) \tag{130}$$

$$= \frac{u}{16}\sum_{k=0}^{\infty}(-1)^k\frac{(4k+1)!}{(3k+2)!k!}\left(\frac{u^2}{4^5}\right)^k . \tag{131}$$

The relation between the constitutive relations *à la Schrödinger* and those in terms of the $I(u)$ function follows from the relations between the T variables and the chiral $T^{\pm}$ variables. This allows to determine $f(u)$ from $I(u)$ and vice versa. Explicitly, given $f(u)$, one obtains $t = t(u)$ from (131); the function $I(u)$ is then given by

$$I(u) = \frac{1}{2}\int t(u)du. \tag{132}$$

4.6 Generalizations

Other non-linear (square-root) BI-like type Lagrangian densities in D dimensions have been suggested in [45, 46], starting from alternative $D = 4$ Goldstino Lagrangians studied in [12, 37, 47].

The first class of Lagrangians is given by D-dimensional generalizations containing pairs field strengths forms of degrees $p+1$ and $D-p-1$, corresponding to gauge fields coupling to $(p-1)$– and $(D-p-3)$– branes, respectively. These Lagrangians generalize the $D = 4$, $p = 2$ case, corresponding to a non-linear Lagrangian for a tensor multiplet regarded as an $\mathcal{N} = 2 \to 1$

Goldstino multiplet in rigid supersymmetry [47, 46]. They read

$$\begin{aligned}
\mathcal{L}\left(X, Y^2; \mu\right) &= \mu^2\left(1-\sqrt{1+\frac{1}{\mu^2}X-\frac{1}{\mu^4}Y^2}\right); & (133)\\
X &:= -\star\left(H_{p+1}\wedge\star H_{p+1}+V_{D-p-1}\wedge\star V_{D-p-1}\right); & (134)\\
Y &:= \star\left(H_{p+1}\wedge V_{D-p-1}\right), & (135)
\end{aligned}$$

and they have the property of being *doubly self-dual* under

$$V'_{D-p-1} = \star H_{p+1}; \quad H'_{p+1} = \star V_{D-p-1}. \tag{136}$$

Moreover, after a single duality the Lagrangian ends up containing two forms of the same degree, and it can be recast in the following manifestly $U(1)_{e.m.}$-invariant form [46]:

$$\mathcal{L}'\left(W, \bar{W}; \mu\right) = \mu^2\left[1-\sqrt{1+\frac{\left(W_{D-p-1}\,\bar{W}_{D-p-1}\right)}{\mu^2}+\frac{\left(W_{D-p-1}\,\bar{W}_{D-p-1}\right)^2-W^2_{D-p-1}\bar{W}^2_{D-p-1}}{4\mu^4}}\right]. \tag{137}$$

Actions with the full $U(n)$ (where n is the number of Maxwell vector fields) were proposed in [48], but they are currently not available in closed form, even for $n=2$.

On the other hand, we can present three relatively simple extensions of the BI Lagrangian to a pair of curvatures F^1 and F^2 whose field equations possess $U(1)$, $U(1)\times U(1)$ and $SU(2)$ duality symmetries. All these models solve subsets of the non-linear Gaillard-Zumino constraints (85) and (86) in the case $\Lambda, \Sigma = 1, 2$, *i.e.* of :

$$\begin{aligned}
F^1\star F^2+G^1\star G^2 &= 0; & (138)\\
F^1\star F^1+G^1\star G^1 &= 0; & (139)\\
F^2\star F^2+G^2\star G^2 &= 0; & (140)\\
F^1\star G^2-F^2\star G^1 &= 0. & (141)
\end{aligned}$$

The Lagrangian with $U(1)$ invariance reads

$$\mathcal{L} = \mu^2\left[1-\sqrt{1+\frac{(F^1)^2+(F^2)^2}{\mu^2}-\frac{\left(\star\left[F^1\wedge F^2\right]\right)^2}{\mu^4}}\right], \tag{142}$$

and was obtained as an application of double duality between two forms of the same degree. This $U(1)$ corresponds to the constraint (85) for $\Lambda=1$ and $\Sigma=2$, namely (138).

The other two examples are simply different complexifications of the two–field system, whose manifest $U(1)$ invariance corresponds to the single constraint of Eq. (86). In particular, letting

$$F = F^1 + i\,F^2\,, \tag{143}$$

the case with $U(1) \times U(1)$ invariance, where the second $U(1)$ constraint corresponds to the equation

$$F^1 \star F^1 + F^2 \star F^2 + G^1 \star G^1 + G^2 \star G^2 = 0, \tag{144}$$

reads

$$\mathcal{L} = \mu^2 \left[1 - \sqrt{1 + \frac{F\,\overline{F}}{\mu^2} - \frac{(\star\,[F \wedge \overline{F}])\,(\star\,[F \wedge \overline{F}])}{\mu^4}} \right] . \tag{145}$$

This is actually a particular case of the $U(n,n)$ dualities of [48] for $n = 1$, where in the absence of scalars $U(1,1)$ reduces to its maximal compact subgroup $U(1) \times U(1)$. Finally, the case with $SU(2)$ invariance reads

$$\mathcal{L} = \mu^2 \left[1 - \sqrt{1 + \frac{F\,\overline{F}}{\mu^2} - \frac{(\star\,[F \wedge F])\,(\star\,[\overline{F} \wedge \overline{F}])}{\mu^4}} \right] . \tag{146}$$

where the other two constraints correspond to the $SU(2)$ generators

$$F^1 \star F^1 - F^2 \star F^2 + G^1 \star G^1 - G^2 \star G^2 = 0, \tag{147}$$

$$F^1 \star F^2 + G^1 \star G^2 = 0. \tag{148}$$

From these examples it is manifest that the standard BI system admits several types of inequivalent complexifications that differ in their quartic couplings.

A fourth option, with maximal $U(2)$ duality, is naturally captured by the construction of [48], but its Lagrangian, as have been mentioned before, is not known in closed form.

All the above non-linear actions can be *massive* by introducing Green-Schwarz terms [45], *i.e.* couplings to another gauge field of the form

$$m H_{p+1} \wedge A_{D-p-1}. \tag{149}$$

A Stückelberg mechanism is then generated, and the non-linear mass terms turn out to take the same functional form as the original non-linear curvature terms; in other words, the $(D-p-1)$-form gauge field has been eaten to give mass to the original p-form gauge field (whose field strength is H_{p+1}). The simplest example of this class of Lagrangians is the $D = 4$ BI action used to make an antisymmetric field $B_{\mu\nu}$ massive : starting point is thus the master action

$$\begin{aligned} \mathcal{L} \;=\; & -\frac{k^2}{12}\, H_{\mu\nu\rho}\, H^{\mu\nu\rho} \;-\; \frac{m}{4}\, \epsilon^{\mu\nu\rho\sigma} B_{\mu\nu}\, F_{\rho\sigma} \\ & + \frac{\mu^2}{8\,g^2} \left[1 - \sqrt{1 + \frac{4}{\mu^2}\, F_{\mu\nu} F^{\mu\nu} - \frac{4}{\mu^4}\, (F_{\mu\nu} \star F^{\mu\nu})^2} \right] , \end{aligned} \tag{150}$$

where we have introduced a dimensionless parameter g, the counterpart of the parameter k that accompanies the two–form kinetic term. The parameter μ is the BI scale factor, with mass–squared dimension, which sizes the non–linear corrections.

As above, a massive variant of the BI action principle would be obtained by eliminating H after moving to a first–order form where it is unconstrained. However, as we have seen, the additional field is just a standard Stückelberg mode, so that for brevity we can just display the gauge–fixed Proca–like Lagrangian for the massive BI vector,

$$\mathcal{L} \;=\; -\frac{m^2}{2\,k^2}\, A_\mu\, A^\mu \;+\; \mathcal{L}_{BI}\,(g, \mu, F_{\mu\nu}) \;, \tag{151}$$

where

$$\mathcal{L}_{BI}\,(g, \mu, F) \;=\; \frac{\mu^2}{8\,g^2}\left[1 \;-\; \sqrt{1 \;+\; \frac{4}{\mu^2}\; F_{\mu\nu}\, F^{\mu\nu} \;-\; \frac{4}{\mu^4}\; (F_{\mu\nu} \,\star\, F^{\mu\nu})^2}\;\right] \;. \tag{152}$$

In the massless case, the self–duality of the BI Lagrangian would translate into the condition that

$$\mathcal{L}_{BI}\Big(g\,,\, \mu\,,\, F_{\mu\nu}(A)\Big) \;=\; \mathcal{L}_{BI}\left(g' = \frac{1}{g}\,,\, \mu' = \frac{\mu}{g^2}\,,\, F_{\mu\nu}(C)\right) \;, \tag{153}$$

where C is the dual gauge field. On the other hand, in the presence of the Green–Schwarz term $(m \neq 0)$, one can eliminate the vector altogether and work, in the dual formulation, solely in terms of the two–form $B_{\mu\nu}$. The self–duality of the massless BI theory then implies that the dual action involves a BI mass term and reads

$$\begin{aligned}\mathcal{L} \;&=\; -\,\frac{k^2}{12}\; H_{\mu\nu\rho}(B)\, H^{\mu\nu\rho}(B) \\ &+\; \frac{\mu^2}{8\,g^2}\left[1 \;-\; \sqrt{1 \;+\; \frac{4\,m^2\,g^4}{\mu^2}\; B_{\mu\nu}\, B^{\mu\nu} \;-\; \frac{4\,m^4\,g^8}{\mu^4}\; (B_{\mu\nu} \,\star\, B^{\mu\nu})^2}\;\right] \;. \end{aligned} \tag{154}$$

The massless limit can be recovered reintroducing the gauge invariant combination $m\,B \;+\; dC$ before letting $m \to 0$. In this fashion, the limiting Lagrangian describes a massless two–form, dual to a scalar, and a dual massless vector C.

This procedure could be relevant when the system is coupled to $\mathcal{N} = 2$ supergravity, in which case one of the two gravitino would belong to a massive spin-$\frac{3}{2}$ multiplet.

Acknowledgements

We are grateful to L. Andrianopoli, P. Aschieri, R. D'Auria, M. Porrati, R. Stora, S. Theisen, M. Trigiante, and especially to A. Sagnotti for useful discussions and/or collaborations.

References

[1] M. K. Gaillard and B. Zumino, *Duality Rotations for Interacting Fields*, Nucl. Phys. **B193**, 221 (1981).

[2] E. Majorana, *Teoria simmetrica dell'elettrone e del positrone*, Nuovo Cimento **14**, 171 (1937).

[3] R. D'Auria, S. Ferrara, M.A. Lledó, and V.S. Varadarajan, *Spinor Algebras*, J. Geom. Phys. **40** (2001) 101, hep-th/0010124.

[4] P. Minkowski, *$e \to \mu\gamma$ at a Rate of One Out of 1-Billion Muon Decays?*, Phys. Lett. **B67**, 421 (1977).

[5] M. Gell-Mann, P. Ramond, and R. Slansky, in : *"Supergravity"*, ed. by D. Freedman and P. Van Nieuwenhuizen, North Holland, Amsterdam (1979), pp. 315-321.

[6] W. Nahm, *Supersymmetries and their representations*, Nuclear Physics **B138**, 149 (1978).

[7] V. G. Kac, *Lie superalgebras*, Advances in Math. **26**, no. 1, 8 (1977).

[8] S. Ferrara and B. Zumino, *Transformation Properties of the Supercurrent*, Nucl. Phys. **B87**, 207 (1975)

[9] A. Salam and J. A. Strathdee, *On Goldstone Fermions*, Phys. Lett. **B49**, 465 (1974).

[10] P. Fayet and J. Iliopoulos, *Spontaneously Broken Supergauge Symmetries and Goldstone Spinors*, Phys. Lett. **B51**, 461 (1974).

[11] J. Bagger and A. Galperin, *A New Goldstone multiplet for partially broken supersymmetry*, Phys. Rev. **D55**, 1091 (1997), hep-th/9608177.

[12] J. Bagger and A. Galperin, *The Tensor Goldstone multiplet for partially broken supersymmetry*, Phys. Lett. **B412**, 296 (1997), hep-th/9707061.

[13] J. Hughes and J. Polchinski, *Partially Broken Global Supersymmetry and the Superstring*, Nucl. Phys. **B278**, 147 (1986). J. Hughes, J. Liu and J. Polchinski, *Supermembranes*, Phys. Lett. **B180**, 370 (1986).

[14] S. Ferrara, L. Girardello and M. Porrati, *Spontaneous breaking of $\mathcal{N}=2$ to $\mathcal{N}=1$ in rigid and local supersymmetric theories*, Phys. Lett. **B376**, 275 (1996), hep-th/9512180.

[15] I. Antoniadis, H. Partouche and T. R. Taylor, *Spontaneous breaking of $\mathcal{N}=2$ global supersymmetry*, Phys. Lett. **B372**, 83 (1996), hep-th/9512006.

[16] L. Andrianopoli, R. D'Auria, S. Ferrara and M. Trigiante, *Observations on the partial breaking of $\mathcal{N}=2$ rigid supersymmetry*, Phys. Lett. **B744**, 116 (2015) 116, arXiv:1501.07842 [hep-th].

[17] L. Andrianopoli, P. Concha, R. D'Auria, E. Rodriguez, and M. Trigiante, *Observations on BI from $\mathcal{N}=2$ Supergravity and the General Ward Identity*, arXiv:1508.01474 [hep-th].

[18] G.R. Dvali and M. A. Shifman, *Domain walls in strongly coupled theories*, Phys. Lett. **B396**, 64 (1997), Phys. Lett. **B407**, 452 (1997), hep-th/9612128.

[19] S. Ferrara and M. Porrati, *Central extensions of supersymmetry in four-dimensions and three-dimensions*, Phys. Lett. **B423**, 255 (1998), hep-th/9711116.

[20] D. V. Volkov and V. P. Akulov, *Is the Neutrino a Goldstone Particle?*, Phys. Lett. **B46**, 109 (1973).

[21] S. Ferrara, M. Porrati, and A. Sagnotti, *$\mathcal{N}=2$ Born-Infeld attractors*, JHEP **1412**, 065 (2014), arXiv:1411.4954 [hep-th].

[22] S. Ferrara and A. Sagnotti, *Some Pathways in non-Linear Supersymmetry: Special Geometry Born-Infeld's, Cosmology and dualities*, arXiv:1506.05730 [hep-th].

[23] S. Ferrara, M. Porrati, A. Sagnotti, R. Stora, and A. Yeranyan, *Generalized Born–Infeld Actions and Projective Cubic Curves*, Fortsch. Phys. **63**, 189 (2015), arXiv:1412.3337 [hep-th].

[24] P. Aschieri, S. Ferrara, and B. Zumino, *Duality Rotations in Nonlinear Electrodynamics and in Extended Supergravity*, Riv. Nuovo Cim. **31**, 625 (2008), arXiv:0807.4039 [hep-th].

[25] P. Aschieri, S. Ferrara and S. Theisen, *Constitutive relations, off shell duality rotations and the hypergeometric form of Born-Infeld theory*, Springer Proc. Phys. **153**, 23 (2014), arXiv:1310.2803 [hep-th].

[26] S. Deser and C. Teitelboim, *Duality Transformations of Abelian and Nonabelian Gauge Fields*, Phys. Rev. **D13**, 1592 (1976).

[27] M. Born and L. Infeld, *Foundations of the New Field Theory*, Proc. R. Soc. London sez. **A 144**, 425 (1934).

[28] S. Deser and R. Puzalowski, *Supersymmetric Nonpolynomial Vector Multiplets and Causal Propagation*, J. Phys. **A13**, 2501 (1980).

[29] S. Cecotti and S. Ferrara, *Supersymmetric Born-Infeld Lagrangians*, Phys. Lett. **B187**, 335 (1987).

[30] J. Broedel, J. J. M. Carrasco, S. Ferrara, R. Kallosh, and R. Roiban, *$\mathcal{N}=2$ Supersymmetry and U(1)-Duality*, Phys. Rev. **D85**, 125036 (2012), arXiv:1202.0014 [hep-th].

[31] W. Chemissany, S. Ferrara, R. Kallosh, and C. S. Shahbazi, *$\mathcal{N}=2$ Supergravity Counterterms, Off and On Shell*, JHEP **1212**, 089 (2012), arXiv:1208.4801 [hep-th].

[32] E. Bergshoeff, F. Coomans, R. Kallosh, C.S. Shahbazi, and A. Van Proeyen, *Dirac-Born-Infeld-Volkov-Akulov and Deformation of Supersymmetry*, JHEP **1308**, 100 (2013), arXiv:1303.5662 [hep-th].

[33] S. Bellucci, S. Krivonos, A. Shcherbakov, and A. Sutulin, *On the road to $\mathcal{N}=2$ supersymmetric Born-Infeld action*, Phys. Lett. **B721**, 353 (2013), arXiv:1212.1902 [hep-th].

[34] L. Andrianopoli, R. D'Auria and M. Trigiante, *On the dualization of Born-Infeld theories*, Phys. Lett. **B744**, 225 (2015), arXiv:1412.6786 [hep-th].

[35] R. Casalbuoni, S. De Curtis, D. Dominici, F. Feruglio and R. Gatto, "Nonlinear Realization of Supersymmetry Algebra From Supersymmetric Constraint," Phys. Lett. B **220** (1989) 569.

[36] Z. Komargodski and N. Seiberg, "From Linear SUSY to Constrained Superfields," JHEP **0909** (2009) 066 [arXiv:0907.2441 [hep-th]].

[37] M. Rocek and A. A. Tseytlin, *Partial breaking of global $D = 4$ supersymmetry, constrained superfields, and three-brane actions*, Phys. Rev. **D59**, 106001 (1999), hep-th/9811232.

[38] A. Tseytlin, *Selfduality of Born-Infeld action and Dirichlet three-brane of type IIB superstring theory*, Nucl. Phys. **B469**, 51 (1996), hep-th/9602064.

[39] E. A. Ivanov and B. M. Zupnik, *New approach to nonlinear electrodynamics: Dualities as symmetries of interaction*, Phys. Atom. Nucl. **67**, 2188 (2004) [Yad. Fiz. **67**, 2212 (2004)] hep-th/0303192.

[40] G. Bossard and H. Nicolai, *Counterterms vs. Dualities*, JHEP **1108**, 074 (2011) arXiv:1105.1273 [hep-th].

[41] J. J. M. Carrasco, R. Kallosh, and R. Roiban, *Covariant procedures for perturbative non-linear deformation of duality-invariant theories*, Phys. Rev. **D85**, 025007 (2012), arxiv:1108.4390 [hep-th]. W. Chemissany, R. Kallosh and T. Ortin, *Born-Infeld with Higher Derivatives*, Phys. Rev. **D85**, 046002 (2012), arXiv:1112.0332 [hep-th].

[42] P. Aschieri and S. Ferrara, *Constitutive relations and Schroedinger's formulation of nonlinear electromagnetic theories*, JHEP **1305**, 087 (2013), arXiv:1302.4737 [hep-th].

[43] E. Schrödinger, *Contributions to Born's New Theory of the Electromagnetic Field*, Proc. R. Soc. London, sez. **A 150**, 465 (1935).

[44] M. K. Gaillard and B. Zumino, *Nonlinear electromagnetic self-duality and Legendre transformations*, hep-th/9712103.

[45] S. Ferrara and A. Sagnotti, *Massive Born–Infeld and Other Dual Pairs*, JHEP **1504**, 032 (2015), arXiv:1502.01650 [hep-th].

[46] S. Ferrara, A. Sagnotti and A. Yeranyan, *Doubly Self-Dual Actions in Various Dimensions*, JHEP **1505**, 051 (2015), arXiv:1503.04731 [hep-th].

[47] S. M. Kuzenko and S. Theisen, *Supersymmetric duality rotations*, JHEP **0003**, 034 (2000), hep-th/0001068.

[48] P. Aschieri, D. Brace, B. Morariu, and B. Zumino, *Nonlinear selfduality in even dimensions*, Nucl. Phys. **B574**, 551 (2000), hep-th/9909021.

LECTURER: S. FERRARA

Majorana Fermions, Supersymmetry Breaking and Born-Infeld Theory

Scientific Secretaries: Nikita Sopenko, Anton Sokolov

DISCUSSION I

- *Campbell:*
In your talk you mentioned that the number of gravitino degrees of freedomin D dimension is given by the formula (D-3)*2^{(D-3)/2}. Could you please explain what happens in D=3 ?

- *Ferrara:*
Well, it is a very good question. In three dimensions the graviton has zero degrees of freedom. The same is implied by supersymmetry for the gravitino, the spin-3/2 supersymmetric partner of the graviton. That is why the formula of the gravitino degrees of freedom is proportional to D-3.

- *Campbell:*
Is there no gravity in D=3 ?

- *t'Hooft:*
It's only topological!

- *Ferrara:*
Exactly! There are no graviton degrees of freedom pertaining to a massless particle of spin 2. Gravity in three dimensions is only topological.

- *Ferreira:*
I would like to ask a really basic question: why doesn't anyone ever deal with particles of spin higher than 2 ?

- *Ferrara:*
Well, in gravity there is the problem of making a consistent coupling of a spin-2 particle (graviton) with spin-3/2 particles. In fact, the second paper on supergravity by Deser and Zumino dealt with "consistent supergravity", exactly because they considered a consistent coupling of gravity with spin-3/2 particles. In electromagnetism, in order to couple matter to the electromagnetic field one must have a conserved U(1) current, which comes from the equation of motion of the Maxwell field as an identity. In gravity, the analogous constraint is provided by the divergence-less nature of the stress-energy tensor. In this morning lecture, I have been mentioning that these equations play an important role in the Friedman-Lemaître metric for the Universe. When one tries to couple particles with spin bigger than 2, one finds that there is an inconsistency in the sense that the constraints given by gauge invariance impose too many conditions on the matter sector, and the theory becomes inconsistent.

String theory apparently is the way out, because it is not made of point-like objects, but rather it is a theory of extended objects.

- *Sagnotti:*

I would like to add the following comment. If one wants to couple higher spin particles, some form of infrared cutoff is needed. In string theory, this infrared cutoff is provided by masses, sized by the string tension itself.

- *Ferrara*

Yes. Let me point out that if one is obliged to have only particles such as a massless graviton and lower spin particles, then from the analysis of the representation of supersymmetry, initiated some time ago by Werner Nahm, one obtains the following bound on the number of dimensions of spacetime: they cannot exceed 11. So, the nice thing is that there is a finite number of theories in higher dimensions because of this constraint.

- *Duff:*

As you know, one can still keep the bound of 128 fermionic degrees of freedom by uplifiting to twelve dimensions with signature (10, 2). Do you think there is any physics associated with that?

- *Ferrara:*

The problem is that the supersymmetry algebra is not the standard one in that case. In fact, there is no momentum on the right-hand side of the anticommutator of the supersymmetric generators. Therefore, the physical meaning of such a non-standard supersymmetry is unclear.

- *Korthals-Altes:*

In your lecture of this morning, you mentioned the Lagrangian structure of the eleven-dimensional supergravity. Could you please comment a little bit on that?

- *Ferrara:*

First of all, in any theory of supergravity the number of bosons and the number of fermions must match, due to supersymmetry. In four dimensions, the graviton has two helicity states (namely, +2 and -2), while the gravitino has two helicity states (i.e., +3/2 and -3/2), and therefore the balance of degrees of freedom is reached. Note that such a balance is not a sufficient but rather a necessary condition. If the same counting is worked out in eleven dimensions, one finds that the graviton has much less degrees of freedom of the gravitino, which has 128 components. Thus, one must have further bosonic fields accounting for the missing degrees of freedom. As first discovered by Cremmer, Julia and Scherk, these are precisely accounted for by a 3-form potential, whose field strength 4-form enter the supergravity Lagrangian. This 3-form potential is essential in order to have M2-brane and M5-brane solutions in the equations of motion of eleven dimensional supergravity.

LECTURER: S. FERRARA

Majorana fermions, Supersymmetry Breaking and Born-Infeld Theory

Scientific Secretaries: Ilya Vilkoviskiy, David Frenklakh

DISCUSSION II

- *Prochazka:*

At some point in the lecture of this morning, you introduced a cutoff "f" for the Born-Infield Lagrangian. I was wondering what sort of cutoff is that, since it is not in the momentum space. Could you please explain it ?

- *Ferrara:*

The presence of such a cutoff implies that the Lagrangian provides a reliable description for energies smaller than the cutoff. In this sense, Born-Infeld theory is similar to the Fermi theory of weak interactions: at energies much smaller than the Fermi scale or than the vector boson masses, the usual theory becomes effective and non-renormalizable (since the operator of the Fermi interaction has dimension 6, there exists a scale $1/m^2$ dubbed the Fermi scale). When inserted in a supersymmetric context, the Born-Infeld cutoff "f" is related to the supersymmetry breaking scale.

- *Prochazka:*

So, could one guess that the ultraviolet completion of these non-linear theories should be provided by the full-fledged supergravity and superstring theory ?

- *Ferrara :*

Yes.

- *Popov :*

Is the Born-Infeld theory renormalizable?

- *Ferrara:*

No, it is not. It is effective theory, just like the Fermi theory of weak interactions.

- *Zichichi:*

In my view, effective theories are used by theorists when there is not a precise knowledge of the physical phenomenon to be described.

- *Ferrara:*

Well, I would say that effective theories are a powerful tool in contemporary high-energy physics. Indeed, we know the physical only within a precise range of energies. Thus, it is very important to have a theoretical description of physics which is energy-dependent, being valid only for energies smaller than some energy scale ("cutoff").

- *Zichichi:*

As I sad in my opening speech, sixty years ago the reason why the anomalous magnetic moment of the muon needed to be measured at a Fermi level was because only an effective theory of electromagnetism was available at that time, assuming that the muon, 200 times heavier than the electron, had some interaction. Do you know how much was the cutoff of such a theory? 1 Gev. The young fellows attending the school should not get confused; when we do not know what to do, we say we have an effective theory.

- *'t Hooft:*

But, all in all, all our theories are effective theories.

- *Zagoskin:*

According to your lecture of this morning, in the Born-Infeld theory D=2 ∂L/∂E and H=-2 ∂L/∂E, but for the simplest electromagnetic Lagrangian this formula yield D=2E, H=-2E. Should we rather leave out '2' from both equations, and substitute 'B' by 'E' in the second one?

- *Ferrara:*

Sorry, there is a typo in the second formula, but the overall factor 2 depends on the normalization of the Lagrangian itself.

- *Sola:*

What's the status of the Cosmological Constant Problem in supergravity? In the case of global supersymmetry, there is exactly zero vacuum energy. On the other hand, it is known that in supergravity some amount of fine-tune can be performed, and in principle the situation can be improved. However, I would like you to explain why this is not so easy.

- *Ferrara:*

In order to explain the situation, let me consider what happens to the transformation law of the fermions in supergravity. Ward identities for spin-1/2 fields "χ " consist of a term which is δ^A_5 times the potential plus another term, which in fact has a minus sign, and is related to the gravitino mass matrices :

$$\delta\chi^A \delta\chi_5 = \delta^A_5\, V - f^C_5 f^A_C$$

Thus, unlike in global supersymmetry, in supergravity a cancellation between these two terms in the right-hand side can occur, if superymmetry is spontaneously broken or broken. Such a cancellation allows for a zero vacuum energy even if supersymmetry is spontaneously broken, because there is a negative contribution due to gravitino spin-3/2 fields; in other words, in supergravity one can have vanishing cosmological constant also in presence of spontaneous supersymmetry breaking. This is actually realized by the so-called no-scale supergravity models, in which the cosmological constant is zero from the beginning because there is an exact cancellation between a negative term and a positive term in the scalar potential. In general, the problem of the cosmological constant is not solved in supergravity, due to the presence of fine-tuning issues, whose mathematical solution is currently unknown.

- *Sola:*

But, in fact, it's not only that cosmological constant vanishes, because in the formula you showed there is this second term with the negative sign. Thus, it could be that the ground state does not coincide with the point at which the cosmological constant vanishes. Even if the two terms are adjusted to reciprocally cancel, it could still be that the vacuum state is lower, implying that it is not a physical point.

- *Ferrara:*

Well, this does depend on the very model under consideration. Criticality of the point must be checked case by case. In the no-scale the scalar potential is semi-positive definite, so the point of zero potential is indeed a critical point.

- *Minkowski:*

As far as I understood, the scale of supersymmetry breaking might be determined by studying the contributions of supersymmetric particles in scattering processes, such as neutrino – gamma scattering. Could you please comment on this ?

- *Ferrara:*

Well, as to my knowledge, I think this is what was investigated in the early times after the discovery of supersymmetry, using low-energy theorems to see processes involving gravitinos.

- *Knoops:*

If there is a cancellation between the F-term and D-term to obtain a vanishing cosmological constant, is this invariant under quantum corrections?

- *Ferrara:*

In supergravity a cancellation between F- and D-terms is not needed for a vanishing cosmological constant; in fact, D-term can vanish itself, and F-term doesn't give a quantitative contribution. Local supersymmetry may allow anti-deSitter vacua, and unbroken supersymmetry can coexist with non-zero cosmological constant, provided that the cosmological constant is negative; this is what is called anti-deSitter supersymmetry. On the other hand, a theorem prevents de Sitter vacua to be supersymmetric. Thus,there exist two classes of supersymmetric vacua:one is anti-deSitter (with negative cosmological constant) and the other is Minkowski (with vanishing cosmological constant).

- *'t Hooft:*

In presence of supersymmetry breaking terms, such as soft supersymmetry breaking terms, couldn't the cosmological constant be softer than them ?

- *Ferrara:*

Indeed, this could happen, because supersymmetry does not explain the value of the cosmological constant.

- *'t Hooft:*

But didn't you just say that the cosmological constant should be negative ?

- *Ferrara:*

No, the unique constraint implied by local supersymmetry is that the cosmological constant will never exceed the fourth power of the Planck mass, generally being proportional to the scale of the soft supersymmetry breaking masses. But the observed "almost zero" value of the cosmological constant cannot be predicted from supersymmetry. Therefore, supersymmetry does not solve the cosmological constant problem.

- *'t Hooft:*

So a priori there is no constraint on the sign of the cosmological constant ?

- *Ferrara:*

If supersymmetry is broken, the three possible classes of vacua (de Sitter, Minkowski, Anti de Sitter) can exist. When supersymmetry is unbroken, deSitter vacua cannot occur. Of course, in inflationary models, supersymmetry may be consistent with a deSitter phase during inflation. But it can be shown that, outside inflation, supersymmetry is not consistent with de Sitter vacua.

Gravity as the square of Yang-Mills?

L. Borsten[1] and M. J. Duff[2]

[1]*School of Theoretical Physics, Dublin Institute for Advanced Studies, 10 Burlington Road, Dublin 4, Ireland*

[2]*Theoretical Physics, Blackett Laboratory, Imperial College London, London SW7 2AZ, United Kingdom*

leron@stp.dias.ie
m.duff@imperial.ac.uk

ABSTRACT

In these lectures we review how the symmetries of gravitational theories may be regarded as originating from those of "Yang-Mills squared". We begin by motivating the idea that certain aspects of gravitational theories can be captured by the product, in some sense, of two distinct Yang-Mills theories, particularly in the context of scattering amplitudes. We then introduce a concrete dictionary for the covariant fields of (super)gravity in terms of the product of two (super) Yang-Mills theories. The dictionary implies that the symmetries of each (super) Yang-Mills factor generate the symmetries of the corresponding (super)gravity theory: general covariance, p-form gauge invariance, local Lorentz invariance, local supersymmetry, R-symmetry and U-duality.

Contents

1 Introduction

The idea that gravitational physics can be understood in terms of gauge theory has reoccured a number of times, in a variety of guises. The most conceptually straight-forward approach is to regard gravity as the gauge theory of Lorentz, Poincaré or de Sitter symmetries [1–5]. The holographic principle [6, 7], concretely realised through the AdS/CFT correspondence [8–10], represents a more subtle realisation of this notion, with profound consequences for our understanding of both gauge and gravity theories. Here we appeal to a third and, at least superficially, independent incarnation:

$$\text{gravity} = \text{Yang-Mills} \times \text{Yang-Mills}. \tag{1.1}$$

At first sight this is a radical proposal; Einstein's general relativity describes gravity as the dynamics of spacetime, while Yang-Mills theories, as used to describe the strong, weak and electromagnetic forces, play out *on* spacetime. General relativity and Yang-Mills theory are seemingly worlds apart in almost every regard, from their fundamental degrees freedom to their basic symmetries. In particular, the Yang-Mills theories underlying the standard model are renormalisable, predictive quantum field theories, in stark contrast to perturbative quantum gravity.

Despite their differences, however, there already exist some fascinating hints that gravity, at least in some regimes, may be related to the square of Yang-Mills theory. String theory provided the first example in the form of the Kawai-Lewellen-Tye (KLT) relations, which connect tree-level amplitudes of closed strings to sums of products of open string amplitudes [11]. More recently, invoking Bern-Carrasco-Johansson (BCJ) colour-kinematic duality [12] it has been conjectured [13] that the on-mass-shell momentum-space scattering amplitudes for gravity are the "double-copy" of gluon scattering amplitudes in Yang-Mills theory to all orders in perturbation theory.

The recent renaissance in amplitude calculations has been principally driven by the "on-shell paradigm". Starting with Lagrangian field theory we learnt how to compute simple amplitudes to low orders in perturbation theory. The factorial growth in complexity with loop order quickly renders traditional approaches impractical. Searching for computational efficiency, over time various generic amplitude structures (on-shell recursion relations, generalised unitarity cuts, Grassmannians, scattering equations ...) were uncovered, eventually allowing the Lagrangian ladder to be kicked away. For an overview of these developments see [14]. This freedom led to the discovery of new features of amplitudes, not visible from the original Lagrangian perspective. BCJ colour-kinematic duality falls into this class of surprises. Conversely, having climbed so high we can no longer see where we can from; the full significance and implications of BCJ duality remain unclear. Can we climb back down by some other route and understand the origin of these remarkable dualities? The basic idea reviewed here is to build a dictionary expressing the covariant fields of (super) gravity as the product, in a well-defined sense, of two arbitrary (super) Yang-Mills theories.

1.1 Motivation

We begin by sketching the BCJ colour-Kinematic duality and the double copy procedure [12, 13, 15]. For a more detailed account of this topic the reader is referred to the reviews [14, 16]. This will not only better motivate (1.1), but also inform our field theory constructions in the subsequent sections.

Let us consider the n-point L-loop amplitude of Yang-Mills theory with an arbitrary gauge group. Converting all four-point contact terms into s, t or u channel trivalent pole diagrams by inserting propagators $1 = s/s = t/t = u/u$, we have,

$$A_n^L = i^L g^{n-2+2L} \sum_{i \in \text{trivalent graphs}} \int \prod_{l=1}^{L} \frac{d^D p_l}{(2\pi)^D S_i} \frac{c_i n_i}{\prod_{a_i} p_{a_i}^2}. \tag{1.2}$$

The sum is over all n-point L-loop graphs i with only trivalent vertices. c_i denotes the kinematic factor of graph i, composed of gauge group structure constants. n_i denotes the kinematic factor of graph i. It is a polynomial of Lorentz-invariant contractions of polarisation vectors and momenta. The $p_{a_i}^2$ are the propagators for each graph i. S_i is the dimension of the automorphism group of graph i.

The set of n-point trivalent graphs can be organised into triples i, j, k such that they differ in only one propagator. For such a triple the three disctint propagators are embedded in the same graph, connected to the same four incoming edges, but in the s, t, u channel for (say) i, j, k, respectively. For such a triple the colour factors will obey a Jacobi identity

$$c_i + c_j + c_k = 0 \tag{1.3}$$

and consequently the generalised gauge transformations

$$n_i \to n_i + s\Delta, \qquad n_j \to n_j + t\Delta, \qquad n_k \to n_k + u\Delta, \tag{1.4}$$

leave the amplitude (1.2) invariant [12]. It was proposed in [12] that one can arrange the diagrams, using the generalised gauge transformations if necessary, to display a colour-kinematic duality:

$$\begin{aligned} c_i = -c_j \quad &\Rightarrow \quad n_i = -n_j \\ c_i + c_j + c_k = 0 \quad &\Rightarrow \quad n_i + n_j + n_k = 0 \end{aligned} \tag{1.5}$$

A reorganisation admitting this surprising relationship between colour and kinematic data was shown to exist for all n-point tree-level amplitudes in [15]. Although there is as yet no proof, the colour-kinematic duality is conjectured to hold, with highly non-trivial evidence [17, 18], at any loop level, thus going beyond the KLT relations [12, 13]. While it is clear that the colour factors should obey Jacobi identities (by definition), it is not at all obvious that the kinematic factors should play by the same rules!

This suggests that there is in fact some underlying kinematic algebra mirroring the properties of conventional Lie algebras, as described in [19,20]. In general, this hidden algebra cannot be made manifest at the Lagrangian level, however for the self-dual sector it can be identified as a diffeomorphism Lie algebra, which determines the kinematic numerators of generic tree-level maximally helicity violating amplitudes [19]. Important features of the BCJ construction can also be derived from string theory. In particular, the BCJ relations [12] (which we have not discussed) have been obtained via monodromy relations [21–23]. Moreover, explicit expressions for colour-kinematic duality respecting local tree-level numerators at n-points have been derived using the pure spinor approach to string theory amplitudes and dimensional reduction [24,25]. The string theoretic approach is also suited to loop level calculations. For example, colour-kinematic duality respecting numerators at five points with one and two loops were obtained in [26,27], while at arbitrary multiplicity one-loop maximally helicity violating colour-kinematic duality respecting numerators have been constructed in [28].

More remarkable still is the double-copy prescription [12, 13, 15]. Assuming one has found a colour-kinematic duality respecting representation of the n-point L-loop gluon amplitude, the equivalent n-point L-loop graviton amplitude is obtained by simply replacing each colour factor, c_i, with a second kinematic factor, $\tilde{n}_i$, as depicted in Figure 1. Examining the unitary cuts of the gravity amplitude obtained via the double-copy is sufficient to prove it reproduces the correct result, assuming colour-kinematic duality is satisfied in one of the Yang-Mills factors. These ideas are seamlessly extended to supersymmetric theories. In particular, the square of the amplitudes of the maximally supersymmetric $\mathcal{N} = 4$ super Yang-Mills theory yield amplitudes of the maximally supersymmetric $\mathcal{N} = 8$ supergravity theory.

The double-copy picture is not only conceptually compelling but also computationally powerful, bringing previously intractable calculations with in reach. This has pushed forward dramatically our understanding of divergences in perturbative quantum gravity, revealing a number of unexpected features and calling into question previously accepted arguments regarding finiteness. For instance, the four-point graviton amplitude in $\mathcal{N} = 8$ supergravity has been shown to be finite to four loops [17], contradicting some early expectations [29]. In particular, the existence of a supersymmetric R^4 counter-term in $\mathcal{N} = 1$ supergravity at three loops was established in [30], although it was already noted there that this result may

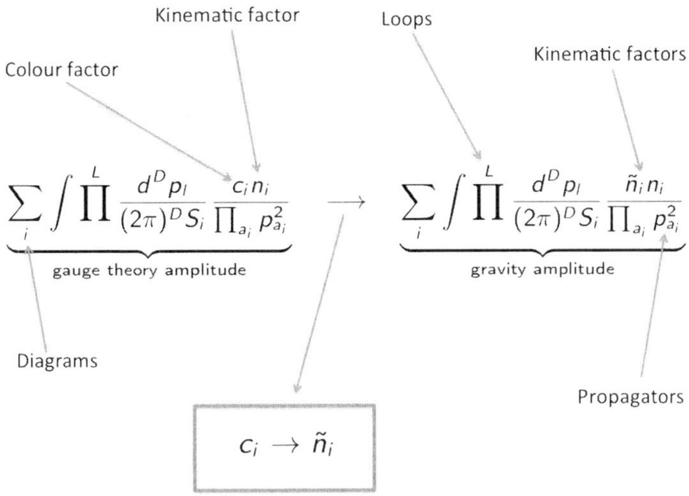

Figure 1: The double-copy procedure. Assuming the gauge theory amplitude on the left has been arranged to display colour-kinematic duality then the gravity amplitude on the right is straight-forwardly obtained by replacing the colour factors with a second copy of the kinematic factors. Note, the second factor does not have correspond to the same Yang-Mills theory. The (supressed) Yang-Mills coupling constants must be replaced by the gravitational coupling constant $g \to \kappa/2$, where $\kappa^2 = 16\pi G_N$.

not hold when more supersymmetry is considered. It has since been shown that the four-loop cancellation can be accounted for by supersymmetry and $E_{7(7)}$ U-duality [31–34]. The consensus, however, is that at seven loops any would-be cancellations cannot be "consequences of supersymmetry in any conventional sense" [31]. Unfortunately, seven loops in $\mathcal{N} = 8$ supergravity remains beyond reach but by decreasing the amount of supersymmetry these arguments apply at lower loop order. Indeed, the four-point amplitude of $D = 4, \mathcal{N} = 5$ supergravity has been shown to be finite to four loops, contrary to all expectations based on standard symmetry arguments [18]. There are "enhanced cancellations" at work and the conclusion that $\mathcal{N} = 8$ supergravity will diverge at seven loops is thrown into doubt. Although the majority opinion is that $\mathcal{N} = 8$ supergravity will diverge at some loop order, there is something deeper at work we have yet to understand fully and question remains very much open.

2 Covariant field dictionary

These developments raise the question: to what extent, or in what sense, can one regard gravity as the square of Yang-Mills? Is there a deeper connection underlying the amplitude relations? One approach to addressing such questions is to build a dictionary at the level of fields, as opposed to on-shell states or amplitudes. In a sense this runs contrary to the "on-shell paradigm" that took us here. Going back off-shell may nonetheless be instructive. This approach has been examined at the level of Lagrangians in [13, 19] and classical solutions in [35]. Here instead we focus on expressing the covariant fields of (super)gravity in terms of the product of (super) Yang-Mills fields. The first consistency check such a dictionary must pass is at the level of symmetries. As we shall review, the gravitational symmetries of general covariance, p-form gauge invariance, local supersymmetry and local chiral symmetry, R-symmetry and U-duality follow from those of Yang-Mills at linearised approximation.

Much of the squaring literature invokes a mysterious product:

$$A_\mu(x)\text{“}\otimes\text{”}\tilde{A}_\nu(x). \tag{2.1}$$

Here, A_μ and $\tilde{A}_\nu$ are the gauge potentials of two distinct Yang-Mills theories, which we will refer to as left (no tilde) and right (tilde), respectively. They can have arbitrary and independent non-Abelian gauge groups G and $\tilde{G}$. Reading off the meaning of "$\otimes$" from the tensor product branching rules of the appropriate spacetime little group representations or corresponding string states one can consistently match the symmetries. See in particular [36, 37]. Here, we are instead seeking a concrete definition of "$\otimes$" at the level of field theory which is valid whether or not there is an underlying string interpretation. This raises two immediate questions: (i) where do the gauge indices go? (ii) does it obey the Leibnitz rule?

Guided by the structure of the amplitude relations and requirements of symmetry we introduced a covariant product rule in [38]:

$$f\text{“}\otimes\text{”}g := f\star\Phi\star g. \tag{2.2}$$

Let us review the ingredients in (2.2). The $\star$ product denotes a convolutive inner tensor product with respect to the Poincaré group combined with a Killing form $\langle\ ,\ \rangle : \mathfrak{g}\otimes\mathfrak{g}\to\mathbb{R}$,

$$[f\star g](x) = \int d^Dy\langle f(y), g(x-y)\rangle. \tag{2.3}$$

We have further introduced the "spectator" field Φ, a $G\times\tilde{G}$ bi-adjoint valued scalar. The convolution reflects the fact that the amplitude relations are multiplicative in momentum space. It turns out to be essential for reproducing the local symmetries of (super)gravity from those of the two (super) Yang-Mills factors. The Killing form accounts for the gauge groups, while the spectator field allows for arbitrary and independent G and $\tilde{G}$. It fact, the appearance of Φ is quite natural from the perspective of amplitude relations. Its necessity was identified by Hodges in the context of twistor-theory [39]. From the perspective presented in subsection 1.1, rather than sending $c_i\to\tilde{n}_i$, doubling the kinematics and removing the colour, one could also send $n_i\to\tilde{c}_i$, doubling the colour and removing the kinematics. In [40, 41] this was shown at tree-level to yield the amplitudes of a global $G\times\tilde{G}$ bi-adjoint scalar field theory with cubic interaction term,

$$\mathcal{L}_{\text{int}} = -f_{ijk}\tilde{f}_{i'j'k'}\Phi^{ii'}\Phi^{jj'}\Phi^{kk'}. \tag{2.4}$$

The transformation rules of Φ are fixed by this theory. A scalar field also appeared independently, but in close analogy to our spectator field, in the double-copy construction of Kerr-Schild gravity solutions from Yang-Mills solutions in [35].

3 $\mathcal{N}=1$ supergravity

Having introduced the covariant product, let us now work through the simplest example exhibiting all the local symmetries of interest. We consider the product of a left $\mathcal{N}=1$ and a right $\tilde{\mathcal{N}}=0$ theories at linearized level:

- Off-shell $\mathcal{N}=1$ Yang-Mills multiplet with (4+4) bosonic + fermionic degrees of freedom and gauge group G:

$$A_\mu,\qquad \psi,\qquad D \tag{3.1}$$

- Off-shell off-shell $\tilde{\mathcal{N}}=0$ Yang-Mills multiplet with (3+0) bosonic + fermionic degrees of freedom and gauge group $\tilde{G}$:

$$\tilde{A}_\nu \tag{3.2}$$

Without making any assumptions regarding the dynamics this yields the (12+12) new-minimal $\mathcal{N}=1$ supergravity multiplet [42]:

$$g_{\mu\nu}, \qquad B_{\mu\nu}, \qquad \psi_\mu, \qquad V_\mu \tag{3.3}$$

where general covariance, 2-form gauge invariance, local supersymmetry and local chiral symmetry follows from the left/right gauge symmetries.

The "gravity=Yang-Mills×Yang-Mills" dictionary and symmetry transformations are most concisely expressed in the superfield formalism. Hence, we consider:

1. A left $\mathcal{N}=1$ real vector superfield,

$$\begin{aligned} V(x,\theta,\bar\theta) =& C + i\theta\chi - i\bar\theta\bar\chi + i\theta^2 F - i\bar\theta^2\bar F - \theta\sigma^\mu\bar\theta A_\mu \\ &+ i\theta^2\bar\theta\left(\bar\psi + \frac{i}{2}\bar\sigma^\rho\partial_\rho\chi\right) - i\bar\theta^2\theta\left(\psi + \frac{i}{2}\sigma^\rho\partial_\rho\bar\chi\right) \\ &+ \frac{1}{2}\bar\theta^2\theta^2\left(D + \frac{1}{2}\Box C\right) \end{aligned} \tag{3.4}$$

transforming under local supergauge, non-Abelian global G and global super-Poincaré:

$$\delta V = \underbrace{\Lambda + \bar\Lambda}_{\text{local Abelian supergauge}} + \overbrace{[V, X]}^{\text{global non-Abelian } G} + \underbrace{\delta_{(a,\lambda,\epsilon)} V}_{\text{global super-Poincaré}} \tag{3.5}$$

where $\Lambda(x,\theta,\bar\theta)$ is a chiral superfield of supergauge parameters

$$\Lambda(x,\theta,\bar\theta) = B + \sqrt{2}\theta\zeta + \theta^2 K + i\theta\sigma^\rho\bar\theta\partial_\rho a + \frac{i}{\sqrt{2}}\theta^2\bar\theta\bar\sigma^\rho\partial_\rho\zeta + \frac{1}{4}\theta^2\bar\theta^2\Box B \tag{3.6}$$

2. A right $\tilde{\mathcal{N}}=0$ Yang-Mills potential $\tilde A_\nu$ transforming under local gauge, non-Abelian global $\tilde G$ and global Poincaré:

$$\delta\tilde A_\nu = \underbrace{\partial_\nu\tilde\sigma}_{\text{local Abelian gauge}} + \overbrace{[\tilde A_\nu, \tilde X]}^{\text{global non-Abelian } \tilde G} + \underbrace{\delta_{(a,\lambda)} A_\nu}_{\text{global Poincaré}}$$

3. The spectator bi-adjoint scalar Φ field transforming under non-Abelian global $G\times\tilde G$ and global Poincaré:

$$\delta\Phi = \overbrace{-[\Phi, X] - [\Phi, \tilde X]}^{\text{global non-Abelian } G\times\tilde G} + \underbrace{\delta_a\Phi}_{\text{global Poincaré}} \tag{3.7}$$

The gravitational symmetries are reproduced here from those of Yang-Mills by invoking the gravity/Yang-Mills dictionary for fields and supergauge parameters:

$$\begin{array}{l|lllllll l} \text{Fields} & \varphi_\nu & = & V & \star & \Phi & \star & \tilde A_\nu & \text{real superfield} \\ \hline \text{Paras} & \phi & = & V & \star & \Phi & \star & \tilde\lambda & \text{real superfield} \\ & S_\nu & = & \Lambda & \star & \Phi & \star & \tilde A_\nu & \text{chiral superfield} \end{array} \tag{3.8}$$

Varying the gravitational superfield

$$\begin{aligned}\varphi_\nu(x,\theta,\bar\theta) =& C_\nu + i\theta\chi_\nu - i\bar\theta\bar\chi_\nu + i\theta^2 F_\nu - i\bar\theta^2\bar F_\nu - \theta\sigma^\mu\bar\theta(g_{\mu\nu}+B_{\mu\nu})\\ &+ i\theta^2\bar\theta\left(\bar\psi_\nu + \frac{i}{2}\bar\sigma^\rho\partial_\rho\chi_\nu\right) - i\bar\theta^2\theta\left(\psi_\nu + \frac{i}{2}\sigma^\rho\partial_\rho\bar\chi_\nu\right)\\ &+ \frac{1}{2}\bar\theta^2\theta^2\left(V_\nu + \frac{1}{2}\Box C_\nu\right)\end{aligned} \tag{3.9}$$

via the dictionary

$$\delta\varphi_\nu = \delta V\star\Phi\star\tilde A_\nu + V\star\delta\Phi\star\tilde A_\nu + V\star\Phi\star\delta\tilde A_\nu \tag{3.10}$$

we obtain

$$\delta\varphi_\nu = S_\nu + \bar S_\nu + \partial_\nu\phi + \delta_{(a,\lambda,\epsilon)}\varphi_\nu. \tag{3.11}$$

This is the complete set of transformation rules for the new-minimal superfield at linearised approximation. Note, this derivation makes use of

$$\langle [X,Y],Z\rangle = \langle X,[Y,Z]\rangle \tag{3.12}$$

and, crucially, the convolution property

$$\partial_\mu(f\star g) = (\partial_\mu f)\star g = f\star(\partial_\mu g). \tag{3.13}$$

To summarise, we have obtained the field content (3.9) and transformation rules (3.11) at linearised approximation of new-minimal $\mathcal{N}=1$ supergravity [43,44]. Hence, the local gravitational symmetries of general covariance, 2-form gauge invariance, local supersymmetry and local chiral symmetry follow from those of Yang-Mills at linear level.

Introducing field equations we should match the on-shell content of the tensor product of spacetime little group representations. This is done covariantly by including the ghost sector in the dictionary [36,37]. The 12 + 12 multiplet splits with respect to superconformal transformations into an 8 + 8 conformal supergravity multiplet plus a 4 + 4 conformal tensor multiplet,

$$\underbrace{\begin{pmatrix}\mathbf{5}+\mathbf{3}+\mathbf{1}+\mathbf{3}\\ \mathbf{4}+\mathbf{2}+\mathbf{4}+\mathbf{2}\end{pmatrix}}_{\text{new-minimal}} \to \underbrace{\begin{pmatrix}\mathbf{5}+\mathbf{3}\\ \mathbf{4}+\mathbf{4}\end{pmatrix}}_{\text{conformal}} + \underbrace{\begin{pmatrix}\mathbf{3}+\mathbf{1}\\ \mathbf{2}+\mathbf{2}\end{pmatrix}}_{\text{tensor}} \tag{3.14}$$

in terms of SO(3) represention s. Since the left (anti)ghost is a chiral superfield the ghost-antighost sector gives a compensating 4 + 4 chiral (dilaton) multiplet [36,37], yielding old-minimal 12 + 12 supergravity [45,46] coupled to a tensor multiplet, which, with the conventional 2-derivative Lagrangian, correctly corresponds to the on-shell content obtained by tensoring left/right helicity states.

4 Extended supersymmetry and U-duality

This minimally supersymmetric example does not fully address the issue of U-duality [47], which, in context of string/M-theory, is of fundamental importance. U-duality manifests itself in supergravity, the low energy effective limit of string/M-theory, in the form of non-compact global symmetries, $\mathcal{G}$, acting non-linearly on the scalar fields [48]. In all cases obtained from "Yang-Mills×Yang-Mills" the scalars parametrise a symmetric space $\mathcal{G}/\mathcal{H}$, where $\mathcal{H}$ is the maximal compact subgroup of $\mathcal{G}$ [49]. The U-dualities and corresponding global symmetries for M-theory compactified on a n-torus are summarised in Table 1. Note, we will also use the term U-duality to refer to $\mathcal{G}$. The question of global symmetries from squaring Yang-Mills has also been addressed in [50–53], particularly in the context of scattering amplitudes.

n-torus	U-duality	$\mathcal{G}$	$\mathcal{H}$
1	$SO(1,1,\mathbb{Z})$	$SO(1,1,\mathbb{R})$	–
2	$SL(2,\mathbb{Z}) \times SO(1,1,\mathbb{Z})$	$SL(2,\mathbb{R}) \times SO(1,1,\mathbb{R})$	$SO(2,\mathbb{R})$
3	$SL(2,\mathbb{Z}) \times SL(3,\mathbb{Z})$	$SL(2,\mathbb{R}) \times SL(3,\mathbb{R})$	$SO(2,\mathbb{R}) \times SO(3,\mathbb{R})$
4	$SL(5,\mathbb{Z})$	$SL(5,\mathbb{R})$	$SO(5,\mathbb{R})$
5	$SO(5,5,\mathbb{Z})$	$SO(5,5,\mathbb{R})$	$SO(5,\mathbb{R}) \times SO(5,\mathbb{R})$
6	$E_{6(6)}(\mathbb{Z})$	$E_{6(6)}(\mathbb{R})$	$USp(8)$
7	$E_{7(7)}(\mathbb{Z})$	$E_{7(7)}(\mathbb{R})$	$SU(8)$
8	$E_{8(8)}(\mathbb{Z})$	$E_{8(8)}(\mathbb{R})$	$SO(16,\mathbb{R})$

Table 1: U-dualities (global symmetries) of M-theory ($D = 11, \mathcal{N} = 1$ supergravity) compactified on an n-torus.

As made clear by Table 1, U-duality becomes increasingly manifest as one descends in dimension[1]. Thus, to fully expose the structure of U-duality with respect to squaring we should consider the product in $D = 3$ of left Yang-Mills theories with $\mathcal{N} = 1, 2, 4, 8$ and right Yang-Mills theories with $\tilde{\mathcal{N}} = 1, 2, 4, 8$. This was done in [58]. The result revealed a rather intriguing mathematical structure. The U-duality algebras obtained make up the Freudenthal-Rosenfeld-Tits magic square [59–61] as given in Table 2. As we shall explain this surprise has an elegant explanation, but first we must spend some time on the magic square itself.

Note, the real forms appearing in Table 2 are not unique; there are numerous possibilities as described in [62]. They also play a role in supergravity. In particular, the $\mathbb{C}$, $\mathbb{H}$, and $\mathbb{O}$ rows of one such magic square (distinct from Table 2) describe the U-dualities of the aptly named magic supergravities in $D = 5, 4, 3$ respectively [63–65]. It should be emphasised, however, that the appearance of the magic square here is unrelated to these constructions.

$\mathcal{N} \otimes \tilde{\mathcal{N}}$	1	2	4	8
1	$\mathfrak{sl}(2,\mathbb{R})$	$\mathfrak{su}(2,1)$	$\mathfrak{sp}(4,2)$	$\mathfrak{f}_{4(-20)}$
2	$\mathfrak{su}(2,1)$	$\mathfrak{su}(2,1) \times \mathfrak{su}(2,1)$	$\mathfrak{su}(4,2)$	$\mathfrak{e}_{6(-14)}$
4	$\mathfrak{sp}(4,2)$	$\mathfrak{su}(4,2)$	$\mathfrak{so}(8,4)$	$\mathfrak{e}_{7(-5)}$
8	$\mathfrak{f}_{4(-20)}$	$\mathfrak{e}_{6(-14)}$	$\mathfrak{e}_{7(-5)}$	$\mathfrak{e}_{8(8)}$

Table 2: The magic square of U-duality algebras obtained from the product of two Yang-Mills theories in $D = 3$ spacetime dimensions.

4.1 Division algebras and the magic square

In this section we follow closely [66, 67]; we refer the reader to these works for more detailed explanations and proofs. An algebra $\mathbb{A}$ defined over $\mathbb{R}$ with identity element e_0, is said to be *composition* if it has a

[1]We stop at $D = 3$, which has $E_{8(8)}$ U-duality, the largest finite dimensional exceptional Lie algebra. One can continue to $D = 2, 1, 0$, invoking the infinite dimensional extended algebras $E_{9(9)}, E_{10(10)}, E_{11(11)}$ [54–57]. Although we will not discuss theses cases here, it would be interesting to investigate whether they can be understood from the perspective of Yang-Mills squared.

non-degenerate quadratic form[2] $\mathbf{n} : \mathbb{A} \to \mathbb{R}$ such that,

$$\mathbf{n}(ab) = \mathbf{n}(a)\mathbf{n}(b), \quad \forall \ a, b \in \mathbb{A}, \tag{4.1}$$

where we denote the multiplicative product of the algebra by juxtaposition. Regarding $\mathbb{R} \subset \mathbb{A}$ as the scalar multiples of the identity $\mathbb{R}e_0$ we may decompose $\mathbb{A}$ into its "real" and "imaginary" parts $\mathbb{A} = \mathbb{R} \oplus \mathbb{A}'$, where $\mathbb{A}' \subset \mathbb{A}$ is the subspace orthogonal to $\mathbb{R}$. An arbitrary element $a \in \mathbb{A}$ may be written $a = \mathrm{Re}(a) + \mathrm{Im}(a)$. Here $\mathrm{Re}(a) \in \mathbb{R}e_0$, $\mathrm{Im}(a) \in \mathbb{A}'$ and

$$\mathrm{Re}(a) = \frac{1}{2}(a + \bar{a}), \qquad \mathrm{Im}(a) = \frac{1}{2}(a - \bar{a}), \tag{4.2}$$

where we have defined conjugation using the bilinear form,

$$\bar{a} := \langle a, e_0 \rangle e_0 - a, \qquad \langle a, b \rangle := \mathbf{n}(a + b) - \mathbf{n}(a) - \mathbf{n}(b). \tag{4.3}$$

A composition algebra $\mathbb{A}$ is said to be *division* if it contains no zero divisors,

$$ab = 0 \quad \Rightarrow \quad a = 0 \quad \text{or} \quad b = 0,$$

in which case $\mathbf{n}$ is positive semi-definite and $\mathbb{A}$ is referred to as a normed division algebra. Hurwitz's celebrated theorem states that there are exactly four normed division algebras [68]: the reals, complexes, quaternions and octonions, denoted respectively by $\mathbb{R}, \mathbb{C}, \mathbb{H}$ and $\mathbb{O}$. They may be constructed via the Cayley-Dickson doubling procedure, $\mathbb{A}' = \mathbb{A} \oplus \mathbb{A}$ with multiplication in $\mathbb{A}'$ defined by

$$(a, b)(c, d) = (ac - d\bar{b}, \bar{a}d + cb). \tag{4.4}$$

With each doubling a property is lost as summarised here:

$\mathbb{A}$	*Construction*	*Dim*	*Division*	*Associative*	*Commutative*	*Ordered*
$\mathbb{R}$	$\mathbb{R}$	1	*yes*	*yes*	*yes*	*yes*
$\mathbb{C}$	$\mathbb{R} \oplus \mathbb{R}$	2	*yes*	*yes*	*yes*	*no*
$\mathbb{H}$	$\mathbb{C} \oplus \mathbb{C}$	4	*yes*	*yes*	*no*	*no*
$\mathbb{O}$	$\mathbb{H} \oplus \mathbb{H}$	8	*yes*	*no*	*no*	*no*
$\mathbb{S}$	$\mathbb{O} \oplus \mathbb{O}$	16	*no*	*no*	*no*	*no*

On doubling the octonions, $\mathbb{S} \cong \mathbb{O} \oplus \mathbb{O}$, the division property fails and we will not consider such cases here. Note that, while the octonions are not associative they are alternative:

$$[a, b, c] := (ab)c - a(bc) \tag{4.5}$$

is an alternating function under the interchange of its arguments. This property is crucial for supersymmetry.

An element $a \in \mathbb{O}$ may be written $a = a^a e_a$, where $a = 0, \ldots, 7$, $a^a \in \mathbb{R}$ and $\{e_a\}$ is a basis with one real e_0 and seven $e_i, i = 1, \ldots, 7$, imaginary elements. The octonionic multiplication rule is,

$$e_a e_b = (\delta_{a0}\delta_{bc} + \delta_{0b}\delta_{ac} - \delta_{ab}\delta_{0c} + C_{abc})\, e_c, \tag{4.6}$$

where C_{abc} is totally antisymmetric and $C_{0bc} = 0$. The non-zero C_{ijk} are given by the Fano plane. See Figure 2.

There are three symmetry algebras on $\mathbb{A}$ that we will make use of:

[2] A *quadratic norm* on a vector space V over a field $\mathbb{R}$ is a map $\mathbf{n} : V \to \mathbb{R}$ such that: (1) $\mathbf{n}(\lambda a) = \lambda^2 \mathbf{n}(a), \lambda \in \mathbb{R}, a \in V$ and (2) $\langle a, b \rangle := \mathbf{n}(a + b) - \mathbf{n}(a) - \mathbf{n}(b)$ is bilinear.

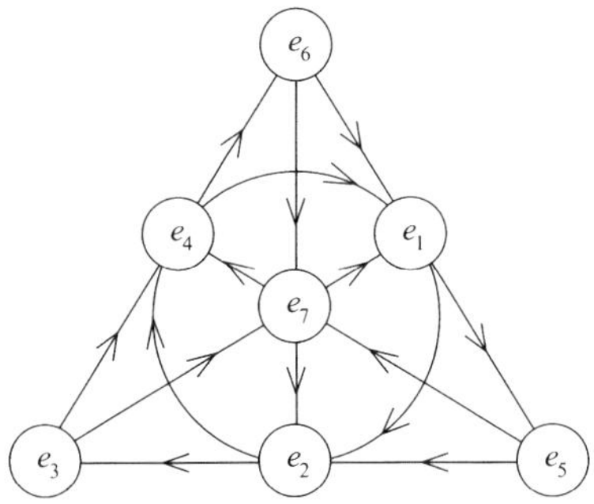

Figure 2: The Fano plane. The structure constants are determined by the Fano plane, $C_{ijk} = 1$ if ijk lies on a line and is ordered according as its orientation. Each oriented line follows the rules of quaternionic multiplication. For example, $e_2e_3 = e_5$ and cyclic permutations; odd permutations go against the direction of the arrows on the Fano plane and we pick up a minus sign, e.g. $e_3e_2 = -e_5$.

1. The *norm preserving* algebra is defined as,

$$\mathfrak{so}(\mathbb{A}) := \{A \in \mathrm{Hom}_{\mathbb{R}}(\mathbb{A}) | \langle Aa, b\rangle + \langle a, Ab\rangle = 0, \ \forall a, b \in \mathbb{A}\}, \tag{4.7}$$

yielding,

$$\begin{aligned} \mathfrak{so}(\mathbb{R}) &\cong \emptyset, \\ \mathfrak{so}(\mathbb{C}) &\cong \mathfrak{so}(2), \\ \mathfrak{so}(\mathbb{H}) &\cong \mathfrak{so}(3) \oplus \mathfrak{so}(3), \\ \mathfrak{so}(\mathbb{O}) &\cong \mathfrak{so}(8). \end{aligned} \tag{4.8}$$

2. The *triality* algebra of $\mathbb{A}$ is defined as,

$$\mathfrak{tri}(\mathbb{A}) := \{(A, B, C) \in \mathfrak{so}(\mathbb{A}) \oplus \mathfrak{so}(\mathbb{A}) \oplus \mathfrak{so}(\mathbb{A}) | A(ab) = B(a)b + aC(b), \ \forall a, b \in \mathbb{A}\}, \tag{4.9}$$

yielding,

$$\begin{aligned} \mathfrak{tri}(\mathbb{R}) &\cong \emptyset, \\ \mathfrak{tri}(\mathbb{C}) &\cong \mathfrak{so}(2) \oplus \mathfrak{so}(2), \\ \mathfrak{tri}(\mathbb{H}) &\cong \mathfrak{so}(3) \oplus \mathfrak{so}(3) \oplus \mathfrak{so}(3), \\ \mathfrak{tri}(\mathbb{O}) &\cong \mathfrak{so}(8). \end{aligned} \tag{4.10}$$

3. One can regard the triality algebra as a generalised form of the *derivation* algebra defined as,

$$\mathfrak{der}(\mathbb{A}) = \{A \in \mathrm{Hom}_{\mathbb{R}}(\mathbb{A}) | A(ab) = A(a)b + aA(b)\}, \tag{4.11}$$

which for $\mathbb{A} = \mathbb{O}$ gives the smallest exceptional Lie algebra,

$$\begin{aligned} \mathfrak{der}(\mathbb{R}) &\cong \emptyset, \\ \mathfrak{der}(\mathbb{C}) &\cong \emptyset, \\ \mathfrak{der}(\mathbb{H}) &\cong \mathfrak{so}(3), \\ \mathfrak{der}(\mathbb{O}) &\cong \mathfrak{g}_{2(-14)}. \end{aligned} \tag{4.12}$$

This provides the first example of a division algebraic description of an exceptional Lie algebra. In fact, the entire Freudenthal magic square can be realised in terms of the division algebras. The magic square was the result of an effort to give a unified and geometrically motivated description of Lie algebras, including the remaining exceptional cases of $\mathfrak{f}_4, \mathfrak{e}_6, \mathfrak{e}_7, \mathfrak{e}_8$. The classical Lie algebras $\mathfrak{so}(n), \mathfrak{su}(2), \mathfrak{sp}(n)$ are very naturally captured by $\mathbb{R}, \mathbb{C}, \mathbb{H}$ geometrical structures, respectively. There are a number of ways of articulating this idea, but perhaps the most concise is in terms of the isometries of projective geometries:

$$\mathfrak{Isom}(\mathbb{RP}^n) \cong \mathfrak{so}(n+1), \quad \mathfrak{Isom}(\mathbb{CP}^n) \cong \mathfrak{su}(n+1), \quad \mathfrak{Isom}(\mathbb{HP}^n) \cong \mathfrak{sp}(n+1). \tag{4.13}$$

This sequence is rather suggestive; we might expect octonionic projective geometries to yield exceptional Lie algebras. Despite non-associativity it was shown by Moufang [69] that one can consistently construct the octonionic projective line and plane, denoted $\mathbb{OP}^1$ and $\mathbb{OP}^2$, respectively. The latter is often referred to as the Cayley plane. We cannot go beyond $n = 2$ for the octonions[3], which in this context reflects the fact that there is indeed just a finite set of exceptional Lie algebras not belonging to any countably infinite family. The $\mathbb{OP}^1$ example is constructed in direct analogy with the real, complex and quaternionic cases[4]. It is diffeomorphic to S^8. The octonionic plane has a more intricate structure. An element $(a, b, c) \in \mathbb{O}^3$ with $\mathbf{n}(a) + \mathbf{n}(a) + \mathbf{n}(c) = 1$ and $(ab)c = a(bc)$ gives a point in $\mathbb{OP}^2$ (the line through the origin containing (a, b, c) in $\mathbb{O}^3$). It is not difficult to show the space of such elements is a 16-dimensional real manifold embedded in $\mathbb{O}^3$ through eight real constraints: $\mathbf{n}(a) + \mathbf{n}(a) + \mathbf{n}(c) = 1$ and $(ab)c = a(bc)$. The lines in $\mathbb{OP}^2$ are copies of $\mathbb{OP}^1$ and there is a duality relation sending lines/points into points/lines preserving the incidence structure. Borel showed that $F_{4(-52)}$ is the isometry group of a 16-dimensional projective plane, which is none other than $\mathbb{OP}^2$. One can show that the points and lines in $\mathbb{OP}^2$ are in one-to-one incidence preserving correspondence with trace 1 and 2 projectors in the Jordan algebra of 3×3 octonionic Hermitian matrices $\mathfrak{J}_3(\mathbb{O})$ (treating projectors as propositions the incidence relation in $\mathfrak{J}_3^{\mathbb{O}}$ is given by implication) [70]. Then $F_{4(-52)} = \mathrm{Isom}(\mathbb{OP}^2)$ follows automatically from the result of Chevalley and Schafer that $F_{4(-52)} = \mathrm{Aut}(\mathfrak{J}_3(\mathbb{O}))$, the group preserving the Jordan product with Lie algebra $\mathfrak{der}(\mathfrak{J}_3(\mathbb{O}))$ [71]. In summary, the sequence in (4.13) is continued to include,

$$\mathfrak{Isom}(\mathbb{OP}^2) \cong \mathfrak{der}(\mathfrak{J}_3(\mathbb{O})) \cong \mathfrak{f}_{4(-52)}. \tag{4.14}$$

Since $F_{4(-52)}$ acts transitively on the space of trace 1 projectors and the stabiliser of a given trace 1 projector is isomorphic to Spin(9) we have,

$$\mathbb{OP}^2 \cong F_{4(-52)}/\,\mathrm{Spin}(9). \tag{4.15}$$

The Cayley plane is a homogenous symmetric space with $T_p(\mathbb{OP}^2) \cong \mathbb{O}^2$, which carries the spinor representation of Spin(9); under $F_{4(-52)} \supset \mathrm{Spin}(9)$ we have

$$\mathbf{52} \to \mathbf{36} + \mathbf{16}, \tag{4.16}$$

or in a more division algebraic form,

$$\begin{aligned} \mathfrak{f}_{4(-52)} &\cong \mathfrak{so}(9) + \mathbf{16} \\ &\cong \mathfrak{so}(\mathbb{R} \oplus \mathbb{O}) + \mathbb{O}^2 \\ &\cong \mathfrak{so}(\mathbb{O}) + \mathbb{O} + \mathbb{O} + \mathbb{O}. \end{aligned} \tag{4.17}$$

The three $\mathbb{O}$ terms in the final line transform in the three triality related 8-dimensional representations of $\mathfrak{so}(8)$, the vector, spinor and conjugate spinor. It is this triality relation which implies that $\mathfrak{tri}(\mathbb{O}) \cong \mathfrak{so}(\mathbb{O})$.

[3] One way to understand this is in terms of Jordan algebras. Points in $\mathbb{OP}^2$ are bijectively identified with trace 1 projectors in $\mathfrak{J}_3^{\mathbb{O}}$, the Jordan algebra of 3×3 octonionic Hermitian matrices. However, for $m > 3$, $m \times m$ octonionic Hermitian matrices do not form a Jordan algebra.

[4] Non-associativity, however, implies that the line through the origin containing the point (a, b) is not given by $\{(\alpha a, \alpha b) | \alpha \in \mathbb{O}\}$, unless $x = 1$ or $y = 1$. This obstacle is easily avoided as all non-zero octonions have an inverse; (a, b) is equivalent to $(b^{-1}a, 1)$ or $(1, a^{-1}b)$ for $b \neq 0$ or $a \neq 0$, giving two charts with a smooth transition function on their overlap. See [67].

Seemingly inspired by the trivial identity $\mathbb{O} \cong \mathbb{R}\otimes\mathbb{O}$ Boris Rosenfeld [61] proposed a natural extension of this construction,

$$\mathfrak{Isom}((\mathbb{C}\otimes\mathbb{O})\mathbb{P}^2) \cong \mathfrak{e}_{6(-78)}, \qquad \mathfrak{Isom}((\mathbb{H}\otimes\mathbb{O})\mathbb{P}^2) \cong \mathfrak{e}_{7(-133)}, \qquad \mathfrak{Isom}((\mathbb{O}\otimes\mathbb{O})\mathbb{P}^2) \cong \mathfrak{e}_{8(-248)}, \tag{4.18}$$

thus giving a uniform geometric description for all Lie algebras. The would-be tangents spaces $(\mathbb{A}\otimes\mathbb{O})^2$ have the correct dimensions and representation theoretic properties. However, it is not actually possible to construct projective spaces over $\mathbb{H}\otimes\mathbb{O}$ and $\mathbb{O}\otimes\mathbb{O}$ using the logic applied to $\mathbb{OP}^2$, essentially because they do not yield Jordan algebras, unlike $\mathbb{C}\otimes\mathbb{O}$. They nonetheless can be identified with Riemannian geometries with isometries $E_{7(-133)}$ and $E_{8(-248)}$, respectively. Indeed, the Lie algebra decompositions[5],

$$\begin{aligned}
\mathfrak{f}_{4(-52)} &\cong \mathfrak{so}(\mathbb{R}\oplus\mathbb{O}) + (\mathbb{R}\otimes\mathbb{O})^2 \\
\mathfrak{e}_{6(-78)} &\cong \mathfrak{so}(\mathbb{C}\oplus\mathbb{O}) \oplus \mathfrak{u}(1) + (\mathbb{C}\otimes\mathbb{O})^2 \\
\mathfrak{e}_{7(-133)} &\cong \mathfrak{so}(\mathbb{H}\oplus\mathbb{O}) \oplus \mathfrak{sp}(1) + (\mathbb{H}\otimes\mathbb{O})^2 \\
\mathfrak{e}_{8(-248)} &\cong \mathfrak{so}(\mathbb{O}\oplus\mathbb{O}) + (\mathbb{O}\otimes\mathbb{O})^2
\end{aligned} \tag{4.19}$$

naturally suggest the identifications

$$\begin{aligned}
\text{Isom}((\mathbb{R}\otimes\mathbb{O})\mathbb{P}^2) &= F_{4(-52)}/\,\text{Spin}(9) \\
\text{Isom}((\mathbb{C}\otimes\mathbb{O})\mathbb{P}^2) &= E_{6(-78)}/[(\text{Spin}(10)\times \text{U}(1))/\mathbb{Z}_4] \\
\text{Isom}((\mathbb{H}\otimes\mathbb{O})\mathbb{P}^2) &= E_{7(-133)}/[(\text{Spin}(10)\times \text{Sp}(1))/\mathbb{Z}_2] \\
\text{Isom}((\mathbb{O}\otimes\mathbb{O})\mathbb{P}^2) &= E_{8(-248)}/[\text{Spin}(16)/\mathbb{Z}_2]
\end{aligned} \tag{4.20}$$

with tangent spaces $(\mathbb{R}\otimes\mathbb{O})^2, (\mathbb{C}\otimes\mathbb{O})^2, (\mathbb{H}\otimes\mathbb{O})^2, (\mathbb{O}\otimes\mathbb{O})^2$ carrying the appropriate spinor representations. Using the Tits' construction [60] the isometry algebras are given by the natural generalisation of (4.14).

$$\begin{aligned}
\mathfrak{f}_{4(-52)} &\cong \mathfrak{Der}(\mathbb{R}) \oplus \mathfrak{Der}(\mathfrak{J}_3(\mathbb{O})) + \text{Im}\mathbb{R}\otimes\mathfrak{J}_3'(\mathbb{O}) \\
\mathfrak{e}_{6(-78)} &\cong \mathfrak{Der}(\mathbb{C}) \oplus \mathfrak{Der}(\mathfrak{J}_3(\mathbb{O})) + \text{Im}\mathbb{C}\otimes\mathfrak{J}_3'(\mathbb{O}) \\
\mathfrak{e}_{7(-133)} &\cong \mathfrak{Der}(\mathbb{H}) \oplus \mathfrak{Der}(\mathfrak{J}_3(\mathbb{O})) + \text{Im}\mathbb{H}\otimes\mathfrak{J}_3'(\mathbb{O}) \\
\mathfrak{e}_{8(-248)} &\cong \mathfrak{Der}(\mathbb{O}) \oplus \mathfrak{Der}(\mathfrak{J}_3(\mathbb{O})) + \text{Im}\mathbb{O}\otimes\mathfrak{J}_3'(\mathbb{O}),
\end{aligned} \tag{4.21}$$

where $\mathfrak{J}'$ denotes the subset of traceless elements in $\mathfrak{J}$. Generalising further, the Tits' construction defines a Lie algebra,

$$\mathfrak{M}(\mathbb{A}_1,\mathbb{A}_2) := \mathfrak{Der}(\mathbb{A}_1) \oplus \mathfrak{Der}(\mathfrak{J}_3(\mathbb{A}_2)) + \text{Im}\mathbb{A}_1\otimes\mathfrak{J}_3'(\mathbb{A}_2), \tag{4.22}$$

for an arbitrary pair $\mathbb{A}_1,\mathbb{A}_2 = \mathbb{R},\mathbb{C},\mathbb{H},\mathbb{O}$, which yields the (compact) magic square given in Table 3. The "magic" is that Table 3 symmetric about the diagonal despite the apparent asymmetry of (4.22). To obtain a magic square with the non-compact real forms that follow from squaring Yang-Mills, as given in Table 2, one can use a Lorentzian Jordan algebra [62],

$$\mathfrak{M}'(\mathbb{A}_1,\mathbb{A}_2) := \mathfrak{Der}(\mathbb{A}_1) \oplus \mathfrak{Der}(\mathfrak{J}_{1,2}(\mathbb{A}_2)) + \text{Im}\mathbb{A}_1\otimes\mathfrak{J}_{1,2}'(\mathbb{A}_2). \tag{4.23}$$

Later we shall see that Yang-Mills squared gives an alternative form of (4.23), based on the Barton-Sudbery triality construction [66], that is manifestly symmetric in $\mathbb{A}_1,\mathbb{A}_2$ [58, 72]. This symmetric form reflects the fact that the squaring procedure is itself symmetric on interchanging the left and right theories.

[5]Note, the additional factors are given by intermediate algebras: $\mathfrak{tri}(\mathbb{A})/\mathfrak{int}(\mathbb{A}) = \emptyset, \mathfrak{u}(1), \mathfrak{sp}(1), \emptyset$ for $\mathbb{A} = \mathbb{R},\mathbb{C},\mathbb{H},\mathbb{O}$ [66].

$\otimes$	$\mathbb{R}$	$\mathbb{C}$	$\mathbb{H}$	$\mathbb{O}$
$\mathbb{R}$	$\mathfrak{su}(2)$	$\mathfrak{su}(3)$	$\mathfrak{sp}(6)$	$\mathfrak{f}_{4(-52)}$
$\mathbb{C}$	$\mathfrak{su}(3)$	$\mathfrak{su}(3)\times\mathfrak{su}(3)$	$\mathfrak{su}(6)$	$\mathfrak{e}_{6(-78)}$
$\mathbb{H}$	$\mathfrak{sp}(6)$	$\mathfrak{su}(6)$	$\mathfrak{so}(12)$	$\mathfrak{e}_{7(-133)}$
$\mathbb{O}$	$\mathfrak{f}_{4(-52)}$	$\mathfrak{e}_{6(-78)}$	$\mathfrak{e}_{7(-133)}$	$\mathfrak{e}_{8(-248)}$

Table 3: The magic square given by the Tits' construction.

4.2 Division algebras and Yang-Mills theories

In the two previous sections we saw that the "square" of $D = 3$ super Yang-Mills theories and the "square" of division algebras both led to the magic square of Freudenthal. Surely this is no coincidence. Indeed, there is a long history of work connecting supersymmetry, spacetime and the division algebras [?, 62–65, 67, 73–101], which as we shall review underlies this magical meeting.

Perhaps the most direct link from division algebras to spacetime symmetries comes via the Lie algebra isomorphism of Sudbery [78],

$$\mathfrak{sl}(2,\mathbb{A}) \cong \mathfrak{so}(1, 1+\dim\mathbb{A}), \tag{4.24}$$

which identifies $D = 3, 4, 6, 10$ as algebraically special. This is itself tied to the earlier observation of Kugo and Townsend [77] that the existence of minimal super Yang-Mills mulitplets in only $D = 3, 4, 6, 10$ is related to the uniqueness of $\mathbb{R}, \mathbb{C}, \mathbb{H}, \mathbb{O}$. This was followed-up by a number of authors [102–106], sharpening the correspondence between supersymmetry and division algebras. The final case of $D = 10, \mathbb{A} = \mathbb{O}$ was developed most carefully in [95], where the link between supersymmetry and the alternativity of $\mathbb{O}$ was emphasised.

Pulling together these ideas, it was shown in [107] that $\mathcal{N}$-extended super Yang-Mills theories in $D = n + 2$ dimensions are completely specified (the field content, Lagrangian and transformation rules) by selecting an ordered pair of division algebras: $\mathbb{A}_n$ for the spacetime dimension and $\mathbb{A}_{n\mathcal{N}}$ for the degree of supersymmetry, where the subscripts denote the dimension of the algebras.

Consequently, the dual appearances of the magic square in $D = 3$, or equivalently for $\mathbb{A}_n = \mathbb{R}$, can be explained by the observation that $D = 3, \mathcal{N} = 1, 2, 4, 8$ Yang-Mills theories can be formulated with a single Lagrangian and a single set of transformation rules, using fields valued in $\mathbb{R}, \mathbb{C}, \mathbb{H}$ and $\mathbb{O}$, respectively [58]. Tensoring an $\mathbb{A}$-valued $D = 3$ super Yang-Mills multiplet with an $\tilde{\mathbb{A}}$-valued $D = 3$ super Yang-Mills multiplet yields a $D = 3$ supergravity mulitplet with fields valued in $\mathbb{A}\otimes\tilde{\mathbb{A}}$, making a magic square of U-dualities appear rather natural.

Let us now review in some more detail these constructions. The Lagrangian for $(n + 2)$-dimensional $\mathcal{N} = 1$ super Yang-Mills with gauge group G over the division algebra $\mathbb{A}_n$ is given [107] by

$$\mathcal{L}(\mathbb{A}_n) = -\frac{1}{4}F^A_{\mu\nu}F^{A\mu\nu} - \mathrm{Re}(i\lambda^{\dagger A}\bar{\sigma}^\mu D_\mu\lambda^A), \qquad \lambda\in\mathbb{A}^2_n, \tag{4.25}$$

where the covariant derivative and field strength are given by the usual expressions

$$\begin{aligned} D_\mu\lambda^A &= \partial_\mu\lambda^A + gf_{BC}{}^A A^B_\mu\lambda^C, \\ F^A_{\mu\nu} &= \partial_\mu A^A_\nu - \partial_\nu A^A_\mu + gf_{BC}{}^A A^B_\mu A^C_\nu, \end{aligned} \tag{4.26}$$

with $A = 0, \dots, \dim[G]$. The $\{\sigma^\mu\}$ are a basis for $\mathbb{A}_n$-valued Hermitian matrices - the straightforward generalisation of the usual complex Pauli matrices [87, 88, 107] to all four normed division algebras, satisfying the usual Clifford algebra relations. We can use these to write the supersymmetry transformations:

$$\delta A^A_\mu = \mathrm{Re}(i\lambda^{\dagger A}\bar{\sigma}_\mu\epsilon), \qquad \delta\lambda^A = \frac{1}{4}F^A_{\mu\nu}\sigma^\mu(\bar{\sigma}^\nu\epsilon). \tag{4.27}$$

Note, since the octonions are non-associative the ordering of the parentheses is important. Moreover, the components λ^{Aa} are anti-commuting; we are dealing with the algebra of octonions defined over the Grassmanns and we cannot rely on the usual spinor identities to hold automatically. However, everything goes through, thanks principally to alternativity.

By dimensionally reducing these theories using the Dixon-halving techniques of [107], we arrive at the Lagrangian for super Yang-Mills in $D = n + 2$ with $\mathcal{N}$ supersymmetries written over the division algebra $\mathbb{A}_{n\mathcal{N}}$. The division algebra associated with spacetime $\mathbb{A}_n$ is viewed as a subalgebra of $\mathbb{A}_{n\mathcal{N}}$. The resulting Lagrangian is:

$$\begin{aligned}\mathcal{L}(\mathbb{A}_n, \mathbb{A}_{n\mathcal{N}}) = &-\frac{1}{4}F^A_{\mu\nu}F^{A\mu\nu} - \frac{1}{2}\langle D_\mu\phi^A|D^\mu\phi^A\rangle - \text{Re}(i\lambda^{\dagger A}\bar{\sigma}^\mu D_\mu\lambda^A) \\ &- gf_{BC}{}^A\text{Re}\left(i\lambda^{\dagger A}\varepsilon\phi^B\lambda^C\right) - \frac{1}{16}g^2 f_{BC}{}^A f_{DE}{}^A\langle\phi^B|\phi^D\rangle\langle\phi^C|\phi^E\rangle,\end{aligned} \tag{4.28}$$

where $\lambda \in \mathbb{A}^2_{n\mathcal{N}}$ (so we have $\mathcal{N}$ spacetime spinors, each valued in $\mathbb{A}^2_n$) and ϕ is a scalar field taking values in $\phi \in \mathbb{A}^{\complement}_n$, the subspace of $\mathbb{A}_{n\mathcal{N}}$ orthogonal to the $\mathbb{A}_n$ subalgebra. The $\{\bar{\sigma}^\mu\}$ are still a basis for $\mathbb{A}_n$-valued Hermitian matrices, again, with $\mathbb{A}_n$ viewed as a division subalgebra of $\mathbb{A}_{n\mathcal{N}}$. As noted in [107], the overall (spacetime little group plus internal) symmetry of the $\mathcal{N} = 1$ theory in $D = n + 2$ dimensions is given by the triality algebra, $\mathfrak{tri}(\mathbb{A}_n)$. If we dimensionally reduce these theories we obtain super Yang-Mills with $\mathcal{N}$ supersymmetries whose overall symmetries are given by,

$$\mathfrak{sym}(\mathbb{A}_n, \mathbb{A}_{n\mathcal{N}}) := \left\{(A, B, C) \in \mathfrak{tri}(\mathbb{A}_{n\mathcal{N}}) | [A, \mathfrak{so}(\mathbb{A}_n)_{ST}] = 0, \quad \forall A \notin \mathfrak{so}(\mathbb{A}_n)_{ST}\right\}, \tag{4.29}$$

where $\mathfrak{so}(\mathbb{A}_n)_{ST}$ is the subalgbra of $\mathfrak{so}(\mathbb{A}_{n\mathcal{N}})$ that acts as orthogonal transformations on $\mathbb{A}_n \subseteq \mathbb{A}_{n\mathcal{N}}$. The division algebras used in each dimension and the corresponding $\mathfrak{sym}$ algebras are summarised in Table 4.

$\mathbb{A}_n \backslash \mathbb{A}_{n\mathcal{N}}$	$\mathbb{O}$	$\mathbb{H}$	$\mathbb{C}$	$\mathbb{R}$
$\mathbb{O}$	$\mathfrak{so}(8)_{ST}$			
$\mathbb{H}$	$\mathfrak{so}(4)_{ST} \oplus \mathfrak{sp}(1) \oplus \mathfrak{sp}(1)$	$\mathfrak{so}(4)_{ST} \oplus \mathfrak{sp}(1)$		
$\mathbb{C}$	$\mathfrak{so}(2)_{ST} \oplus \mathfrak{su}(4)$	$\mathfrak{so}(2)_{ST} \oplus \mathfrak{sp}(1) \oplus \mathfrak{so}(2)$	$\mathfrak{so}(2)_{ST} \oplus \mathfrak{so}(2)$	
$\mathbb{R}$	$\mathfrak{so}(8)$	$\mathfrak{so}(4) \oplus \mathfrak{sp}(1)$	$\mathfrak{so}(2) \oplus \mathfrak{so}(2)$	$\emptyset$

Table 4: A table of algebras: $\mathfrak{sym}(\mathbb{A}_n, \mathbb{A}_{n\mathcal{N}})$. This lets us read off the spacetime and internal symmetries in each Yang-Mills theory. For example, one can see the familiar R-symmetries in $D = 4$: U(1), U(2) and SU(4) for $\mathcal{N} = 1, 2, 4$, respectively. Note that the symmetries in $D = 3$ are entirely internal and that they include the R-symmetry as a subgroup (these are actually the symmetries of the theories after dualising the vector to a scalar).

Let us take $D = 3$ as a concrete example. The $\mathcal{N} = 8$ Lagrangian is given by

$$\begin{aligned}\mathcal{L} = &-\tfrac{1}{4}F^A_{\mu\nu}F^{A\mu\nu} - \tfrac{1}{2}D_\mu\phi^A_i D^\mu\phi^A_i + i\bar{\lambda}^A_a\gamma^\mu D_\mu\lambda^A_a \\ &- \tfrac{1}{4}g^2 f_{BC}{}^A f_{DE}{}^A\phi^B_i\phi^D_i\phi^C_j\phi^E_j \\ &- gf_{BC}{}^A\phi^B_i\bar{\lambda}^{Aa}\Gamma^i_{ab}\lambda^{Cb},\end{aligned} \tag{4.30}$$

where Γ^i_{ab}, $i = 1, \ldots, 7$, $a, b = 0, \ldots, 7$, belongs to the SO(7) Clifford algebra. The key observation is that this gamma matrix can be represented by the octonionic structure constants,

$$\Gamma^i_{ab} = i(\delta_{bi}\delta_{a0} - \delta_{b0}\delta_{ai} + C_{iab}), \tag{4.31}$$

which allows us to rewrite the action over octonionic fields. If we replace $\mathbb{O}$ with a general division algebra $\mathbb{A}$, the result is $\mathcal{N} = 1, 2, 4, 8$ over $\mathbb{R}, \mathbb{C}, \mathbb{H}, \mathbb{O}$:

$$\begin{aligned}\mathcal{L} = &-\tfrac{1}{4}F^{A}_{\mu\nu}F^{A\mu\nu} - \tfrac{1}{2}D_\mu\phi^{*A}D^\mu\phi^A + i\bar{\lambda}^A\gamma^\mu D_\mu\lambda^A \\ &-\tfrac{1}{4}g^2 f_{BC}{}^{A} f_{DE}{}^{A}\langle\phi^B|\phi^D\rangle\langle\phi^C|\phi^E\rangle \\ &+\tfrac{i}{2}g f_{BC}{}^{A}\left((\bar{\lambda}^A\phi^B)\lambda^C - \bar{\lambda}^A(\phi^{*B}\lambda^C)\right),\end{aligned} \tag{4.32}$$

where $\phi = \phi^i e_i$ is an $\mathrm{Im}\mathbb{A}$-valued scalar field, $\lambda = \lambda^a e_a$ is an $\mathbb{A}$-valued two-component spinor and $\bar{\lambda} = \bar{\lambda}^a e^*_a$.

The supersymmetry transformations in this language are given by

$$\begin{aligned}\delta\lambda^A &= \frac{1}{2}(F^{A\mu\nu} + \varepsilon^{\mu\nu\rho}D_\rho\phi^A)\sigma_{\mu\nu}\epsilon - \frac{1}{4}g f_{BC}{}^{A}\phi^B(\phi^C\epsilon), \\ \delta A^A_\mu &= -\frac{i}{2}(\bar{\epsilon}\gamma_\mu\lambda^A - \bar{\lambda}^A\gamma_\mu\epsilon), \\ \delta\phi^A &= \frac{i}{2}e_i[(\bar{\epsilon}e_i)\lambda^A - \bar{\lambda}^A(e_i\epsilon)],\end{aligned} \tag{4.33}$$

where ϵ is an $\mathbb{A}$-valued two-component spinor and $\sigma_{\mu\nu}$ are the generators of $\mathrm{SL}(2, \mathbb{R}) \cong \mathrm{Spin}(1, 2)$. The form of the first term in the λ^A transformation also highlights the vector's status as the missing real part of the $\mathrm{Im}\mathbb{A}$-valued scalar field. Indeed, in the free $g = 0$ theory one may dualise the vector to a scalar to obtain a full $\mathbb{A}$-valued field.

Now consider the product of two division algebraic multiplets:

1. A left $\mathcal{N} = \dim\mathbb{A}$ multiplet
$$\{A_\mu \in \mathrm{Re}\mathbb{A}, \quad \phi \in \mathrm{Im}\mathbb{A}, \quad \lambda \in \mathbb{A}\} \tag{4.34}$$
2. A right $\tilde{\mathcal{N}} = \dim\tilde{\mathbb{A}}$ multiplet
$$\{\tilde{A}_\nu \in \mathrm{Re}\tilde{\mathbb{A}}, \quad \tilde{\phi} \in \mathrm{Im}\tilde{\mathbb{A}}, \quad \tilde{\lambda} \in \tilde{\mathbb{A}}\} \tag{4.35}$$

We obtain the field content of an $(\mathcal{N} + \tilde{\mathcal{N}})$-extended supergravity theory valued in both $\tilde{\mathbb{A}}$ and $\tilde{\mathbb{A}}$:

$$g_{\mu\nu} \in \mathbb{R}, \quad \Psi_\mu \in \begin{pmatrix}\mathbb{A} \\ \tilde{\mathbb{A}}\end{pmatrix}, \quad \varphi, \chi \in \begin{pmatrix}\mathbb{A}\otimes\tilde{\mathbb{A}} \\ \mathbb{A}\otimes\tilde{\mathbb{A}}\end{pmatrix}. \tag{4.36}$$

The $\mathbb{R}$-valued graviton and $\mathbb{A}\oplus\tilde{\mathbb{A}}$-valued gravitino carry no degrees of freedom. The $(\mathbb{A}\otimes\tilde{\mathbb{A}})^2$-valued scalar and Majorana spinor each have $2(\dim\mathbb{A}\times\dim\tilde{\mathbb{A}})$ degrees of freedom.

The $\mathcal{H}$ algebra then follows immediately in this division algebraic language. The left and right factors each come with a commuting copy of the triality algebra, $\mathfrak{tri}(\mathbb{A})\oplus\mathfrak{tri}(\tilde{\mathbb{A}})$. However, the $\mathbb{A}\otimes\tilde{\mathbb{A}}$ doublets in (4.36) form irreducible representations of R-symmetry. The corresponding generators must themselves transform under $\mathfrak{tri}(\mathbb{A})\oplus\mathfrak{tri}(\tilde{\mathbb{A}})$ consistently, implying they are elements of $\mathbb{A}\otimes\tilde{\mathbb{A}}$. This follows, formally, from the left/right supersymmetries. The conventional infinitesimal supersymmetry variation of the $left \otimes right$ states correctly gives the infinitesimal supersymmetry variation on the corresponding supergravity states [37,38,50]. Seeking, instead, internal *bosonic* transformations on the supergavity multiplet suggests starting from the rather unconventional tensor product of the left and right supercharges, $Q\otimes\tilde{Q}$. See Figure 3. This follows, at least formally, from the observation $Q\otimes\tilde{Q}\in\mathbb{A}\otimes\tilde{\mathbb{A}}$, where we are explicitly suppressing the spacetime representation space. Note, these are "pseudo-supersymmetry" transformations since they do not change the mass dimension of the component fields. For an explicit construction see [49]. In summary, we have in total:

$$\mathfrak{h}(\mathbb{A}, \tilde{\mathbb{A}}) := \mathfrak{tri}(\mathbb{A})\oplus\mathfrak{tri}(\tilde{\mathbb{A}}) + \mathbb{A}\otimes\tilde{\mathbb{A}}. \tag{4.37}$$

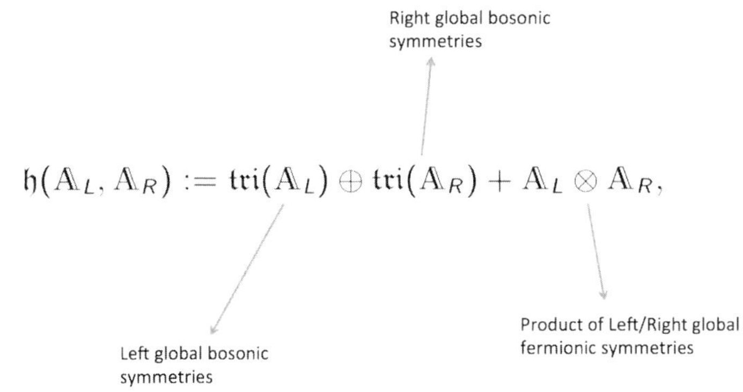

Figure 3: The $\mathcal{H}$ algebra in terms of the left/right super Yang-Mills theories in a division algebraic language.

$\mathbb{A}_L \backslash \mathbb{A}_R$	$\mathbb{R}$	$\mathbb{C}$	$\mathbb{H}$	$\mathbb{O}$
$\mathbb{R}$	$\mathfrak{so}(2)$	$\mathfrak{so}(3) \times \mathfrak{so}(2)$	$\mathfrak{so}(5) \times \mathfrak{so}(3)$	$\mathfrak{so}(9)$
$\mathbb{C}$	$\mathfrak{so}(3) \times \mathfrak{so}(2)$	$[\mathfrak{so}(3) \times \mathfrak{so}(2)]^2$	$\mathfrak{so}(6) \times \mathfrak{so}(3) \times \mathfrak{so}(2)$	$\mathfrak{so}(10) \times \mathfrak{so}(2)$
$\mathbb{H}$	$\mathfrak{so}(5) \times \mathfrak{so}(3)$	$\mathfrak{so}(6) \times \mathfrak{so}(3) \times \mathfrak{so}(2)$	$\mathfrak{so}(8) \times \mathfrak{so}(4)$	$\mathfrak{so}(12) \times \mathfrak{so}(3)$
$\mathbb{O}$	$\mathfrak{so}(9)$	$\mathfrak{so}(10) \times \mathfrak{so}(2)$	$\mathfrak{so}(12) \times \mathfrak{so}(3)$	$\mathfrak{so}(16)$

Table 5: Magic square of maximal compact subalgebras.

This Lie algebra, see [72] for the commutators, yields the maximal compact subalgebras of the corresponding U-dualities, given in Table 5.

The U-dualities $\mathcal{G}$ are realised non-linearly on the scalars, which parametrise the symmetric spaces $\mathcal{G}/\mathcal{H}$. This can be understood using the identity relating $(\mathbb{A} \otimes \tilde{\mathbb{A}})^2$ to $\mathcal{G}/\mathcal{H}$,

$$(\mathbb{A} \otimes \tilde{\mathbb{A}})\mathbb{P}^2 \cong \mathcal{G}/\mathcal{H}. \tag{4.38}$$

The scalar fields may be regarded as points in division-algebraic projective planes. The tangent space $T_p(\mathcal{G}/\mathcal{H}) \cong \mathfrak{p} = \mathfrak{g} \ominus \mathfrak{h}$ implies the scalars carry the $\mathfrak{p}$-representation of $\mathcal{H}$. The tangent space at any point of $(\mathbb{A} \otimes \tilde{\mathbb{A}})\mathbb{P}^2$ is just $(\mathbb{A} \otimes \tilde{\mathbb{A}})^2$, the required representation space of $\mathcal{H}$. Since $\mathcal{G}/\mathcal{H}$ is a symmetric space, the U-duality Lie algebra is given by adjoining the scalar representation space $(\mathbb{A} \otimes \tilde{\mathbb{A}})^2$ to Figure 3,

$$\mathfrak{m}(\mathbb{A}, \tilde{\mathbb{A}}) := \underbrace{\mathfrak{tri}(\mathbb{A}) \oplus \mathfrak{tri}(\tilde{\mathbb{A}}) + (\mathbb{A} \otimes \tilde{\mathbb{A}})}_{\mathfrak{h}(\mathbb{A}, \tilde{\mathbb{A}})} + \underbrace{(\mathbb{A} \otimes \tilde{\mathbb{A}})^2}_{\text{"scalars"}}. \tag{4.39}$$

This has a $\mathbb{Z}_2 \times \mathbb{Z}_2$ graded Lie algebra structure uniquely determined by the left/right super Yang-Mills factors and yields precisely the magic square. See [72] for a full account of the commutation relations. The triality construction described in [66] is isomorphic to (4.39) as a vector space, but has a different Lie algebra structure, as reflected in the distinct real forms appearing in each case. In conclusion, the product of division algebras and super Yang-Mills theories both lead to the magic square, as depicted in Figure 4.

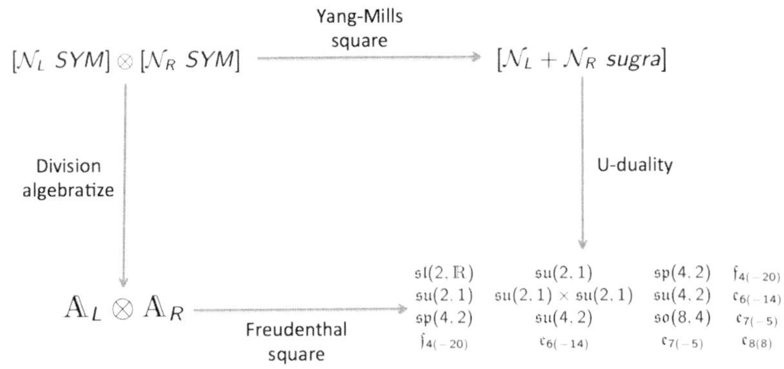

Figure 4: All roads lead to the magic square.

For $D = n+2$, we begin with a pair of Yang-Mills theories with $\mathcal{N}$ and $\tilde{\mathcal{N}}$ supersymmetries written over the division algebras $\mathbb{A}_{n\mathcal{N}}$ and $\mathbb{A}_{n\tilde{\mathcal{N}}}$, respectively, as described by (4.28). In terms of spacetime little group representations we may then write all the bosons of the left (right) theory as a single element $b \in \mathbb{A}_{n\mathcal{N}}$ ($\tilde{b} \in \mathbb{A}_{n\tilde{\mathcal{N}}}$), and similarly for the fermions $f \in \mathbb{A}_{n\mathcal{N}}$ ($\tilde{f} \in \mathbb{A}_{n\tilde{\mathcal{N}}}$). After tensoring we arrange the resulting supergravity fields into a bosonic doublet and a fermionic doublet,

$$B = \begin{pmatrix} b \otimes \tilde{b} \\ f \otimes \tilde{f} \end{pmatrix}, \quad F = \begin{pmatrix} b \otimes \tilde{f} \\ f \otimes \tilde{b} \end{pmatrix}, \tag{4.40}$$

just as we did in $D = 3$. The algebra (4.37) acts naturally on these doublets. However, a diagonal $\mathfrak{so}(\mathbb{A}_n)_{ST}$ subalgebra of this corresponds to spacetime transformations, so we must restrict $\mathfrak{h}(\mathbb{A}_{n\mathcal{N}}, \mathbb{A}_{n\tilde{\mathcal{N}}})$ to the subalgebra that commutes with $\mathfrak{so}(\mathbb{A}_n)_{ST}$. Heuristically, we identify a diagonal spacetime subalgebra $\mathbb{A}_n$ in $\mathbb{A}_{n\mathcal{N}} \otimes \mathbb{A}_{n\tilde{\mathcal{N}}}$ and require that it is preserved by the global isometries, which picks out a subset in $\mathfrak{Isom}((\mathbb{A}_{n\mathcal{N}} \otimes \mathbb{A}_{n\tilde{\mathcal{N}}})\mathbb{P}^2)$. Imposing this condition selects the U-duality algebra of the $D = n+2$, $(\mathcal{N}+\tilde{\mathcal{N}})$-extended supergravity theory obtained by tensoring $\mathcal{N}$ and $\tilde{\mathcal{N}}$ super Yang-Mills theories. The Lie algebras are given by the *magic pyramid formula*:

$$\mathfrak{MPyr}(\mathbb{A}_n, \mathbb{A}_{n\mathcal{N}}, \mathbb{A}_{n\tilde{\mathcal{N}}}) := \left\{ u \in \mathfrak{m}(\mathbb{A}_{n\mathcal{N}}, \mathbb{A}_{n\tilde{\mathcal{N}}}) - \mathfrak{so}(\mathbb{A}_n)_{ST} \middle| [u, \mathfrak{so}(\mathbb{A}_n)_{ST}] = 0 \right\}. \tag{4.41}$$

The terminology is made clear by the pyramid of corresponding U-dualities groups presented in Figure 5. The base of the pyramid in $D = 3$ is the 4×4 Freudenthal magic square, while the higher levels are comprised of a 3×3 square in $D = 4$, a 2×2 square in $D = 6$ and Type II supergravity at the apex in $D = 10$. Note, in [54] the oxidation of $\mathcal{N}$-extended $D = 3$ dimensional supergravity theories was shown to generate a partially symmetric "trapezoid" of non-compact global symmetries for $D = 3, 4, \ldots 11$ and $0, 2^0, 2^1, \ldots 2^7$ supercharges. A subset of algebras in the trapezoid with $D = 3, 4, 5$ and $2^5, 2^6, 2^7$ supercharges matches the $D = 3, 4, 5$ and $\mathbb{A} = \mathbb{C}, \mathbb{H}, \mathbb{O}$ exterior wall of the pyramid of Figure 5.

Let us conclude with some comments on the product of theories other than super Yang-Mills. Particularly interesting examples are provided by the superconformal multiplets in $D = 3, 4, 6$. In a manner directly analogous to the magic pyramid the tensor product of left and right superconformal theories yields the "conformal pyramid", described in [72]. It has the remarkable property that its faces are also given by the Freudenthal magic square, as depicted in Figure 6. In particular, ascending up the maximal spine one encounters the famous exceptional sequence $E_{8(8)}, E_{7(7)}, E_{6(6)}$, but where $E_{6(6)}$ belongs to the $D = 6, (4,0)$ theory proposed by Hull as the superconformal limit of M-theory compactified on a 6-torus [108–110]. This pattern suggests the existence of some highly exotic $D = 10$ theory with $F_{4(4)}$

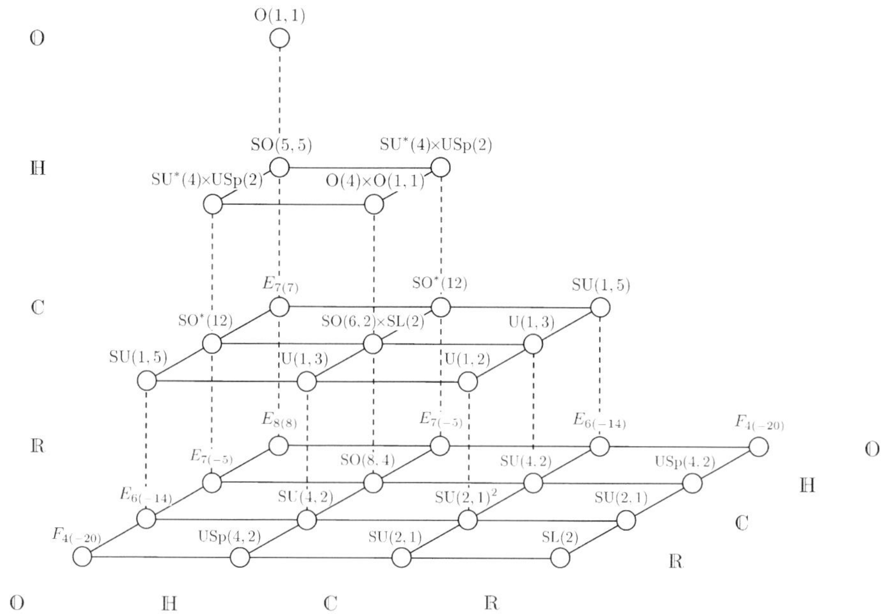

Figure 5: A magic pyramid of supergravities. The vertical axis labels the spacetime division algebra $\mathbb{A}_n$, while the horizontal axes label the algebras associated with the number of supersymmetries $\mathbb{A}_{n\mathcal{N}}$ and $\mathbb{A}_{n\tilde{\mathcal{N}}}$.

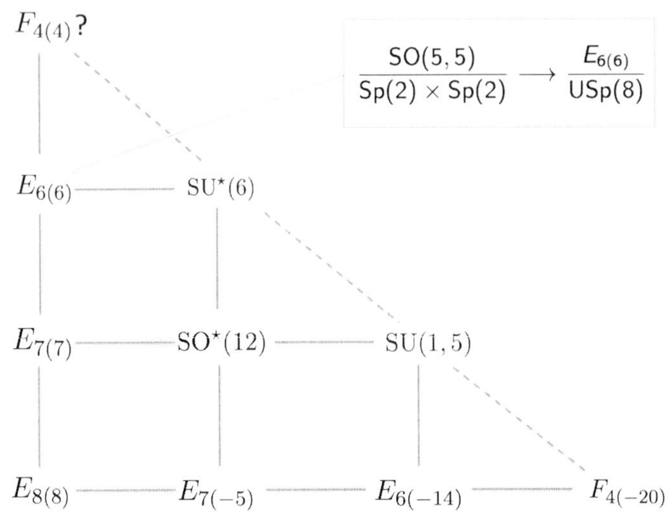

Figure 6: The magic faces of the conformal pyramid. We have highlighted the the maxiamally supersymmetric case in $D = 6$. The $\mathcal{N} = (2,2)$ supergravity theory of the magic pyramid with U-Duality SO(5,5) is replaced by the $\mathcal{N} = (4,0)$ superconfromal theory, proposed by Hull [108–110], with U-duality $E_{6(6)}$. As a consequence, the outer faces of the conformal pyramid are diagonal slices of the Freudenthal magic square, upto the $F_{4(4)}$ slot, which currently has does not have a theory associated to it.

U-duality group. The existence of such a theory would be more than a little surprising and there is a (slightly) more conventional interpretation of the conformal pyramid, including its $F_{4(4)}$ tip, but for theories in $D = 3, 4, 5, 6$, as described [72].

The product of conformal theories in the context of amplitudes has been considered previously in, for example, [51, 111–114]. In particular, the maximally supersymmetric $D = 3$, $\mathcal{N} = 8$ Bagger-Lambert-Gustavsson (BLG) Chern-Simons-matter theory [115–117] has been shown to enjoy a colour-kinematic duality reflecting its three-algebra structure [113]. The "square" of BLG amplitudes yields those of $\mathcal{N} = 16$ supergravity. Since $\mathcal{N} = 16$ supergravity is the unique theory with 32 supercharges in three dimensions it is also the "square" of the $\mathcal{N} = 8$ Yang-Mills theory. The square of the amplitudes in both cases agree, despite their distinct structures [114].

In $D = 6$ one might expect relations between the "square" of the $\mathcal{N} = (2,0)$ tensor multiplet and the $\mathcal{N} = (4,0)$ theory proposed by Hull [108–110], as discussed in [51]. Of course, amplitudes are generically not well-defined in these cases, but one can make some precise statements in terms of the tree-level S-matrix in particular regimes, as discussed in [111, 112]. For example, in the absence of additional degrees of freedom all tree-level amplitudes of the (2,0) tensor multiplet vanish [111]. The $D = 5, \mathcal{N} = 4$ super Yang-Mills theory squares to give the amplitudes of $D = 5, \mathcal{N} = 8$ supergravity. However, being non-renormalisable it ought to be regarded as a superconformal $D = 6, \mathcal{N} = (2,0)$ theory compactified on a circle of radius $R = g^2_{YM}/4\pi^2$. At linearised level Hull's (4,0) theory follows from the square the (2,0) theory [118] and gives $\mathcal{N} = 8, D = 5$ supergravity when compactified on a circle [109]. From this perspective the $(2,0) \times (2,0) = (4,0)$ identity constitutes an, as yet ill-defined, M-theory up-lift of the maximally supersymmetric $D = 5$ squaring relation.

Acknowledgments

MJD is grateful to Prof Zichichi for his hospitality. Based on work done in collaboration with A. Anastasiou, M. J. Hughes and S. Nagy. We are very grateful for their numerous and essential contributions. We would also like to thank A. Marrani for his insights on the symmetries of supergravity. The work of LB was supported by a Schödinger Fellowship. The work of MJD is supported by the STFC under rolling grant ST/G000743/1.

References

[1] R. Utiyama, "Invariant Theoretical Interpretation of Interaction," *Physical Review* **101** (Mar., 1956) 1597–1607.

[2] T. W. B. Kibble, "Lorentz Invariance and the Gravitational Field," *Journal of Mathematical Physics* **2** (Mar., 1961) 212–221.

[3] S. W. MacDowell and F. Mansouri, "Unified geometric theory of gravity and supergravity," *Physical Review Letters* **38** (Apr., 1977) 739–742.

[4] A. H. Chamseddine and P. C. West, "Supergravity as a gauge theory of supersymmetry," *Nuclear Physics B* **129** (Oct., 1977) 39–44.

[5] K. S. Stelle and P. C. West, "de Sitter gauge invariance and the geometry of the Einstein-Cartan theory," *Journal of Physics A Mathematical General* **12** (Aug., 1979) L205–L210.

[6] G. Hooft, "Dimensional reduction in quantum gravity," *Salamfestschrift: a collection of talks* **4** (1994) no. A, 1–13, `gr-qc/9310026`.

[7] L. Susskind, "The World as a hologram," *J.Math.Phys.* **36** (1995) 6377–6396, `arXiv:hep-th/9409089 [hep-th]`.

[8] J. M. Maldacena, "The large N limit of superconformal field theories and supergravity," *Adv. Theor. Math. Phys.* **2** (1998) 231–252, `arXiv:hep-th/9711200`.

[9] E. Witten, "Anti-de Sitter space and holography," *Adv.Theor.Math.Phys.* **2** (1998) 253–291, `arXiv:hep-th/9802150 [hep-th]`.

[10] S. Gubser, I. R. Klebanov, and A. M. Polyakov, "Gauge theory correlators from noncritical string theory," *Phys.Lett.* **B428** (1998) 105–114, `arXiv:hep-th/9802109 [hep-th]`.

[11] H. Kawai, D. Lewellen, and S. Tye, "A Relation Between Tree Amplitudes of Closed and Open Strings," *Nucl.Phys.* **B269** (1986) 1.

[12] Z. Bern, J. Carrasco, and H. Johansson, "New Relations for Gauge-Theory Amplitudes," *Phys.Rev.* **D78** (2008) 085011, `arXiv:0805.3993 [hep-ph]`.

[13] Z. Bern, J. J. M. Carrasco, and H. Johansson, "Perturbative Quantum Gravity as a Double Copy of Gauge Theory," *Phys.Rev.Lett.* **105** (2010) 061602, `arXiv:1004.0476 [hep-th]`.

[14] H. Elvang and Y.-t. Huang, "Scattering Amplitudes," `arXiv:1308.1697 [hep-th]`.

[15] Z. Bern, T. Dennen, Y.-t. Huang, and M. Kiermaier, "Gravity as the Square of Gauge Theory," *Phys.Rev.* **D82** (2010) 065003, `arXiv:1004.0693 [hep-th]`.

[16] J. J. M. Carrasco, "Gauge and Gravity Amplitude Relations," `arXiv:1506.00974 [hep-th]`.

[17] Z. Bern, J. J. Carrasco, L. J. Dixon, H. Johansson, and R. Roiban, "The Ultraviolet Behavior of N=8 Supergravity at Four Loops," *Phys. Rev. Lett.* **103** (2009) 081301, arXiv:0905.2326 [hep-th].

[18] Z. Bern, S. Davies, and T. Dennen, "Enhanced ultraviolet cancellations in $\mathcal{N} = 5$ supergravity at four loops," *Phys.Rev.* **D90** (2014) no. 10, 105011, arXiv:1409.3089 [hep-th].

[19] R. Monteiro and D. O'Connell, "The Kinematic Algebra From the Self-Dual Sector," *JHEP* **1107** (2011) 007, arXiv:1105.2565 [hep-th].

[20] R. Monteiro and D. O'Connell, "The Kinematic Algebras from the Scattering Equations," *JHEP* **1403** (2014) 110, arXiv:1311.1151 [hep-th].

[21] N. E. J. Bjerrum-Bohr, P. H. Damgaard, and P. Vanhove, "Minimal Basis for Gauge Theory Amplitudes," *Phys. Rev. Lett.* **103** (2009) 161602, arXiv:0907.1425 [hep-th].

[22] S. Stieberger, "Open & Closed vs. Pure Open String Disk Amplitudes," arXiv:0907.2211 [hep-th].

[23] S. H. Henry Tye and Y. Zhang, "Dual Identities inside the Gluon and the Graviton Scattering Amplitudes," *JHEP* **06** (2010) 071, arXiv:1003.1732 [hep-th]. [Erratum: JHEP04,114(2011)].

[24] C. R. Mafra, O. Schlotterer, and S. Stieberger, "Explicit BCJ Numerators from Pure Spinors," *JHEP* **07** (2011) 092, arXiv:1104.5224 [hep-th].

[25] C. R. Mafra and O. Schlotterer, "Berends-Giele recursions and the BCJ duality in superspace and components," arXiv:1510.08846 [hep-th].

[26] C. R. Mafra and O. Schlotterer, "Towards one-loop SYM amplitudes from the pure spinor BRST cohomology," *Fortsch. Phys.* **63** (2015) no. 2, 105–131, arXiv:1410.0668 [hep-th].

[27] C. R. Mafra and O. Schlotterer, "Two-loop five-point amplitudes of super Yang-Mills and supergravity in pure spinor superspace," *JHEP* **10** (2015) 124, arXiv:1505.02746 [hep-th].

[28] S. He, R. Monteiro, and O. Schlotterer, "String-inspired BCJ numerators for one-loop MHV amplitudes," *JHEP* **01** (2016) 171, arXiv:1507.06288 [hep-th].

[29] P. S. Howe and K. Stelle, "The Ultraviolet Properties of Supersymmetric Field Theories," *Int.J.Mod.Phys.* **A4** (1989) 1871.

[30] S. Deser, J. H. Kay, and K. S. Stelle, "Renormalizability Properties of Supergravity," *Phys. Rev. Lett.* **38** (1977) 527, arXiv:1506.03757 [hep-th].

[31] M. B. Green, J. G. Russo, and P. Vanhove, "String theory dualities and supergravity divergences," *JHEP* **1006** (2010) 075, arXiv:1002.3805 [hep-th].

[32] G. Bossard, P. Howe, and K. Stelle, "On duality symmetries of supergravity invariants," *JHEP* **1101** (2011) 020, arXiv:1009.0743 [hep-th].

[33] N. Beisert, H. Elvang, D. Z. Freedman, M. Kiermaier, A. Morales, *et al.*, "E7(7) constraints on counterterms in N=8 supergravity," *Phys.Lett.* **B694** (2010) 265–271, arXiv:1009.1643 [hep-th].

[34] G. Bossard, P. Howe, K. Stelle, and P. Vanhove, "The vanishing volume of D=4 superspace," *Class.Quant.Grav.* **28** (2011) 215005, arXiv:1105.6087 [hep-th].

[35] R. Monteiro, D. O'Connell, and C. D. White, "Black holes and the double copy," *JHEP* **1412** (2014) 056, arXiv:1410.0239 [hep-th].

[36] W. Siegel, "SUPERSTRINGS GIVE OLD MINIMAL SUPERGRAVITY," *Phys.Lett.* **B211** (1988) 55.

[37] W. Siegel, "Curved extended superspace from Yang-Mills theory a la strings," *Phys.Rev.* **D53** (1996) 3324–3336, arXiv:hep-th/9510150 [hep-th].

[38] A. Anastasiou, L. Borsten, M. J. Duff, L. J. Hughes, and S. Nagy, "Yang-Mills origin of gravitational symmetries," *Phys.Rev.Lett.* **113** (2014) no. 23, 231606, arXiv:1408.4434 [hep-th].

[39] A. Hodges, "New expressions for gravitational scattering amplitudes," *Journal of High Energy Physics* **1307** (2013) , arXiv:1108.2227 [hep-th].

[40] F. Cachazo, S. He, and E. Y. Yuan, "Scattering of Massless Particles: Scalars, Gluons and Gravitons," *JHEP* **1407** (2014) 033, arXiv:1309.0885 [hep-th].

[41] L. Dolan and P. Goddard, "Proof of the Formula of Cachazo, He and Yuan for Yang-Mills Tree Amplitudes in Arbitrary Dimension," *JHEP* **1405** (2014) 010, arXiv:1311.5200 [hep-th].

[42] M. F. Sohnius and P. C. West, "An Alternative Minimal Off-Shell Version of N=1 Supergravity," *Phys.Lett.* **B105** (1981) 353.

[43] S. Cecotti, S. Ferrara, M. Porrati, and S. Sabharwal, "NEW MINIMAL HIGHER DERIVATIVE SUPERGRAVITY COUPLED TO MATTER," *Nucl.Phys.* **B306** (1988) 160.

[44] S. Ferrara and S. Sabharwal, "Structure of New Minimal Supergravity," *Annals Phys.* **189** (1989) 318–351.

[45] K. Stelle and P. C. West, "Minimal Auxiliary Fields for Supergravity," *Phys.Lett.* **B74** (1978) 330.

[46] S. Ferrara and P. van Nieuwenhuizen, "The Auxiliary Fields of Supergravity," *Phys.Lett.* **B74** (1978) 333.

[47] C. M. Hull and P. K. Townsend, "Unity of superstring dualities," *Nucl. Phys.* **B438** (1995) 109–137, arXiv:hep-th/9410167.

[48] E. Cremmer and B. Julia, "The $SO(8)$ supergravity," *Nucl. Phys.* **B159** (1979) 141.

[49] A. Anastasiou, L. Borsten, M. J. Hughes, and S. Nagy, "Global symmetries of Yang-Mills squared in various dimensions," *(to appear in JHEP)* (2015) , arXiv:1502.05359 [hep-th].

[50] M. Bianchi, H. Elvang, and D. Z. Freedman, "Generating Tree Amplitudes in N=4 SYM and N = 8 SG," *JHEP* **0809** (2008) 063, arXiv:0805.0757 [hep-th].

[51] M. Chiodaroli, M. Gunaydin, and R. Roiban, "Superconformal symmetry and maximal supergravity in various dimensions," *JHEP* **1203** (2012) 093, arXiv:1108.3085 [hep-th].

[52] J. J. M. Carrasco, M. Chiodaroli, M. Günaydin, and R. Roiban, "One-loop four-point amplitudes in pure and matter-coupled N = 4 supergravity," *JHEP* **1303** (2013) 056, arXiv:1212.1146 [hep-th].

[53] M. Chiodaroli, M. Günaydin, H. Johansson, and R. Roiban, "Scattering amplitudes in $\mathcal{N} = 2$ Maxwell-Einstein and Yang-Mills/Einstein supergravity," *JHEP* **01** (2015) 081, arXiv:1408.0764 [hep-th].

[54] B. Julia, "Group disintegrations," in *Superspace and Supergravity*, S. Hawking and M. Rocek, eds., vol. C8006162 of *Nuffield Gravity Workshop*, pp. 331–350. Cambridge University Press, 1980.

[55] H. Nicolai, "The Integrability of $N = 16$ Supergravity," *Phys.Lett.* **B194** (1987) 402.

[56] P. C. West, "E(11) and M theory," *Class.Quant.Grav.* **18** (2001) 4443–4460, `arXiv:hep-th/0104081 [hep-th]`.

[57] T. Damour, M. Henneaux, and H. Nicolai, "E(10) and a 'small tension expansion' of M theory," *Phys.Rev.Lett.* **89** (2002) 221601, `arXiv:hep-th/0207267 [hep-th]`.

[58] L. Borsten, M. J. Duff, L. J. Hughes, and S. Nagy, "A magic square from Yang-Mills squared," *Phys.Rev.Lett.* **112** (2014) 131601, `arXiv:1301.4176 [hep-th]`.

[59] H. Freudenthal, "Beziehungen der E_7 und E_8 zur oktavenebene I-II," *Nederl. Akad. Wetensch. Proc. Ser.* **57** (1954) 218–230.

[60] J. Tits, "Interprétation géométriques de groupes de Lie simples compacts de la classe E," *Mém. Acad. Roy. Belg. Sci* **29** (1955) 3.

[61] B. A. Rosenfeld, "Geometrical interpretation of the compact simple Lie groups of the class E (Russian)," *Dokl. Akad. Nauk. SSSR* **106** (1956) 600–603.

[62] S. L. Cacciatori, B. L. Cerchiai, and A. Marrani, "Squaring the Magic," `arXiv:1208.6153 [math-ph]`.

[63] M. Günaydin, G. Sierra, and P. K. Townsend, "Exceptional supergravity theories and the magic square," *Phys. Lett.* **B133** (1983) 72.

[64] M. Günaydin, G. Sierra, and P. K. Townsend, "The geometry of $N = 2$ Maxwell-Einstein supergravity and Jordan algebras," *Nucl. Phys.* **B242** (1984) 244.

[65] M. Günaydin, G. Sierra, and P. K. Townsend, "Gauging the $d = 5$ Maxwell-Einstein supergravity theories: More on Jordan algebras," *Nucl. Phys.* **B253** (1985) 573.

[66] C. H. Barton and A. Sudbery, "Magic squares and matrix models of Lie algebras," *Adv. in Math.* **180** (2003) no. 2, 596–647, `arXiv:math/0203010`.

[67] J. C. Baez, "The octonions," *Bull. Amer. Math. Soc.* **39** (2001) 145–205, `arXiv:math/0105155`.

[68] A. Hurwitz, "Uber die komposition der quadratishen formen von beliebig vielen variabeln," *Nachr. Ges. Wiss. Gottingen* (1898) 309–316.

[69] R. Moufang, "Alternativkörper und der satz vom vollsta ndigen vierseit," *Abh. Math. Sem. Hamburg* **9** (1933) 207–222.

[70] P. Jordan, "Über eine nicht-desarguessche ebene projektive geometrie," *Abh. Math. Sem. Hamburg* **16** (1949) 74–76.

[71] C. Chevalley and R. D. Schafer, "The exceptional simple lie algebras f_4 and e_6," *Proc. Nat. Acad. Sci. USA* (1950) 137–141.

[72] A. Anastasiou, L. Borsten, M. J. Duff, L. J. Hughes, and S. Nagy, "A magic pyramid of supergravities," *JHEP* **1404** (2014) 178, `arXiv:1312.6523 [hep-th]`.

[73] M. Gunaydin, "Exceptional Realizations of Lorentz Group: Supersymmetries and Leptons," *Nuovo Cim.* **A29** (1975) 467.

[74] M. Gunaydin, "Octonionic Hilbert Spaces, the Poincare Group and SU(3)," *J.Math.Phys.* **17** (1976) 1875.

[75] F. Gursey, "Octonionic Structures in Particle Physics," *Lect.Notes Phys.* **94** (1979) 508–521.

[76] M. Gunaydin, "Quadratic Jordan Formulation Of Quantum Mechanics And Construction Of Lie (Super)Algebras From Jordan (Super)Algebras," in *International colloquium on group theoretical methods in physics*, vol. 10, p. 18. Israel Grp.Th.Meth., 1979.

[77] T. Kugo and P. K. Townsend, "Supersymmetry and the division algebras," *Nucl. Phys.* **B221** (1983) 357.

[78] A. Sudbery, "Division algebras, (pseudo)orthogonal groups, and spinors," *J. Phys.* **A17** (1984) no. 5, 939–955.

[79] F. Gursey, "Superpoincare Groups and Division Algebras," *Mod.Phys.Lett.* **A2** (1987) 967.

[80] M. B. Green, J. H. Schwarz, and E. Witten, *Superstring Theory vol. 1: Introduction*. Cambridge Monographs on Mathematical Physics. Cambridge University Press, Cambridge, UK, 1987. 469 p.

[81] J. M. Evans, "Supersymmetric Yang-Mills theories and division algebras," *Nucl. Phys.* **B298** (1988) 92.

[82] M. J. Duff, "Supermembranes: The First Fifteen Weeks," *Class.Quant.Grav.* **5** (1988) 189.

[83] M. Blencowe and M. J. Duff, "Supermembranes and the Signature of Space-time," *Nucl.Phys.* **B310** (1988) 387.

[84] M. Gunaydin, "Generalized conformal and superconformal group actions and Jordan algebras," *Mod.Phys.Lett.* **A8** (1993) 1407–1416, `arXiv:hep-th/9301050 [hep-th]`.

[85] N. Berkovits, "A Ten-dimensional superYang-Mills action with off-shell supersymmetry," *Phys.Lett.* **B318** (1993) 104–106, `arXiv:hep-th/9308128 [hep-th]`.

[86] C. A. Manogue and J. Schray, "Finite Lorentz transformations, automorphisms, and division algebras," *J. Math. Phys.* **34** (1993) 3746–3767, `arXiv:hep-th/9302044`.

[87] J. M. Evans, "Auxiliary fields for superYang-Mills from division algebras," *Lect.Notes Phys.* **447** (1995) 218–223, `arXiv:hep-th/9410239 [hep-th]`.

[88] J. Schray and C. A. Manogue, "Octonionic representations of Clifford algebras and triality," *Found. Phys.* **26** (1996) no. 1, 17–70, `arXiv:hep-th/9407179`.

[89] F. Gürsey and C.-H. Tze, *On the role of division, Jordan and related algebras in particle physics*. World Scientific, London, 1996.
`http://www.worldscientific.com/doi/abs/10.1142/9789812819857`.

[90] C. A. Manogue and T. Dray, "Dimensional reduction," *Mod.Phys.Lett.* **A14** (1999) 99–104, `arXiv:hep-th/9807044 [hep-th]`.

[91] M. Günaydin, K. Koepsell, and H. Nicolai, "Conformal and quasiconformal realizations of exceptional Lie groups," *Commun. Math. Phys.* **221** (2001) 57–76, `arXiv:hep-th/0008063`.

[92] F. Toppan, "On the octonionic M-superalgebra," in *Sao Paulo 2002, Integrable theories, solitons and duality*. 2002. `arXiv:hep-th/0301163`.

[93] M. Günaydin and O. Pavlyk, "Generalized spacetimes defined by cubic forms and the minimal unitary realizations of their quasiconformal groups," *JHEP* **08** (2005) 101, arXiv:hep-th/0506010.

[94] L. Borsten, D. Dahanayake, M. J. Duff, H. Ebrahim, and W. Rubens, "Black Holes, Qubits and Octonions," *Phys. Rep.* **471** (2009) no. 3–4, 113–219, arXiv:0809.4685 [hep-th].

[95] J. C. Baez and J. Huerta, "Division algebras and supersymmetry i," in *Superstrings, Geometry, Topology, and C*-Algebras, eds. R. Doran, G. Friedman and J. Rosenberg, Proc. Symp. Pure Math*, vol. 81, pp. 65–80. 2009. arXiv:0909.0551 [hep-th].

[96] J. C. Baez and J. Huerta, "Division Algebras and Supersymmetry II," *Adv.Theor.Math.Phys.* **15** (2011) 1373–1410, arXiv:1003.3436 [hep-th].

[97] J. Huerta, "Division Algebras, Supersymmetry and Higher Gauge Theory," arXiv:1106.3385 [math-ph].

[98] J. Huerta, "Division Algebras and Supersymmetry III," *Adv.Theor.Math.Phys.* **16** (2012) 1485–1589, arXiv:1109.3574 [hep-th].

[99] A. Anastasiou, L. Borsten, M. J. Duff, L. J. Hughes, and S. Nagy, "An octonionic formulation of the M-theory algebra," *JHEP* **1411** (2014) 022, arXiv:1402.4649 [hep-th].

[100] J. Huerta, "Division Algebras and Supersymmetry IV," arXiv:1409.4361 [hep-th].

[101] A. Marrani and P. Truini, "Exceptional Lie Algebras, SU(3) and Jordan Pairs Part 2: Zorn-type Representations," *J.Phys.* **A47** (2014) 265202, arXiv:1403.5120 [math-ph].

[102] K. Chung and A. Sudbery, "Octonions and the Lorentz and Conformal Groups of Ten-dimensional Space-time," *Phys.Lett.* **B198** (1987) 161.

[103] D. B. Fairlie and C. A. Manogue, "A Parametrization of the Covariant Superstring," *Phys.Rev.* **D36** (1987) 475.

[104] C. A. Manogue and A. Sudbery, "General Solutions of Covariant Superstring Equations of Motion," *Phys.Rev.* **D40** (1989) 4073.

[105] J. Schray, "The General classical solution of the superparticle," *Class.Quant.Grav.* **13** (1996) 27–38, arXiv:hep-th/9407045 [hep-th].

[106] T. Dray, J. Janesky, and C. A. Manogue, "Octonionic hermitian matrices with non-real eigenvalues," *Advances in Applied Clifford Algebras* **10** (2000) no. 2, 193–216.

[107] A. Anastasiou, L. Borsten, M. J. Duff, L. J. Hughes, and S. Nagy, "Super Yang-Mills, division algebras and triality," *JHEP* **1408** (2014) 080, arXiv:1309.0546 [hep-th].

[108] C. Hull, "Strongly coupled gravity and duality," *Nucl.Phys.* **B583** (2000) 237–259, arXiv:hep-th/0004195 [hep-th].

[109] C. Hull, "Symmetries and compactifications of (4,0) conformal gravity," *JHEP* **0012** (2000) 007, arXiv:hep-th/0011215 [hep-th].

[110] C. Hull, "Conformal nongemetric gravity in six-dimensions and M theory above the Planck energy," *Class.Quant.Grav.* **18** (2001) 3233–3240, arXiv:hep-th/0011171 [hep-th].

[111] Y.-t. Huang and A. E. Lipstein, "Amplitudes of 3D and 6D Maximal Superconformal Theories in Supertwistor Space," *JHEP* **1010** (2010) 007, arXiv:1004.4735 [hep-th].

[112] B. Czech, Y.-t. Huang, and M. Rozali, "Chiral three-point interactions in 5 and 6 dimensions," *JHEP* **1210** (2012) 143, arXiv:1110.2791 [hep-th].

[113] T. Bargheer, S. He, and T. McLoughlin, "New Relations for Three-Dimensional Supersymmetric Scattering Amplitudes," *Phys.Rev.Lett.* **108** (2012) 231601, arXiv:1203.0562 [hep-th].

[114] Y.-t. Huang and H. Johansson, "Equivalent D=3 Supergravity Amplitudes from Double Copies of Three-Algebra and Two-Algebra Gauge Theories," *Phys.Rev.Lett.* **110** (2013) 171601, arXiv:1210.2255 [hep-th].

[115] J. Bagger and N. Lambert, "Modeling Multiple M2's," *Phys.Rev.* **D75** (2007) 045020, arXiv:hep-th/0611108 [hep-th].

[116] A. Gustavsson, "Algebraic structures on parallel M2-branes," *Nucl.Phys.* **B811** (2009) 66–76, arXiv:0709.1260 [hep-th].

[117] J. Bagger and N. Lambert, "Gauge symmetry and supersymmetry of multiple M2-branes," *Phys.Rev.* **D77** (2008) 065008, arXiv:0711.0955 [hep-th].

[118] L. Borsten and et al, "On $d = 6, \mathcal{N} = (4,0)$ theories," to appear, 2015.

LECTURER: M.J. DUFF

Gravity as the square of Yang-Mills?

Scientific Secretaries: Y. Klopot, O.O. Novikov

DISCUSSION

- *Novikov:*

I have two questions, probably one depending on your answer to the first one. This relation between supergravity and two super Yang-Mills models, as I knew it for the first time at the level of scattering amplitudes, is somewhat disconnected from the details about gauge groups of these theories. In your result the gauge transformations were instrumental in getting the diffeomorphism invariance of the supergravity. So my question is: is there some influence of the specific choice of the gauge group of super Yang-Mills models that can arise in this relation?

- *Duff:*

Well, we wanted to do the most general thing we could do in order to get the right answer. And we found we were able to take both the left and right-handed gauge groups to be arbitrary. At least in the calculation that we did we didn't find any constraint on the gauge groups. When we come to the extended supergravities, there is an interesting question however that I don't think we've fully answered. And that is that the supergravity theories have electric-magnetic dualities that transform the field strengths into their duals. And you might say: ``What is Yang-Mills origin of these gravitational dualities?'' The most obvious answer you might think of is that it comes from the Yang-Mills dualities. However, the Yang-Mills dualities are rather subtle because they are non-abelian and you don't see them in perturbation theory. N=4 Yang-Mills for example is believed to be self-dual and there is symmetry between the magnetic monopoles in the theory and the fundamental particles in theory. But that is a non-perturbative effect and it's a bit of mystery how that fits into the gravitational self-duality if indeed that's the reason for it. Now in the case of Yang-Mills duality, I believe that if the gauge group is simply laced - the dual theory gives you back the same gauge group but if it isn't, the dual theory gives you back a different gauge group. So that sounds like it might be imposing some constraint on the choice of gauge group. It might tell you for example that the G_L has to be the dual of G_R, I don't know. So there are some issues that haven't been resolved that may constraint the gauge groups but at the level of the calculation I was doing this morning we don't see these subtleties.

- *Novikov:*

Thanks for the answer, that was very interesting, but I still have the second question. What if we consider one or both of these Yang-Mills theories in the planar limit? What could you say about the result of such a product?

- *Duff:*

Well, we haven't looked at the dynamics of the Yang-Mills. Everything I said this morning was addressing a simpler question: how to derive the gravitational symmetries starting from the Yang-Mills symmetries. The whole issue of the dynamics (which is a very important one) is something we will address after we've understood the symmetries. An interesting question but we haven't looked at that.

- 't Hooft:

It's just related to this question. As you know Zvi Bern was giving lectures last year at Erice and he was addressing these dynamics questions very much. Are you talking really about the same thing or is it is a different kind of conjecture that he is considering?

- Duff:

I think we are talking about the same thing but the connection is not at all obvious. He is doing on-shell momentum-space scattering amplitudes: we're doing off-shell fields in coordinate space. Somewhere along the line they must meet but we haven't yet made a contact with what he's doing. For example one thing we worried about was taking care of these indices and the way we chose to do it is to put a scalar field between two Yang-Mills. You might say maybe you can put on the other side of this equation. And if you look at the work of Cachazo and others, what they find is indeed that the product of the two spin-one amplitudes is a spin-two amplitude times the spin-zero amplitude or rather

M(2)=M_L(1)M^(-1)(0)M_R(1)

where M(s) is the spin s amplitude. So we're a bit mystified if our field is the field that Cachazo and his collaborators use, or its inverse. What we do believe is that these four fields (gravity, right YM, left YM and the scalar) live in the different worlds. Maximal supergravity can't couple to anything else, it's unique. Similarly, in the maximal case the Yang-Mills fields can't couple to anything even if you wanted them to. Things don't have to be maximal but I'm picking the maximal case to illustrate my point. ^{3} interaction according to Cachazo and Hodges who first proposed it. So it's not that we have gravity coupled to Yang-Mills or Yang-Mills coupled to a scalar. Each of these fields has its own dynamics. Now if that is true, if I look at the correlations of field products, they will factorize into correlation functions of the individual fields and that factorization should be to all orders. So if you have correlation functions which factorize to all orders that may go someway towards explaining why M(2)M(0)=M_L(1)M(1). So to answer your question, we believe there is a connection but there's a lot of work still to be done.

- Sopenko:

My first question is this: is there any conceptual reason for such an unusual product?

- Duff:

Well, what is the right product? That's the question we struggled with for a long time. First of all you must do something with the gauge indices; you can't leave them just floating around because the gravity theory has no Yang-Mills quantum numbers. So we have to do something and that's the easiest thing to do. And this scalar field... We weren't the first people to introduce it; Hodges I think was the first. He found it useful in his amplitude calculations. And we find it fills the bill in explaining where the indices go. So our field is in the bi-adjoint representation G_L x G_R and so was his. The next question is why we have these convolutions. And that seems to be necessary in order to avoid the Leibnitz rule because if you have the Leibnitz rule there are too many derivatives and you don't derive the correct gravitational transformation rules from those of Yang-Mills. So our justification in a way is a posteriori - it worked. At some deeper level why is it so, we are not sure, but if we use it we get the right answer. That's the best reply I can give.

- *Sopenko:*

The second question is: We can in principle consider the reduction of the 11-dimensional gravity to the 2 dimensions and in this case we get infinite-dimensional symmetry group E9. Is it possible to obtain magic square for D=2 supergravity for affine Lie algebras?

- *Duff:*

Very good question! The magic pyramid focused on dimensions 3,4,6 and 10 but if you like you can fill in the gaps and do it for the other dimensions. And you get the so-called U-duality groups: E8 in 3, E7 in 4 and so on. Now as you correctly pointed out, if you go down to two dimensions this U-duality group becomes the infinite-dimensional group as Cremmer and Julia first noted. So if what we are saying is true we ought to be able to derive those infinite-dimensional symmetries from Yang-Mills as well. We haven't addressed that yet but you were right to say that it ought to follow if what we say is true.

- *Marrani:*

I also have two questions for you. The first question is this: you mentioned that not all supergravity theories have off-shell formulations. I was wondering whether you might get some linearized off-shell formulations of supergravity starting from Yang-Mills theories by your tensoring procedure?

- *Duff:*

Good. Off-shell. Well, before I answer the supergravity question let me first make a remark about Yang-Mills, because something interesting happens here. It's been known for many years that if you look at the Yang-Mills theories, you can find off-shell formulations for all except the maximally supersymmetric ones, strangely enough. So, you can't do it at all in D=10, you can do it for N=1 in D=6, you can do it for N=1,2 in D=4, and you can do it for N=1,2,4 in D=3. But it's well known that in the maximal cases (N=1,D=10), (N=2,D=6), (N=4,D=4) and (N=8,D=2), the algebra only closes on-shell, not off-shell. Now, we reformulate these theories in terms of R,C,H and O and go through the business of closing the algebra. Sure enough, we find the old result but in the new framework. The reason why the maximal theories fail to close off-shell in this language is precisely because of the non-associativity of the octonions! Now we take the square, and, of course, we can't get off-shell supergravity theories by squaring unless we have the off-shell Yang-Mills theory to start with. So, we don't get anything that wasn't already known.

- *Marrani:*

My second question concerns that scalar in bi-adjoint, that we use for your product that does not satisfy Leibnitz rule. If I understand well, bi scalar is neither in supergravity theory nor in the Yang-Mills, so it's some external spectator scalar field. So, how can you constrain it in the sense that you said, it should have a cubic theory, but why?

- *Duff:*

In fact to the linearized order, we don't need to know that it is based on a phi-cubed theory. I believe that when we do the non-linear case, that issue should arise. If you think about the two structure constants for G_L and G_R, and you have a field in a bi-adjoint, you can form a cubic invariant – the phi-cube theory. It cannot be part of Yang-Mills nor can it be part of the gravity.

- *Marrani*:

So it's not a physical field, it is a technical tool?

- *Duff:*

So far yes.

- *Tokareva:*

Can we construct gravity from 2 vector fields without supersymmetry?

- *Duff:*

Can we square non-supersymmetric Yang-Mills to get non-supersymmetric gravity? Yes. Supersymmetry does not seem essential. It just keeps you on the straight and narrow. There are a lot of consistency checks you can do in supersymmetric theories that helps you to keep track of what's going on, but it doesn't have to be supersymmetric.

- *Salakh:*

Are there geometrical or physical reasons for using the division algebras in the supersymmetrical Yang-Mills theory apart from their use in magic square?

- *Duff:*

Well, I think there are, yes. But let us first look at the squaring. Because if you go back to the magic square, historically, the mathematicians were trying to give a geometric picture of the exceptional groups, the E7, the E6, the E8. That was the reason. And they succeeded in doing that with this construction which involves in some way squaring the division algebras to get the Lie algebras, and the exceptional groups appear along this final row and column. When we come to do the physics, something similar happens. The Yang-Mills theories have this RCHO interpretation and when we square them to get the gravity theory we get the same exceptional groups that mathematicians got. And so that same geometrical relation between division algebras and lie algebras persists in physics as it did in mathematics. So, that's, presumably, the underlying reason why we are finding what we are finding. We find it for the same reason as mathematicians find it. If you want more geometrical interpretation, there is something else that I mentioned only briefly. The scalar fields in our supergravity theory belong to coset G/H. G is the non-compact symmetries disappearing in magic square and H are the maximal compact subgroups. It turns out that those cosets have geometrical interpretation in terms of projective planes, and I refer you to these papers which explain this in great detail. We've only just started to look at this, but I think that the message would be that there is indeed much more geometry behind what I was telling you than we've so fare been able to uncover. So, the short answer to your question is: yes, it's very geometrical. If we forget the magic square for a moment, however, we present a master Lagrangian which encapsulates all cases. Each possibility is obtained from the unique (O,O) (D = 10, N = 1) theory by a combination of Cayley-Dickson halving, which amounts to dimensional reduction, and removing points, lines and quadrangles of the Fano plane, which amounts to consistent truncation. The so-called triality algebras associated with the division algebras allow for a novel formula for the overall (spacetime plus internal) symmetries of the on-shell degrees of freedom of the theories.

Strong Coupling and Classicalization

Gia Dvali[a,b,c],

[a]*Arnold Sommerfeld Center for Theoretical Physics*
Department für Physik, Ludwig-Maximilians-Universität München
Theresienstr. 37, 80333 München, Germany
[b]*Max-Planck-Institut für Physik*
Föhringer Ring 6, 80805 München, Germany
[c]*Center for Cosmology and Particle Physics*
Department of Physics, New York University,
4 Washington Place, New York, NY 10003, USA

Abstract

Classicalization is a phenomenon by which a theory shields itself from the strong coupling regime by redistributing a high energy of few initial hard quanta among many weakly-interacting soft quanta. In this way, the high-energy scattering process splits into large number of soft elementary processes, each with very low momentum exchange. The outcome of such a scattering experiment is a production of states of large occupation number that are approximately classical. It is evident that black hole creation in particle collision at super-Planckian energies is a result of classicalization, but there is no *a priory* reason why this phenomenon must be limited to gravity. If the hierarchy problem is solved by classicalization, the LHC has a chance of detecting the tower of resonances. The heavier resonances are expected to be longer-lived and to decay into larger number of the softer quanta.

1 Motivation for New Physics Around TeV: The Hierarchy Problem

There are several motivations for new physics beyond the standard model:

1) More simplicity, elegance and predictivity of possible extensions of the Standard Model;
2) The need to account for the phenomena of nature that Standard Model cannot explain (e.g., dark matter, dark energy, inflation, baryogenesis);
3) The unification principle (e.g., unification of gauge forces and/or unification with quantum gravity);
4) Naturalness.

I shell split the naturalness problems into the following two categories:

I. Problems of UV-sensitivity. An example is the Hierarchy Problem: the quadratic sensitivity of the Higgs mass to an UV-cutoff.

II. Problems of Vacuum (Super)Selection. An example is the Strong-CP Problem originating from the θ-vacuum in QCD.

The hierarchy problem has a clear physical meaning because of gravity. Gravity contains a fundamental scale, the Planck length, defined in the following way,

$$L_P^2 \equiv \hbar G_N \,, \tag{1}$$

where G_N is the Newton's constant. The corresponding mass scale, $M_P \equiv \frac{\hbar}{L_P}$, is the Planck mass. Planck mass represents a boundary between the world of elementary particles and the world of black holes. The elementary particles heavier than M_P do not exist, they are black holes!

In order to understand this, consider a particle of mass m. There come the two associated length-scales with it. The first one is the Compton wavelength, $L_C \equiv \frac{\hbar}{m}$. This is a distance at which the energy of quantum fluctuations exceeds the mass of the particle. The second length-scale is the gravitational radius, $L_g \equiv \hbar \frac{m}{M_P^2}$. An object of mass m localized beyond this length-scale becomes a black hole. Notice, this length-scale is *classical*, because it is independent of $\hbar$.

For $m < M_P$, we have, $L_C > L_P > L_g$, and thus, the Compton wavelength of a particle is the dominant length-scale. For example, if we think of a particle of mass $m \ll M_P$ as a gravitating source, we discover that the classical Newtonian gravitational potential created by such a particle at distance $r \sim L_C$ is very weak,

$$\phi_{Newton} \sim \frac{mG_N}{L_C} = \frac{L_g}{L_C} \ll 1 \,. \tag{2}$$

This means that if we approach particle from infinity, the quantum effects become important way before we have a chance to probe the scale L_g, i.e., the gravitational radius is completely shielded by quantum effects. This is why the elementary particles - despite being treated as point-like - cannot be considered as black holes.

For $m > M_P$, the story changes dramatically. We now have $L_C < L_P < L_g$ and the dominant length is the gravitational radius. We are dealing with a black hole!

The three length-scales meet, $L_C = L_P = L_g$, for $m = M_P$. Hence, the Planck mass is an absolute UV-cutoff in the sense that it represents an upper bound on the mass of any elementary particle.

The particles with $m \sim M_P$ are on the boundary of the two worlds and combine properties of both. On one hand, they are strongly gravitating at their Compton wavelength, i.e., the Newtonian potential (2) created by them at the distance $r \sim L_C$ is order one. On the other hand they are far from being classical black holes, since quantum fluctuations are 100 percent important and cannot be ignored. So, I shall refer to such particles as *quantum black holes* .

The above makes the hierarchy problem real. We have understood that Higgs cannot be heavier than M_P. However, since it is 17 orders of magnitude lighter than M_P, it is fully legitimate to ask, what keeps the Higgs particle so light?

In contrast, it would not be a hard question to answer, for example, why Higgs is much lighter than the Earth. Such a question could be legitimate in quantum field theory without gravity, but with gravity the answer is obvious: the Higgs particle cannot be as heavy as the Earth, because in such a case it would not be an elementary particle, but instead a classical black hole of ~cm size. Thus, we know that non-perturbative gravity by consistency prevents Higgs from renormalization of its physical mass beyond M_P. But, why is the Higgs mass-square about 34 orders of magnitude smaller than the scale of its natural stability? Putting it differently we can summarize the Hierarchy Problem as a question:

Why is the Higgs particle so far from being a quantum black hole?

2 Classicalization

If the hierarchy problem is not a problem of vacuum-selection, there must exist some new physics around TeV energies. This new physics can be weakly-coupled or strongly-coupled. Let me focus on a strongly-coupled case. The usual attitude towards the strong coupling is that this is something we should be scared of. I shall take a different attitude. First, the strong-coupling probably is the best thing that could happen to LHC: if LHC reaches the strong-coupling scale we cannot miss new physics. But, this is not the only reason for my focus on the strong coupling in this lecture. Strong coupling will force us to think about new possibilities for UV-completion that until recently have been overlooked.

Let me explain this. In quantum field theory, the strength of the interaction among elementary degrees of freedom is controlled by a quantum couplings, which can be chosen to be dimensionless and which I shall denote by α. I shall assume that the coupling is properly normalized in such a way that for $\alpha \ll 1$ the interactions are weak and the scattering rate is small. For $\alpha \gtrsim 1$ the interaction becomes strong and the naively-evaluated scattering amplitude violates unitarity. In quantum field theory the value of the coupling $\alpha(E)$ depends on the energy scale E at which it is evaluated. Thus, the strength of the coupling is a scale-dependent notion.

What happens when a theory hits the strong coupling, $\alpha(\Lambda) \sim 1$, at some energy scale Λ? Above the scale Λ the theory becomes a theory of something else: new degrees of freedom enter the game. Here we can distinguish the two possible scenarios.

In the first scenario the new degrees of freedom can coexist with the old ones. That is, for $E < \Lambda$ the relevant degrees of freedom are X, whereas for $E > \Lambda$ the relevant degrees of freedom are X, Y. A well known example is the Higgs particle in the standard model. In the standard model without Higgs, the interaction of the longitudinal W-bosons becomes strong for energies above the scale Λ not far above the weak scale. This strong coupling is resolved by inclusion of the Higgs boson, which unitarizes the $W - W$ scattering at high energies. All the low energy degrees of freedom of the standard model continue to exist as legitimate weakly-interacting degrees of freedom also in the high-energy domain with the sole addition of the Higgs boson.

The second scenario is that the theory fully changes across the scale Λ and the old IR degrees of freedom X are replaces by completely different UV degrees of freedom Y.

The examples of the latter behavior are QCD and gravity. In QCD the role of the strong coupling scale is assumed by the QCD scale Λ_{QCD}. For energies below Λ_{QCD}, or equivalently, for distances much larger than $L_{QCD} \equiv \hbar \Lambda_{QCD}^{-1}$, the relevant degrees of freedom are mesons, glueballs and baryons. (Despite the fact that some of these are heavier than Λ_{QCD}, nevertheless they are the correct asymptotic states that a low energy observer can detect). Whereas, for distances smaller than L_{QCD} the good degrees of freedom are quarks and gluons.

In gravity, the role of Λ is taken-up by the Planck mass M_P. As we have discussed above, for energies below M_P gravity represents a theory of graviton and of other elementary particles, whereas above the scale M_P, gravity becomes a theory of black holes, since there exist no elementary particles with the mass exceeding M_P.

In both of these examples the universal effect is that when we approach the scale Λ from either side, the degrees of freedom become strongly coupled. Due to this, the scale Λ is an universal regulator in the sense that the weakly-coupled degree of freedom "inhabiting" a given domain cannot cross over into a domain of the strong coupling.

Consider a theory with a four-point interaction of some degrees of freedom X that gets strong above the scale Λ. Consider a head-on collision of two X-quanta with the center of mass energy $\sqrt{s} \gg \Lambda$. Then, the coupling evaluated at such a high energy is strong, $\alpha(\sqrt{s}) \gg 1$. So, the unitarity seems to be violated. This violation means that perturbation theory in α breaks down. If the theory is to make any sense, this perturbative expansion must be replaced by something else.

We shall now try to make a guess. Our task is to design a scenario that will avoid violation of unitarity, but the rules are:

1) We are not allowed to invent new elementary quanta above the scale Λ;

2) We must respect all the basic rules of the quantum field theory (conservatio of energy-momentum, no negative norm, etc...)

Our proposed solution, following [1, 2], is the following. The problem comes from the fact that the energy per X-quantum is too high and if this energy momentum will get exchanged among the two quanta the quanta will be in a strong-coupling regime. So, in order to avoid this, the system must turn the two-particle scattering process $2X \to 2X$ into a multi-particle process $2X \to NX$ consisting of many elementary processes in such a way that the momentum exchange per-quantum in each elementary process is small-enough so that the corresponding coupling α is weak. In this way, the system has a chance to never leave the weak-coupling regime throughout the scattering process.

That is, the entire $\sqrt{s}$ energy should be re-distributed among N quanta, such that $\alpha(\sqrt{s}/N) < 1$. Equivalently, the number N must satisfy

$$\sqrt{s}/N < \Lambda\,. \tag{3}$$

Obviously, higher is $\sqrt{s}$, more quanta must be produced in order to satisfy this requirement. Hence, such a theory has a chance to unitarize at the expense of creating more and more soft quanta with an increasing center of mass energy. But, the states of high-occupation number are behaving approximately *classically*. Hence, the name *classicalization* [1, 2].

Thus, classicalization is a phenomenon describing a situation in which a theory prevents itself from entering the strong-coupling regime at high energy, $\sqrt{s} \gg \Lambda$, by means of redistributing the high center of mass energy among N soft quanta, such that they satisfy (3), and therefore are weakly interacting.

As we shall explain later, this is precisely what happens in gravity in very high energy particle collisions. However, there is no *a priory* reason why this phenomenon should be restricted to gravity. A theory with a genuine strong coupling scale Λ - and with no Wilsonian UV-completion available above this scale - can save unitarity by classicalization.

3 Hierarchy Problem and Classicalization

An interesting question - both from fundamental as well as the phenomenological points of view - is whether the hierarchy problem can be solved via classicalization due to some new interaction of the standard model Higgs? This could happen, if the Higgs particle is subjected to some fundamental interaction that becomes strong above a cutoff scale $\Lambda \sim$ TeV. Let us focus on a situation when this new classicalizing interaction does not imply introduction of new elementary particles.

In such a case we have to endow the Standard Model sector with interaction that becomes strong around TeV energies, but we must forbid the resolution of this interaction by some new weakly-coupled degrees of freedom. In other words, the only possibility for the system to unitarize is due to production of multi-particle states exclusively using its low energy quanta of the Standard Model.

For example, let us consider a Higgs self-coupling trough some simple derivative interaction,

$$G\,(\partial_\mu H \partial^\mu H)^2 \;=\; \frac{1}{\hbar}\frac{1}{\Lambda^4}\,(\partial_\mu H \partial^\mu H)^2\,. \tag{4}$$

Here G is the coupling constant of a *classical* theory, which has a dimensionality $[G] = \frac{(length)^3}{energy}$. We have rewritten it in terms of a quantum cutoff scale as $G = \frac{1}{\hbar\Lambda^4}$. In $\hbar \to 0$ limit, $\Lambda \to 0$ in such a way that G is finite, as it should. The dimensionless *quantum* coupling for a characteristic momentum exchange p, is defined as

$$\alpha \equiv \hbar G p^4 = \frac{p^4}{\Lambda^4}\,. \tag{5}$$

We shall assume that the above interaction is *fundamental*, in the sense that it is not obtained via integrating-out some weakly-coupled elementary degrees of freedom. To put it differently, we demand that the strong coupling is *genuine*.

The above interaction gives $2 \to 2$ scattering of the Higgs bosons, with the naive effective coupling strength that grows as $\alpha \sim \frac{s^2}{\Lambda^4}$. If system classicalizes, then at energies $\sqrt{s} \gg \Lambda$, we expect the scattering to be dominated by production of many soft Higgs quanta. Can we estimate their number N?

It is interesting that this number can be deduced in a model independent way, from the requirement that the characteristic wavelength of the soft quanta is independent of $\hbar$. As shown in [1], there exists an unique classical length-scale that can be built out of the center of mass energy $\sqrt{s}$ and the coupling constants of the theory and it is given by,

$$L_* \equiv \frac{\hbar}{\Lambda}\left(\frac{\sqrt{s}}{\Lambda}\right)^{\frac{1}{3}} = (G\sqrt{s})^{\frac{1}{3}} . \tag{6}$$

In [1] this scale was called a *classicalization radius* (notation used there was r_*). From here we can easily derive the number of quanta of the above wavelength among which the initial energy $\sqrt{s}$ must be redistributed. We get,

$$N = \sqrt{s}\frac{L_*}{\hbar} = \left(\frac{\sqrt{s}}{\Lambda}\right)^{\frac{4}{3}} . \tag{7}$$

In the same time evaluating the effective coupling α given by (5) for the soft quanta of wavelength (6), we discover,

$$\alpha = \left(\frac{\sqrt{s}}{\Lambda}\right)^{-\frac{4}{3}} , \tag{8}$$

or equivalently,

$$\alpha N = 1 . \tag{9}$$

To put it shortly, we discover that the requirement that the wavelength of the soft quanta be given by the unique classical scale existing in the problem, fixes the number of quanta to be given by the inverse of their quantum coupling! What is the significance of this fact? The answer is: the quantum criticality of the quantum state of N soft bosons [7]. It is because of this quantum criticality that the production of the N-particle state in scattering of two energetic quanta can potentially avoid the suppression by the semiclassical factor, e^{-N}, that would be expected in usual weakly-coupled theories.

4 Lesson from Gravity

.

One theory for which we have excellent arguments indicating that the classicalization works, is gravity [2,10]. This is because the necessary creation of multi-particle states is guaranteed by matching the fact that the high-energy scattering in gravity is dominated by black holes. It is an old idea that in gravity in very high energy particle collisions black holes are produced [3–5].

The qualitative argument goes as follows. Consider a collision of two elementary particles at trans-Planckian center of mass energy, $\sqrt{s} \gg M_P$. We shall assume that there is no long-range gravity-competing repulsive force acting between these particles. Then, if the impact parameter is less than the gravitational radius corresponding to the center of mass energy, $L_g = \sqrt{s}L_P^2$, the initial energy will get localized within the region smaller than its own gravitational radius. Hence a black hole of mass $\sqrt{s}$ is expected to form.

Following [2, 10], let us now explain that - from the microscopic point of view - the black hole creation in high energy particle collision is due to classicallization.

There exists a strong evidence [10] that from the point of view of microscopic theory the amplitude capturing the essence of this process is the one describing a transition of two high energy quanta into N soft gravitons, of momenta $\hbar/L_g$. Correspondingly, their number is given by $N = sL_P^2$. Note that for this process we have $N = \alpha^{-1}$, where $\alpha = M_P^2/s$, is the gravitational coupling of final gravitons. This is exactly the same relation as described by (9).

The rate of the $2 \to N$ process in this kinematic regime scales as (see [10])

$$\gamma_{2\to N} \sim \alpha^N N! \sim e^{-N}, \tag{10}$$

where we have used Stirling's formula in order to capture the leading-order exponential scaling, ignoring the power-law pre-factors and numerical coefficients.

Few observations are in order. First, notice that the number of gravitons N of the wavelength L_g - that can make-up the initial center of mass energy $\sqrt{s}$ - scales as the Bekenstein entropy [8] of the corresponding mass black hole,

$$N = \alpha^{-1} = S_{Bek}\,. \tag{11}$$

Consequently, the exponential suppression factor in (10) matches the expected entropy suppression factor $\sim e^{-S_{Bek}}$ that one could get - from very general semi-classical considerations - for a transition rate from an initial two-particle state to any given black hole micro-state.

Secondly, notice that the black hole entropy scales as the inverse coupling of the final soft gravitons, $S_{Bek} \sim \alpha^{-1}$.

From here we are learning the two things.

First, we clearly see glimpses of black hole entropy in $2 \to N$ amplitude. Secondly, we see that this glimpses appear in the kinematic regime in which the number and the coupling of outgoing gravitons satisfy the relation (9). According to the black hole portrait of [6, 7] this is not an accident. The microscopic significance of the relation (9) is that it corresponds to the critical point of quantum phase transition, at which N gravitons form a self-sustained bound-state, a black hole. Because, of quantum criticality this point is characterized by appearance of large number of collective Bogoliubov modes, with the tiny energy gap $\sim 1/N$. These are the modes that carry black hole entropy. They result into an exponentially-large number of nearly-degenerate micro-states. It is this quantum criticality that defies the usual intuition that creation of multi-particle classical configurations in two particle collision are expected to be exponentially-suppressed.

Can we generalize the lesson from gravity to the other strongly-coupled theories? If yes, the guideline would be to look for quantum-criticality of N-particle state. We saw that the classicalization point in non-gravitational scalar theory satisfies the same relation (9) as the gravitons do in black hole formation process. Hence, this indicates an intrinsic connection between classicalization and quantum criticality.

One case when such criticality is achieved is if the interaction among bosons is attractive. In such a case, the quantum critical point $N\alpha = 1$ produces a near-gapless excitations and correspondingly a high-density of states. Moreover, if the interaction among bosons is momentum-dependent the number of modes can scale as the power-law with N and correspondingly the number of states can scale exponentially with N [9]. This may produce an exponentially-large density of micro-states, just as in black hole case, that could potentially compensate the naive suppression factor of production of multi-particle states.

We finish the discussion by briefly commenting on experimental signatures of classicalization. What we learn from above is that a classicalizing theory in scattering above the scale Λ must create states with many soft quanta. The number as well as the softness of the constituent quanta must increase with the center of mass energy. This must be seen as the production of the tower of resonances. At any given center of mass energy, $\sqrt{s} \gg \Lambda$, the scattering is expected to be dominated by the production of the heaviest energetically-allowed resonances. The heavier resonances should be longer lived and decay into larger number of the soft quanta.

Acknowledgements

It is a pleasure to thank Prof. Antonino Zichichi for exciting discussions and for invitation to Erice summer school "Future of Our Physics Including New Frontiers".

References

[1] G. Dvali, G.F. Giudice, C. Gomez and A. Kehagias, "UV-Completion by Classicalization," JHEP **1108**, 108 (2011) [arXiv:1010.1415 [hep-ph]].

[2] G. Dvali, C. Gomez and A. Kehagias, "Classicalization of Gravitons and Goldstones," JHEP **1111**, 070 (2011) [arXiv:1103.5963 [hep-th]].

[3] G. 't Hooft, "Graviton Dominance in Ultrahigh-Energy Scattering," Phys. Lett. **B198**, 61-63 (1987);

[4] D. Amati, M. Ciafaloni and G. Veneziano, "Superstring Collisions at Planckian Energies," Phys. Lett. B **197**, 81 (1987);

"Classical and Quantum Gravity Effects from Planckian Energy Superstring Collisions," Int. J. Mod. Phys. A **3**, 1615 (1988); "Can Space-Time Be Probed Below the String Size?," Phys. Lett. B **216**, 41 (1989); "Higher Order Gravitational Deflection and Soft Bremsstrahlung in Planckian Energy Superstring Collisions," Nucl. Phys. B **347**, 550 (1990); "Effective action and all order gravitational eikonal at Planckian energies," Nucl. Phys. B **403**, 707 (1993).

[5] D.J. Gross and P.F. Mende, "The High-Energy Behavior of String Scattering Amplitudes," Phys. Lett. B **197**, 129 (1987); "String Theory Beyond the Planck Scale," Nucl. Phys. B **303**, 407 (1988);

D.J. Gross and J.L. Manes, "The High-energy Behavior of Open String Scattering," Nucl. Phys. B **326**, 73 (1989).

[6] G. Dvali and C. Gomez, "Black Hole's Quantum N-Portrait", Fortsch.Phys. 61 (2013) 742-767, arXiv:1112.3359 [hep-th].

[7] G.Dvali and C. Gomez, "Black Holes as Critical Point of Quantum Phase Transition" Eur.Phys.J. C74 (2014) 2752, arXiv:1207.4059 [hep-th]; "Black Hole Macro-Quantumness", arXiv:1212.0765 [hep-th].

D. Flassig, A. Pritzel and N. Wintergerst, "Black Holes and Quantumness on Macroscopic Scales," Phys. Rev. D **87**, 084007 (2013) [arXiv:1212.3344].

G. Dvali, D. Flassig, C. Gomez, A. Pritzel and N. Wintergerst, "Scrambling in the Black Hole Portrait," Phys. Rev. D **88**, no. 12, 124041 (2013) [arXiv:1307.3458 [hep-th]].

[8] J. D. Bekenstein, "Black Holes and Entropy," Phys. Rev. D **7**, 2333 (1973).

[9] G. Dvali, A. Franca, C. Gomez, N. Wintergerst, "Nambu-Goldstone Effective Theory of Information at Quantum Criticality", Phys.Rev. D92 (2015) 125002, arXiv:1507.02948 [hep-th].

[10] G. Dvali, C. Gomez, R.S. Isermann, D. Lüst, S. Stieberger, "Black hole formation and classicalization in ultra-Planckian $2 \to N$ scattering", Nucl. Phys. B893 (2015) 187-235, arXiv:1409.7405 [hep-th]

Texture Zero Mass Matrices and Flavor Mixing of Quarks and Leptons

Harald Fritzsch
Department für Physik
Universität München
Theresienstraße 37
D-80333 München, Germany

Abstract

We discuss mass matrices with four texture zeros for the quarks and leptons. The three mixing angles for the quarks and leptons are functions of the fermion masses. The results agree with the experimental data. The ratio of the masses of the first two neutrinos is given by the solar mixing angle. The neutrino masses are calculated:
$m_1 \approx 0.004$ eV , $m_2 \approx 0.010$ eV , $m_3 \approx 0.070$ eV.

The flavor mixing of the quarks is parametrized by the CKM-matrix. There are several ways to describe the CKM-matrix in terms of three angles and one phase parameter. I prefer the parametrization, which Z. Xing and I introduced years ago (ref.(1)), given by the angles θ_u, θ_d and θ:

$$U = \begin{pmatrix} c_u & s_u & 0 \\ -s_u & c_u & 0 \\ 0 & 0 & 1 \end{pmatrix} \times \begin{pmatrix} e^{-i\phi} & 0 & 0 \\ 0 & c & s \\ 0 & -s & c \end{pmatrix} \times \begin{pmatrix} c_d & -s_d & 0 \\ s_d & c_d & 0 \\ 0 & 0 & 1 \end{pmatrix}, \qquad (1)$$

where $c_{u,d} \sim \cos\theta_{u,d}$, $s_{u,d} \sim \sin\theta_{u,d}$, $c \sim \cos\theta$ and $s \sim \sin\theta$. The phase parameter ϕ describes CP-violation. These three angles have been determined by the experiments: $\theta_u \simeq 5.4°$, $\theta_d \simeq 11.7°$, $\theta \simeq 2.4°$.

Relations between the quark masses and the mixing angles can be derived, if the quark mass matrices have "texture zeros", as shown by S. Weinberg and me in 1977 (ref.(2)). For six quarks the mass matrices have four "texture zeros":

$$M = \begin{pmatrix} 0 & A & 0 \\ A^* & 0 & B \\ 0 & B^* & C \end{pmatrix}. \qquad (2)$$

We can calculate the angles θ_u and θ_d as functions of the mass eigenvalues:

$$\theta_u \simeq \sqrt{m_u/m_c}, \qquad \theta_d \simeq \sqrt{m_d/m_s}. \qquad (3)$$

Using the observed mass values for the quarks, we find:

$$\theta_d \simeq (13.0 \pm 0.4)°, \qquad \theta_u \simeq (5.0 \pm 0.7)°.$$

The experimental values agree with the theoretical results:

$$\theta_d \simeq (11.7 \pm 2.6)°, \qquad \theta_u \simeq (5.4 \pm 1.1)°.$$

We can calculate the Cabibbo angle as a function of the quark masses and of the phase ϕ:

$$\theta_c \cong |\sqrt{m_d/m_s} + e^{i\phi}\sqrt{m_u/m_c}|. \qquad (4)$$

Taking into account the observed quark masses, the phase angle ϕ must be close to 90°. Using the representation of the CKM-matrix, given in eq. (1), one finds that this angle is the angle α of the unitarity triangle, which is measured to about 90°.

The CKM-element V_{cb} has been determined by the experiments:

$$V_{cb} \cong 0.041. \tag{5}$$

There is a relation between the four heavy quark masses, a phase angle ψ and V_{cb}:

$$V_{cb} \cong |\sqrt{m_s/m_b} + e^{i\psi}\sqrt{m_c/m_t}|. \tag{6}$$

This relation seems to be a problem, if the mass of the top quark is as large as 174 GeV. For example, let us assume that the phase angle is 180° and let us take the following values for the quark masses: $m_s \simeq 0.11\ GeV$, $m_b \simeq 4.3\ GeV$, $m_c \simeq 1.1\ GeV$, $m_t \simeq 174\ GeV$. The result disagrees with the experimental result:

$$V_{cb} \cong \sqrt{m_s/m_b} - \sqrt{m_c/m_t} \cong 0.08. \tag{7}$$

The "texture zero" - matrices are related to specific symmetries, which are present at the energy scale of the Grand Unification. But the flavor mixing angles and the quark masses are measured at relatively low energies. The observed quark masses receive also a small contribution from electromagnetic radiative corrections, which are proportional to the fine structure constant. An observed quark mass is the sum of a "bare quark mass", given by the "texture zero" mass matrix, and a radiative correction. The relations between the flavor mixing angles and the quark masses are relations, involving the bare quark masses - they do not include the radiative corrections.

Let us consider relation (6). The observed value of the CKM element V_{cb} is about 0.041. Taking into account radiative corrections, the two terms in relation (6) vary as follows:

$$0.11 \leq \sqrt{m_s/m_b} \leq 0.21, \qquad 0.06 \leq \sqrt{m_c/m_t} \leq 0.10. \tag{8}$$

Thus for certain values of the bare quark masses the relation (6) is valid, e.g. for the values:

$$m_t \simeq 172\ GeV,\ m_c \simeq 1.5\ GeV,\ m_b \simeq 4.2\ GeV,\ m_s \simeq 0.08\ GeV.$$

Now we consider the flavor mixing of the leptons. It is described by a 3×3 unitary matrix U, similar to the CKM mixing matrix for the quarks. It can be parametrized in terms of three angles and three phases. I use a parametrization, introduced by Z. Xing and me (ref.(3)):

$$U = \begin{pmatrix} c_l & s_l & 0 \\ -s_l & c_l & 0 \\ 0 & 0 & 1 \end{pmatrix} \times \begin{pmatrix} e^{-i\phi} & 0 & 0 \\ 0 & c & s \\ 0 & -s & c \end{pmatrix} \times \begin{pmatrix} c_\nu & -s_\nu & 0 \\ s_\nu & c_\nu & 0 \\ 0 & 0 & 1 \end{pmatrix} \times P_\nu \qquad (9)$$

where $c_{l,\nu} \sim \cos\theta_{l,\nu}$, $s_{l,\nu} \sim \sin\theta_{l,\nu}$, $c \sim \cos\theta$, $s \sim \sin\theta$. The angle θ_ν is the solar angle θ_{sun}, the angle θ is the atmospheric angle θ_{at}, and the angle θ_l is the "reactor angle". The phase matrix $P_\nu = \mathrm{Diag}\{e^{i\rho}, e^{i\sigma}, 1\}$ is relevant only for Majorana masses.

The neutrino oscillations are described by the three angles, which have been measured: $\theta_{sun} = \theta_\nu \simeq 34°$, $\theta_{at} = \theta \simeq 45°$, $\theta_l \simeq 13°$.

We assume that the mass matrices of the leptons also have four texture zeros:

$$M = \begin{pmatrix} 0 & A & 0 \\ A^* & 0 & C \\ 0 & C^* & D \end{pmatrix}. \qquad (10)$$

In this case we can calculate the solar and the reactor angle as functions of the lepton masses:

$$\tan\theta_l \simeq \sqrt{m_e/(m_e + m_\mu)}, \qquad \tan\theta_\nu \simeq \sqrt{m_1/(m_1 + m_2)}. \qquad (11)$$

From the solar mixing angle we obtain for the neutrino mass ratio:

$$m_1/m_2 \simeq 0.44. \qquad (12)$$

We use this relation and the experimental results for the mass differences of the neutrinos to determine the three neutrino masses:

$$\begin{aligned} m_1 &\simeq 0.004\ eV, \\ m_2 &\simeq 0.010\ eV, \\ m_3 &\simeq 0.070\ eV. \end{aligned} \qquad (13)$$

The atmospheric mixing angle is determined by the lepton masses:

$$sin\ \theta_{at} \cong |\sqrt{\frac{m_3}{m_2+m_3}} \times \sqrt{\frac{m_\mu}{m_\mu+m_\tau}} + e^{i\alpha}\sqrt{\frac{m_2}{m_2+m_3}} \times \sqrt{\frac{m_\tau}{m_\mu+m_\tau}}| \quad (14)$$

Here the phase parameter α is not known. Thus the atmospheric angle should be in the range

$$13^\circ < \theta_{at} < 40^\circ. \quad (15)$$

If α vanishes, one has $\theta_{at} = 40^\circ$, in agreement with the experimental value. It is interesting to observe that here the phase parameter must be close to zero, in eq. (5) it must be close to 180°, and in eq. (4) close to 90°. Perhaps this follows from a specific symmetry of the "texture zero" mass matrices.

The mass matrices of the quarks and leptons are not exactly given by texture zero matrices. Radiative corrections of the order of the fine-structure constant α will contribute - the zeros will be replaced by small numbers.

The ratios of the masses of the quarks with the same electric charge and of the masses of the neutrinos seem to be universal:

$$\frac{m_u}{m_c} \simeq \frac{m_c}{m_t} \simeq 0.005,$$

$$\frac{m_d}{m_s} \simeq \frac{m_s}{m_b} \simeq 0.044, \quad (16)$$

$$\frac{m_1}{m_2} \simeq \frac{m_2}{m_3} \simeq 0.25.$$

The dynamical reason for this universality is unclear. But the mass ratios of the charged leptons are not universal:

$$\frac{m_e}{m_\mu} \simeq 0.005,$$

$$\frac{m_\mu}{m_\tau} \simeq 0.06. \quad (17)$$

If the ratio of the muon mass and the electron mass would be equal to the ratio of the tauon mass and the muon mass, the mass of the electron would have to be about ten times larger ($\simeq 6\ MeV$).

The universality is not expected to be exact, due to radiative corrections. A radiative correction of the order of $\pm(\alpha/\pi)m_\tau \simeq \pm 4\ MeV$ would have to be added to the charged lepton masses. Such a contribution is relatively small for the muon and the tauon, but it is large for the electron. One expects that the physical electron mass is the sum of a bare electron mass $\widetilde{m}_e$, due to the texture zero mass matrix, and a radiative correction r_e: $m_e = \widetilde{m}_e - r_e$. The ratio of the bare electron mass and the muon mass should be equal to the ratio of the muon mass and the tauon mass: $\widetilde{m}_e \simeq 5.51$ MeV, $r_e \simeq 5.00$ MeV.

Radiative corrections also contribute to the muon and the tauon mass, but here the corrections are small in comparison to the bare masses and can be neglected.

We calculate the angle θ_l (see eq.(11)) and obtain a result, which agrees with the experimental value:

$$\begin{aligned} \tan\theta_l \simeq \sqrt{\widetilde{m}_e/m_\mu} \simeq 0.23, \\ \theta_l \simeq 13^\circ. \end{aligned} \qquad (18)$$

In this paper we have shown that mass matrices with four "texture zeros" describe well the flavor mixing. The mixing angles for the quarks and the leptons are functions of the fermion masses. The results agree with the experimental data.

References

[1] H. Fritzsch and Z. Z. Xing, Phys. Lett. B 413, 396 (1997), H. Fritzsch and Z. Z. Xing, Phys. Rev. D 57, 594 (1998).

[2] S. Weinberg, Trans. New York Acad. Sci. 38, 185 (1977), H. Fritzsch, Phys. Lett. 70B, 436 (1977), H. Fritzsch, Phys. Lett. 73B, 317 (1978).

[3] H. Fritzsch and Z. Z. Xing, Phys. Lett. B 682, 220 (2009).

[4] H. Fritzsch, Mod. Phys. Lett. A, 27, 1250079 (2012).

LECTURER: H. FRITZSCH

Texture Zero Mass Matrices and Flavour Mixing of Quarks and Leptons

Scientific Secretaries: I. Ivanova, L. Varnhorst

DISCUSSION

- *Avdoshkin:*

What lays behind the relations which the quark masses happen to satisfy?

- *Fritzsch:*

The texture zero conditions come from the symmetry which is related to the grand unification at very high energies.

- *Dunin-Barkowski:*

How are the results for tau neutrinos related to the results from OPERA experiment?

- *Fritzsch:*

The results of our work are in agreement with the results from OPERA experiment and other experiments.

- *'t Hooft:*

Why do texture zeros come about in SO(10) and not in SU(5) ?

- *Fritzsch:*

In SO(10) the symmetry breaking is arranged by multiplets of SO(10). Texture zeros arise when the 16 representations of SO(10) are absent.

- *Tannenbaum:*

When you mentioned $m_c/m_t=1/207$, I thought, isn't that the same as m_e/m_μ? Why did you change m_e?

- *Fritzsch:*

m_e/m_μ is the same as the ratios for the quarks. It is probably an accident. It would be better if in nature the electron should have a mass of about 5 MeV.

- *Korthals-Altes:*

Symmetry causes the texture zeros for quarks. What is the reason for the texture zero in the see-saw mechanism?

- *Fritzsch:*

It is an assumption. The fermions are in a 16-dimentional representation of SO(10). The mass matrix is assumed to be in a 10-dimensional representation.

Roadmap at the LHC to the Higgs boson and beyond

(Including highlights from ATLAS)

Peter Jenni
Albert-Ludwigs-University Freiburg, Germany, and CERN, Geneva, Switzerland

Abstract

The journey for the search of the Standard Model Higgs boson at the LHC started more than two decades ago. During 2010-2012, experiments at the Large Hadron Collider (LHC) have started exploring physics at the high-energy frontier. Thanks to the superb turn-on of the LHC already a rich harvest of initial physics results has been obtained by the general-purpose experiments ATLAS and CMS. The initial data have allowed one to test, at the highest collision energies ever reached in a laboratory, the Standard Model (SM) of elementary particles, and to make many searches Beyond the Standard Model (BSM). Clearly the most important results have been obtained with the discovery of a (or 'the') Higgs boson, which establishes the postulated electro-weak symmetry breaking mechanism in the SM.

1 Introduction

The discovery of a scalar boson, which shows within the present statistical precision achieved all the expected properties for the famous Higgs boson, announced by the ATLAS and CMS Collaborations on 4th July 2012 [1,2], was the culminating experimental triumph for the Standard Model [3-11]. The Standard Model (SM) of particle physics is one of the most remarkable achievements of physics over the past 50 years. Its descriptive and predictive power has been experimentally demonstrated with unprecedented accuracy in many generations of experiments ranging from low to high energies. The SM comprises the fundamental building blocks of all visible matter, with the three fermion families of quarks and leptons, and their interactions via three out of the four fundamental interactions mediated by bosons, namely the massless photon for the electromagnetism, the heavy W and Z bosons for the weak force, these two interactions unified in the electroweak theory, and the massless gluons for the strong interaction.

In order to solve the mystery of generation of mass, a spontaneous symmetry-breaking mechanism was proposed introducing a complex scalar field that permeates the entire Universe. This mechanism, known as Brout-Englert-Higgs (BEH) mechanism [6-9], gives the W and Z their large masses and leaves the photon massless. Interaction with the scalar field generates masses to the quarks and leptons in proportion to the strength of their couplings to it. This field leads to an additional massive scalar boson as its quantum, called the Higgs boson. After the discovery of the W and Z bosons in the early 1980s, the hunt for the Higgs boson, considered to be the keystone of the SM, became a central theme in particle physics, and also a primary motivation for the LHC. Finding the Higgs boson would establish the existence of the postulated BEH field, and thereby marking a crucial step in the understanding of Nature.

There is a vast body of literature available for 'telling the story' of the Higgs discovery at the LHC. Somehow unjustified and selectively, the following accounts are heavily based on a few didactic articles co-authored by the lecturer with longtime colleagues since the early years of the LHC adventure [12,13] and hadron collider experiments even before [14]. For scientifically rigorous accounts and full references the reader is referred to the original publications [1,2] and the full references therein, as well as to the updated public reference lists of publications by ATLAS [15] and CMS [16].

2 The LHC Project

The LHC project must be seen as a global scientific adventure, combining the accelerator complex, the experiment collaborations with their detectors, a worldwide computing grid, and a motivating theory community, that started more than 30 years ago. Obviously the main issue is the fundamental physics it addresses, but the project itself is truly 'a marvel of technology', a detailed account of which can be found in [17], and specific technological highlights are featured in [18], from which some of the technical discussion is reproduced here.

2.1 Historical Setting and Time Line

A comprehensive history of the years leading to the LHC can be found in [19]. The great success in making the experimentally "clean" W and Z boson discoveries, despite the huge hadronic backgrounds, at the CERN SppbarS Collider in the early 1980s was crucial for the community to dare to even dream of a future powerful high-energy hadron collider in order to make a decisive search for the Higgs boson. The idea that the tunnel for the, at that time, still future Large Electron-Positron (LEP) machine should be able to house, at some distant time, the LHC, was already in the air in the late 1970s [20], not the least thanks to the foresight of Professor Zichichi chairing an ECFA-LEP Working Group. Thankfully, those leading CERN at the time had the vision to plan for a tunnel with dimensions that could accommodate it. Enthusiasm for an LHC surfaced strongly in 1984 at a CERN-ECFA workshop in Lausanne entitled "LHC in the LEP Tunnel", which brought together working groups that comprised machine experts, theorists and experimentalists.

With the promise of great physics at the LHC, several motivating workshops and conferences followed, where the formidable experimental challenges started to appear manageable, provided that enough R&D work on detectors would be carried out. Highlights of these "LHC experiment preliminaries" were the 1987 Workshop in La Thuile of the so-called "Rubbia Long-Range Planning Committee" and the large Aachen ECFA LHC Workshop in 1990. Finally, in March 1992, the famous conference entitled "Towards the LHC Experimental Programme", took place in Evian-les-Bains, where several proto-collaborations presented their designs in "Expressions of Interest". This was also the time when CERN created an international peer review committee for the LHC experiments (LHCC). Moreover, from the early 1990s, CERN's LHC Detector R&D Committee (DRDC), which reviewed and steered R&D collaborations, greatly stimulated innovative developments in detector technology.

It cannot be stressed enough how important the many years of R&D were that preceded the final detector construction for the LHC experiments. Technologies had to be taken far beyond their state-of-the-art of the late 1980s in terms of performance criteria in the anticipated harsh LHC environment, like granularity and speed of readout, radiation resistance, reliability, but also considering buildable sizes of the detector components and number of units, and very importantly at an affordable cost. For many detector subsystems there were initially a few parallel developments pursued as options, because it was not guaranteed from the onset that a given proposed technology would finally fulfil all the necessary requirements. Increasingly more realistic prototypes were developed, in a learning process for both the detector communities and the industries involved.

Some of the major technology decisions were taken by the Collaborations before the submission of the Technical Proposals to the LHCC end of 1994, which were finally approved early in 1996. For other choices the R&D needed more time, and they could only be made in the subsequent years from 1996 to the early 2000s, thereby defining the timing for the final Technical Design Reports of the various detector components.

The long duration of the LHC project until the Higgs discovery is illustrated in Table 1, with a few selected milestones concerning the LHC and the general-purpose experiments.

Table 1: The LHC Timeline

1984	Workshop on a Large Hadron Collider in the LEP tunnel, Lausanne, Switzerland.
1987	Workshop on the Physics at Future Accelerators, La Thuile, Italy. The Rubbia "Long-Range Planning Committee" recommends the Large Hadron Collider as the right choice for CERN's future.
1990	LHC Workshop, Aachen, Germany (discussion of physics, technologies and detector design concepts).
1992	General Meeting on LHC Physics and Detectors, Evian-les-Bains, France (with four general-purpose experiment designs presented).
1993	Three Letters of Intent evaluated by the CERN peer review committee LHCC. ATLAS and CMS selected to proceed to a detailed technical proposal.
1994	The LHC accelerator approved for construction, initially in two stages.
1996	ATLAS and CMS Technical Proposals approved.
1997	Formal approval for ATLAS and CMS to move to construction (materials cost ceiling of 475 MCHF).
1997	Construction commences (after approval of detailed Technical Design Reports of detector subsystems).
2000	Assembly of experiments commences, LEP accelerator is closed down to make way for the LHC.
2008	LHC experiments ready for pp collisions. LHC starts operation. An incident stops LHC operation.
2009	LHC restarts operation, pp collisions recorded by LHC detectors.
2010	LHC collides protons at high energy (centre of mass energy of 7 TeV).
2012	LHC operates at 8 TeV: discovery of a Higgs boson.
2015	LHC starts Run-2 at 13 TeV collision energy.

2.2 The Collider

In the LHC the two counter-rotating beams travel in separate vacuum pipes in the arcs around the ring and consist of many particle bunches separated by a small distance. They are made to collide at the centre of the detectors with a small crossing angle. In the case of the LHC, the nominal number of bunches is 2808, the distance between bunches is 7.5 m and the crossing angle 285 μrad. The layout of the collider is shown in Fig.1. There are eight arcs (bending radius of 2804 m), where the beams are bent, and eight straight-sections used for utilities or collision insertions. Four insertions are equipped with experimental detectors, where the two counter-rotating beams can be brought to collision.

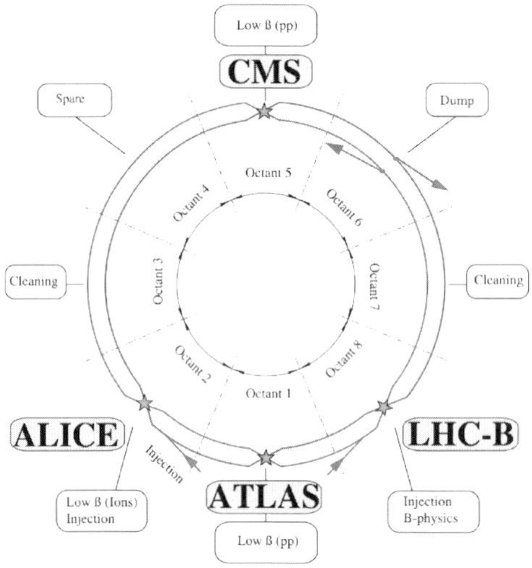

Fig. 1: Layout of the Large Hadron Collider

The two most important parameters of a proton-proton collider are the energy of the collisions and the luminosity, a parameter proportional to the number of collision events per second. The energy is related to the discovery potential of new particles of higher mass and the luminosity to the production of a relevant number of the desired events in a reasonable running time. The luminosity is a measure of the quality of the beams and their collisions, which is the result of careful design and mastering of several phenomena.

The ***beam energy*** is proportional to the product of B (the magnetic field of the main dipole magnets) and ρ (the bending radius of the arcs). Obviously, since ρ was fixed by the dimensions of the LEP tunnel, it was important to aim at the highest possible field B. The choice for LHC was to use NbTi superconductors cooled at a temperature of 1.9 K provided by superfluid helium (He-II). It is interesting to note that the solution adopted for the LHC reconciled two requirements, namely the quest for the highest possible magnetic field and a design such that the magnets could be constructed in existing industry. This design with 8.3 Tesla magnetic field in the dipoles resulted in a nominal beam energy of 7 TeV. Out of the almost ten thousands magnets forming the collider, the 1232 dipole magnets (length 16 m, mass 27 tons) represented a major design and constructional effort.

The solution of NbTi conductors at 1.9 K offers another important advantage at the expense of a more complex cryogenic system. It is due to the peculiar transport properties of pressurized superfluid helium, such as high heat capacity, low viscosity, and good effective thermal conductivity: it permits to cool the long string of magnets with a very small flow of liquid. The total cold mass of 35000 tonnes operating at 1.9 K is the coldest spot in the Universe.

The considerable beam power (362 MJ) and the large electromagnetic energy stored in the magnets (11 GJ for the complete system) require a very sophisticated protection system to prevent damage in case of beam guiding problems or resistive transition (quench). In fact each of the two beams have an energy sufficient to melt 500 kg of copper and the electromagnetic energy of the magnets, if not properly discharged, could provoke large damages.

The ***luminosity L*** is related to the properties of the colliding beams at the collision point and it is measured in $cm^{-2}\ s^{-1}$. It is proportional to the square of number of particle per bunch (N), the number of bunches around the ring (n_b) and inversely proportional to the transverse dimensions of the beams at the collision point. A small correction factor F takes into account the small angle of the beams at crossing. The product of L with the cross-section σ of the process to be investigated gives the average number of events produced per second. L varies with time, since the stored beams degrade during the collision run. The integrated luminosity is the integral L(t)dt over a certain time period. The luminosity can be considered as a figure of merit of the global quality of the machine. During the 2012 run, with beam energy of 4 TeV, the highest initial luminosity was $7.7\ 10^{33}\ cm^{-2}\ s^{-1}$. The integrated luminosity collected by ATLAS and CMS in 2012 was 21.5 fb^{-1} (delivered 23 fb^{-1}).

2.3 Overview and Motivation for the LHC Detectors

The enthusiasm for the great physics potential of an LHC arose in the community at the Lausanne ECFA-CERN workshop in 1984 already mentioned before. Finding the Higgs boson, the direct experimental manifestation of the Brout-Englert-Higgs mechanism, was clearly a priority for the future of particle physics, but was also expected to be very challenging. Its mass is not predicted by the Standard Model and could have been as high as 1 TeV. This required a search over a broad range of mass, hence ideally suited at a high-energy pp collider where the energy spectrum of the constituents of the protons (quarks and gluons) allow all possible Higgs masses to be looked for at the same time. Because of the predicted low detectable cross-sections a very high-luminosity collider was mandatory.

But the Higgs search was by far not the only reason to stimulate great interest for the LHC. Already in the 1980s there were clear motivations from speculative theories predicting physics Beyond the Standard Model (BSM), the most popular one being Supersymmetry (SUSY) with its characteristic missing transverse energy signatures due to the escaping lightest neutral SUSY particle (LSP). Other hypothetical models predicted new heavy resonances, or leptoquarks (particles containing both quarks and leptons), or even substructure to quarks, and many other exotic ideas were around. It was also realized early on that the LHC would produce huge numbers of heavy flavour particles, opening a new frontier in precision flavour physics. Furthermore, unprecedented exploratory steps could be made in studying the quark-gluon plasma at high density and temperatures when colliding heavy ions, like fully ionized lead nuclei. Motivated by these physics prospects, CERN opted ultimately for an experimental programme with two very large general-purpose detectors (ATLAS and CMS), two large apparatus optimized respectively for flavour physics (LHCb) and for heavy ion collisions (ALICE), complemented later by three much smaller specialized experiments (TOTEM, total cross-section and forward physics; LHCf, measuring forward neutral particle production; MoEDAL, monopole search).

The detection of the Higgs boson played a particularly important role in the design of the ***general-purpose experiments ATLAS and CMS***. In the region of low mass ($114 < m_H < 150$ GeV), the two channels considered mostly suited for unambiguous discovery were the decay to two photons and the decay to two Z bosons, each decaying in turn into e^+e^- or $\mu^+\mu^-$, where one or both of the Z bosons could be virtual. As the natural width of the low-mass Higgs boson is < 10 MeV, the width of any observed peak would be entirely dominated by the instrumental mass-resolution. This meant that in designing the general-purpose detectors, considerable care was placed on the value of the magnetic field strength, on the precision tracking systems and on high-resolution em calorimeters. The high-mass region, as well as the signatures from supersymmetry, drove the need for good resolution for jets and missing transverse energy (E_T^{miss}), implying also almost full 4π hadronic calorimetry coverage.

The choice of the field configuration determined the overall design for these experiments. It was also well understood that to stand the best chance of making discoveries at the new energy scale of the LHC - and in the harsh conditions generated by about a billion pairs of protons interacting every second, several tens every bunch crossing - would require the invention of new technologies while at the same time pushing existing ones to their limits. In fact, a prevalent saying was "We think we know how to build a high-energy, high-luminosity hadron collider – but we don't have the technology to build a detector for it". In reality of course both turned out to be difficult and demanded technological breakthroughs. Early on it was realized that detectors will have to face eventually even luminosities beyond the initial LHC design to reach some of the ultimate physics goals. That the general-purpose experiments have worked so marvellously well since the start-up of the LHC is a testament to the difficult technology-choices made by the conceivers and the critical decisions made during the construction of these experiments. It is noteworthy that indeed the very same elements mentioned above were crucial in the discovery of a Higgs boson.

Very different challenges were faced for the two ***special-purpose experiments LHCb and ALICE***, which are reflected in their very different specific designs. The only aspects they have in common is their operation at lower luminosity, typically at $2\ 10^{32}$ cm $^{-2}$ s $^{-1}$ or lower, avoiding basically multiple events per bunch crossing, and their specific optimizations for particle identifications, as will be discussed later.

It is far beyond the scope of this lecture note to give a comprehensive description of these four sophisticated instruments that have been developed with very considerable R&D efforts, culminating in many large-scale prototype measurement campaigns in particle beams at CERN and other accelerator laboratories, over the 1990s and early 2000s. The construction of the various components took place over about 10 years, starting in the second half of the 1990s, in universities, national laboratories, and industries. Typically, after local initial testing, the components were sent to CERN, where after reception tests they were assembled and installed in the experimental caverns, followed by commissioning tests. During all this time, from the first Letters of Intent in 1992 to the operation phase, CERN's LHCC played an important role closely guiding and monitoring the experiments. Figure 2 shows photographs of the four detectors during their late installation phase, before the detectors were completely closed for operation.

Fig. 2: The four LHC detectors during the installation phase. Upper row left (a) ATLAS, right (b) CMS, lower row left (c) LHCb, and right (d) ALICE

In the following the four detectors are very briefly introduced separately. The four collaborations have published very detailed and comprehensive technical descriptions of their detectors as finally built and operated during the first years of LHC in [21-24]. The interested reader is invited to consult these major documentations for detailed information.

2.3.1 ATLAS [21]

The design of the ATLAS detector is based on a large toroid magnet system for the muon spectrometer complemented with a small superconducting solenoid around the inner tracking cylinder, centred at the interaction point. The novel and challenging superconducting air-core toroid magnet system, contains about 80 km of superconductor cable in eight separate barrel coils (each 25×5 m^2 in a 'racetrack' shape) and two matching end-cap toroids. A field of ~0.5 Tesla is generated over a large volume. The toroids are complemented with a thin solenoid (2.4 m diameter, 5.3 m length), which provides an axial magnetic field of 2 Tesla.The momentum of the muons can be precisely measured as they travel unperturbed by material for more than 5 m in the air-core toroid field. About 1200 large muon drift tube chambers of various shapes, with a total area of 5000 m^2, measure the impact positions with an accuracy of better than 0.1

mm. Another set of about 4200 fast chambers are used to provide a muon trigger (resistive plate chambers in the barrel, thin gap chambers in the end-caps).

In the field-free region between solenoid and toroids there is a highly granular em calorimeter using a novel Lead - liquid Argon (LAr) sampling calorimeter complemented by full-coverage hadronic sampling calorimeters. For the latter a plastic scintillator – iron sampling technique, also with a novel geometry, is used in the barrel part of the experiment, whereas LAr calorimeters cover the end-cap regions near the beam axis where particle fluxes, and thereby radiation exposures, are highest. The em and hadronic calorimeters have 200000 and 10000 cells respectively.

The reconstruction of all charged particles, including those of displaced secondary vertices, is done in the inner tracking detector, which combines highly granular pixel (50 μm x 400 μm, total 80 million channels) and microstrip (13 cm x 80 μm, total 6 million channels) silicon semiconductor sensors close to the beam axis, and a 'straw tube' gaseous detector (4 mm diameter, 350000 channels) that provides about 35 signal hits per track. The latter also helps in the identification of electrons using information from the effect of transition radiation.

Figure 2a shows one end of the cylindrical barrel detector after about 4 years of in-situ installation work in the underground cavern, 1.5 years before completion. The ends of four of the barrel toroid coils are visible, illustrating the eightfold symmetry of the structure. The relatively lightweight overall structure of the detector adds up to 7000 tonnes, spanning over a large volume of 22 m diameter with a length of 46 m.

2.3.2 CMS [22]

The CMS detector features prominently a state-of-the-art superconducting high-field solenoid of 4 Tesla. This single magnet solution serves both the high precision inner tracking chambers as well as the external muon detector, which is instrumented with large gaseous drift chambers in the barrel and cathode strip chambers in the end-caps complemented by resistive plate trigger chambers, embedded in the return yoke. This configuration allowed one to achieve a compact overall design, limiting the diameter to 15 m. The magnet yoke makes up for most of the total detector weight of 12500 tonnes.

Muon detection, and their most accurate measurement, was a priority criterion for the CMS design, followed by precision measurements for photons and electrons, achieved with a new type of radiation hard em crystal calorimeter, the largest ever built. The challenging, but very successful development and manufacture of the 75848 lead tungstate scintillating crystals, in the final set-up pointing to the interaction point, took more than a decade.

The solution to charged particle tracking was to opt for a small number of precise position measurements for each charged particle trajectory (13 layers with a position resolution of ~15 μm per measurement) leading to a large number of cells distributed inside a cylindrical volume 5.8 m long and 2.5 m in diameter: 66 million 100 × 150 μm^2 silicon pixels and 9.3 million silicon microstrips ranging from about 10 cm × 80 μm to 20 cm × 180 μm. With 198 m^2 of silicon detector area the CMS tracker is by far the largest silicon tracker ever built.

Finally the hadron calorimeter, comprising ~3000 small solid angle projective towers covering almost the full solid angle, is built from alternate plates of ~5 cm brass absorber and ~4 mm thick scintillator plates that sample the shower energy. The scintillation light is detected by photodetectors (hybrid photodiodes) that can operate in the strong magnetic field, as the calorimeters are placed inside the solenoid coil.

The iron yoke of the CMS detector is sectioned into five barrel wheels and three endcap disks at each end. The sectioning enabled the detector to be assembled and tested in a large surface hall while the underground cavern was being prepared. The sections, weighing between 350 and 2000 tonnes, were then lowered into the cavern (Fig. 2b) between October 2006 and January 2008, using a dedicated gantry crane system equipped with strand jacks: a pioneering use of this technology to simplify the underground assembly of large experiments.

2.3.3 LHCb [23]

The LHCb detector concept exploits the dominant rate of beauty production towards the beam directions, and for practical reasons concentrates just on one of the two sides. Away from the LHC collision region the layout therefore resembles a fixed target spectrometer, but with very unique features.

A silicon strip vertex locator (VELO) detector can be positioned during stable beams very close to the interaction region and beam line in order to measure accurately primary and secondary vertices, selecting events with b-quarks. Particle identification to cleanly identify the various final states is achieved by two ring image Cerenkov detectors (RICH), whereas the momentum measurements are based on a large-aperture warm dipole magnet generating an integrated field path of 4 Tesla-meters for trajectories going through all tracking stations (silicon strips and straw tube drift chambers) of the spectrometer. Calorimetry is provided for by sampling scintillator lead (em) and coarser scintillator iron devices. The muon detector behind the calorimeters, an absorber with iron plates of a total of 20 interaction lengths, sampled by four chamber layers, completes the LHCb spectrometer. Figure 2c shows a picture of the fully installed LHCb detector in LHC Point-8.

2.3.4 ALICE [24]

The ALICE detector has to cope with extremely high multiplicity events, characteristic of heavy ion collisions, including charged particle measurements at an as low as possible momentum threshold. Furthermore, particle identification is needed for many of the specific heavy ion physics signatures.

The ALICE Collaboration has reused the former huge L3 warm solenoid magnet providing a field of 0.5 Tesla over a large central volume. Within the magnet, with its 10000 tonnes heavy iron yoke, is located, as central tracking detector, the world's largest Time Projection Chamber (TPC) with a field cage of 5.6 m diameter and 5.4 m length, which provides precision tracking as well as particle identification by dE/dx ionization information. The innermost region, inside the TPC, facing the most extreme particle density region around the collision point, is covered by an optimized inner tracking system with silicon pixel and silicon drift detectors followed by double-sided silicon strip detectors. Several detector systems dedicated to particle identification over various limited solid angles cover the outside radius of the TPC: transition radiation and Cherenkov radiation detectors, and a state-of-the-art time-of-flight system (TOF). High resolution em calorimetry for photon measurements is implemented with lead tungstate scintillating crystals (similar characteristics as for CMS).

A muon spectrometer starting with a massive 4 m long sophisticated hadron absorber cone, and featuring a classical dipole magnet, covers on one side the solid angle from 2 to 9 degrees with respect to the beam direction. A front view of the ALICE detector nearing installation completion is shown in Fig. 2d.

2.4 Triggering and Computing

A particular challenge for ATLAS and CMS are the very high collision rates in the LHC, necessary for the Higgs search and studies, given the small production cross section combined with the need to investigate final states with very small branching fractions. In the first three years of operation the LHC reached a peak instantaneous luminosity of $7x10^{33}$ $cm^{-2}s^{-1}$ with a 50 ns bunch spacing, which meant that the detectors had to simultaneously cope with up to ~ 50 overlapping (pile-up) events in a given bunch crossing. In the years ahead, the instantaneous luminosity is still expected to rise two- to three-fold.

It is technically not possible to store all data for all events, therefore a trigger system is used to reject large numbers of events and retain only the interesting ones from crossings with potential physics processes of interest. This is done in real time by sophisticated integrated trigger and data acquisition systems, involving custom-made fast electronics in a first stage and large computing farms in subsequent stages before the data is transferred to mass storage for further analyses. The initial data rate from up to 40 MHz bunch crossings with multiple pile-up events is thereby reduced to a few hundreds of Hz for offline analysis. A description of these systems is far beyond the scope of these lectures, see [21,22] for details, as well as [23,24] for the specific data selection and data flow challenges for LHCb and ALICE.

The LHC experiments generate huge amounts of data (tens of petabytes of data per year) requiring a fully distributed computing model. The worldwide LHC Computing Grid (wLCG) was developed to deal with this task [17,25]. With its hierarchical architecture of tiered centres it allows any user anywhere access to any data recorded or produced in the analyses steps during the lifetime of the experiments. The centre at CERN receives the raw data, carries out prompt reconstruction, almost in real time, and exports the raw and reconstructed data to the Tier-1 centres and also to Tier-2 centres for physics analysis. The Tier-0 must keep pace with the event rate of several hundred Hz of typically 1 MB of raw data per event from each experiment. The large Tier-1 centres provide also long-term storage of raw data and reconstructed data outside of CERN (as a second copy). They carry out, for example, second-pass reconstruction, when better calibration constants are available. The large number of events simulated by Monte Carlo methods and necessary for quantifying the expectations are produced mainly in Tier-2 centres.

2.5 Comment on Testing and Commissioning

The Individual detector components (e.g. chambers) were built and assembled in a distributed way all around the globe in the numerous participating institutes and were typically first tested at their production sites, then after delivered to CERN, and finally again after their installation in the underground caverns. The collaborations also invested enormous efforts in testing representative samples of the detectors in test beams at CERN and other accelerator laboratories around the world. These test beam campaigns not only verified that performance criteria were met over the several years of production of detector components, but were also used to prepare the calibration and alignment data for LHC operation. Very important were the so-called large combined test beam set-ups, which represented whole 'slices' of the different detector layers of the final detectors.

During the progressing installation the experiments made extensive use of the constant flow of cosmic rays impinging on Earth providing a reasonable flux of muons even at a depth of 100 m underground, typically a few hundred per second traversing the detectors. These muons were used to check the whole chain from hardware to analysis programs of the experiments, and to align the detector elements and calibrate their response prior to the pp collisions. In particular, after the LHC incident on 19th September 2008 the experiments used the 15 months LHC down time, before the first collisions on 23rd November 2009, to run the full detectors in very extensive cosmic ray campaigns, collecting many hundreds of millions of muon events. These runs allowed the experiments to be ready for physics operation, with already accurately pre-calibrated and pre-aligned detectors, by the time of the first pp collisions.

An excellent account of this huge and essential commissioning work has been given in lectures to the CERN – Latin American School of High Energy Physics by A. Hoecker, and the reader is highly recommended to consult [26].

3 The Discovery of the Higgs Boson

3.1 Standard Model Measurements to Demonstrate the Performance

Observing, and measuring accurately, at the LHC collision energies, the production of known particles of the SM, was always considered to be a necessary stepping stone towards exploring the full potential of the LHC with its promise of new physics, firstly of the discovery of the Higgs boson. The SM processes, such as W and Z production, are often referred to as 'standard candles' for the experiments. However, there is much more value to measuring SM processes than this: never before could the SM physics be studied at a hadron collider with such sophisticated and highly accurate detectors, ultimately allowing tests of detailed predictions of the SM with unprecedented precision and minimal instrumental systematic errors.

An example of a very early measurement is shown in Fig. 3, produced only after a month or so after first high-energy collisions in spring 2010. ATLAS and CMS observed in such di-muon invariant mass distributions a 'summary' of decades of particle physics, with remarkable mass resolution.

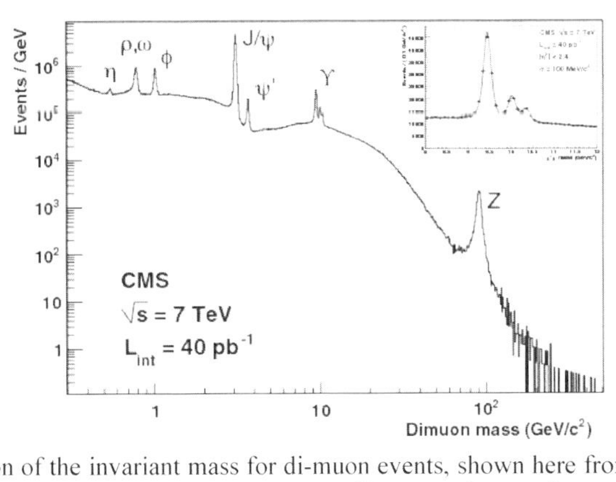

Fig. 3: The distribution of the invariant mass for di-muon events, shown here from CMS, displays the various well-known resonant states of the SM. The inset illustrates the excellent mass resolution for the three states of the Υ family.

In the following paragraphs a few examples will be shown from a very extensive body of publications and publically released conference contributions, which are all available at [15] for ATLAS and [16] for CMS, where detailed specific references can be found.

The charged and neutral Intermediate Vector Bosons (IVB) W and Z are the major benchmark measurements at the LHC for demonstrating the excellent detector performance, as well as for testing model predictions to a high degree of accuracy. The Z decays into electron and muon pairs can be extracted almost free of any backgrounds, as shown in Fig. 4.

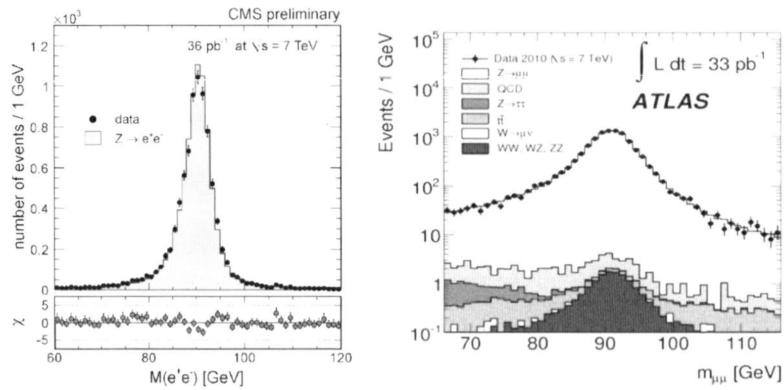

Fig. 4: The CMS electron-pair mass distribution on a linear (left) and the ATLAS muon-pair mass distribution on a logarithmic (right) vertical scale, in the Z mass region. The estimated small background contributions are indicated, as well as the expected signal shape from MC simulations.

The classical W decay signatures into an electron or muon and the associated neutrino are an excellent test for the E_T^{miss} performance of the detector due to the undetected neutrino. E_T^{miss} is inferred from the measured energy imbalance in the transverse projection of all observed signals w.r.t. the beam axis. The E_T^{miss} spectrum for events with a well-identified muon candidate is shown in Fig. 5a, and shows a clear W signal over the expected SM background sources. After applying a selection of events with $E_T^{miss} > 25$ GeV only a small residual background remains present under the W signal, as indicated in the transverse mass distribution given in Fig. 5b.

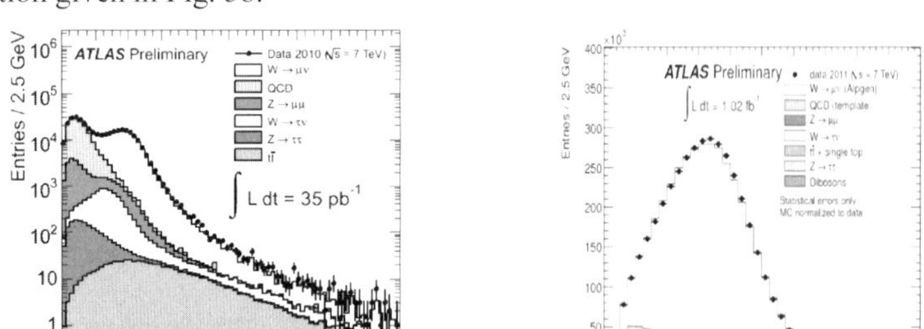

Fig. 5: (a, left) Missing transverse energy distribution for events with a muon candidate. (b, right) Transverse mass distribution for W to muon decays. The expected background contributions are indicated as well (examples from ATLAS).

The good agreement between the measured and expected cross-sections times the leptonic decay branching ratios (which is the expected rate for W and Z bosons to be produced and then decay to leptons) is illustrated in Fig. 6. With the presently available data samples the measurements are expected to already strongly constrain the theoretical model parameters. Figure 6a shows the cross-section measurements and predictions as a function of the collision energy, whereas in (b) the W and Z cross-section results are displayed in a 2-dimensional plot including their correlated error ellipse, and compared to predictions with various parton distribution functions (describing the quark and gluon momentum distributions inside the protons). Detailed measurements of properties for IVB production and decay at the LHC have been published already and are being refined now with the full Run-1 data samples. They include, for example, the lepton charge asymmetry measurements for W decays, which were an important signature of the electro-weak nature of the W at the time of their discovery some 30 years ago.

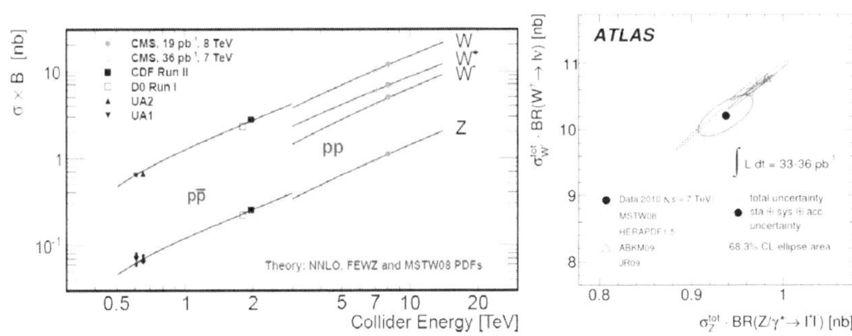

Fig. 6: (a, left) CMS W and Z production cross-sections times leptonic branching ratio as a function of the collision energy, showing also previous measurements at lower energy colliders. (b, right) Correlation of the measured (solid dot) leptonic W and Z cross-section from ATLAS, compared to theoretical expectations with various choices for the parton distribution functions (open symbols).

Hard collisions (characterized by having final state particles with significant transverse energy) at the LHC are dominated by the production of high transverse momentum jets, which are the collimated sprays of particles from the hadronization of the initially scattered partons (quarks, gluons) in the colliding protons. At work is the strong interaction described by Quantum Chromo Dynamics (QCD). Most commonly two jets emerge at opposite azimuth with balanced transverse momenta, from an initially lowest order parton-parton scattering process. However, higher order QCD corrections alter this picture significantly, and detailed measurements of multi-jet configurations are very important to constrain the QCD descriptions of hadronic processes.

The most impressive results at this stage are the inclusive jet and the di-jet cross-section measurements; an example from ATLAS for them is shown in Fig. 7a. These measurements cover unprecedented kinematical ranges spanning typically over jet transverse momenta from 20 GeV to 2 TeV, in many angular (pseudorapidity) bins up to $|\eta| < 4.4$ (i.e. very close to the beam axis). The cross-sections vary over these ranges by up to 12 orders of magnitude. In general the agreement with perturbative QCD calculations, including next to leading order

(NLO) corrections, is well within the systematic uncertainties. This cannot be seen in Fig. 7a directly, only in ratio plots measurement/theory for a given η-interval as shown for CMS data in Fig. 7b. The systematic uncertainties in the ratios are typically less than 20%, which is a great achievement compared to previous such measurements. The systematic uncertainties on the measurements are dominated by the jet energy scale uncertainty (calibration of the detectors for the energy of jets), which, thanks to a considerable effort, has been determined to typically better than 3%.

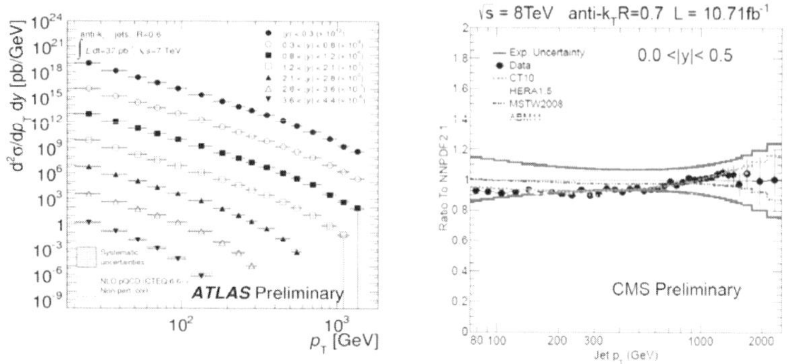

Fig. 7: ATLAS measurements of inclusive jet (a, left) cross-sections, and (b, right) CMS jet measurements compared to NLO perturbative QCD predictions plotted as ratio data/calculation

Jets can also be produced together with W and Z bosons, so-called QCD corrections to the Intermediate Vector Boson production. Many results of these processes have been published. A good understanding of them is particularly important as they are, in many cases, a dominant source of backgrounds to the search for new particles, as well as to the measurements of top quark production discussed next.

The heaviest known particle in the SM is the top quark with its roughly 175 GeV mass. It decays almost exclusively into a W and a bottom quark. The measurement of top quark pair production typically requests that at least one of the W decays leptonically (also needed to trigger the events), and therefore the final states require one or two leptons (electrons or muons), E_T^{miss}, and jets, some of which, coming from the b-quarks, can be tagged by the displaced secondary vertices due to the finite life times of b-hadrons. Whilst it is beyond the scope of these notes to describe the sophisticated analyses employed, the message is that there are clear top pair signals in ATLAS and CMS, both in the single and two-lepton channels, when considering the correct jet topologies. The resulting cross-sections are shown in Fig. 8 which also illustrates the expected large rise of the cross-section with the collision energy increase from 2 TeV at the Tevatron to 7 TeV and 8 TeV at the LHC. Good agreement with NLO QCD calculations is seen within the present few % measurement errors. It can be mentioned that both ATLAS and CMS have also reported first single top measurements (events with just one top quark) at a rate in good agreement with QCD expectations, as well as detailed studies of top properties like its mass.

In summary one can note that the data collected in the first three years of high-energy LHC operation have allowed ATLAS and CMS to make numerous precise measurements of SM processes, including production of bottom and top quarks, W and Z bosons, singly and in

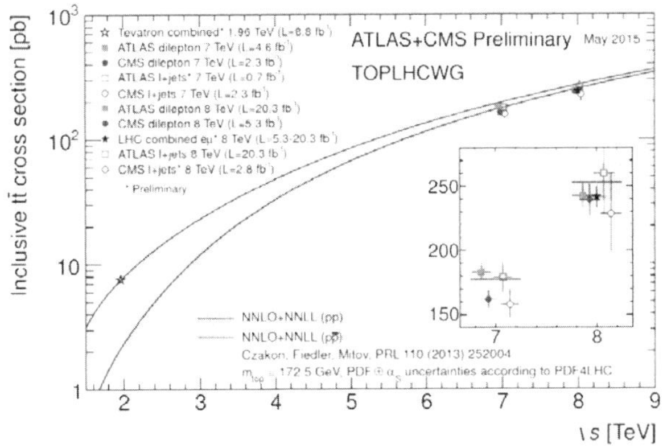

Fig. 8: Top pair production cross-section as a function of the collision energy, showing the Tevatron and LHC measurements

pairs. In particular very detailed measurements of QCD processes have been made. A collection of examples of such studies is shown in Fig. 9 where measurements of cross-sections for various selected electroweak and QCD processes are compared with the SM predictions. These very diverse measurements, probing cross-sections over a range of many orders of magnitude, confirm the predictions of the SM within the errors in all cases. Establishing this agreement is essential before any claims for discoveries can be made, i.e. to demonstrate on the one hand that the detector performance is well understood, and on the other hand that known SM processes are correctly observed in the experiments as they often constitute large backgrounds to signatures of new physics, such as those expected for the Higgs boson. The speed with which the wide range of measurements have shown that SM predictions for known physics have been essentially spot-on is a tribute to a large amount of work done by many particle physics theorists.

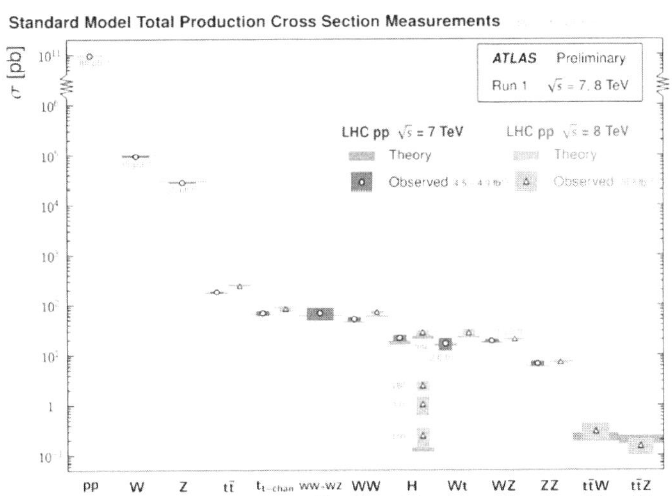

Fig. 9: A comparison of cross-section measurements for electroweak and QCD processes with theoretical predictions from the SM, shown here as example from the ATLAS experiment

3.2 Discovery and Measurements of the Higgs boson

The discovery of a heavy scalar boson was announced jointly by the ATLAS and CMS Collaborations [1,2] on 4th July 2012 with a partial data sample corresponding to about 10 fb^{-1} coming to about equal parts from running at 7 TeV collision energy in 2011 and 8 TeV in 2012 until June. The fantastic performance of the LHC during the second half of 2012 allowed the experiments to more than double their data sets. By the end of 2012 (LHC Run-1) the total amount of data that had been examined corresponded to ~5 fb^{-1} at $\sqrt{s}$ = 7 TeV and ~20 fb^{-1} at $\sqrt{s}$ = 8 TeV, equating to the examination of some 2000 trillion proton-proton collisions. Using these data first measurements of the properties of the new boson were also made. The accumulated luminosity evolution over Run-1 is illustrated in Fig. 10 for ATLAS, showing also that the experiment was very efficient in recording stably delivered luminosity as well as maintaining a high fraction (~90%) of data quality 'good for physics'.

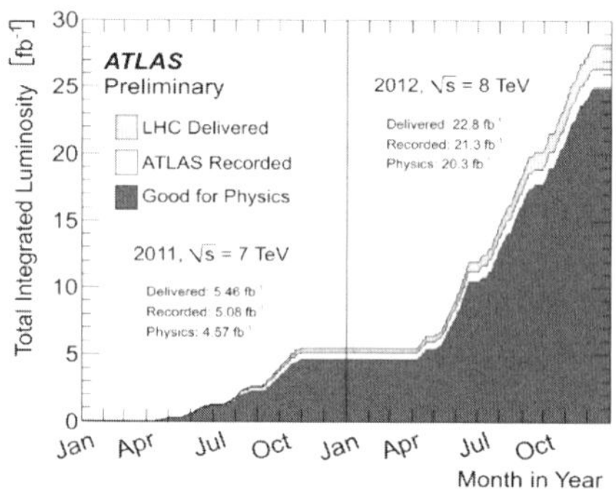

Fig. 10: Integrated luminosity over the high-energy running periods of Run-1 in 2011 and 2012, showing the stably delivered, recorded and finally used data sets for physics (shown is the example for ATLAS, CMS is very similar)

3.2.1 Decays to bosons: the $H\rightarrow\gamma\gamma$, the $H\rightarrow ZZ\rightarrow 4l$ and $H\rightarrow WW\rightarrow 2l2\nu$ decay modes

As examples for the full data sets, the results from the ATLAS experiment are shown for the $H\rightarrow\gamma\gamma$ decay mode (Fig. 11a) and those from the CMS experiment for the $H\rightarrow ZZ\rightarrow 4l$ mode (Fig. 11b). The signal is unmistakable and the significances are summarized in Table 2. The data show a clear excess of events above the expected background around 125 GeV. As for all the Higgs analysis figures shown in the following, the complementary data plots and results from the two experiments can be found in the detailed list of publications available from ATLAS [15] and CMS [16].

The search for H → WW is primarily based on the study of the final state in which both W bosons decay leptonically, resulting in a signature with two isolated, oppositely charged, high p_T leptons (electrons or muons) and large E_T^{miss} due to the undetected neutrinos. The signal sensitivity is improved by separating events according to lepton flavour; into e^+e^-, $\mu^+\mu^-$, and $e\mu$ samples, and according to jet multiplicity into 0-jet and 1-jet samples. The dominant background arises from irreducible non-resonant WW production. Any background arising from Z bosons, with same flavour but opposite sign leptons, is removed by a di-lepton mass cut $(m_Z - 15) < m_{ll} < (m_Z + 15)$ GeV.

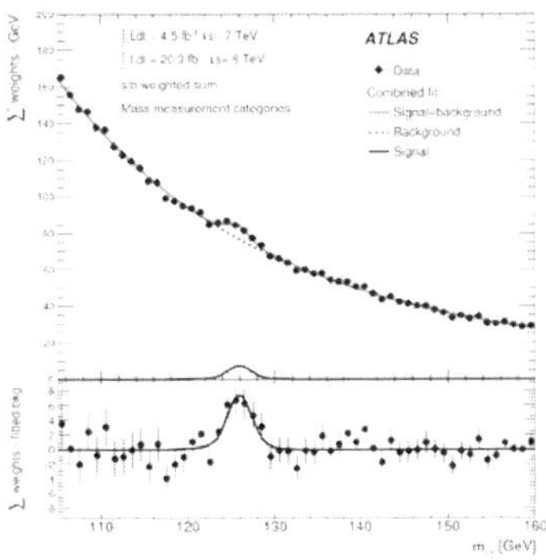

Fig. 11a: Invariant mass distribution of di-photon candidates. The result of a fit to the background described by a polynomial and the sum of signal components is superimposed. The bottom inset displays the residuals of the data with respect to the fitted background component.

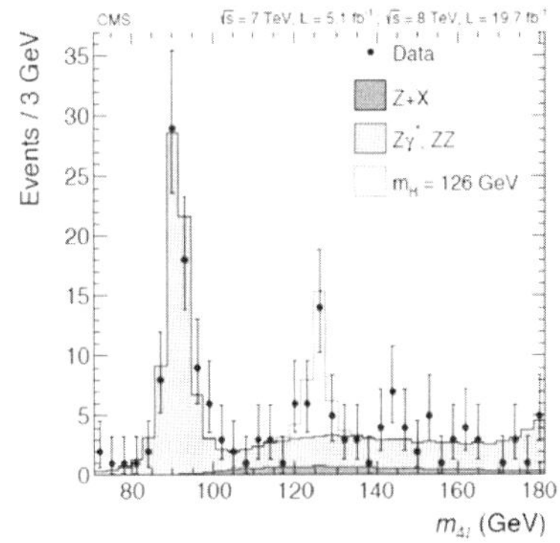

Fig. 11b: The four-lepton invariant mass distribution in the CMS experiment for selected candidates relative to the background expectation. The expected signal contribution is also shown.

The m_{ll} distribution in the 0-jet and eμ final state is shown for CMS in Fig. 12a. The expected contribution from a SM Higgs boson with m_H = 125 GeV is also indicated. The transverse mass, m_T, distribution is shown in Fig. 12b from ATLAS, as well as the background-subtracted distribution. Both show a clear excess of events compatible with a Higgs boson with mass ~125 GeV. The observed (expected) significances of the excess with respect to the background-only hypothesis are shown in Table 2.

3.2.2 Decays to fermions: the H → ττ and the H → bb decay modes

It is important to establish whether this new particle also couples to fermions, and in particular to down-type fermions, since the measurements above mainly constrain the couplings to the up-type top quark. Determination of the couplings to down-type fermions requires direct measurement of the Higgs boson decays to bottom quarks and τ leptons.

The H → $\iota\iota$ search is typically performed using the final-state signatures eμ, $\mu\mu$, eτ_h, $\mu\tau_h$, $\tau_h\tau_h$, where electrons and muons arise from leptonic τ-decays and τ_h denotes a τ lepton decaying hadronically. Each of these categories is further divided into two exclusive sub-categories based on the number and the type of the jets in the event: (i) events with one forward and one backward jet, consistent with the Vector-Boson-Fusion (VBF) topology, (ii) events with at least one high p_T hadronic jet but not selected in the previous category. In each of these categories, a search is made for an excess in the reconstructed $\tau\tau$ mass distribution. The main irreducible background, Z → $\tau\tau$ production, and the largest reducible backgrounds (W + jets, multijet production, Z → ee) are evaluated from various control samples in data.

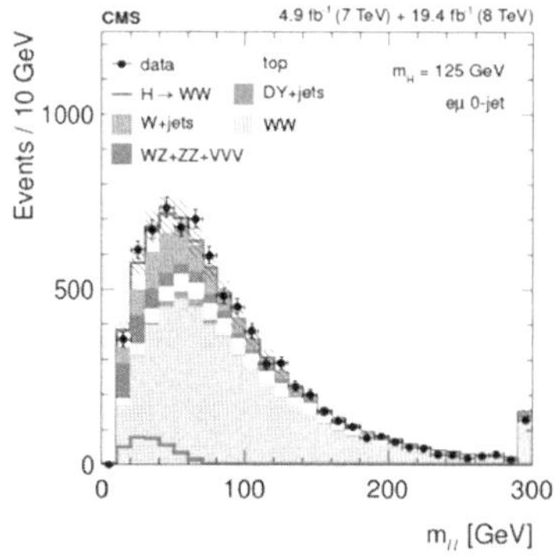

Fig. 12a: Distribution of dilepton mass in the 0-jet, eμ final state in CMS for a m_H = 125 GeV SM Higgs boson decaying via $H \rightarrow WW \rightarrow l\nu l\nu$ and for the main backgrounds

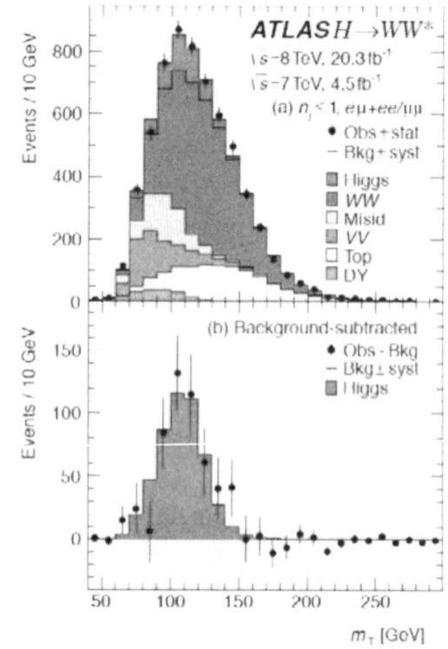

Fig. 12b: The transverse mass distributions for events passing the full selection of the $H \rightarrow WW \rightarrow l\nu l\nu$ analysis in ATLAS summed over all lepton flavours for final states with $N_{\text{jet}} \leq 1$. In the lower part the residuals of the data with respect to the estimated background are shown, compared to the expected m_T distribution for a SM Higgs boson.

The H→bb decay mode has by far the largest branching ratio (~54%). However since σ_{bb} (QCD) ~ $10^7 \times \sigma$(H→ bb) the search concentrates on Higgs boson production in association with a W or Z boson using the following decay modes: W → eν/$\mu\nu$ and Z → ee/$\mu\mu$/ $\nu\nu$. The Z → $\nu\nu$ decay is identified by the requirement of a large missing transverse energy. The Higgs boson candidate is reconstructed by requiring two b-tagged jets.

Evidence for a Higgs boson decaying to a $\tau\tau$ lepton pair is reported by the CMS and ATLAS Collaborations. The results are given in Table 2. The CMS results reported in Table 2 include both the H → $\tau\tau$ and H → WW contributions, considered as signal in the $\tau\tau$ decay-tag analysis. This treatment leads to an increased sensitivity to the presence of a Higgs boson that decays into both $\tau\tau$ and WW.

The CMS measurements in the H → $\tau\tau$ and VH with H → bb searches are mutually consistent, within the precision of the present data, and with the expectation for the production and decay of the SM Higgs boson. CMS has combined these two results, requiring the simultaneous analysis of the data selected by the two individual measurements. Figure 13 shows the scan of the profile likelihood as a function of the signal strength relative to the expectation for the production and decay to fermions (bb and ττ) of a standard model Higgs boson for m_H = 125 GeV. The evidence against the background-only hypothesis is found to have a maximum of 3.8σ for m_H = 125 GeV.

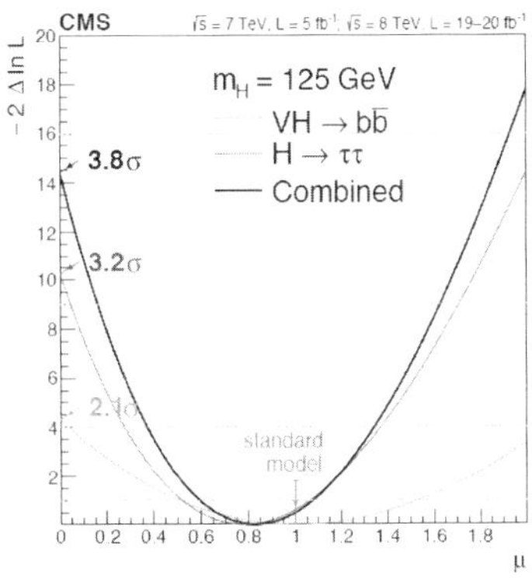

Fig. 13: Scan of the profile likelihood as a function of the signal strength relative to the expectation for the production and decay of a standard model Higgs boson, for m_H = 125 GeV

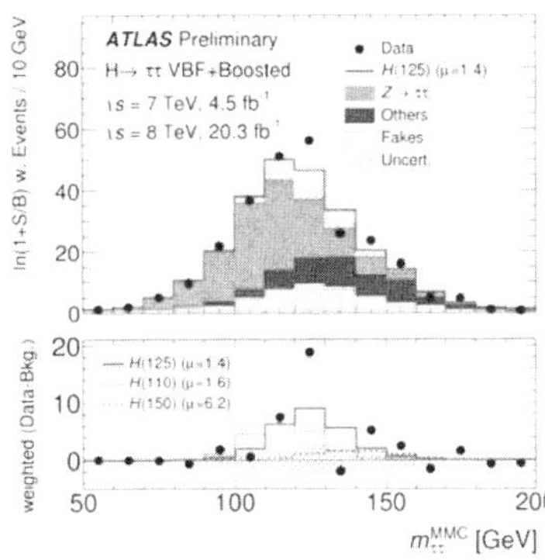

Fig. 14: Observed and expected weighted di-tau mass distributions in ATLAS. The bottom panel shows the difference between weighted data events and weighted background events (points) compared to signal events yields for various masses, with signal strengths set to their best-fit values.

Figure 14 shows the observed and expected $\tau\tau$ mass distributions from the ATLAS experiment, weighing all sub-distributions in each category of each channel by the ratio between the expected signal and background yields for the respective category in a di-tau mass interval containing 68% of the signal. The plot also shows the difference between the observed data and expected background distributions, together with the expected distribution for a SM Higgs boson signal with m_H = 125 GeV. The observed (expected) significance of the excess with respect to the background-only hypothesis at this mass is 4.5 (3.4) standard deviations in the ATLAS experiment.

The Tevatron experiments, CDF and D0, have also reported a combined observed significance of 3.0σ [27], where the H→ bb mode is the dominant one. All these results establish the existence of the fermionic decays of the new boson, consistent with the expectation from the SM.

3.2.3 Higgs Boson Properties

3.2.3.1 The mass of the Higgs boson

Both ATLAS and CMS experiments have separately combined their measurements of the mass of the Higgs bosons from the two channels that have the best mass resolution, namely $H\rightarrow\gamma\gamma$ and $H\rightarrow ZZ\rightarrow 4l$. The signal in all channels is assumed to be due to a state with a unique mass. The obtained values are from ATLAS m_H = 125.36 ± 0.37(stat) ± 0.18(syst) GeV and from CMS m_H = 125.02 ± 0.27(stat) ± 0.14 (syst) GeV, in excellent agreement.

3.2.3.2 Significance of the observed excess

Table 2 summarises the median expected and observed local significances for a SM Higgs boson mass hypothesis of 125 GeV from the individual decay modes in ATLAS and CMS. Both experiments confirm independently the discovery of a new particle with a mass near 125 GeV.

Table 2: The expected and observed local p-values in ATLAS and CMS expressed as the corresponding number of standard deviations of the observed excess from the background-only hypothesis, for $m_H = 125$ GeV, for various decay modes

Experiment	ATLAS		CMS	
Decay mode	Expected (σ)	Observed (σ)	Expected (σ)	Observed (σ)
$\gamma\gamma$	4.6	5.2	5.3	5.6
ZZ	6.2	8.1	6.3	6.5
WW	5.8	6.1	5.4	4.7
bb	2.6	1.4	2.6	2.0
$\tau\tau$	3.4	4.5	3.9	3.8

3.2.3.3 Signal strength

To establish whether or not the newly found state is the Higgs boson of the SM, one needs to precisely measure its other properties and attributes. Several tests of compatibility of the observed excesses with those expected from a standard model Higgs boson have been made.

In one comparison labelled as the signal strength $\mu = \sigma/\sigma_{SM}$, the measured production times decay rate of the signal is compared with the SM expectation, determined for each decay mode individually and for the overall combination of all channels. A signal strength of one would be indicative of a SM Higgs boson.

Both the ATLAS and CMS experiments have measured μ values, by decay mode and by additional tags used to select preferentially events from a particular production mechanism. The best-fit value for the common signal strength μ, obtained in the different sub-combinations and the overall combination of all search channels in the ATLAS and CMS experiments is shown in Fig. 15. The observed μ value is 1.00 ± 0.09 (stat) ± 0.08 (theory) for CMS for a Higgs boson mass of 125.0 GeV and 1.18 ± 0.15 in ATLAS for a Higgs boson mass of 125.4 GeV. In both the experiments the μ-values are consistent with the value expected for the SM Higgs boson ($\mu = 1$). The Tevatron has also measured the value of this signal strength, primarily using the bb channel and find it to be 1.44 ± 0.59 [27].

3.2.3.4 Couplings of the Higgs boson

Figure 16 illustrates the dependence of the Higgs boson couplings on the mass of the decay particles (τ, b-quark, W, Z and t-quark). The couplings are plotted in terms of λ or $\sqrt{(g/2v)}$. The

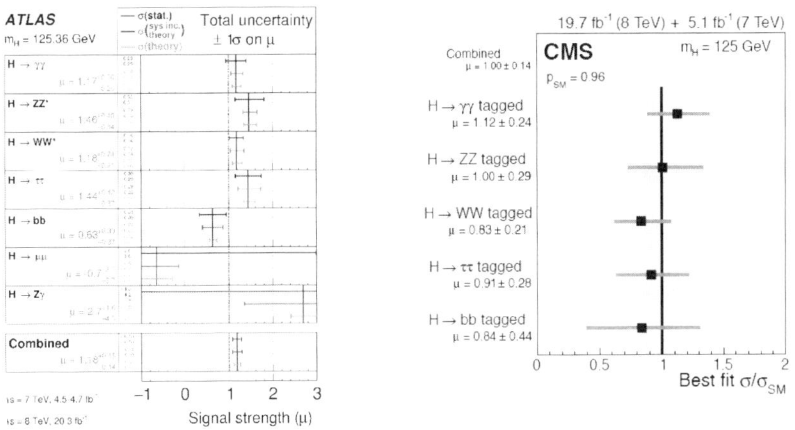

Fig. 15: Values of μ for sub-combinations by decay mode in (left) ATLAS and (right) in CMS

line is the expectation from the SM. For the fermions, the values, λ, of the fitted Yukawa couplings *Hff* are shown, while for vector bosons the square-root of the coupling for the *HVV* vertex divided by twice the vacuum expectation value of the Higgs boson field (√(g/2v). For a Higgs boson with a mass of 125 GeV decaying to μμ CMS has found that the observed (expected) upper limit on the production rate is 7.4 (6.5 +2.8, -1.9). This corresponds to an upper limit on the branching fraction of 0.0016. The couplings are indeed proportional to mass, as expected for a SM Higgs boson, over a broad mass range, from the τ-lepton mass (about 1.8 GeV) to that of the top quark (mass about one hundred times larger).

3.2.3.5 Spin and parity

Another key to the identity of the new boson is its quantum numbers amongst which is the spin-parity (J^P). The angular distributions of the decay particles can be used to test various spin hypotheses.

In the decay mode H → *ZZ* → 4*l* the full final state is reconstructed, including the angular variables sensitive to the spin-parity. The information from the five angles and the two di-lepton pair masses are combined to form boosted decision tree (BDT) discriminants. A decision tree is a set of cuts employed to classify events as "signal-like" or "background-like".

In the decay mode H → WW → *lνlν*, for example, in the ATLAS experiment the discriminants used in the fit are outputs of two different BDTs, trained separately against all backgrounds to identify 0^+ and 2^+ events, respectively. For the BDT the kinematic variables used are the transverse mass m_T, the azimuthal separation of the two leptons, $\Delta\phi_{ll}$, m_{ll} and dilepton p^T_{ll}.

A first study has been presented by CMS in the *ZZ* → 4*l* channel with the data already disfavouring the pure pseudo-scalar hypothesis (Fig. 17). The CMS experiment has combined

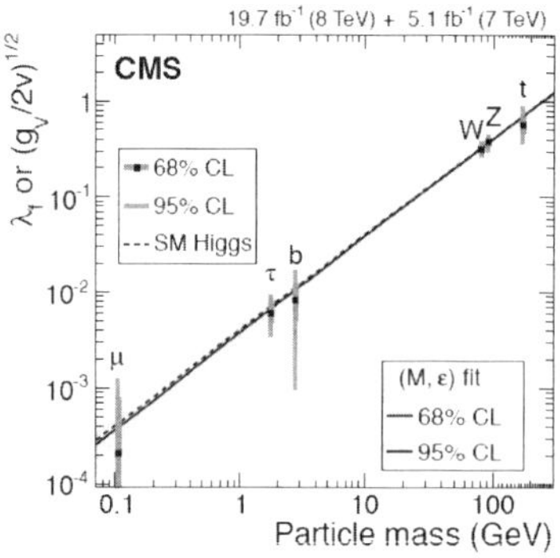

Fig. 16: Summary of the fits from the CMS experiment for deviations in the couplings λ or √(g/2v) as function of particle mass for a Higgs boson with a mass of 125 GeV (see text)

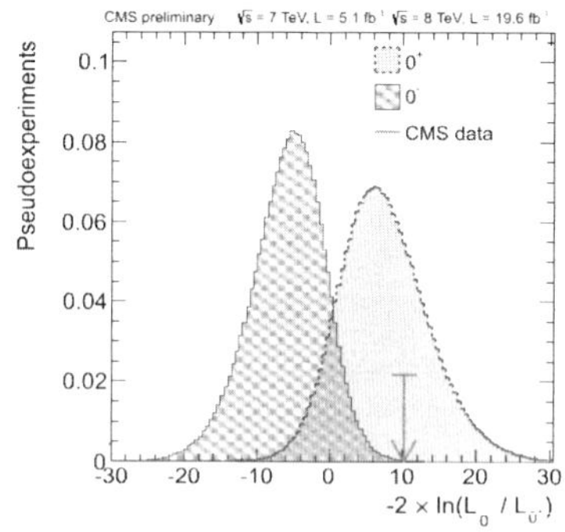

Fig. 17: Distribution of q=$-2ln(L_{JP}/L_{SM})$ for two signal types, 0^+ (yellow histogram) and 0^- hypothesis (blue histogram) for m_H = 126 GeV for a large number of generated experiments. The arrow indicates the observed value.

the ZZ → 4*l* and WW → *lνlν* spin analyses. Under the assumption that the observed boson has J^P=0^+, the data disfavour the hypothesis of a graviton-like boson with minimal couplings produced in gluon fusion, $J^P = 2^+$, with a CLs value of 0.60%.

ATLAS has also presented a combined study of the spin of the Higgs boson candidate using the H → γγ, H → WW → *lνlν* and H → ZZ → 4*l* decays to discriminate between the SM assignment of J^P=0^+ and a specific model of $J^P = 2^+$. The data strongly favour the J^P=0^+ hypothesis (see Fig. 18). The specific J^P=2^+ hypothesis is excluded with a confidence level above 99.9%, independently of the assumed contributions of gluon fusion and quark-antiquark annihilation processes in the production of the spin-2 particle.

The above-mentioned example analyses show that the spin-parity J^P=0^+ hypothesis is strongly favoured by both experiment, with the alternatives $J^P = 0^-, 1^+, 1^-, 2^+$ hypotheses rejected with confidence levels larger than 97.8%.

4 Beyond the Standard Model at the LHC

Besides the quest to elucidate the mechanism of the electro-weak symmetry breaking by searching for the Higgs boson, the major excitement for LHC comes from the great potential to explore uncharted territory of physics Beyond the Standard Model (BSM), thanks to its highest collision energy ever available in a laboratory so far. Since the beginning of the project, the search for Supersymmetry (SUSY) was a strong motivation, and besides the H boson it has been the other main benchmark physics that was guiding the detector designs. However, many other hypothetical new processes can be searched for, and indeed ATLAS and CMS have already reported in many publications a very broad spectrum of searches for BSM signatures (mass peaks for new particles or kinematical distributions with deviations from the expectations of known physics processes). No such new effect has yet been found, and all of these searches

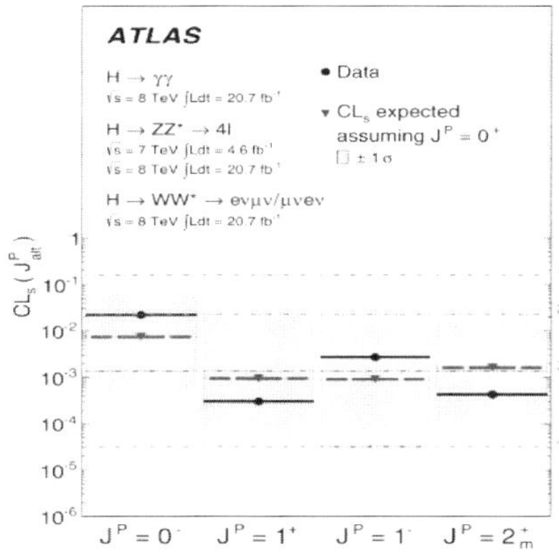

Fig. 18: Expected (blue triangles/dashed lines) and observed (black circles/solid lines) confidence level CL_S for alternative spin–parity hypotheses assuming a $J^P=0^+$ signal. The green band represents the 68% $CL_S(J_{alt}^P)$ expected exclusion range for a signal with assumed $J^P=0^+$.

result in highly-improved, stringent exclusion limits, often well beyond the one TeV scale already. Only a few examples are mentioned below.

The most popular searches concern SUSY, which predicts additional fundamental particles. The search for SUSY is motivated in part by the prospect that the lightest stable neutral SUSY particle (LSP) could be an excellent candidate for explaining the Dark Matter (DM) in the Universe. The mysterious existence of DM was postulated by Fritz Zwicky, and rather convincingly evidenced by Vera Rubin, both astronomers, in the 1930s and 1970s respectively.

The SUSY searches at LHC are very complex as they must be sensitive to many (model-dependent) decay chains, implying a large variety of possible final state topologies. A common feature for most of them is the existence of significant missing transverse energy, E_T^{miss}, due to the escaping LSPs (an experimental signature similar to that of the neutrinos in the W decays). Furthermore, the SUSY signatures often include high transverse momentum jets, some tagged as b-jets for third-generation squarks as particularly motivated by naturalness arguments developed in other lectures, and leptons. The expected topologies depend not only on the model parameters, but also on the mass relations between various squarks and gluinos (the SUSY partners of the SM quarks and the gluons). A summary of 95% CL mass exclusion regions from many SUSY searches is shown in Fig. 19 from ATLAS. Similar results are available from CMS.

Many other searches aimed at exploring BSM physics have been conducted, at this stage all without finding anywhere an excess of observed event rates over the expected backgrounds from the known SM processes. However, much more stringent limits and constraints could be established than what were available up to now. A non-exhaustive summary of 95% CL limits is displayed in Fig. 20 from CMS, with very similar results being reported by ATLAS.

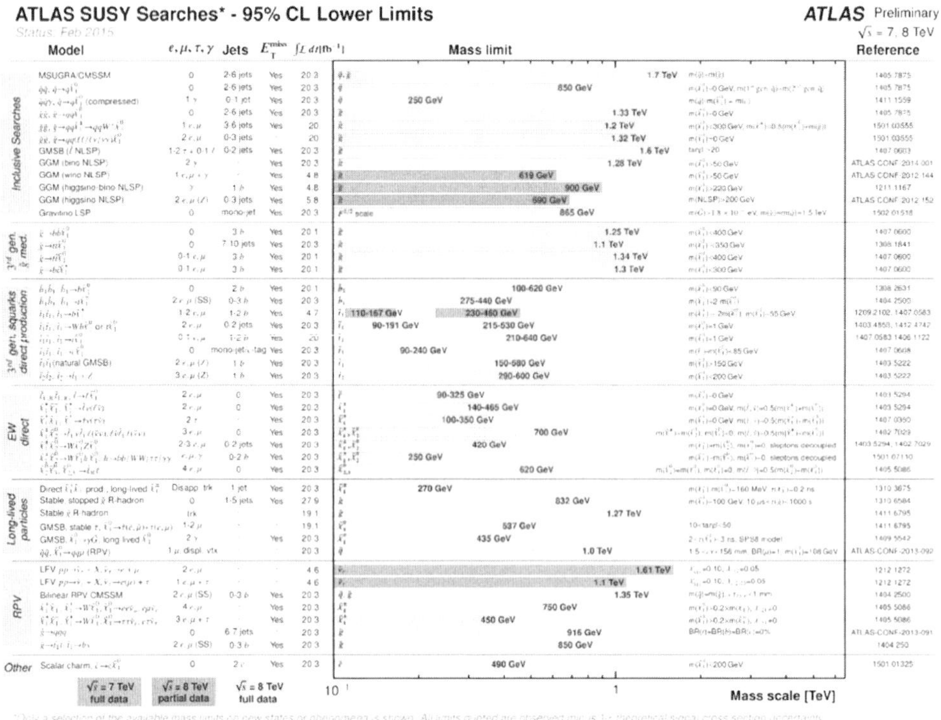

Fig. 19: A summary of 95% CL mass exclusion limits from many SUSY searches as obtained by ATLAS (similar results are available from CMS as well)

5 Outlook

The LHC is just starting operation again with Run-2 at the collision energy of 13 TeV and eventually 14 TeV in the coming years. Together with the increase in energy there will be also an increase in the luminosity, bringing the LHC to its full design performance (14 TeV, 2×10^{34} $cm^{-2}s^{-1}$). The projected integrated luminosity by 2022 is about 300 fb^{-1}.

The increased energy means larger cross-sections, particularly striking for heavy objects, as can be seen in Fig. 21. It is therefore with great expectations that the experiments are looking forward to collect data in the forthcoming Run-2 and Run-3 periods, covering the initial LHC project planning.

5.1 Prospects with the High Luminosity Upgrades (HL-LHC)

The roadmap of physics at the LHC beyond its initial design phase, with typically 300 fb^{-1} integrated luminosity until the early 2020s, has dramatically changed with the discovery of the Higgs-like boson. Not only will there be the unchallenged window for directly observable hypothetical heavy mass particles, messengers of new physics beyond the Standard Model, but

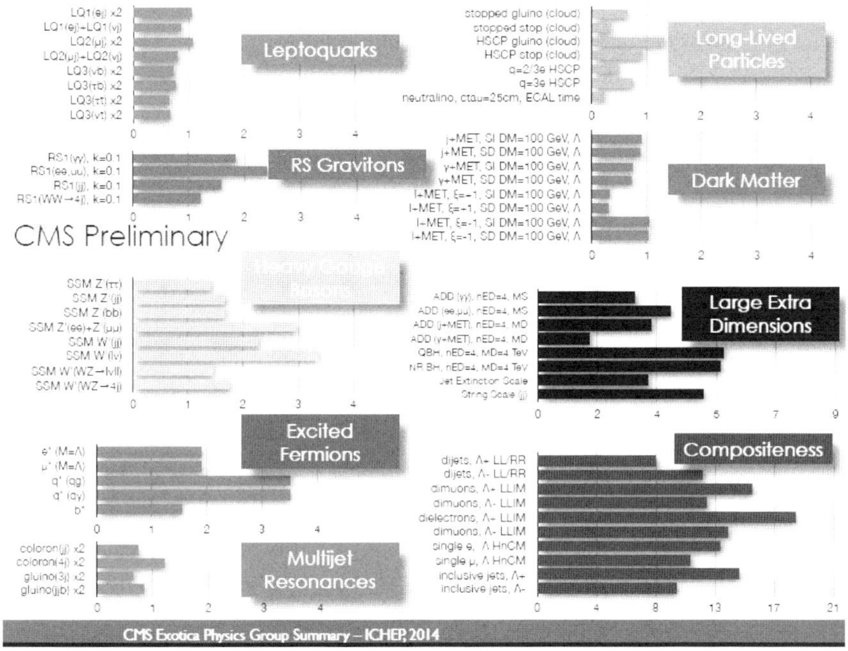

Fig. 20: A summary of 95% CL exclusion limits from many BSM searches other than SUSY as obtained by CMS (very similar results are available from ATLAS as well)

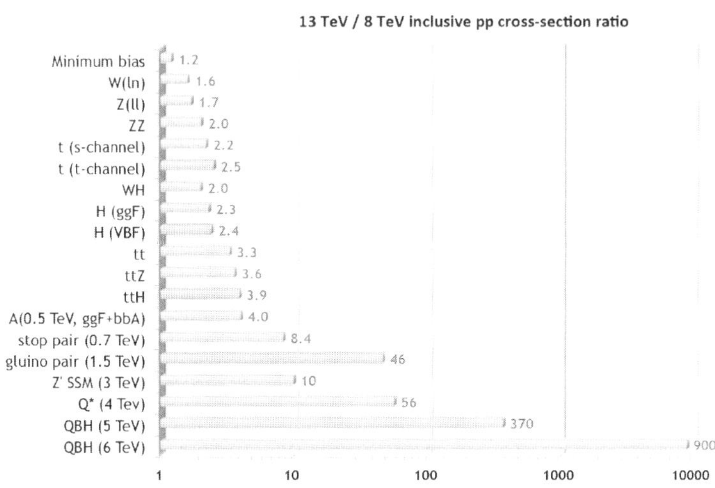

Fig. 21: Examples of inclusive production cross-section ratios for 13 TeV / 8 TeV

also a clear task to investigate in greatest details the properties of the new boson. Needless to say, this basic scenario could well be strongly enriched further if the forthcoming 14 TeV data of the current decade would reveal any new BSM physics, which would then be of course exploited best with the highest available integrated luminosity.

These prospects have strongly motivated to launch a very mayor high luminosity upgrade project planning both for the experiments and the LHC machine, called the HL-LHC, with the goal to integrate a tenfold luminosity (3000 fb^{-1}) by the early 2030s. The importance of this future direction for particle physics has been fully recognized in the Update of the European Strategy for Particle Physics [28] where HL-LHC is singled out as first-priority in the European road map for the decades to come. A few examples are given here based on the studies in this context. The Fig. 22 illustrates the updated anticipated road map for the LHC operation for the coming decades.

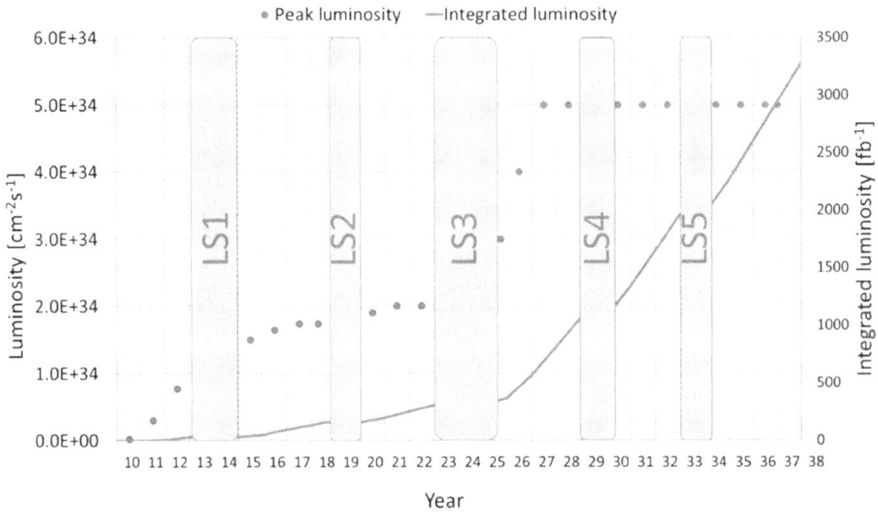

Fig. 22: Updated anticipated road-map for the LHC operation, with the HL-LHC starting after the Long Shut-Down number 3 (LS3) around 2025

The LHC potential for detailed studies of the electroweak symmetry breaking mechanism will be discussed first, namely the precision measurements of the Higgs couplings, the Higgs self-coupling, and vector boson scattering at high energy. In a second part a few examples of extending the reach into exploratory BSM physics will be given, including SUSY and searches for massive heavy resonances. The ATLAS and CMS Collaborations have presented a wealth of evaluations for the physics reach with the anticipated luminosity of 3000 fb^{-1} for the HL-LHC era [29,30]. These estimates, given here per single experiment, are based on a very substantial simulation effort taking into account realistic pile-up conditions. Both Collaborations work on very substantial detector upgrade projects that will maintain similar detector performances as at present, and which are needed in any case for critical components to allow operation beyond the initial LHC design era. Note that similarly LHCb and ALICE have engaged into very major upgrade projects as well.

5.1.1 Measurements of Higgs boson couplings

While measurements of the Higgs boson couplings have already begun by ATLAS and CMS, these will remain a central topic within the approved LHC programme. The luminosity of the HL-LHC will provide further substantially improved statistical precision for all established channels. However, most importantly, it will also allow one to study crucial rare Higgs boson production and decay modes.

Two examples for families of channels that will only become accessible in a quantitative way with the HL-LHC are mentioned here for illustration:

- WH / ZH, H → $\gamma\gamma$ and ttH, H → $\gamma\gamma$. These channels have a low signal rate at the LHC, but one can expect to observe more than 100 events at the HL-LHC. The ttH initial state gives the cleanest signal with a signal-to-background ratio (S/B) of ~ 20%. It also provides a measurement of the top-Yukawa coupling, which is not easily accessible elsewhere. Figure 23a shows the expected signal.

- H → $\mu\mu$. The S/B of this low-rate channel is only ~ 0.2% but the narrow peak allows one to extract a more than 6 σ significant signal for an inclusive measurement, see Fig. 23b. The exclusive ttH, H → $\mu\mu$ would yield a clean (S/B > 1) sample of 30 events providing information on both top- and μ-Yukawa couplings.

An overview of the expected measurement precisions on the signal rate in each channel is given in Fig. 24 from ATLAS, but very similar results are available from CMS as well, comparing 300 and 3000 fb^{-1}. It should be stressed that only a limited selection of channels (initial and final states) were studied so far, and further improvements can be expected with future studies.

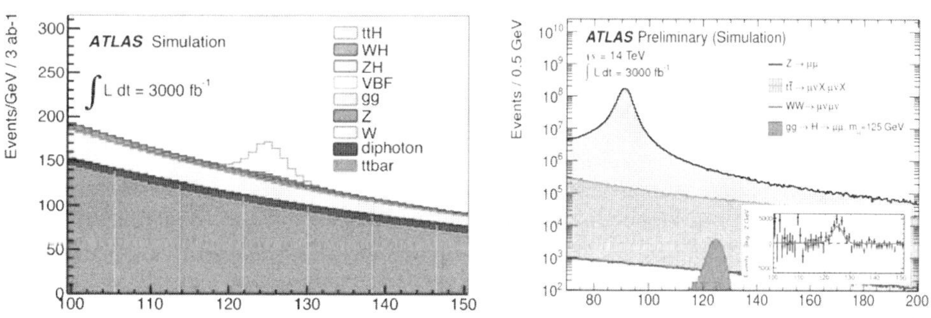

Fig. 23: Examples of expected invariant mass distributions for 3000 fb^{-1}, (a) for ttH, H → $\gamma\gamma$ selected with 1 lepton, and (b) inclusive H → $\mu\mu$

5.1.2 Observation of the Higgs self-coupling

In order to fully determine the parameters of the SM and to establish the EW symmetry breaking mechanism, the measurement of the Higgs self-coupling is important. A direct analysis of the Higgs trilinear self-coupling λ_{HHH} can be done via the detection of Higgs boson pair production, through interference effects with the dominant pair production at LHC by gluon-gluon fusion.

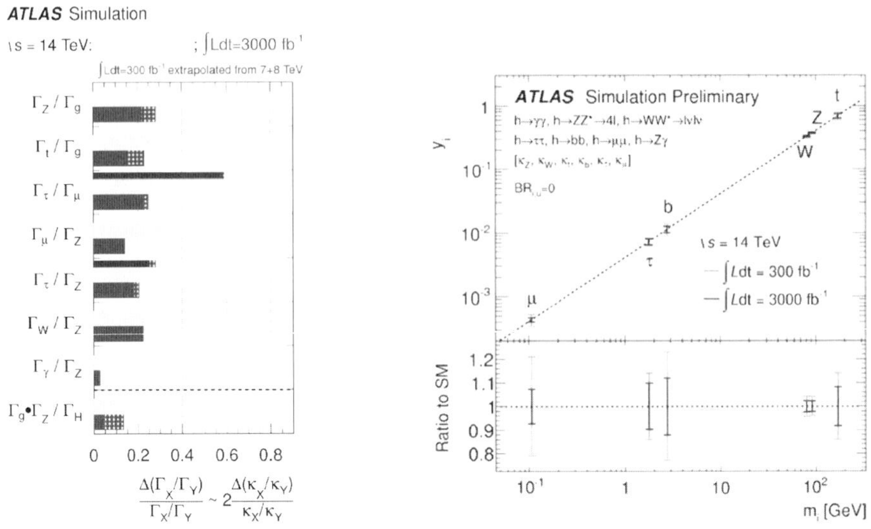

Fig. 24: (a) Expected precisions on ratios of Higgs boson partial widths. The bars give the expected relative uncertainty for a SM Higgs with mass 125 GeV (dashed are current theory uncertainty from QCD scale and PDFs). The thin bar for ττ show extrapolations from current analysis to 300 fb^{-1}, instead of the dedicated studies for VBF channels. (b) Expected precisions for the couplings (see also Fig. 16).

Initial sensitivity studies have been performed only on two channels so far, HH → bbγγ and bbWW, for their clean signature and high branching ratio, respectively. Only the bbγγ final state has been found to be accessible for 3000 fb^{-1}, yielding a 3 σ observation per experiment. Additionally, promising channels like bbττ in the final state are under investigation. The expectation is that a 30% measurement on λ_{HHH} can be achieved by combining the HL-LHC measurements.

5.1.3 Vector boson scattering

If the new boson discovered at LHC is fully confirmed to be the SM Higgs, then unitarity of scattering amplitudes in longitudinal Vector Boson Scattering (VBS) should be preserved at high energy. It is important to confirm this prediction experimentally. It would also be important to look for new physics contributing to the regularization of the cross section or else enhancing it. For example, Technicolour or little Higgs models, postulate TeV scale resonances to become observable.

At the LHC the VBS are tagged with two forward jets on either side, the remnants of the quarks that have emitted the vector bosons involved in the scattering process. Studies of several channels have been reported for different VB decay final states for WW+2jets, WZ+2jets, and ZZ+2jets events. As an example the clean channel ZZ+2jets → 4 charged leptons + 2jets, which would allow one to fully reconstruct a hypothetical 1 TeV mass ZZ resonance peak over the SM VBS events and non-VBS di-boson background has been reported in [29].

5.1.4 Exploratory Beyond Standard Model physics at HL-LHC

Exploratory physics reach for BSM has always been a great motivation for the LHC, and that remains true more than ever also for the HL-LHC. Many quantitative studies exist, and have been refined now with sophisticated simulations by ATLAS and CMS with their realistic detector understanding, gained by the current LHC running in already very challenging pile-up conditions.

Considering first Supersymmetry (SUSY) searches, the new studies have confirmed that the mass reach in the generic searches for gluinos and squarks of the first two generations will be extended from typically 2.6 TeV to 3.2 TeV when adding the HL-LHC data. These results remain essentially unchanged for lightest supersymmetry particle (LSP) masses up to 1/3 of the mass of the strongly produced sparticles.

Naturalness arguments suggest the top squark to be light, preferably below 1 TeV. At 14 TeV the direct stop pair production cross section for 600 GeV (1 TeV) stops is 240 fb (10 fb). An increase in the luminosity from 300 to 3000 fb^{-1} increases therefore the sensitivity significantly for heavy stop in the interesting region or, if stop candidates are found, will enable to measure their properties. As an illustrative example of a new detailed study the Fig. 25 summarizes the results in the stop-LSP plane for two decay chains. Both the 5σ discovery range and the 95 % CL exclusion limits are shown. The cross sections for electroweak gaugino searches are small at the LHC, and the discovery potential will get strongly enhanced by the ten-fold luminosity increase. For example, the discovery potential for associated production of charginos and neutralinos extends to scenarios with chargino masses of about 800 GeV for neutralino masses below 300 GeV.

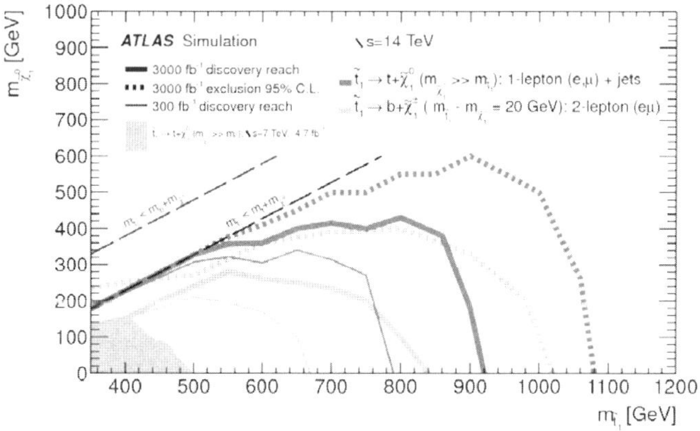

Fig. 25: 5σ discovery reach and 95% CL exclusion limits in the stop-LSP mass plane for two decay channels, as indicated, for direct stop pair production

A broad variety of resonances and other exotic signatures are sought for at the LHC. The reach for direct observations extends deep into the TeV mass scale, as a typical example one can quote the straight-forward searches for new sequential standard model like Z' decaying into charged lepton pairs. The mass reach of typically 6.5 TeV with 300 fb^{-1} will increase to 7.8 TeV with 3000 fb^{-1}. This improved reach of about 20% is very typical for many other searches.

A notable area of exotic physics that will benefit particularly from an HL-LHC phase is the sector of final states with top quarks. Strongly and weakly produced top-antitop resonances have been studied as an interesting benchmark. For example, strongly-produced Kaluza-Klein gluons in extradimension models could result in broad top-antitop resonance signals. The mass reach for them will increase very significantly from 4.3 TeV at 300 fb^{-1} to 6.7 TeV with 3000 fb^{-1}.

5.2 Prospects beyond the LHC

In the lectures an outlook for ambitious, future facilities beyond the LHC project was given. A rich variety of ambitious project dreams are pursed in the community to explore further the High Energy Frontier for many decades to come. They include hadron colliders as well as e+e- colliders (linear colliders and circular storage rings), a field too vast to describe here in this limited write-up. Studies are evolving fast, and the students are encouraged to consult the updated web information available for the CERN Future Circular Collider (FCC) studies [31], the CERN Compact Linear Collider (CLIC) studies [32], the International Linear Collider (ILC) studies [33] and the Chinese collider ring complex (CEPC-SPPC) studies [34].

With the LHC the journey into new physics territory at the high energy frontier has only just begun, and rarely before have we enjoyed such an exciting time in particle physics with great promises for discoveries. But the long LHC story, still only at the beginning of its exploitation, has also told us that timely plans and courageous decisions on a global scale have to be made by the world community of particle physicists, in order to 'plant the right seed' for the future of our field.

Acknowledgement

I warmly thank Professor Antonino (Nino) Zichichi, Director of the Ettore Majorana Foundation and Centre for Scientific Culture, for having invited me to this motivating school, and for his generous hospitality in the beautiful town of Erice. I also thank all the brilliant students for their great stimulating discussions, which make me confident that our field will be in good hands for the future. For this write-up I have significantly benefitted from summary articles that I had the pleasure to co-author with long-standing colleagues in the field of hadron colliders, namely Tejinder S. Virdee and Giorgio Brianti.

References

[1] ATLAS Collaboration, Observation of a new particle in the search for the Standard Model Higgs boson with the ATLAS detector at the LHC, *Phys. Lett.* **B 716** (2012) 1.

[2] CMS Collaboration, Observation of a new boson at a mass of 125 GeV with the CMS experiment at the LHC, *Phys. Lett.* **B 716** (2012) 30.

[3] S.L. Glashow, *Nucl. Phys.* **22** (1961) 579.

[4] S. Weinberg, *Phys. Rev. Lett.* **19** (1967) 1264.

[5] A. Salam, Proceedings of the eighth Nobel symposium, ed. N. Svartholm, p. 367. Almqvist & Wiskell, 1968.

[6] F. Englert and R. Brout, *Phys. Rev. Lett.* **13** (1964) 321.

[7] P.W. Higgs, *Phys. Lett.* **12** (1964) 132.

[8] P.W. Higgs, Phys. *Rev. Lett.* **13** (1964) 508.

[9] G.S. Guralnik, C.R. Hagen, T.W.B. Kibble, *Phys. Rev.* **155** (1967) 1554.

[10] P.W. Higgs, *Phys. Rev.* **145** (1966) 1156.

[11] T.W.B. Kibble, *Phys. Rev.* **155** (1967) 1554.

[12] M. Della Negra, P. Jenni, T. S. Virdee, Journey in the Search for the Higgs Boson: The ATLAS and CMS Experiments at the Large Hadron Collider, Science Vol. **338** no. 6114 pp. 1560-1568 (2012), http://www.sciencemag.org/content/338/6114/1560.full.html.

[13] P. Jenni and T. S. Virdee, The Discovery of the Higgs Boson at the LHC, in *60 Years of CERN Experiments*, Eds. H. Schopper and L. Di Lella (World Scientific, 2015), 978-981-4644-14-3.

[14] P. Grannis and P. Jenni, The Evolution of Hadron-Collider Experiments, *Phys. Today* **66(6)** (2013) 38.

[15] ATLAS publications: https://twiki.cern.ch/twiki/bin/view/AtlasPublic

[16] CMS publications: http://cms-results.web.cern.ch/cms-results/public-results/publications/

[17] *The Large Hadron Collider, a Marvel of Technology*, Ed. L. Evans, EPFL Press, 2009.

[18] G. Brianti and P. Jenni, The Large Hadron Collider, in *Technology Meets Research, 60 Years of Technology Achievements at CERN*, Eds. C. Benvenutti et al. (World Scientific, 2015).

[19] C. Llewellyn Smith, *Nature* **448** (2007) 281.

[20] *ECFA-LEP Working Group, 1979 Progress Report*, Ed. A. Zichichi, ECFA/79/39 (1979).

[21] ATLAS Collaboration, The ATLAS Experiment at the LHC, *JINST* **3**, S08003 (2008).

[22] CMS Collaboration, The CMS Experiment at the LHC, *JINST* **3**, S08004 (2008).

[23] LHCb Collaboration, The LHCb Detector at the LHC, *JINST* **3**, S08005 (2008).

[24] ALICE Collaboration, The ALICE Experiment at the LHC, *JINST* **3**, S08002 (2008).

[25] WLCG comprehensive information: http://wlcg.web.cern.ch/

[26] A. Hoecker, Commissioning and early physics analysis with the ATLAS and CMS experiments, in *5th CERN – Latin-American School of High Energy Physics*, 2009, CERN-2010-001, 449, arXiv:1010.5976.

[27] CDF and D0 Collaborations, Higgs boson studies at the Tevatron, *Phys. Rev.* **D88** (2013) 052014.

[28] The European Strategy for Particle Physics Update 2013, *CERN-Council-S/106* (2013), http://cds.cern.ch/record/1567258/files/esc-e-106.pdf

[29] ATLAS Collaboration, Physics at a High-Luminosity LHC with ATLAS, *ATL-PHYS-PUB-2013-007* (2013), arXiv:1307.7292v2[hep-ex].

[30] CMS Collaboration, Projected Performance of an Upgraded CMS Detector at the LHC and HL-LHC, *CMS NOTE-13-002* (2013), arXiv:1307.7135v2[hep-ex].

[31] CERN Future Circular Collider (FCC) web site: https://fcc.web.cern.ch/Pages/default.aspx

[32] CERN Compact Linear Collider (CLIC) web site: http://clic-study.web.cern.ch/

[33] International Linear Collider (ILC) web site: https://www.linearcollider.org/ILC

[34] Chinese Circular Electron Positron Collider web site: http://cepc.ihep.ac.cn/

LECTURER: P. JENNI

Roadmap at the LHC to the Higgs boson and beyond
(Including highlights from ATLAS)

Scientific Secretaries: L. Barak, A. Salakh

DISCUSSION I

- *Ievlev:*

You mentioned that the LHCb experiment might help us to understand why most observed matter consists of particles rather than antiparticles. Can you please explain how this is made possible and which detector features are responsible for it?

- *Jenni:*

Tomorrow there will be a lecture by the LHCb spokesperson, Professor G. Wilkinson, so it is probably better to leave this question to him. He will be able to give you a competent answer and cover the subject of matter/antimatter asymmetry detection more thoroughly.

- *Campbell:*

How do you convince people to expand the scope and the budget of a project? How do you balance the risk of asking too much and getting nothing?

- *Jenni:*

You can convince people about the value of the extra cost by highlighting the richness of the additional physics. The risk of asking too much is real and the SSC is an example of a project that was stopped because it asked too much. Part of balancing this risk is having a good sense of the funding climate. The current climate isn't quite as good as it was when the LHC experiments were first being proposed. So for future experiment upgrades, compromises may be necessary.

LECTURER: P. JENNI

Roadmap at the LHC to the Higgs boson and beyond
(Including highlights from ATLAS)

Scientific Secretaries: S. Chen, T. Nooney

DISCUSSION II

- *Campbell:*

You showed the different Higgs decay measurements including the significance you projected and the significance you obtained (slide 100). In the case of $H \rightarrow ZZ \rightarrow 4l$, you report 6.2 σ projected and 8.1σ obtained. What is the source of this difference?

- *Jenni:*

This is just a question of statistics in this particular case. In the 4 lepton channel, the number of events we see is still very low, of the order of 30 in each experiment, so you're exposed to large statistical fluctuations.

- *Puccio:*

Could you comment why the ATLAS systematic uncertainty for the mass on *Higgs → 4 leptons* is so much smaller than the equivalent CMS systematic uncertainty (slide 169)?

- *Jenni:*

The two groups have worked together to produce errors for the joint analysis. Both groups do what they can to get the best result and then publish it.

- *Landsberg:*

ATLAS has more aggressive systematic errors because that was a precursor for the W mass measurement. Since you're completely dominated by statistical errors at CMS, you're not so aggressive getting the muon momentum scale. To minimise the systematic uncertainties you need to do a serious amount of work. You can see that overall the error from CMS is smaller due to the slightly higher efficiency for low p^T leptons. At the time of the publication, we were dominated by statistical uncertainties so we decided not to go for the best possible systematic uncertainties because it wasn't significant to the final result. CMS is working on the W mass measurement as ATLAS does and these systematics will decrease in the future.

- *Sokolov:*

Some theorists predicted the appearance of microscopic black holes at the LHC. What are the ways of detecting them and what are the current results from these searches?

- *Jenni:*

There are several model dependent analysis being done. One of the earliest and most popular ones is a model where the micro-black holes decay in a democratic way into many Standard Model (SM) particles like leptons and jets. People look for events with many leptons and jets and compare the distributions with what one expects from the SM. Two examples of this are shown in the backup (slide 205). The first from CMS looks at the distribution of the scalar energy sum of objects at high p^T. These are model dependent micro-black hole models which can be used to set upper limits, typically at the few TeV mass range. Alternatively you could have many similar objects in a democratic way; same sign muons, for example, which are shown by ATLAS. Once again these are model dependent and set limits in the few TeV mass range.

- *Dvali:*

Unfortunately these models tested in ATLAS and CMS do not make very much sense. They often assume many energetic particles in the final state, however this sort of decay in which the micro-black holes produce multi-particle decays, multi jets and many soft decays is simply impossible due to unitarity. You can't immediately produce resonances that decay into many soft quanta before first producing resonances that decay into a moderate number, e.g. two jets. The properties mentioned could of course be properties once you go above the threshold, but you must produce first a resonance, which is the first precursor of what you would call a 'black hole tower'. This first resonance would be literally like an elementary particle, decaying for example into two jets. It would look like a very short lived resonance. If we are incredibly lucky and can go above the threshold, gradually we'll see you're producing resonances that are longer and longer lived and decay into many softer and softer quanta. This could be a signal, once you've passed the first resonance. What we understand is that this democracy that you mentioned must be a transient phenomena. You should see more and more democracy with higher mass. You cannot immediately hit a particle which is immediately democratic. By unitarity this is impossible and I think it's extremely important.

- *'t Hooft:*

LHC is far off the Planck scale so it has inner contradictions, or there are extra-dimensions. In that case, the Planck scale could be low enough for the LHC to reach, but we would first see extra dimensions (Kaluza-Klein tower), then black holes.

- *Dvali:*

No, a black hole like resonance cannot be missed once we reach the Planck scale. This is a model independent feature, thanks to the universal scale of gravity. The others are model dependent, for example the Kaluza-Klein tower might not be observed if the string coupling is too weak etc.

- *Landsberg:*

Yes, we actually also look for black holes which decay into two particles like di-jets, di-muons or decays which involve flavour violating decays. The magenta line in slide 169 is exactly the case of a BH $\rightarrow$ di-jets. The reason why you still see 3 or 4 jets events is due to pileups from LHC.

- *Jenni:*

Yes we indeed look for all kinds of signatures and so called "exotics", so don't worry.

- *Savitskyi:*

All the top mass measurements presented on slide 61 tend to agree, which is good. However, all measurements are strongly dependent on the understanding of jet energy scales and on the input top mass used for Monte Carlo (MC) generations. This means that these results are measuring the top mass driven by MC rather than the pole mass which should be used in theory and can be obtained via cross section measurements. Is it fair to compare measurements from experiments which use different MC generators and do experiments such as ATLAS and CMS plan to develop a common MC generator?

- *Jenni:*

Yes, the measurements shown here are comparable. The measurements make templates with the MC and compare these to the reconstructed top mass. But the experiments use different methods and there are different ways that are pursued and published. For example the inclusive cross section is a function of the pole mass but the precision isn't as good. The data is combined and there is a strong coherence and exchange between the experiments. The code may not be identical, but this is good for getting an estimate on the systematic errors.

- *Landsberg:*

We did exercises comparing the same MC generator next to leading order and found that the results are not dependent on this. This is a very small part of the systematics.

- *Tannenbaum:*

Can you explain the plot shown for $H \rightarrow \mu^+ \mu^-$ on slide 78.

- *Jenni:*

This has not been measured by anyone yet. This is an important example of why we need the high luminosity LHC. With 300 fb^{-1} we will get a first signal, but 3000 fb^{-1} are needed to get significantly above 5σ, so we really need the high luminosity LHC for this measurement.

- *Li:*

You present the expected and observed significance of Higgs decays, but some channels are not consistent (slide 143). Do you have any comparison of branching ratios from experimental results and theoretical calculations?

- *Jenni:*

In terms of signal strengths, the results are normalised to the cross section times branching ratio. In the analysis for the coupling strengths, there are ratios. We have to do ratios because we don't measure Higgs decays into invisible channels with high precision. There are constraints, no more than of the order of 25%. For the fits of the coupling strengths, one typically normalises to the decay widths of the vector bosons.

- *Li:*

Why is the error bar of the inclusive tt cross section using 8 TeV data larger than that for 7 TeV, even though both the cross section and luminosity are higher for 8 TeV?

- *Jenni:*

Good question. I will have to check.

- *Klapot:*

When do you expect ATLAS to collect enough data to exclude TeV scale supersymmetry (SUSY)? For example a minimal SUSY model at 5σ?

- *Jenni:*

I don't think one can give a simple answer to that. If, for example, you look at third generation SUSY particles which are very close to the kinematic limits, then it is quite difficult. If you take the simple 1st and 2nd generation classical decays with just a few jets plus missing E_T, there are already exclusions larger than 1 TeV. Experimentally the most favourable result is currently at 1.7 TeV, but this is not the favoured model of our theory colleagues.

IceCube: the Discovery of High-Energy Cosmic Neutrinos

Francis Halzen
University of Wisconsin–Madison, Madison, WI

By transforming a cubic kilometer of natural Antarctic ice into a neutrino detector, the IceCube project created the opportunity to observe cosmic neutrinos. We describe the experiment and the complementary methods presently used to study the flux of the recently discovered cosmic neutrinos. In one method, events are selected in which neutrinos interacted inside the instrumented volume of the detector, yielding a sample of events dominated by neutrinos of electron and tau flavor. Alternatively, another method detects secondary muons produced by neutrinos selected for having traveled through the Earth to reach the detector, providing a pure sample of muon neutrinos. We will summarize the results obtained with the enlarged data set collected since the initial discovery and appraise the current status of high-energy neutrino astronomy. Continued observations are closing in on the source candidates. In this context, we highlight the potential of multimessenger analyses as well as the compelling case for constructing a next-generation detector larger by one order of magnitude in volume.

1 Introduction

Soon after the 1956 observation of the neutrino [1], the idea emerged that it represented the ideal astronomical messenger [2, 3, 4]. Neutrinos can reach us from the edge of the Universe without absorption and with no deflection by magnetic fields. They can escape unscathed from the inner neighborhood of black holes and from the accelerators where cosmic rays are born. The weak interactions that turn neutrinos into ideal astronomical messengers also make them notoriously difficult to detect. It has been anticipated for some time that kilometer-scale particle detectors are required to collect neutrinos from the cosmos in statistically significant numbers [5, 6, 7, 8]. Already by 1969, it had been understood [9] that such an instrument would create the possibility of observing the so-called Greisen-Zatsepin-Kuzmin (GZK) neutrinos produced in the interactions of cosmic rays with background microwave photons [10].

Given the detector's required size, early efforts concentrated on instrumenting large volumes of natural water with photomultipliers [11] that detect the Cherenkov light emitted by secondary particles produced when neutrinos interact with nuclei inside or near the detector [12, 13, 14, 15, 16, 17]. In contrast, the IceCube detector [18, 19, 20, 21] transforms deep natural Antarctic ice into a Cherenkov detector. The instrument shown in Fig. 1 consists of 86 strings, each instrumented with 60 ten-inch photomultipliers spaced 17 m apart over a total length of one kilometer. The deepest modules are located at a depth of 2.45 km so that the instrument is shielded from the large background of cosmic rays at the surface by approximately 1.5 km of ice. Strings are arranged at apexes of

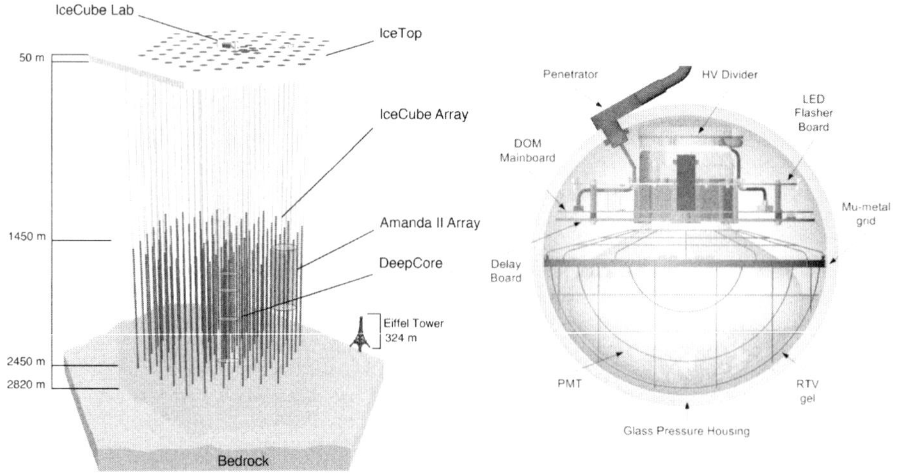

Figure 1 – Sketch of the IceCube observatory (left) and the digital optical module, or DOM (right).

equilateral triangles that are 125 m on a side. The instrumented detector volume is a cubic kilometer of dark and highly transparent [22] Antarctic ice.

Neutrino events may be broadly classified in two groups that reflect the patterns of Cherenkov light emitted, muon tracks and particle showers (often referred to as cascades). Tracks are produced by charged current interactions of muon neutrinos, while cascades are produced by charged current interactions of electron and tau neutrinos as well as neutral current interactions of all flavors. The travel range of ν_μ-induced muons is on the order of kilometers, while the length characteristic of the electromagnetic showers that dominate the cascade events is only tens of meters.

A detailed understanding of the propagation of photons in the Antarctic ice [22] is essential in order to relate light generated to light observed in the digital optical modules (DOMs) that house and record the signals of the photomultipliers [20]: the arrival time of photons at the DOMs determines the trajectory [23], while the number of photons determine the deposited energy. For typical kilometer-long tracks, the angular resolution is better than 0.4°. Reconstruction of the direction of cascade events is still in the development stage in IceCube [24]. The direction of cascades can be reconstructed to within 15° based on photon timing patterns collected by individual DOMs. Determining the deposited energy from the observed light pool is relatively straightforward, and a resolution of better than 15 % is possible; the same value holds for the reconstruction of the energy deposited by a muon track inside the detector. At this point, IceCube's precision in reconstruction is limited by computing, not by the properties of the Cherenkov medium, guaranteeing improved performance in the future.

Another way to classify neutrino events is to distinguish between events that start inside the detector from those in which the neutrino interacts outside the detector. By specializing to events starting inside the detector, one detects neutrinos of all flavors. For $E_\nu < 1\,\text{PeV}$, the interaction of a ν_τ in IceCube will look much like that of a ν_e because the track length of the ν_τ of less than 50 m is smaller than the 125 m string spacings and difficult to identify. The largest neutrino sample consists of ν_μ-induced muon tracks

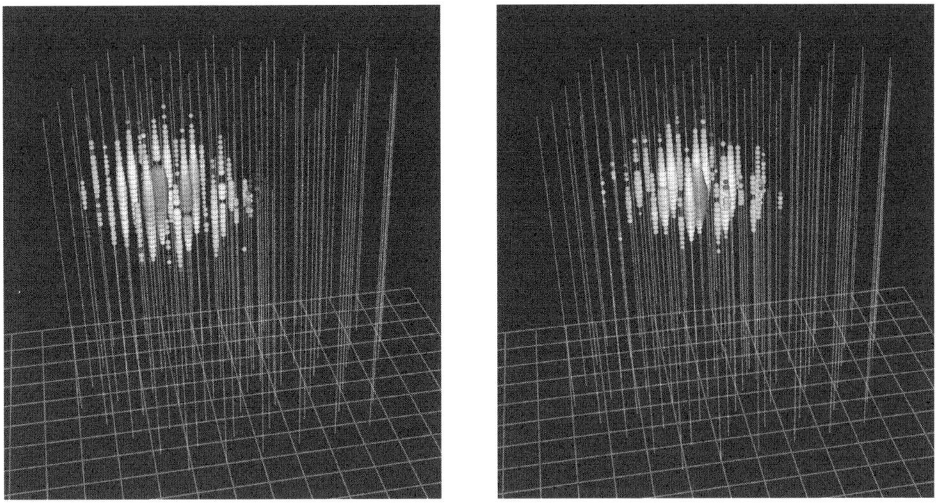

Figure 2 – Light pool produced in IceCube by the two cascades "Bert" (left) and "Ernie" (right). Their measured energy is 1.04 PeV and 1.14 PeV, respectively, which represents a lower limit on the energy of the neutrino that initiated each shower. White dots represent sensors with no signal. For the colored dots, color indicates arrival time, from red (early) to purple (late) following the rainbow, and size reflects the number of photons detected.

entering the detector from zenith angles too large to be atmospheric in origin, typically $\theta \geq 85°$. The muons are secondaries produced by neutrinos that have penetrated the Earth. The rate of such events in the full IceCube detector is approximately 200 per day or more, depending on the threshold for a particular analysis. Their mean energy is $1 \sim 10$ TeV.

For neutrino astronomy, the first challenge is to select sufficiently pure samples of neutrinos, and the second is to identify the small fraction of the neutrinos that are astrophysical in origin, expected to be at most tens of events per year. Atmospheric neutrinos are an overwhelming background for cosmic neutrinos, at least at energies below ~ 100 TeV. Above this energy, however, their flux is too small to produce events in a kilometer-scale detector and every event is a discovery. This is how IceCube made its initial serendipitous observation of cosmic neutrinos.

2 The Status of Neutrino Astronomy

GZK neutrinos were the target of a dedicated search using IceCube data collected between May 2010 and May 2012. Two interesting events were found [25]. However, their energies, rather than EeV as expected for GZK neutrinos, were in the PeV range: 1,040 TeV and 1,140 TeV. The events are particle showers initiated by neutrinos interacting inside the instrumented detector volume. Their light pool of roughly one hundred thousand photoelectrons extends over more than 500 meters; see Fig. 2. With no evidence of a muon track, they are showers initiated by electron or tau neutrinos. With PeV energy, they are unlikely to be atmospheric in origin.

Previous to this serendipitous discovery, neutrino searches had almost exclusively specialized to the observation of muon neutrinos that interact primarily outside the detector to produce kilometer-long muon tracks passing through the instrumented volume. Al-

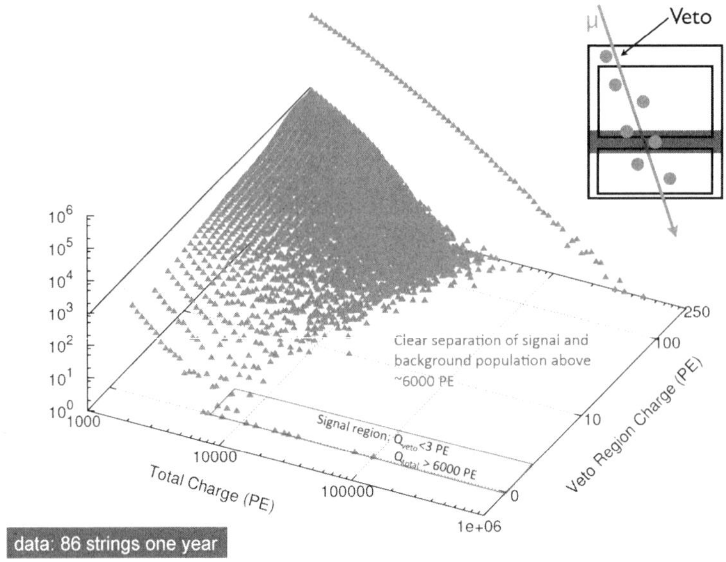

Figure 3 – One year of IceCube data from its final 86-string configuration showing number of events as a function of the total number of photoelectrons and the number present in the veto region. The signal region is defined as having more than 6000 photoelectrons with fewer than three of the first 250 in the veto region of the detector. The signal, including nine events with reconstructed energy in excess of 100 TeV, is clearly separated from the background.

though creating the opportunity to observe neutrinos interacting outside the detector, it is necessary to use the Earth as a filter to remove the huge background flux of muons produced by cosmic ray interactions in the atmosphere. This limits the neutrino view to a single flavor and half the sky. Inspired by the observation of the two PeV events, a filter was designed that exclusively identifies neutrinos interacting inside the detector. It divides the instrumented volume of ice into an outer veto shield and a 420 megaton inner fiducial volume. The separation between veto and signal regions was optimized to reduce the background of atmospheric muons and neutrinos to a handful of events per year while keeping 98% of the signal. The great advantage of specializing to neutrinos interacting inside the instrumented volume of ice is that the detector functions as a total absorption calorimeter measuring energy with a 10-15 % resolution. Also, neutrinos from all directions in the sky can be identified, including both muon tracks produced in ν_μ charged-current interactions and secondary showers produced by neutrinos of all flavors.

Analyzing the data covering the same time period as the GZK neutrino search, 28 candidate neutrino events were identified with in-detector deposited energies between 30 and 1140 TeV; see Fig. 3. Of these, 21 are showers. The remaining seven events are muon tracks, which do allow for subdegree angular reconstruction; only a lower limit on their energy can be established because of the unknown fraction carried away by the exiting muon track. Furthermore, with the present statistics, these are difficult to separate from a remnant of the competing atmospheric muon background. The 28 events include the two PeV events previously revealed in the GZK neutrino search.

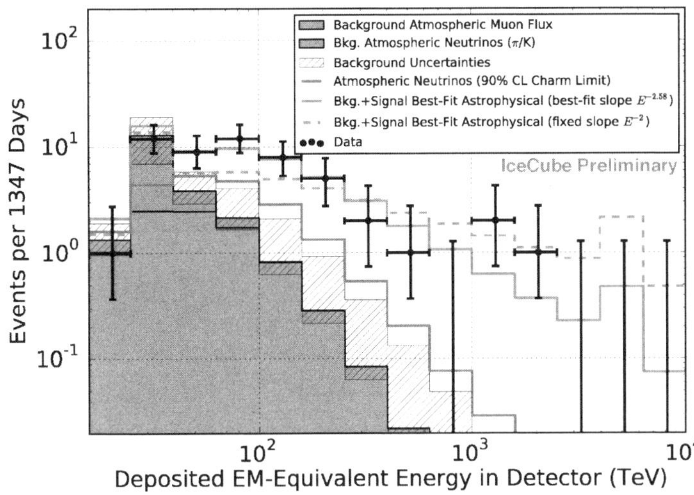

Figure 4 – Deposited energies of muons observed in four years of data. The hashed region shows uncertainties on the sum of all backgrounds. The atmospheric muon flux (red) and its uncertainty is computed from simulation to overcome statistical limitations in our background measurement and scaled to match the total measured background rate. The atmospheric neutrino flux is derived from previous measurements of both the π, K and charm components of the atmospheric spectrum [26]. Also shown are two power-law fits to the spectrum.

Fitting the data to a superposition of extraterrestrial neutrinos on an atmospheric background yields a cosmic neutrino flux of

$$E_\nu^2 \frac{\mathrm{d}N}{\mathrm{d}E_\nu} = 3.6 \times 10^{-11}\,\mathrm{TeV\,cm^{-2}\,s^{-1}\,sr^{-1}} \tag{1}$$

for the sum of the three neutrino flavors. The energy and zenith angle dependence observed is consistent with what is expected for a flux of neutrinos produced by cosmic accelerators; see Fig. 4. The flavor composition of the flux is, after corrections for the acceptances of the detector to the different flavors, consistent with $\nu_e : \nu_\mu : \nu_\tau \sim 1 : 1 : 1$ as anticipated for a flux originating in cosmic sources.

The major uncertainty in the spectrum of atmospheric neutrinos at high energy is the flux of charm origin that must be present at some level. Short-lived charmed hadrons produced by cosmic rays in the atmosphere promptly decay, producing muons and neutrinos with the same spectrum as their parent cosmic rays. This "prompt" flux of leptons has not been identified. Existing limits [27, 28] allow a charm component at a level predicted by a dipole model calculation for the high-energy charm cross section [29]. In this context, it is important to point out that the muon produced in the same decay as the neutrino, in the case of muon neutrinos from above, is guaranteed to reach the detector provided that the neutrino energy is sufficiently high and the zenith angle sufficiently small [30, 31]. In this case, the atmospheric neutrino provides its own self-veto. This self-veto is routinely applied to the IceCube data sample and suppresses any potential charm component.

In the meantime, two additional years of data have been analyzed, doubling the statistics of the discovery publication [32]. A purely atmospheric explanation can be excluded at 7σ. The four-year data set contains a total of 54 neutrino events with deposited energies ranging from 30 to 2000 TeV. In the fourth year, muon neutrinos were found that

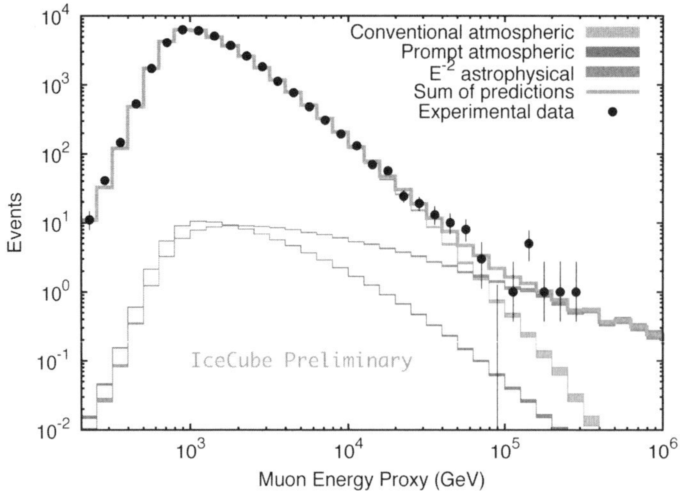

Figure 5 Spectrum of secondary muons initiated by muon neutrinos that have traversed the Earth, i.e., with zenith angle less than 5° above the horizon, as a function of the energy they deposit inside the detector. The highest energy muons are, on average, initiated by PeV neutrinos.

deposited $\sim 500\,\text{TeV}$ energy inside the detector, indicating PeV-energy parent neutrinos. One of them reconstructs through IceTop, IceCube's surface array, with no evidence for an air shower. Combining the absence of an air shower in IceTop with large deposited muon energy results in a high significance for astrophysical origin from a single event.

Recently, an independent analysis has also revealed an excess over atmospheric background of high-energy ν_μ-induced muons penetrating the Earth from the Northern Hemisphere [33, 34] confirming the signal observed in the starting event analyses; see Fig. 5, which shows the muon neutrino flux as a function of the energy deposited by the muons inside the detector. Standard Model physics allows one to infer the energy spectrum of the parent neutrinos; for instance, the highest energies in Fig. 5 correspond, on average, to parent neutrinos with energy in excess of PeV. A best fit to the spectrum that includes a conventional, charm and astrophysical component with free normalizations yields the results shown in the figure. With one additional year of data, the statistical significance exceeds $4.3\,\sigma$; in the 2014 sample, an event was found with energy $2.6 \pm 0.3\,\text{PeV}$, representing the highest energy neutrino ever recorded, with energy in the $5 \sim 10\,\text{PeV}$ energy range [35].

Where do the cosmic neutrinos originate? Figure 6 shows the arrival directions of the four-year starting-event sample in Galactic coordinates. Cascade events are labeled (+), track events (×). The color scale indicates the value of the test statistic (TS) of an unbinned maximum likelihood test searching for anisotropies of the neutrino arrival directions. No significant local excess in the sky was found when compared to randomized pseudo-experiments. The correlation of neutrino events with the Galactic plane is not significant. Letting the width of the plane float freely, the best fit returned a correlation for a value of $\pm 7.5°$ with a post-trial chance probability of 3.3%. Neither probability decreased after doubling the data. The IceCube Collaboration also searched for clustering of the events in time [36] and investigated a possible correlation with the times of observed GRBs [37]. No statistically significant correlation was found.

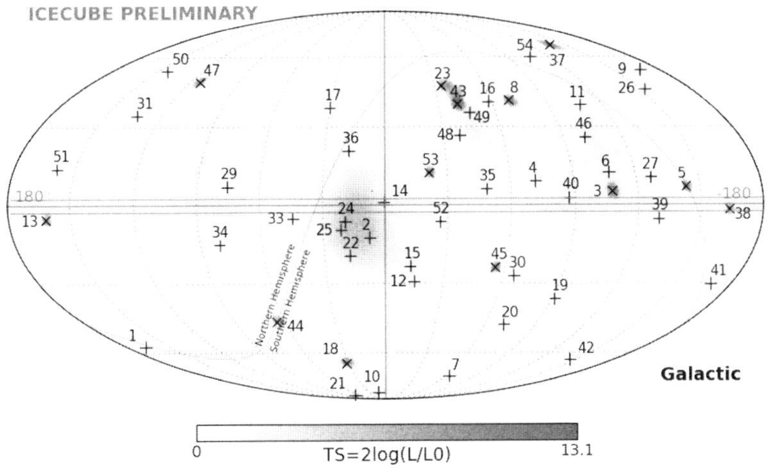

Figure 6 – Arrival directions of neutrinos in the four-year starting-event sample in Galactic coordinates. Shower-like events are shown with "+" and those containing muon tracks with "×". The color scale indicates the value of the test statistic (TS) of an unbinned maximum likelihood test searching for anisotropies of the event arrival directions. The horizontal red lines indicate the minimum width (and step-size) ±2.5° used in the search for extended emission along the Galactic Plane. Note that the track-like event 28 has been omitted following the discussion in Ref. [32].

In summary, the observed neutrino flux is consistent with an isotropic distribution of arrival directions and equal contributions of all neutrino flavors [38], suggesting the observation of extragalactic sources whose flux has equilibrated in the three flavors after propagation over cosmic distances [39, 40]. Increasingly, a variety of analyses [38, 41] suggest that the cosmic neutrino flux dominates the atmospheric background above an energy that may be as low as 30 TeV with an energy spectrum that is not described as a single power. This is reinforced by the fact that fitting the excess flux in different ranges of energy yields different values for the power.

3 Closing in on the Sources

Rather than speculate, we will focus on the multimessenger connection of cosmic neutrinos to cosmic rays and gamma rays; this will turn out to be revealing. The overall energy density of the observed neutrino flux is close to the limit for neutrino production in the sources of ultra-high-energy (UHE) extragalactic CRs [42]. This may be a coincidence, but it may also indicate a multimessenger relation. The neutrino and CR nucleon (N) emission rates Q are related by

$$\frac{1}{3}\sum_{\alpha} E_\nu^2 Q_{\nu_\alpha}(E_\nu) \simeq \frac{1}{4}\frac{f_\pi K_\pi}{1+K_\pi} E_N^2 Q_N(E_N), \tag{2}$$

where $f_\pi(<1)$ is the efficiency for pion production, K_π the ratio of charged to neutral pions, and $E_\nu \simeq 0.05 E_N$. The emission rate density of UHE CRs depends on their spectrum and composition, for a E^{-2} flux of protons $E_p^2 Q_p(E_p) \simeq (1-2)\times 10^{44}\,\mathrm{erg\,Mpc^{-3}\,yr^{-1}}$ [43]. Unlike cosmic rays that are collected from sources within the GZK radius of $\sim 100\,\mathrm{Mpc}$, neutrinos reach us from all redshifts. Their relative fluxes therefore depend on the evolution of the sources with redshift, which can be parametrized by a factor $\rho(z)$. For no evolution and $\gamma = 2$, $\xi_z \simeq 0.5$, whereas an evolution of sources following the star-formation rate (SFR) yields $\xi_z \simeq 2.4$. In the end, we obtain a diffuse neutrino flux given by

$$E_\nu^2 \phi_\nu(E_\nu) \simeq \frac{\xi_z f_\pi K_\pi}{1+K_\pi}(2-4) \times 10^{-8}\,\mathrm{GeV\,cm^{-2}\,s^{-1}\,sr}\,. \tag{3}$$

With $f_\pi < 1$ this reproduces the *Waxman-Bahcall* bound [42]. Interestingly, the observed flux is close to the bound requiring $f_\pi \simeq 1$ requiring a very efficient production of neutrinos in sources with optical thickness $\tau > 1$. These are often referred to as CR calorimeters or CR reservoirs [44], such as starburst galaxies [45, 46] and clusters of galaxies [47, 48, 49].

The production of PeV neutrinos is inevitably associated with the production of PeV gamma rays [50, 51, 52, 53, 54, 55, 56]: hadronic accelerators produce fluxes of both neutral and charged pions that are the parents of gamma rays and neutrinos, respectively. The production rates of neutrinos and gamma rays are related by

$$\frac{1}{3}\sum_\alpha E_\nu^2 Q_{\nu_\alpha}(E_\nu) \simeq \frac{K_\pi}{4} E_\gamma^2 Q_\gamma(E_\gamma)\,. \tag{4}$$

Note that this relation does not depend on the pion production efficiency but only on the relative charged-to-neutral pion rate, K_π. On the other hand, the production rate of gamma rays described by Eq. (4) is not necessarily the emission rate observed. For instance, in hadronic sources that efficiently produce neutrinos via $p\gamma$ interactions, the target photon field can also efficiently reduce the pionic gamma rays via pair production. Also, inverse-Compton scattering and synchrotron emission in magnetic fields will shift the emitted gamma-ray spectrum to lower energies. This is a calorimetric process that will, however, conserve the total energy of hadronic gamma rays.

It is straightforward to apply the multimessenger relation to the cosmic neutrino flux observed by IceCube. Figure 7 shows the associated gamma-ray flux for two illustrative *pp* emission scenarios that accommodate the observed neutrino flux. The black and red lines show the neutrino and gamma-ray spectra after accounting for cosmic evolution and cascading in cosmic radiation backgrounds. The thick solid line shows an $E^{-2.15}$ neutrino spectrum with an exponential cutoff around PeV. This scenario actually matches the extragalactic isotropic diffuse gamma-ray background (IGRB) measured by *Fermi* [58]. The alternative scenario illustrates that even for harder emission spectra the cascaded flux will still make a significant contribution to the Fermi IGRB. For illustration, we show the effect of hadronic emission that produces a peaked neutrino spectrum in the 10 TeV to 1 PeV energy region (thin black line). Note that such a spectrum is not expected for a *pp* scenario. The observed gamma-ray spectrum (thin red line) is in this case dominated by secondary cascaded photons. The contribution to the *Fermi* IGRB between 100 GeV to 1 TeV is still at the level of 10% even though the input neutrino spectrum underestimates the neutrino flux substantially. Soft emission spectra ($\Gamma \gtrsim 2.2$) are inconsistent with the Fermi observations [57, 59] and point at obscured or hidden sources.

In general, this exercise indicates that the diffuse gamma-ray contribution to the *Fermi* IGRB is large for *pp* cosmic beam dumps and suggests a common origin of some of the sources. This is intriguing because a recent analysis shows that blazars dominate the Fermi flux. Are blazars the final answer? The good news is that IceCube, by accumulating more events, can eventually identify blazars by observing multiple neutrinos from the same sources [60] . On the other hand, a search for neutrinos from *Fermi*'s *identified*

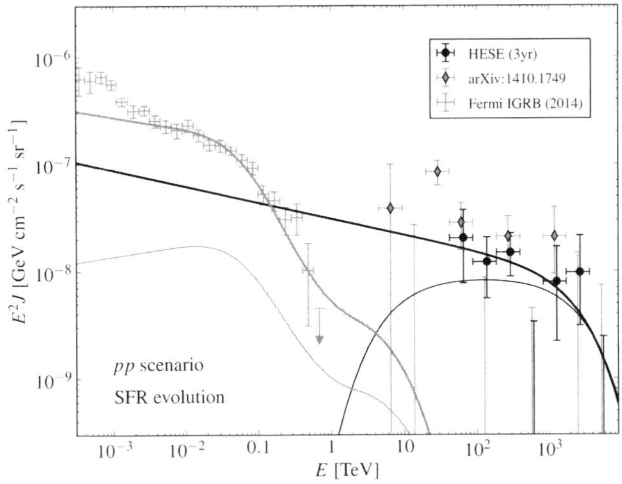

Figure 7 – Two models of the astrophysical neutrino flux (black lines) observed by IceCube and the corresponding cascaded gamma-ray flux (red lines) observed by *Fermi*. The models assume that the decay products of neutral and charged pions from pp interactions are responsible for the nonthermal emission in the Universe [57]. The thin lines represent an attempt to minimize the contribution of the pionic gamma-ray flux to the *Fermi* observations. It assumes an injected flux of E^{-2} with exponential cutoff at low and high energy. The black data points are measured by the IceCube three-year "high-energy starting event" (HESE') analysis [32], the gray data points are from an IceCube analysis lowering the energy threshold for events starting in the detector even further [41].

blazars [61] has come up empty. However, these represent only 50% of the total diffuse background, allowing for the possibility that a partially different source population produces the neutrinos. Although no definite identification of the sources of cosmic neutrinos has emerged, it is rather clear that a multiwavelength path to the neutrino sources looks very promising.

By building IceCube, it was possible to map the optical properties of natural ice over large distances, and the IceCube Collaboration made the surprising discovery that the absorption length of the Cherenkov light to which the DOMs are sensitive exceeds 100 m. In fact, in the lower half of the detector it exceeds 200 m. The absorption length dictates the distance by which one can space the strings of sensors without spoiling the uniformity of the detector. Modeling indicates that spacings of 250 m, possibly larger, are acceptable. One can therefore instrument a ten-times-larger volume of ice with the same number of strings used to build IceCube. The project would be free of risk since the performance of the DOMs is well understood and we already know how to deploy them. Furthermore, the costs are also understood. Designed before 1999, all of the components of IceCube can be significantly updated for improved performance.

The larger spacings do of course result in a higher threshold, but this is not necessarily bad. While the 100,000 or so atmospheric neutrinos that IceCube collects above a threshold of 100 GeV every year were useful for calibration, they represent a severe background for isolating the cosmic component of the flux. The peak sensitivity to an E^{-2} spectrum is reached at 40 TeV [62]. While the detector has to be efficient below that energy, a threshold much lower than this value introduces background without a gain in signal. Designs for a next-generation instrument are in progress [63].

4 The Search for Dark Matter

IceCube was very much designed as a discovery instrument, and it covers a range of science, besides the search for the nature of the cosmic particle accelerators, whose interests are multidisciplinary. Examples includes the search for Galactic supernova explosions and the study of neutrinos themselves, at higher energies and with improved statistics. Also, for some time now, neutrino "telescopes" have been recognized as powerful tools in the search for the particle nature of dark matter. IceCube's DeepCore infill, which lowers IceCube's threshold over a significant part of the detector volume, was initially proposed by Per Olof Hulth as a way to enhance IceCube's capabilities for detecting lower mass dark matter particles. In this context ,it is worth noting that the AMANDA detector, the forerunner and proof of concept for IceCube, received a significant fraction of its initial funding from the Berkeley Center for Particle Astrophysics to search for dark matter.

IceCube searches indirectly for dark matter by looking for neutrinos from concentrations of weakly interacting massive particles (WIMPs) in the Sun [64], in the Milky Way [65, 66], and in nearby galaxies [67]. The neutrinos are secondary products of annihilation of pairs of WIMPs into Standard Model particles, which include decays into neutrinos. The evidence that yet-to-be-detected weakly interacting massive particles (WIMPs) make up dark matter is compelling. WIMPs are swept up by the Sun as the Solar System moves about the galactic halo. Though interacting weakly, they will occasionally scatter elastically with nuclei in the Sun and lose enough momentum to become gravitationally bound. Over the lifetime of the Sun, a sufficient density of WIMPs may accumulate in its center so that an equilibrium is established between their capture and annihilation. The annihilation products of these WIMPs represent an indirect signature of halo dark matter, their presence revealed by neutrinos that escape the Sun with minimal absorption. The neutrinos are, for instance, the decay products of heavy quarks and weak bosons resulting from the annihilation of WIMPs into $\chi\chi \rightarrow \tau\bar{\tau}$, $b\bar{b}$ or W^+W^-. Neutrino telescopes have are sensitive to such neutrinos because of the relatively large neutrino energy reflecting the mass of the WIMP.

The beauty of the indirect detection technique using neutrinos originating in the Sun is that the astrophysics of the problem is understood. The source in the Sun has built up over solar time, sampling the dark matter throughout the galaxy; therefore, any possible structure in the halo has been averaged out. Given a WIMP mass and properties, one can unambiguously predict the signal in a neutrino telescope. If not observed, the model is ruled out. This is in contrast to indirect searches for photons from WIMP annihilation, whose sensitivity depends critically on the structure of halo dark matter; observation requires cuspy structure near the galactic center or clustering on appropriate scales elsewhere. Observation not only necessitates appropriate WIMP properties but also favorable astrophysical circumstances.

Extensions of the Standard Model of quarks and leptons, required to solve the hierarchy problem, naturally yield dark matter candidates. For instance, the neutralino, the lightest stable particle in supersymmetric models, has been intensively studied as a possible dark matter candidate. Detecting it has become the benchmark by which experiments are evaluated and mutually compared. We will follow this tradition here. Supersymmetric models allow for a large number of free parameters and, unfortunately, for a variety of parameter sets that are able to generate the observed dark matter density in the context of standard big bang cosmology. The neutralino interacts with ordinary matter by spin-independent (e.g., Higgs exchange) and by spin-dependent (e.g., Z-boson exchange) interactions. The first mechanism favors direct detection experiments [68] because the WIMP interacts coherently, resulting in an increase in sensitivity proportional to the square of the atomic number of the detector material. Given the rapid improvement in the sensitivity of direct experiments, indirect detection experiments are therefore not competitive for models in which dark matter has a large spin-independent coupling. We will show that this is not the case for spin-dependent models, however. Within the context of supersymmetry, direct and indirect experiments are complementary.

Although IceCube detects neutrinos of all flavors, sensitivity to neutrinos produced by WIMPs in the Sun is achieved by exploiting the degree accuracy with which muon neutrinos can be pointed back to the Sun.

IceCube is most sensitive to WIMPs with significant spin-dependent cross sections for interactions with protons because this would lead to strong concentrations in the Sun, a nearby and readily identifiable source. The signal would simply be an excess of neutrinos of GeV energy and above from the direction of the Sun over the atmospheric neutrino background in the same angular window. There would be no alternative astrophysical explanation of such a signal, which represents a smoking gun for dark matter particles. In most WIMP scenarios, the cross sections for WIMP capture ($\sigma_{\chi,p}$) and for WIMP annihilation ($\sigma_{\chi,\chi}$) are large enough so that an equilibrium between capture and annihilation would have been achieved within the age of the solar system [69]. In this case, limits on neutrinos from the Sun can be expressed in terms of the capture cross section, $\sigma_{\chi,p}$. If equilibrium is not reached, weaker limits can still be derived.

A WIMP may interact with ordinary nuclei by spin-independent (e.g. Higgs exchange) and by spin-dependent (e.g. Z-boson exchange) interactions. The first mechanism favors direct detection experiments [68] because the WIMP interacts coherently, resulting in an increase in sensitivity proportional to the square of the atomic number of the detector material. The latter favors indirect experiments in which the rates are dominated by spin-dependent interactions with protons. In the case of the Sun, the WIMPS have accumulated over solar time scales, sampling the dark matter throughout the galaxy and averaging out any structure in the halo that theory may not have accounted for. Within the context of supersymmetry, direct and indirect experiments are complementary.

Quantitative interpretation of the IceCube limits is complicated, depending on detailed analysis of capture rates and annihilation channels. The analysis uses DarkSUSY [69], which builds on the classic calculations of Gould [70] for capture and annihilation rates. The neutrino spectrum from annihilation of pairs of WIMPs depends on the dominant channel for coupling to Standard Model particles. The neutrino spectrum at production is also modified by subsequent interactions and oscillations of neutrinos as they propagate out from the solar core. Three flavor oscillations with matter effects are included, as in Ref. [71]. The range of possibilities is bracketed by calculating two extremes,

$$\chi\chi \to b\bar{b}\,(\text{soft}) \text{ and } \chi\chi \to W^+W^-,\ \tau\bar{\tau}\,(\text{hard}). \tag{5}$$

In the hard channel, $W^\pm$ are replaced by $\tau^\pm$ for $m_\chi < m_W$.

The current IceCube limits [64, 72, 73] are shown in Figs. 8 and 9. Also shown is a sampling of the WIMP parameter space that is not ruled out by other experiments. IceCube has produced the most stringent limits on the cross section for spin-dependent interactions of dark matter particles with ordinary matter. Since the exact branching ratios of WIMP annihilation into different channels is model-dependent, experiments usually

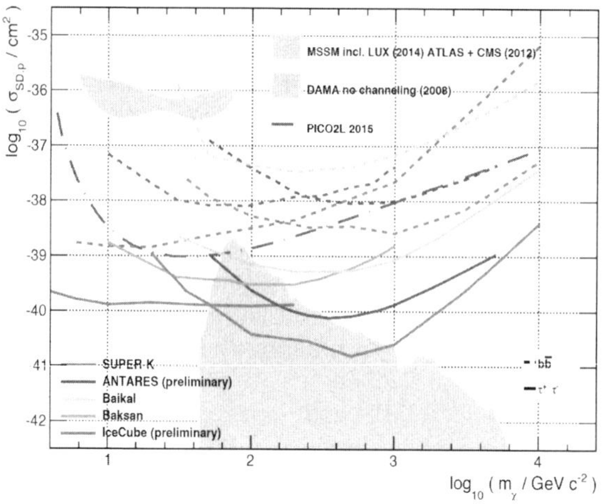

Figure 8 Upper limits at 90% confidence level on the spin-dependent neutralino-proton cross section assuming that the neutrinos are produced by $b\bar{b}$, $\tau\bar{\tau}$ and W^+W^- annihilation. Limits from IceCube [64, 72, 73], Super-K [75], ANTARES [76], Baikal [77] and Baksan [78] are shown. Full lines refer to limits on the annihilation channel and dashed lines to the $b\bar{b}$ channel. Direct search results from PICO [79] and LUX [80] and tentative signal regions [81, 82, 83] (gray-shaded areas) are included for comparison. The brown-shaded region indicates the allowed parameter space in the MSSM not ruled out by direct experiments.

choose two annihilation channels which give extreme neutrino spectra to show their results. Annihilation into $b\bar{b}$ is chosen as a representative case producing a soft neutrino spectrum, and annihilation into W^+W^- or $\tau\bar{\tau}$ as a hard spectrum. Assuming a 100% branching ratio to each of these channels brackets the expected neutrino spectrum from any more realistic model with branching to more channels. IceCube or SuperK reach bounds at the $10^{-40} - 10^{-41}\text{cm}^2$ level, covering the WIMP mass range, between the two experiments, from a few GeV to 100 TeV. Because of the A^2 coherence factor for scattering on heavy nuclei, the indirect experiments have an advantage over IceCube for the case of spin-independent interactions; see Fig. 9.

Current IceCube limits are based on observations of ν_μ-induced muons with the nearly complete detector, with 79 strings of digital optical modules (DOMs) and including for the first time the DeepCore subarray. Data were collected from May 2010 to May 2011, including the austral summer (October-March) when the Sun is above the horizon. By including events that start inside DeepCore (summer and winter), the mass range for the WIMP search could be extended down to 20 GeV, which overlaps some of the allowed region from the DAMA experiment [74]. By including muons from below the horizon entering IceCube from outside (winter only because of the high background of cosmic-ray muons from above), the mass range is extended to 5 TeV.

Note that the limits are derived from a single year of data with a detector that was not quite completed. Not only are the analysis methods subject to improvements, five years of data is now available and will be analysed in the near future. This will yield a discovery or limits improved well beyond the square root of time expected from statistics alone.

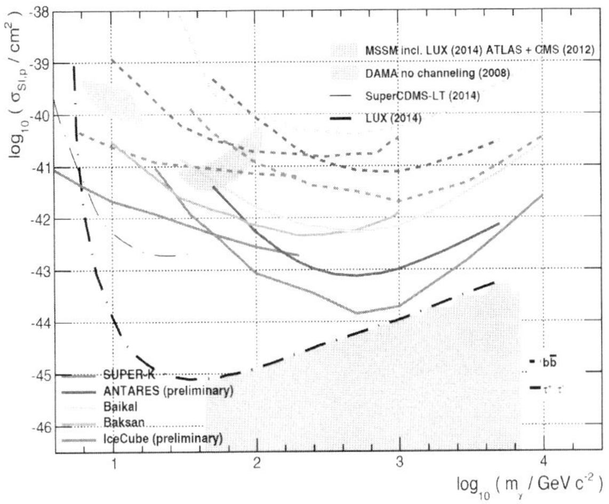

Figure 9 – Same as the previous figure for spin-independent interactions of dark matter with ordinary matter.

Acknowledgments

Discussion with collaborators inside and outside the IceCube Collaboration, too many to be listed, have greatly shaped this presentation. Thanks. This research was supported in part by the U.S. National Science Foundation under Grants No. ANT-0937462 and PHY-1306958 and by the University of Wisconsin Research Committee with funds granted by the Wisconsin Alumni Research Foundation.

References

1. Frederick Reines and Clyde L. Cowan. The neutrino. *Nature*, 178:446–449, 1956.
2. K. Greisen. Cosmic ray showers. *Ann.Rev.Nucl.Part.Sci.*, 10:63–108, 1960.
3. Frederick Reines. Neutrino interactions. *Ann.Rev.Nucl.Part.Sci.*, 10:1–26, 1960.
4. M.A. Markov. On high energy neutrino physics. *Proc. of the 10th Intl. Conf. on High-Energy Physics*, pages 578–581, 1960.
5. Thomas K. Gaisser, Francis Halzen, and Todor Stanev. Particle astrophysics with high-energy neutrinos. *Phys.Rept.*, 258:173–236, 1995.
6. J.G. Learned and K. Mannheim. High-energy neutrino astrophysics. *Ann.Rev.Nucl.Part.Sci.*, 50:679–749, 2000.
7. Francis Halzen and Dan Hooper. High-energy neutrino astronomy: The Cosmic ray connection. *Rept.Prog.Phys.*, 65:1025–1078, 2002.
8. Julia K. Becker. Status of neutrino astronomy. *J.Phys.Conf.Ser.*, 136:022055, 2008.
9. A. Roberts. The Birth of high-energy neutrino astronomy: A Personal history of the DUMAND project. *Rev.Mod.Phys.*, 64:259–312, 1992.
10. V.S. Berezinsky and G.T. Zatsepin. Cosmic rays at ultrahigh-energies (neutrino?). *Phys.Lett.*, B28:423–424, 1969.
11. Zheleznykh, I., Int. J. Mod. Phys. A 21S1 (2006) 1; Markov, M. and Zheleznykh, I., Nucl. Phys. 27 (1961) 385.

12. J. Babson et al. Cosmic Ray Muons in the Deep Ocean. *Phys.Rev.*, D42:3613–3620, 1990.
13. V.A. Balkanov et al. The BAIKAL neutrino project: Status report. *Nucl.Phys.Proc.Suppl.*, 118:363–370, 2003.
14. G. Aggouras et al. A measurement of the cosmic-ray muon flux with a module of the NESTOR neutrino telescope. *Astropart.Phys.*, 23:377–392, 2005.
15. J.A. Aguilar et al. First results of the Instrumentation Line for the deep-sea ANTARES neutrino telescope. *Astropart.Phys.*, 26:314–324, 2006.
16. Emilio Migneco. Progress and latest results from Baikal, Nestor, NEMO and KM3NeT. *J.Phys.Conf.Ser.*, 136:022048, 2008.
17. A. Margiotta. Status of the KM3NeT project. *JINST*, 9:C04020, 2014.
18. P. Bagley et al. *IceCube Preliminary Design Document*. The IceCube Collaboration, 2001. `http://www.icecube.wisc.edu/science/publications/pdd/pdd.pdf`.
19. J. Ahrens et al. Sensitivity of the IceCube detector to astrophysical sources of high energy muon neutrinos. *Astropart.Phys.*, 20:507–532, 2004.
20. Francis Halzen and Spencer R. Klein. IceCube: An Instrument for Neutrino Astronomy. *Rev. Sci. Instrum.*, 81:081101, 2010.
21. F. Halzen. Neutrino astronomy: An update. *Riv.Nuovo Cim.*, 36(03):81–104, 2013.
22. M.G. Aartsen et al. Measurement of South Pole ice transparency with the IceCube LED calibration system. *Nucl.Instrum.Meth.*, A711:73–89, 2013.
23. J. Ahrens et al. Muon track reconstruction and data selection techniques in AMANDA. *Nucl.Instrum.Meth.*, A524:169–194, 2004.
24. M.G. Aartsen et al. Energy Reconstruction Methods in the IceCube Neutrino Telescope. *JINST*, 9:P03009, 2014.
25. M.G. Aartsen et al. First observation of PeV-energy neutrinos with IceCube. *Phys.Rev.Lett.*, 111:021103, 2013.
26. M.G. Aartsen et al. Search for neutrino-induced particle showers with IceCube-40. *Phys.Rev.*, D89:102001, 2014.
27. M. Aglietta et al. Upper limit on the prompt muon flux derived from the LVD underground experiment. *Phys.Rev.*, D60:112001, 1999.
28. Anne Schukraft. A view of prompt atmospheric neutrinos with IceCube. *Nucl.Phys.Proc.Suppl.*, 237-238:266–268, 2013.
29. Rikard Enberg, Mary Hall Reno, and Ina Sarcevic. Prompt neutrino fluxes from atmospheric charm. *Phys.Rev.*, D78:043005, 2008.
30. Stefan Schönert, Thomas K. Gaisser, Elisa Resconi, and Olaf Schulz. Vetoing atmospheric neutrinos in a high energy neutrino telescope. *Phys.Rev.*, D79:043009, 2009.
31. Thomas K. Gaisser, Kyle Jero, Albrecht Karle, and Jakob van Santen. A generalized self-veto probability for atmospheric neutrinos. *Phys.Rev.*, D90:023009, 2014.
32. M.G. Aartsen et al. Observation of High-Energy Astrophysical Neutrinos in Three Years of IceCube Data. *Phys.Rev.Lett.*, 113:101101, 2014.
33. M. G. Aartsen et al. Evidence for Astrophysical Muon Neutrinos from the Northern Sky with IceCube. *Phys. Rev. Lett.*, 115(8):081102, 2015.
34. C. Weaver. Spring APS Meeting, Savannah, Georgia, USA, 2014.
35. Sebastian Schoenen and Leif Rädel. *Proceedings of ICRC2015*, 1079:642, 2015.
36. M. G. Aartsen et al. Searches for Time Dependent Neutrino Sources with IceCube Data from 2008 to 2012. *Astrophys. J.*, 807(1):46, 2015.
37. M. G. Aartsen et al. Search for Prompt Neutrino Emission from Gamma-Ray Bursts

with IceCube. *Astrophys. J.*, 805(1):L5, 2015.
38. M.G. Aartsen et al. Flavor Ratio of Astrophysical Neutrinos above 35 TeV in IceCube. 2015.
39. John G. Learned and Sandip Pakvasa. Detecting tau-neutrino oscillations at PeV energies. *Astropart.Phys.*, 3:267–274, 1995.
40. H. Athar, M. Jezabek, and O. Yasuda. Effects of neutrino mixing on high-energy cosmic neutrino flux. *Phys.Rev.*, D62:103007, 2000.
41. M.G. Aartsen et al. Atmospheric and Astrophysical Neutrinos above 1 TeV Interacting in IceCube. 2014.
42. Eli Waxman and John N. Bahcall. High-energy neutrinos from astrophysical sources: An Upper bound. *Phys.Rev.*, D59:023002, 1999.
43. Markus Ahlers and Francis Halzen. Minimal Cosmogenic Neutrinos. *Phys.Rev.*, D86:083010, 2012.
44. Boaz Katz, Eli Waxman, Todd Thompson, and Abraham Loeb. The energy production rate density of cosmic rays in the local universe is $\sim 10^{44-45}$erg Mpc^{-3} yr^{-1} at all particle energies. 2013.
45. Abraham Loeb and Eli Waxman. The Cumulative background of high energy neutrinos from starburst galaxies. *JCAP*, 0605:003, 2006.
46. Irene Tamborra, Shin'ichiro Ando, and Kohta Murase. Star-forming galaxies as the origin of diffuse high-energy backgrounds: Gamma-ray and neutrino connections, and implications for starburst history. *JCAP*, 1409(09):043, 2014.
47. V.S. Berezinsky, P. Blasi, and V.S. Ptuskin. Clusters of galaxies as a storage room for cosmic rays. *Astrophys J.*, 487:529–535, 1997.
48. Kohta Murase, Susumu Inoue, and Shigehiro Nagataki. Cosmic Rays Above the Second Knee from Clusters of Galaxies and Associated High-Energy Neutrino Emission. *Astrophys.J.*, 689:L105, 2008.
49. Fabio Zandanel, Irene Tamborra, Stefano Gabici, and Shin'ichiro Ando. High-energy gamma-ray and neutrino backgrounds from clusters of galaxies and radio constraints. 2014.
50. J. Alvarez-Muniz and F. Halzen. Possible high-energy neutrinos from the cosmic accelerator RX J1713.7-3946. *Astrophys.J.*, 576:L33–L36, 2002.
51. D.B. Fox, K. Kashiyama, and P. Meszaros. Sub-PeV Neutrinos from TeV Unidentified Sources in the Galaxy. *Astrophys.J.*, 774:74, 2013.
52. M.C. Gonzalez-Garcia, F. Halzen, and V. Niro. Reevaluation of the Prospect of Observing Neutrinos from Galactic Sources in the Light of Recent Results in Gamma Ray and Neutrino Astronomy. *Astropart.Phys.*, 57-58:39–48, 2014.
53. Thomas K. Gaisser. Very high energy photons and neutrinos: Implications for UHECR. *EPJ Web Conf.*, 53:01012, 2013.
54. P. Padovani and E. Resconi. Are both BL Lacs and pulsar wind nebulae the astrophysical counterparts of IceCube neutrino events? *Mon.Not.Roy.Astron.Soc.*, 443:474–484, 2014.
55. Luis A. Anchordoqui, Haim Goldberg, Thomas C. Paul, Luiz H. M. da Silva, and Brian J. Vlcek. Estimating the contribution of Galactic sources to the diffuse neutrino flux. 2014.
56. Anthony M. Brown, Jenni Adams, and Paula M. Chadwick. γ-ray observations of extraterrestrial neutrino track event positions. 2015.
57. Kohta Murase, Markus Ahlers, and Brian C. Lacki. Testing the Hadronuclear Origin of PeV Neutrinos Observed with IceCube. *Phys.Rev.*, D88(12):121301, 2013.

58. M. Ackermann et al. The spectrum of isotropic diffuse gamma-ray emission between 100 MeV and 820 GeV. 2014.
59. Xiao-Chuan Chang and Xiang-Yu Wang. The diffuse gamma-ray flux associated with sub-PeV/PeV neutrinos from starburst galaxies. *Astrophys.J.*, 793(2):131, 2014.
60. Markus Ahlers and Francis Halzen. Pinpointing Extragalactic Neutrino Sources in Light of Recent IceCube Observations. 2014.
61. Thorsten Glüsenkamp. Analysis of the cumulative neutrino flux from Fermi-LAT blazar populations using 3 years of IceCube data. 2015.
62. Francis Halzen, Alexander Kappes, and Aongus O'Murchadha. Prospects for identifying the sources of the Galactic cosmic rays with IceCube. *Phys.Rev.*, D78:063004, 2008.
63. M.G. Aartsen et al. IceCube-Gen2: A Vision for the Future of Neutrino Astronomy in Antarctica. 2014.
64. M.G. Aartsen et al. Search for dark matter annihilations in the Sun with the 79-string IceCube detector. *Phys.Rev.Lett.*, 110(13):131302, 2013.
65. R. Abbasi et al. Search for Dark Matter from the Galactic Halo with the IceCube Neutrino Observatory. *Phys.Rev.*, D84:022004, 2011.
66. R. Abbasi et al. Search for Neutrinos from Annihilating Dark Matter in the Direction of the Galactic Center with the 40-String IceCube Neutrino Observatory. 2012.
67. M.G. Aartsen et al. IceCube Search for Dark Matter Annihilation in nearby Galaxies and Galaxy Clusters. *Phys.Rev.*, D88(12):122001, 2013.
68. B. Sadoulet. Particle dark matter in the universe: At the Brink of discovery? *Science*, 315:61, 2007.
69. P. Gondolo, J. Edsjo, P. Ullio, L. Bergstrom, Mia Schelke, and E. A. Baltz. DarkSUSY: Computing supersymmetric dark matter properties numerically. *JCAP*, 0407:008, 2004.
70. Andrew Gould. Cosmological density of WIMPs from solar and terrestrial annihilations. *Astrophys. J.*, 388:338–344, 1992.
71. Mattias Blennow, Joakim Edsjo, and Tommy Ohlsson. Neutrinos from WIMP annihilations using a full three-flavor Monte Carlo. *JCAP*, 0801:021, 2008.
72. M. G. Aartsen et al. Search for Dark Matter annihilations in the Sun using the completed IceCube neutrino telescope. *PoS*, ICRC2015(1209), 2015.
73. M. G. Aartsen et al. mproved methods for solar Dark Matter searches with the IceCube neutrino telescope. *PoS*, ICRC2015(1099), 2015.
74. Piero Ullio, Marc Kamionkowski, and Petr Vogel. Spin dependent WIMPs in DAMA? *JHEP*, 07:044, 2001.
75. K. Choi et al. Search for neutrinos from annihilation of captured low-mass dark matter particles in the Sun by Super-Kamiokande. *Phys. Rev. Lett.*, 114(14):141301, 2015.
76. C. Tonnis et al. Overview of dark matter searches with ANTARES. *PoS*, ICRC2015, 2015.
77. A. D. Avrorin et al. Search for neutrino emission from relic dark matter in the Sun with the Baikal NT200 detector. *Astropart. Phys.*, 62:12–20, 2014.
78. M. M. Boliev, S. V. Demidov, S. P. Mikheyev, and O. V. Suvorova. Search for muon signal from dark matter annihilations inthe Sun with the Baksan Underground Scintillator Telescope for 24.12 years. *JCAP*, 1309:019, 2013.

79. C. Amole et al. Dark Matter Search Results from the PICO-2L C_3F_8 Bubble Chamber. *Phys. Rev. Lett.*, 114(23):231302, 2015.
80. D. S. Akerib et al. First results from the LUX dark matter experiment at the Sanford Underground Research Facility. *Phys. Rev. Lett.*, 112:091303, 2014.
81. C. Savage, G. Gelmini, P. Gondolo, and K. Freese. Compatibility of DAMA/LIBRA dark matter detection with other searches. *JCAP*, 0904:010, 2009.
82. C. E. Aalseth et al. Search for an Annual Modulation in a P-type Point Contact Germanium Dark Matter Detector. *Phys. Rev. Lett.*, 107:141301, 2011.
83. R. Agnese et al. Silicon Detector Dark Matter Results from the Final Exposure of CDMS II. *Phys. Rev. Lett.*, 111(25):251301, 2013.

New physics is needed to understand black holes

or

Renormalization of the gravitational force in analogy with the electroweak theory

Lectures at the Erice School of Subnuclear Physics
June 26, 27, 2015

Gerard 't Hooft

Institute for Theoretical Physics
EMMEΦ
Centre for Extreme Matter and Emergent Phenomena
Science Faculty
Utrecht University
POBox 80.195
3808TD, Utrecht
The Netherlands

e-mail: g.thooft@uu.nl
internet: http://www.staff.science.uu.nl/~hooft101/

Abstract

It seems to be deceptively easy to turn Einstein's gravity theory into a renormalizable theory by following a procedure that would have successfully turned Fermi's non renormalizable theory of the weak interactions into a renormalizable gauge theory as in the Standard Model. In gravity, the resulting theory is a conformal system, which features a negative metric massive spin 2 particle. Apart from this apparent disease, this theory has many desirable features. Black holes, for instance, turn from exotic oddities into regular topological solitons.

Version: April 2016

1 Introduction

Black holes must be an essential ingredient of any theory that purports to reconcile general relativity with quantum mechanics. As we have argued in several previous Schools here at Erice [?][?], it won't do to simply replace Einstein's equations with something that looks quantum mechanical, such as a functional integral. Black holes force us into a major reconsideration of what space and time themselves are like. As is well-known, the general coordinate transformation relating the physics near and inside black holes as seen by outside observers, to what would be perceived by someone travelling in, does not just involve the time coordinate, but even inverts it at some places, and to handle this situation, entirely new physics is needed.

In this year's lectures, we focus on the role played by local conformal invariance, and some of its consequences. First, we shall see that this invariance is likely to be an *exact* invariance, which is *spontaneously broken*. How this works is well understood.[?]

But there is more: in most known cases, conformal symmetry inevitably leads to fields with indefinite metric and/or negative energy. This is enigmatic; we explain what the field that we get looks like (a negative metric massive spin 2 particle), with some very vague hints as to how it should be treated – it is not at all well-understood.

Finally, on the side of quantum mechanics, there may also be further constraints. It is very tempting to suspect that there is an underlying deterministic system of equations, which do not look like quantum mechanics at all. How this can happen, how such an idea would lead us to believe in *superdeterminism* [?], and how one may well cope with the fear that such a theory leads to what is called "conspiracy", has also been the topic of previous year's lectures.[?]

Physics as a science could not exist without *symmetries*. For example, Galileï invariance tells us how a particle moving with any velocity, behaves, if we know its behaviour when at rest. Inside a solid, for instance, there is no Galileï invariance, or anything that looks like it, so that consequently we do not know a priori how any local disturbance can move about, even if we know exactly what it looks like when at rest. When atoms and molecules move in a vacuum, we do not have such problems. We just need the mass (or energy) when at rest, and we can calculate everything that can happen with them when they move with any velocity.

Galileï invariance, however, turns out not to be exact. It was H. A. Lorentz who discovered what one has to replace it with when you want to take the electromagnetic properties of particles in account when they move: Galileï invariance was turned into Lorentz invariance.

Then this was extended into Poincaré invariance: if you know the properties of a particle at rest at one particular point in space and time, this symmetry dictates how such an object will behave when moving around *elsewhere*.

2 Conformal invariance

Then why is physics still so difficult? The answer to this is: even if we know exactly how particles collide at one given energy, or equivalently, at one time- and distance scale, we still know nothing about what happens at other energies, or other time- and distance scales. Whenever we want to understand a new energy scale, we have to build a new particle accelerator to find out what happens, what nature's laws are like in that new domain.

Clearly, a symmetry is missing: how do we perform scale transformations? [?] Children's toys are based on scale symmetries. By giving them toy cars and small dolls, we make them believe that scale invariance works, but in reality, it works very badly. ++++ There are more examples of symmetries working badly, but sometimes it helps to put the garbage in the transformation laws. More precisely, we could assume that the *vacuum state is not invariant*. This is called spontaneous symmetry breaking. There would be a great advantage if we had a theory telling us how exactly the vacuum state transforms under scaling; such a law could be simple, and it might suggest how to select objects, and states, which indeed are scale invariant, so that they teach us things about domains of this world that we know nothing about: physics at infinite energies, or infinite distances.

Now switch on gravity. It may well dictate that scale invariance must be a local symmetry. Thus we arrive at a *spontaneously broken local conformal invariance*. We call the symmetry conformal, because in scale transformations, typically angles remain the same. There is not so much an urge also to transform angles into different angles, because angles occupy a compact domain: they always stay between 0 and 2π. Actually, angles may also be transformed by gravity, this would be a consequence of general coordinate transformations.

Now here may be a surprise: gravity already is a theory with spontaneously broken conformal invariance [?][?][?]. All we have to do is rename its basic dynamical variables. Transform

$$g_{\mu\nu}(\vec{x},t) \rightarrow \Omega^2(\vec{x},t)\, g_{\mu\nu}(\vec{x},t)\ , \tag{2.1}$$

Since, in gravity, the metric tensor $g_{\mu\nu}(\vec{x},t)$ is a dynamical field, the transformation (??) is a completely legal field transformation that should leave the equations of motion unaltered, but we have to realise that the vacuum state has

$$\langle\emptyset|g_{\mu\nu}(\vec{x},t)|\emptyset\rangle = \eta_{\mu\nu}\ , \tag{2.2}$$

where

$$\eta_{\mu\nu} = diag(-1,+1,+1,+1)\ . \tag{2.3}$$

so, indeed, the vacuum is not invariant.

3 An analogy: The electroweak theory

Of all symmetries, spontaneously broken local symmetries are among the most important ones. Without a spontaneously broken Yang-Mills gauge symmetry, we would have no electroweak theory. Prior to the introduction of this symmetry, the phenomenological theories we had were *non-renormalizable.* This means that the interactions ran wild at too high energies. I will now explain how the introduction of gauge theories could have been done stepwise. It is not how the theory was actually conceived, so this is a non-historical explanation, but its value lies in the fact that clever physicists in the 1950s and 1960s could have argued their way into the modern theories following these lines. Now that we know this, we might try the same procedure to quantum gravity. This may help us to obtain a renormalizable theory for gravity, if nature is kind to us. As explained in the beginning, this is not good enough for a complete theory, but it would help enormously.

So here is how we look upon the situation in the electroweak case. The original Fermi interaction could be written as

$$\mathcal{L}^F = \tfrac{1}{2} g_F^2 J_\mu J_\mu \,; \qquad J_\mu = \overline{\psi}\gamma_\mu\psi \,, \tag{3.1}$$

where g_F is a coupling constant. We could have asked the question: *which renormalizable theory could have produced this action at its low energy limit?* Well, we could have rewritten the action by inserting an auxiliary field $A_\mu(x)$:

$$\mathcal{L} = -\tfrac{1}{2} A_\mu^2 + g_F A_\mu J_\mu \,, \tag{3.2}$$

since one can solve the Euler Lagrange equation for A_μ to yield: $\mathcal{L} = \mathcal{L}^F$. Now, a second auxiliary field could be used, being a scalar field $\phi(x)$ with isospin $1/2$:

$$\mathcal{L} = -\tfrac{1}{2} A_\mu^2 \phi^2 + g_1 A_\mu J_\mu - V(\phi) \,, \qquad V(\phi) = \tfrac{1}{2}\lambda_\phi(\phi^2 - F^2)^2 \,, \tag{3.3}$$

where F is a constant, and $g_1 = g_F F$. All fermionic masses, if not too heavy, could have been written as

$$\mathcal{L}_m = -\lambda_F \overline{\psi}\phi\psi \,, \tag{3.4}$$

where λ_F is a coupling parameter, $\lambda_F = m_F/F$.

Note that all of this does not change any of the physics. The scalar field takes the value $\phi = F$ (assuming that the effects of the terms (??) are negligible), This is because, as yet, we did not include any kinetic terms for the auxiliary fields A_μ and ϕ.

The theory is non renormalizable. Just look at the divergences of the 1-loop diagrams of Fig. ??. When we return to the original Lagrangian, Eq. (??), we see that these are the divergences that ensure the non renormalizability of the Fermi interaction. But now we may also see a possible cure. The divergences here have the same external momentum dependence as the kinetic terms of a gauge theory. If we promote all our "auxiliary fields" into genuine kinetic fields, the divergences can indeed be absorbed into the new kinetic

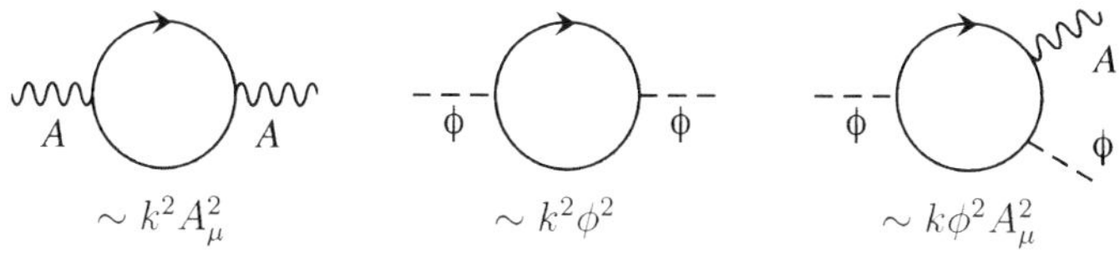

Figure 1: Divergences in the Fermi theory. Counter terms will be unavoidable.

terms. The three diagrams in the Figure have the familiar form (with constraints dictated by the symmetry structures of the system):

$$-\tfrac{1}{4}F_{\mu\nu}F_{\mu\nu} \,, \qquad -\tfrac{1}{2}(\partial_\mu\phi)^2 \,, \qquad -A_\mu\,\phi\,\partial_\mu\phi \,, \tag{3.5}$$

where the last two are parts of

$$-\tfrac{1}{2}(D_\mu\phi)^2 \,, \qquad D_\mu = \partial_\mu + g_F A_\mu \,. \tag{3.6}$$

We see here that our original theory 'automatically' transmutes into a familiar Yang-Mills gauge theory with BEH mechanism. Indeed, local (non-Abelian) gauge invariance emerged; it was already present, but hidden, in Eqs. (??) and (??).

This would have been a successful sequence of arguments and procedures to discover a renormalizable theory that produces the desired effective interactions at low energies. Indeed, at low energies the added kinetic terms become insignificant (they all generate factors k in the induced corrections, which at low energies vanish, almost by definition).

Something else happens as well: in the non-renormalizable theory, there was no place for magnetic monopoles, other than Dirac monopoles, which would be associated with singularities: the Dirac strings. The gauge theory may have locally regular solutions with magnetic charges.

4 Turning the Einstein-Hilbert action into a renormalizable theory

What we saw in the previous section is that all we have to do to turn a non-renormalizable theory into a renormalizable one, is to reformulate the former using sets of auxiliary fields, after which we identify the obnoxious divergences with possible kinetic terms for these auxiliary fields. The original theory wasn't renormalizable because the kinetic terms for the auxiliary fields had been missing, so that they had not been available as counter terms. Once we can identify the kinetic terms that are still required for this purpose, under the condition that their dimensionality is as required for a renormalizable theory, we may be able to complete the program of producing a renormalizable theory. The question will be whether the kinetic terms we arrive at will indeed be permissible, with the correct positivity constraints, as we shall see.

Following the same route for gravity leads to an interesting suggestion. We start with the Einstein-Hilbert action:

$$\begin{aligned} \mathcal{L} &= \mathcal{L}^{\rm EH} + \mathcal{L}^{\rm matter} , & (4.1)\\ \mathcal{L}^{\rm EH} &= \tfrac{1}{16\pi G}\sqrt{-g}\,(R-2\Lambda) , & (4.2)\\ \mathcal{L}^{\rm matter} &= \mathcal{L}^{\rm YM}(A) + \mathcal{L}^{\rm bos}(A,\,\phi,\,g_{\mu\nu}) + \mathcal{L}^{\rm Dirac}(A,\,\psi,\,\phi,\,g_{\mu\nu}) . & (4.3)\end{aligned}$$

This system is non renormalizable, and what is more, in this setting, as is well-known, topologically non-trivial solutions may appear, in the form of black holes. These are associated with horizons and singularities, which create considerable problems.

Interestingly, we can introduce auxiliary fields here as well. Write [?]:

$$g_{\mu\nu} = \omega^2(\vec{x},t)\hat{g}_{\mu\nu} . \qquad \mathcal{L} = \mathcal{L}(\omega,\,\hat{g}_{\mu\nu},\,A_\mu,\,\psi,\,\phi) . \tag{4.4}$$

We find a "trivial" local symmetry:

$$\hat{g}_{\mu\nu} \rightarrow \Omega^2(\vec{x},t)\hat{g}_{\mu\nu} , \qquad \omega \rightarrow \Omega^{-1}\omega , \quad \phi \rightarrow \Omega^{-1}\phi , \quad \psi \rightarrow \Omega^{-3/2}\psi . \tag{4.5}$$

This is a *local, conformal invariance.* The various components of our Lagrangian now read:

$$\begin{aligned} \mathcal{L}^{\rm EH} &= \sqrt{-\hat{g}}\left(\tfrac{1}{16\pi G}(\omega^2\hat{R} + 6\hat{g}^{\mu\nu}\partial_\mu\omega\partial_\nu\omega) - \tfrac{\Lambda}{8\pi G}\omega^4\right) & (4.6)\\ \mathcal{L}^{\rm matter} &= -\tfrac{1}{4}F_{\mu\nu}F_{\mu\nu} + \\ &\quad \sqrt{-\hat{g}}\left(-\tfrac{1}{2}\hat{g}^{\mu\nu}D_\mu\phi D_\nu\phi - \tfrac{1}{2}m^2\omega^2\phi^2 - \tfrac{1}{12}\hat{R}\phi^2 - \tfrac{\lambda}{8}\phi^4\right) + \\ &\quad + \mathcal{L}^{\rm Dirac} , & (4.7)\end{aligned}$$

where we explicitly displayed the contribution of a scalar field in the non-Yang-Mills bosonic part.

The ω field defines the sizes of things, as well as the duration of time. The field $\hat{g}_{\mu\nu}$ defines the shapes of the light cones. The field combinations ϕ/ω and $\psi/\omega^{3/2}$ here play the roles of the usual fields ϕ and ψ. Sizes of atoms are always measured *with respect to* $\omega^{-1}(\vec{x},t)$.

ω is our first auxiliary field. Interestingly, it automatically obtained a kinetic term from the Einstein-Hilbert Lagrangian, eq. (??) – we merely have to do a (finite) renormalization:

$$\omega = \sqrt{\tfrac{16\pi G}{6}}\,\eta(\vec{x},t) , \tag{4.8}$$

except that the sign is still wrong. A replacement $\eta \rightarrow i\eta$ seems to improve the situation, as long as the η or ω field only interacts with even powers; surprisingly, this is indeed the case for the Standard model.

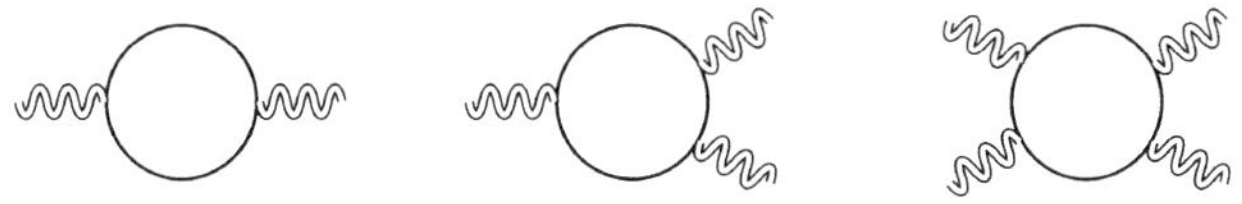

Figure 2: Divergences of graphs containing $n = 2, 3$, or 4 external $\hat{g}_{\mu\nu}$ lines. Counter terms behaving as k^4 will be needed for all n.

Thus we find that the overall scale, the ω field, is a beautifully renormalizable part of the system. The problem lies with the $\hat{g}_{\mu\nu}$ field. Its kinetic term has entirely disappeared. The entire theory is renormalizable *as long as we omit internal lines of the* $\hat{g}_{\mu\nu}$ *field.*

The diagrams of Figure ?? might then seem to be renormalizable, but they are not. The $\hat{g}_{\mu\nu}$ field always couples with derivatives. Writing $\hat{g}_{\mu\nu} \equiv \eta_{\mu\nu} + \delta\hat{g}_{\mu\nu}$, where $\eta_{\mu\nu} = \mathrm{diag}(-1, 1, 1, 1)$, one finds that all these diagrams all diverge as $k^4(\delta\hat{g}_{\mu\nu})^n$, where n is the number of external lines. We see here that, what we need is kinetic terms for the $\hat{g}_{\mu\nu}$ fields that are quartic in the derivatives ∂_μ.

Now comes a very important point. The way we wrote this theory, there is a manifest local symmetry: local conformal invariance, or, invariance under the transformations (??). Therefore, all counter terms needed for the diagrams of Fig. ?? must form a locally conformally invariant field combination. There exists exactly one such combination: the Weyl term,

$$\mathcal{L}^{\mathrm{Weyl}} = -\tfrac{\lambda^W}{2} C_{\mu\nu\alpha\beta} C_{\mu\nu\alpha\beta} \quad \to \quad -\tfrac{\lambda^W}{4} (\partial^2 \hat{g}_{\mu\nu}^{\mathrm{traceless,transverse}})^2 \,. \tag{4.9}$$

Here $C_{\mu\nu\alpha\beta}$ is that part of the Riemann curvature $R_{\mu\nu\alpha\beta}$ that is traceless in all pairs of indices. It transforms as ω^2, but the combination of $\hat{g}^{\mu\nu}$ needed for connecting all index pairs, together with the density factor $\sqrt{-\hat{g}}$ needed in front, is such that this term is entirely invariant under (??). ω drops out altogether.

According to the rule that was perfectly valid for the Standard Model, every counter term that is needed in the Lagrangian, must also be in the "bare" Lagrangian. This certainly should also hold for kinetic terms. Indeed, if we accept $\mathcal{L}^{\mathrm{Weyl}}$ as a kinetic term for gravity [?], the theory becomes renormalizable. The usual perturbation expansion is now an expansion in powers of $1/\lambda^W$

However, this kinetic term is not of the usual form. It has k^4-terms, *including terms of order* k_0^4. Consequently, it is difficult to see how we can ever avoid negative metric states. We return to this in Section ??.

5 Black holes

In black holes, the use of conformal symmetry may be very important. It means that we basically have to follow the structure of the local light cones, while all other fields, in particular also the ω field, may be regarded as matter, to be included at later stages of our theories.

We expect black holes to undergo the same fate as magnetic monopoles in renormalizable gauge theories. Topological objects, quite untreatable in the non-renormalizable scenarios, may become neatly behaved when renormalizability of the theory has been recovered. Our problem has always been how to account for the micro-states of a black hole, when it actually decays [?]. Now, we see that the mass M of a black hole is not invariant under local conformal transformations, simply because it is not dimensionless.

Consequently, while an observer falling in experiences a horizon typical for a black hole with mass M, an outside observer sees the mass M gradually shrinking to zero while the in-going observer is still lingering on the horizon. Now, we do not have to decide who is right and who is wrong. The mass is not invariant. In fact, in black holes, the vacuum state itself is not invariant. Since the two observers disagree about the vacuum state, they may well disagree about the vacuum expectation value of the mass M.

If we write the $\hat{g}_{\mu\nu}$ metric as

$$\mathrm{d}s^2 = M^2(\tilde{t})\left(-\mathrm{d}t^2(1-\tfrac{2}{r}) + \frac{\mathrm{d}r^2}{1-2/r} + r^2(\mathrm{d}\theta^2 + \sin^2\theta\,\mathrm{d}\varphi^2)\right) , \tag{5.1}$$

then we can choose $M^2(\tilde{t})$ to be different for different observers: they use a different way to divide $g_{\mu\nu}$ up in $\hat{g}_{\mu\nu}$ and ω.

For the outside observer, there is no horizon. (S)he also sees no singularity[1].

The basic new feature brought along by local conformal symmetry as introduced here is this: we postulate that the ω field (more precisely, i times the η field, Eq. (??)) should not be fundamentally different from other scalar fields. This means that there should be no singularities when $\omega \to 0$. This is actually the small distance limit when looked at in terms of the original variables. Thus, stating that ω (or rather, the field $\eta(\vec{x},t)$) does not produce any singularity when it tends to zero, will give us new constraints.

What these constraints are is easy to see. The small distance limit of renormalizable theories is determined by the running coupling parameters. In the standard formalisms, we expect logarithmic behaviour at small distances, or, at best, convergence towards some non-trivial fixed point of the renormalization group β functions. Now, we demand such anomalies to be absent: the theory should be *at* a renormalization group fixed point; the β functions, recording anomalous scaling, would violate conformal symmetry unless they all cancel out to zero.

This gives us as many constraints as there are coupling (and mass) parameters in the theory. Therefore, we expect to arrive at a theory where all these parameters are determined by the equations for vanishing β functions; they should therefore be computable. Unfortunately, it seems to be still much too early to attempt to actually compute them. preliminary attempts did not lead to even remotely satisfactory answers.

[1]This situation is clarified a lot more in a publication that came out after the Erice School [?].

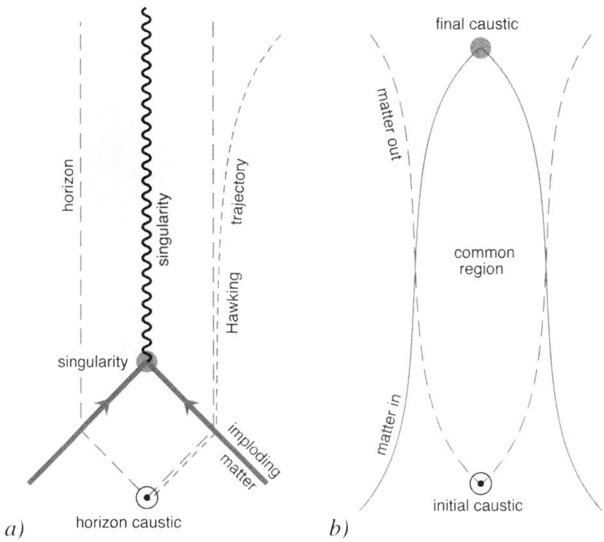

Figure 3: *a*) Black hole as seen by observer falling in. The mass still has the value M after passing the horizon. *b*) Black hole seen from the outside. While the in-going observer is still lingering on the horizon, the outside observer sees that the mass has evaporated. The picture also has become time-reversal symmetric. All this is possible when we have local conformal symmetry.

These equations only depend on the algebra chosen for the Standard Model interactions. Therefore, we are now in the position to suspect that all we have to do is establish Nature's symmetry algebra. The rest follows when we are able to determine the renormalization group fixed points. Again: conformal symmetry mandates that we are *at* a fixed point, not that we should be approaching it from elsewhere, as it happens in other theories [?].

An interesting alley to investigate further is to try algebras with large numbers N in them. These may allow for $1/N$ expansions that take the place of coupling constant expansions. This is technically complex, but physically interesting.

6 The negative metric states

Now let us investigate what exactly goes wrong if we accept the Weyl action, Eq. (??). This is an interesting exercise. To do it right, one must expand the solutions for the resulting equations in plane waves, for instance in the z-direction. One finds physical propagators and ghosts, the latter of course depending on the choice of gauge (we now have a 5-dimensional local symmetry: the local coordinate transformations and the scalar conformal transformation (??)). The surviving poles are of two types:

- a massless pole, of the form $1/(k^2 - i\varepsilon)$, with a 2-fold degeneracy. These are the two spin 2 polarisations of the standard Einstein-Hilbert gravitons. At long distances, we expect only these poles to survive, since no changes in Einstein's gravity equations are expected at long distances.
- a 5-fold pole, of the form $1/(k^2 + M^2 - i\varepsilon)$, all at the same mass $M = 1/\sqrt{\lambda^W}$. It has spin helicities ± 2, ± 1, and 0. Clearly, these are the 5 states of a massive spin 2 particle. However, it has negative metric.

This negative metric is inevitable. Together with the graviton pole, one finds

$$1/k^2 - 1/(k^2 + M^2) = 1/(k^2 + \lambda^W k^4) \ , \tag{6.1}$$

where the k^4 divergence at large k comes from the Weyl term with 4 derivatives, which dominates at large k.

How should we interpret this result? Note, that local conformal symmetry differs in essential ways from ordinary symmetries in field theories. *Energy is not conformally invariant, and the metric $\hat{g}_{\mu\nu}$ is not locally observable.* Therefore, arguments assuming stability, and unitarity, have to be considered with extreme care [?].

Consider a flat background, and write $\hat{g}_{\mu\nu} = \eta_{\mu\nu} + \delta\hat{g}_{\mu\nu}$. All wave-like solutions are in $\delta\hat{g}_{\mu\nu}$ only, so, looking at a specific wave 4-vector k, we just concentrate on $\hat{g}_{\mu\nu}(k)$ and $\omega(k)$. We write $g^{\rm trtr}_{\mu\nu}$ as being the traceless, transverse part of $\hat{g}_{\mu\nu}$:

$$g^{\rm trtr}_{\mu 0} = 0 \ ; \quad k_\mu g^{\rm trtr}_{\mu\nu} = 0 \ ; \tag{6.2}$$

and there should be two vectors, u_μ and v_μ and a scalar w such that

$$\hat{g}_{\mu\nu} = g^{\rm trtr}_{\mu\nu} + \delta_{\mu 0} u_\nu + \delta_{\nu 0} u_\mu + k_\mu v_\nu + k_\nu v_\mu + w\,\eta_{\mu\nu}. \tag{6.3}$$

We gauge ω to be constant. Then, in the helicity 2 domain, one finds that

$$\begin{aligned} \mathcal{L}^{\rm kin} = \mathcal{L}^{\rm EH} - \tfrac{\lambda^W}{2} C_{\mu\nu\alpha\beta} C_{\mu\nu\alpha\beta} \rightarrow \\ -\tfrac{1}{4}(\partial g^{\rm trtr}_{\mu\nu})^2 - \tfrac{\lambda^W}{4}(\partial^2 g^{\rm trtr}_{\mu\nu})^2 \ . \end{aligned} \tag{6.4}$$

Let the $g_{\mu\nu}$ field be coupled to the energy-momentum tensor $T_{\mu\nu}$, written as a source term $J_{\mu\nu}$. Introducing an auxiliary field ϕ_2, one can rewrite all of this in short-hand notation:

$$\begin{aligned} \mathcal{L} &= -\tfrac{1}{2}(\partial g)^2 - \tfrac{1}{2}\lambda(\partial^2 g)^2 + Jg \\ &= -\tfrac{1}{2}(\partial g)^2 - \tfrac{1}{2}\phi_2^2/\lambda + Jg \quad + \eta(\phi_2 - \lambda\partial^2 g) \ , \\ &\quad \eta = \phi_2/\lambda \ , \quad g \equiv \phi_1 + \phi_2 \ , \quad \phi_2 \rightarrow \lambda\partial^2 g \ . \\ \mathcal{L} &= -\tfrac{1}{2}(\partial\phi_1)^2 + \tfrac{1}{2}(\partial\phi_2)^2 + \tfrac{1}{2}M^2\phi_2^2 + J(\phi_1 + \phi_2) \ . \end{aligned} \tag{6.5}$$

Here, the field η in the second line is just a Lagrange multiplier field, and solving the equations for ϕ_2 gives us the third line. The last line follows from some algebra, and shows us how a regular-metric particle ϕ_1 appears (the graviton), and an anomalous metric particle ϕ_2 with mass M. We propose the name *gravitello* for this strange object.

Also, the ω field has wrong metric, but this is a locally unobservable ghost field, so it should be harmless. The ϕ_2 field may be identified with the Weyl curvature field $C_{\mu\nu\alpha\beta}$.

Massive fields with negative metric imply that the unitarity equation for the scattering matrix S take the form

$$S\sum(|\text{light}\rangle\langle\text{light}| - |\text{heavy}\rangle\langle\text{heavy}|)S^\dagger = \mathbb{I}\ , \tag{6.6}$$

but, in second-quantized theories, one may also exchange negative metric with negative energy. Suppose, we have two oscillators, one with frequency $A = |\vec{k}|$ and one with frequency $B = \sqrt{\vec{k}^2 + M^2}$, then the Hamiltonian

$$H = A(p_1^2 + x_1^2) - B(p_2^2 + x_2^2) = A\,a^\dagger a - B\,b^\dagger b + C\ , \tag{6.7}$$

shows negative energies, and therefore negative signs in the Hamiltonian, as well as in the Lagrangian. However, we could interchange b and $b^\dagger$. This interchanges creation and annihilation of particles with energy B, so we return to regular energies. The price one then pays is that there is no ground state: one can always annihilate another particle, while there is a limit to what can be created: the original vacuum state is that limit.

Conceivably, the gravitello field is in a highly excited mode, so that it has no ground state. Whether one can build a more sophisticated theory along such lines is unknown. Note also, that the replacement $b \to b^\dagger$, $b^\dagger \to -b$, which keeps the commutation rule intact but inverts B, corresponds to the replacement $x \to ix$, $p \to -ip$, so that the field of the gravitello may have to be chosen purely imaginary rather than real.

It should be emphasised that these considerations do not solve the problems associated with wrong metric fields; we still would have problems with unitarity violation, but one can conjecture that locally conformally invariant theories of matter may be needed for a more satisfactory treatment. [?]

7 Asymptotic safety

A related theory is known as the "asymptotic safety" idea. [?] When the various interaction parameters are allowed to run as a function of renormalization scale, an ultraviolet fixed point might be reached. Once we are at the fixed point, the effective theory there becomes locally conformally invariant.

In our view, this theory does not resolve the difficulties explained in the previous section: there may be a Landau ghost, playing the role of "gravitello". The question whether we have to stay *exactly* at the fixed point, or move away from it, may be essentially a semantic one, much comparable with the question whether local conformal invariance is spontaneously or explicitly broken.

8 Conclusion

Gravity may well be handled as a candidate - renormalizable theory. Locally conformal invariance then plays the same role as local gauge invariance replacing the older Fermi theory.

Thus, locally conformal symmetry is spontaneously broken in the same sense that the BEH mechanism spontaneously breaks local Yang-Mills gauge invariance: the invariance is still an exact property of natural law, but we have to fix the gauge before writing down equations that can be handled mathematically.

Atoms, and all other particles obtain their physical sizes only *relative to a field* $\omega(\vec{x}, t)$. the new parameters m and coordinates x have no absolute scale, just as the field ω, but now the ratios m/ω, $x\omega$ have absolute scales. The small-distance limit is now the limit $\omega \to 0$. The interesting thing about this is that, if ω is to be treated as an ordinary scalar field, there should be no singularities in points where ω happens to vanish, just as what we expect magnetic monopoles to do in gauge theories: there is no physical singularity at the centre of a monopole, where $\varphi(\vec{x}, t)$ vanishes.

This, however, should mean that, in the new formalism, *all* renormalization group β functions should vanish, or, we are at a renormalization group fixed point. This would require important modifications in the algebras underlying fundamental quantum field theories for the sub-atomic particles.

Thus we came to suspect that, starting from the requirement that black holes should be made consistent with theories of the sub-atomic particles, important constraints will be found on the interactions of the latter, while at the same time one may hope on obtaining superior theories for quantum space-time. Needless to say, these are mere conjectures. We have not yet reached our ambitious goal.

References

[1] 't Hooft, G.: *Canonical Methods in Deterministic Quantum Mechanics*, International School of Subnuclear Physics, Erice, 2014.

[2] 't Hooft, G.: *Probing the small distance structure of canonical quantum gravity using the conformal group*, arXiv:1009.0669; id.: *Quantum Gravity without Space-time Singularities or Horizons*, International School of Subnuclear Physics, Erice, 2009, arXiv:0909.3426.

[3] 't Hooft, G.: *Local Conformal Symmetry: the Missing Symmetry Component for Space and Time*, International Journal of Modern Physics **D 24** No. 12 (2015) 1543001, arXiv:1410.6675 v2 [gr-qc].

[4] Hossenfelder, S.: *Testing super-deterministic hidden variables theories*, Found. Phys. **41** (2011) 1521 [arXiv:1105.4326 [quant-ph]], id., *Testing Superdeterministic Conspiracy*, arXiv:1401.0286.

[5] Mannheim, P. D. and Kazanas, D.: *Exact vacuum solution to conformal Weyl gravity and galactic rotation curves*, Astrophys. J. **342** (1989) 635; Kazanas , D. and Mannheim, P. D.: *General structure of the gravitational equations of motion in conformal Weyl gravity* Astrophys. J. Suppl. **76** (1991) 431; Mannheim, P. D.: *Alternatives to Dark Matter and Dark Energy*, Prog. Part. Nucl. Phys. **56**, 340 (2006), astro-ph/0505266; id.: *Intrinsically Quantum-Mechanical Gravity and the Cosmological Constant Problem*, arXiv:1005.5108 [hep-th]; id.: *Solution to the ghost problem in fourth order derivative theories*, Found. Phys. **37** (2006) 532. arXiv:hep-th/0608154.

[6] Hawking, S.W.: *Particle Creation by Black Holes*, Commun. Math. Phys. **43** 199 (1975); id.: *Breakdown of predictability in gravitational collapse*, Phys. Rev. **D14** (1976) 2460.

[7] 't Hooft, G.: *Black hole unitarity and antipodal entanglement*, arXiv:1601.03447[gr-qc], Found. Phys., to be pub.

[8] Niedermaier, M., and Reuter, M.: *The Asymptotic Safety Scenario in Quantum Gravity*, Living Rev. Relativity, **9** (2006), 5, http:// www.livingreviews.org/lrr-2006-5;
Weinberg, S., *Ultraviolet divergencies in quantum theories of gravitation*, in *General Relativity: An Einstein Centenary Survey*, (Cambridge University Press, Cambridge, U.K.; New York, U.S.A., 1979).

[9] Bekenstein, J.D.: *Black holes and entropy*. Phys. Rev. **D 7** (8), 2333–2346. (1973).

[10] 't Hooft, G. and Nobbenhuis, S.: *Invariance under complex transformations, and its relevance to the cosmological constant problem*, Class. Quantum Grav. **23** (2006) 3819-3832; reprinted in: Mathematical Physics Proceedings of the 12th Regional Conference, Islamabad, Pakistan, Mach-April 2006, M.J. Aslam et al eds., World Scientific 2007, pp. 301-319, arXiv:gr-qc/0602076.

LECTURER: G.'t HOOFT

New physics is needed to understand black holes

Scientific secretaries: L. Bishler, S. Zhidkova

DISCUSSION I

- Baguet:

You start with the Einstein-Hilbert action to which you add a Weyl term, that is conformally invariant. However, unless you are in D=2 the Einstein-Hilbert is not conformally invariant, right? So, wouldn't we need to start from something that is conformally invariant and get the Einstein-Hilbert term from spontaneously broken conformal invariance symmetry?

- 't Hooft:

The answer is no, because I do it differently, and the theory is going to be made first conformal invariant, you could say by brute force, by adding powers of the ω field [see slide 11]). The ω field has a conformal dimension and so transforms non-trivially under the conformal transformation, including the Weyl term and the kinetic term. So whenever you encounter something that doesn't have the right dimensionality here you multiply by powers of ω to get the right dimensionality. We do want to have only terms of positive integer powers of so that the theory looks renormalizable in terms of the ω field itself. Here we just multiply by ω because otherwise, perturbation expansion would generate powers of ω^{-1}. We are not going to get a useful perturbation expansion that way. So we have ω^2 here in the numerator that makes it conformally invariant. I don't have to do anything else than just that, substitute $g_{\mu\nu}$by $\omega^2 g_{\mu\nu}$ thenω^2 automatically appears here. And it is in my next slide [slide 12]. You see what happens: the entire Einstein-Hilbert action becomes renormalizable, but as you see, it has powers of ω in the numerator, that makes it conformally invariant and then the cosmological term is ω^4 and so on and so forth. That is the point why your question is very well posed. However, now the kinetic term for the$g_{\mu\nu}$ field disappears altogether, and that can't be right. So, basically, I'm replacing the kinetic term for the $g_{\mu\nu}$ field by kinetic term for ωfield and that makes the physics conformally invariant. Until this point, the physics has not yet been modified by this operation. But in that procedure I lost the kinetic term for the $g_{\mu\nu}$ field altogether, which is why I subsequently add the quadratic Weyl term, which is also conformally invariant.

-Duff:

Naively, when you make your field redefinition the physics doesn't change. So you could equally not change the variables. But then I would be faced with the fact that world wasn't conformal invariant. So what's wrong with that?

- 't Hooft:

First, the conformal transformation is only in the redefined fields, while the original, "physical" fields are not affected at all, so this conformal transformation is physically trivial. But then, I do change the physics, in a secret way which I have not yet explained. It will be explained tomorrow. The thing that changes is that I use only pure powers of ω and not logarithms or any fractional powers. If you would renormalize the theory first and then add your ω field, you would get logarithms all over the place. I want the theory to stay renormalizable in terms of the ω field. In particular, when $\omega \to 0$,we talk about the small distance limit. But in the so-called renormalizable interaction here$\omega \to 0$ seems to be a totally regular region. That is where I change the physics, I say that when $\omega \to 0$ nothing should happen. That means that our formalism is not only going to be conformally invariant, it is going to be sitting at a fixed point. I'll say more about it tomorrow. I will make a solid departure from the standard physics, but so far, you are quite right, I've changed nothing, this is just a field redefinition.

-Simonello:

I got confused between slides 15-16. I think you started slide 15 by saying that we are going to see the same $g_{\mu\nu}$, but different ω and then we end it up at different $g_{\mu\nu}$ ant I lost the path here.

-'t Hooft:

Ok, maybe I didn't say it quite clearly. It is $\hat{g}_{\mu\nu}$ which transforms together with ω, so that the original $g_{\mu\nu}$ doesn't change, $\hat{g}_{\mu\nu}$ changes. And so, I now make a transition to use $\hat{g}_{\mu\nu}$ as metric instead of old $g_{\mu\nu}$, and then, of course, the black hole is going to look totally different, and then the thing is that your original theory was very singular when $g_{\mu\nu} \to 0$ or when you went to very small distance limit. That means $\omega \to 0$. The answer to the previous question is that $\omega \to 0$ shouldn't be a singular limit at all. So, now the black hole becomes much more regular because of this extra condition. This is where I change the physics and that includes the physics of the black hole.

- Savitskyi:

Today you mentioned that the counter term for the kinetic term should be added to the bare Lagrangian and, basically, this reminds me of the picture in QFT with Fadeev-Popov ghosts which correspond to non-observable, non-physical objects in QFT. In your theory does this correspond to some kind of physically observed objects or not? What can be the case?

- 't Hooft:

I'll tell more about ghosts tomorrow, because, for one thing, there will be negative metric states. I'll explain how to understand negative metric states. There is a problem. So I have not done all my work yet, I still have a problem with those negative metric states. The Fadeev-Popov ghosts only come about when you fix the gauge by introducing derivatives in the gauge fixing condition. Then Fadeev-Popov ghost pops up and does its own thing, just as the other negative metric particles. Of these you know that they should cancel out against the gauge artifacts that you get anyway. I haven't really considered much the Fadeev-Popov ghosts of the theory, they are coming right on the top of the other ghosts, which are much more problematic but these I shall discuss tomorrow.

- Sopenko:

As a warm-up example we could consider 3-dimensional gravity. In that case the theory is much more simple because there is no propagating gravitons and, in fact, due to that it's renormalizable. It is still non-trivial, there are BTZ solutions and it is not clear what quantum gravity is. Does what you were talking about tell us something on that case?

-'t Hooft:

2+1 gravity is a completely different theory from my point of view. The standard way of dealing with 2+1 gravity is still not renormalizable, but you can play some tricks such that it looks renormalizable, though actually there still are problems. When you discuss it in the way many people have been doing, for instance Carlip and Witten, the problem is that you leave out matter. But when I do 2+1 gravity I see that all the physics is in the matter fields because, as you say, the gravitons have zero degrees of freedom. The gravitons just represent the locally flat spacetime and they are not physically observable. They are ghosts altogether. So, when you say that 2+1 D gravity is renormalizable you are talking about a theory that, to my taste, is a pretty empty theory: there is no matter and there is no gravitational force. Although, there are topological non-trivial structures.

- *Sopenko:*

Yes, it is topological but there are non-trivial global solutions. It is not clear how to deal with global structures in quantum gravity.

- *'t Hooft:*

That theory has a quite different physical structure than the 3+1 dimensional case. The procedure I am applying is very much dependent on the dimension. d=4 is very special. In d=5 I would not know how to do it because I would get ω^3or fractional powers rather than ω^2 in my Lagrangian. What I will be talking about tomorrow only applies to d=4. Something very strange is going on with 3D gravity because I say it is very trivial on the one hand and it is impossibly difficult on the other. 2+1 gravity may require its own program of renormalization which Carlip and Witten were trying to do. Now I would like to put matter in that theory as well, and then again ask the same question. Now usually, the first thing they do is throw the matter away which is a too simple way out to my taste.

- *Gulitskiy:*

What is the origin of the problems with the renormalization of gravity? When you want to renormalize the electroweak theory the basic idea is that you can't measure the bare charges, you can only measure the charges modified due to interaction with the vacuum in all orders of the expansion. And this clear and simple physical idea is enough to renormalize the electroweak theory. Why is it not enough in gravity? And why do we have some problems there?

- *'t Hooft:*

I was not handling the electroweak theory, I was discussing the old electroweak theory of Fermi. Fermi knew there were two currents interacting, but he didn't know that there was the Yang – Mills structure behind it at all. So this was before we had the Yang – Mills structure, and then the theory was non-renormalizable; the question was how can we do renormalize the weak interactions. The answer was that we cannot. For the same reason we say now that for gravity it doesn't work. The scheme we are talking about, having bare charges and bare fields and so on that scale logarithmically, doesn't work for gravity because Newton's constant in standard gravity has a dimension. If you look at the one-loop corrections, you not only get Newton's constant squared, but also factor k squared, that means that the integrals are going to diverge more and more when you go to higher energies. Therefore, the counter-terms needed in gravity do not seem to have the same form as the Einstein – Hilbert action. In fact, they have a form of the Weyl curvature squared, and onwards, if you do it perturbatively, you get a Weyl to the third power, Weyl to the fourth power and so on, and as long as you don't look at conformal invariance, you get the other Riemann tensor components raised to very high powers indefinitely. In the standard way of phrasing what renormalizability means we say gravity is non-renormalizable. You could also say it is a question of taste whether gravity is non-renormalizable, because you can also say: yes, ok fine, just add all the counter-terms with unknown coefficients, then gravity is a theory of physics with an infinite number of unknown coefficients. If you are not afraid of that, you could say that you can renormalize gravity. But most people don't like theories which have infinite numbers of freely-adjustable parameters. That's what happens with gravity. I'm not sure that I answered your question.

- *Prochazka:*

I want to ask you a question about 4-dilaton interactions. In slide 14, when you wrote Weyl tensor squared term. Why don't you have also terms which come from the Euler density? Because the Euler density is not conformally invariant, so, in general, you can get 4-dilaton interactions. Why can you neglect them in this picture?

- *'t Hooft:*

Well, first of all the Euler density is a topological term, so it doesn't affect the local equations of motion and it doesn't produce any quantum corrections, it is a topological term. Similarly, one can ask anything about the electroweak theory whether you add for example topological twists. If the topological twist is locally regular and globally obeys the correct boundary conditions there is no objection against them. Now in gravity you can ask: "Do we need topologically nontrivial spaces in ordinary physics?" I suspect that the surrounding space is topologically trivial.

- *Prochazka:*

Yes, but you talk about black holes, so if there was singularity?

- *'t Hooft:*

Well, it seems that black holes also become topologically trivial. There is a danger, because in the large black hole limit they are topologically distinct. Black holes in this picture, certainly, will not contribute to the Pontryagin number, because they will have a trivial Penrose diagram, the same as flat Minkowski space. I don't see topology entering physics in this theory up to now, but the theory is not yet finished at all, so maybe in our finally completed formulas of the theory we will see the topological things happening. The topological winding number doesn't seem to be conformally invariant itself. You can discuss a theory without topological invariance. I'm not sure of this. I first need the theory without any topological numbers and then worry about topology at alatest stage.

- *Prochazka:*

A second question: this is our classical scale invariance and conformal invariance, so apart from couplings there are also beta functions and there will be counter terms which will depend on these local charges...

- *'t Hooft:*

That is a very important point, it is all in the tomorrow lecture. So, yes, there will be beta functions, and we will have to do something about them. And, in fact, the only correct answer will be "look at the fixed points". But that would be a very important constraint. We can say now that the SM does not obey those constraints. It means that we have to do something about the standard model. We should add other things. What we have to do is to find out fixed points of our very special kinds of beta functions, and that's going to be a very complicated and difficult job.

- *Bellantuono:*

The conformally symmetric black hole has not a horizon with a singularity inside, but is like a locally regular soliton. Is there a way of defining the analogue of the Bekenstein – Hawking entropy for this kind of black hole, even if we don't have any more a horizon?

- *'t Hooft:*

Compare the regular, magnetic monopoles that come along in agrand unified theory; those monopoles carry many microstates as well. Microstates in a sense that you can have exitations of those monopoles. For instance, if you give them an electric charge they become dyons. If you give them angle momentum they become rotating monopoles. There will be very many states. You can call them particles. You can also say that they form a collection of microstates for the monopole. The microstates of a black hole enter exactly the same way. Now, of course, it has still to be shown in the theory and I've not been able to show that microstates will build up in the same way as required by the Hawking radiation. That's what has still to be shown, but, for sure, there are microstates. The question is how many, and the answer is that I don't know. If the microstates are as in classical black holes, they imply Hawking radiation via the rules of thermodynamics.

- *Ivanitskiy:*

You mention that in-going and out-going spectators of the black hole can define the vacuum differently. Therefore, it seems that excitation of that vacuum also can be defined differently. Can it help us to solve the problem of information loss when we speak about some microstates which also can be considered as excitations of vacuum?

- *'t Hooft:*

Well, this is nearly the same question as the previous one. Yes, this black hole has microstates, and in the conventional black hole the microstates always seem to diverge in a way similar to one-loop diagrams. So, if you try to reproduce the microstates in a quantum field theory in a vicinity of a black hole you find a limitless number of microstates, which can't be right, because that's in conflict with Hawking's results, and therefore we must have only a finite number of microstates. If we now look at very small black holes in this picture, I just see them as a tiny, regular soliton-like configuration, and therefore, from that point of view it would be very surprising if I had any infinity there. What I really consider as the black hole information problem is the fact that the number of microstates seems to tend to infinity if you apply background quantum field theory. Of course, that is not correct here. The problem is how to do this right; this is the black hole information problem. I suspect the problem will be completely resolved, but that has still to be shown.

- *Varnhost:*

You spoke about negative metric states whose consequences would be problematic...

- *'t Hooft:*

Negative metric states give you a violation of unitarity, so instead of the equation that we have unitarity with some intermediate states, here you generate intermediate states which from a Lagrangian point of view look like completely illegal intermediate states, they contribute to unitarity with the wrong sign. Suddenly the theory generates states which contribute negatively to the equation that says that the sum of all probabilities must be conserved. If you have a scattering event, the sum of all probabilities should add up to one. Instead, we now have the contribution of some states minus the contribution of other states. About that minus sign I'll show tomorrow how it enters in this theory, it will be a disaster for the theory. It will not let you do decent loop calculations. So, something has to be done about it.

- *Tokareva:*

You describe the black hole without singularities in conformal gravity, but the conformal symmetry must be broken. How would black holes look in the theory with spontaneously broken conformal symmetry?

- *'tHooft:*

The whole point of the question is the word "spontaneously". Spontaneously means that the vacuum state doesn't obey that symmetry. Different observers will identify different states as being the vacuum. So the in-going observers will say: well space – time around me is vacuum, I see nothing. And the outer observers will say at the same time: while you were jumping into the black hole I saw all these Hawking particles come out. That's not the vacuum, but a state full of particles. Both disagree about what should be called vacuum state. Now suppose that you normalize ω to be 1 for the vacuum. They say: wait a minute, which vacuum do you mean? And so the in-going observer will use a different vacuum to describe his space-time than the observer outside will do, because he will include Hawking particles in his description of space and time.

He will renormalize something else to be one. So his ω will automatically be some local multiplication, of the previous ω. And that is where this theory of the black hole differs from the previous theory. ω is such that it tends to zero at the origin of the black hole, which means that as the original black hole collapses to a point, we should go back to the previous metric, as seen by the outside observer. Some people call that a firewall. But it is not really a firewall at all, because the inner observer has no problem, he goes into the black hole and still thinks that the singularity line is in front of him, while the outside observer says: no-no, all of internal space – time should be replaced by a point and there is no singularity here any more. The ω field looks very wild, but not singular at that point.

- *Tokareva:*

What may be the effect of non-unitarity for black holes? May it lead to some instability or singularity?

- *'t Hooft:*

To my mind, black holes ought to behave according to the laws of quantum mechanics. And that means they should contribute to unitarity in the usual fashion. So, once we write down the full Schroedinger equations for the theory that Schroedinger equation should only contain states with positive metric. If it doesn't then the theory goes heywire. That's the first step. How to correct that I will explain tomorrow. That's the easy part, but then the problem reappears somewhere else, and we don't know what to do about it unless the theory is further approved. And there is a complication, let me say it right now. When you talk about unitarity you talk about energy. And, in fact, instead of having a non-unitary contribution to the state you could say I have a negative energy contribution. And now energy is a problematic notion in the case of a conformal theory, because energy is transformed to something else by a conformal transformation. Therefore, strictly speaking, we don't have a Hamiltonian any more. So once you say these things, you see how dangerous the hole familiar concept is going to be in the case of conformal quantum gravity. This is why I suspect that taking conformal invariance into account properly might solve the problem. But without that the problem is very serious. It also applies to black holes: they should be described with positive energy and positive metric states. Otherwise, we might be in a deep trouble.

LECTURER: G. 't HOOFT

New physics is needed to understand black holes

Scientific Secretaries: M.V. Gargiulo, M.Zoccali

DISCUSSION II

- *Anastasiou:*

Do you know whether it would be possible to obtain such a super conformal gravity theory from a string theory compactification?

- *'t Hooft:*

So far I have only rarely touched supersymmetry or supergravity in this kind of work. Extending these ideas to supersymmetric or supergravity scenarios is very natural, however it is yet to be done. If Michael Duff is here he might comment on this.

- *Duff:*

The only field theory I am aware of that is conformally invariant at the quantum level is the maximally supersymmetric conformal supergravity, which is not the one you would want for your purposes. However it has the virtue that the anomalies cancel and appear as the low energy limit of some twist or-type string by Witten et al. They didn't like it, however, because it violates unitarity.

- *'t Hooft:*

According to the way of looking at gravity presented here, one can make any theory conformally invariant in a trivial sense at first, and then apply physical conditions to make it non-trivial. Naturally, a valid alternative would be to combine this with known methods and results for super conformal theories, namely adding an extra scalar field to a large N super conformal theory, to break the symmetry, and study how the algebra survives in the 1/N or 1/N^2 approximation. However, I believe we need something smarter than that, albeit super conformal theories remain extremely interesting for model making.

- *Zoller:*

You claimed that conformal symmetry must be spontaneously broken. What sort of mechanism could break this symmetry in your opinion? Do you introduce a new Higgs-like field or could it be something different? How does the breaking scale enter?

- *'t Hooft:*

In this case the ω field introduced has an expectation value equal one, whereas the conformal transformation properties are centred around its value zero: this is equivalent to the Brout-Englert-Higgs mechanism of the SM, where the field φ has value one but its symmetry structure is centred around its zero point. Despite some similarities, there exist a few differences as well: the ω field is conformally invariant at zero, however its vacuum expectation value is at 1/G. In this case, however, there is no potential forcing it in that direction as it would explicitly break the conformal invariance. Moreover the ω field does not split into transverse and radial components as in the case of the Higgs field, and in particular no observable particle survives the symmetry breaking process.

- Zoller:

In your theory you need all the beta functions to vanish, however when you go to the effective theory - to the SM at low scale- you have the running of couplings. So when precisely do the couplings start to run? Does this enter after conformal symmetry is broken or below some specific scale?

- 't Hooft:

The usual beta functions of the SM refer to what we call physical fields. In this formalism they would be the ratio of fields, such as φ/ω or $\psi/\omega^{3/2}$. These scale non-trivially, so the beta functions do apply. In the case of the fundamental conformally invariant fields, this is not true. The conformal fixed point is relative to the new metric $\hat{g}_{\mu\nu}$, used to renormalize the theory, which is fundamentally different from using the ordinary metric $g_{\mu\nu}$. As an example of this difference, the mass terms in the Standard Model usually do not occur directly in the beta functions; in this theory the masses act as dimensionless couplings, therefore they are associated with their own beta functions. In summary, the Standard Model would be modified here not only because one would modify the algebra, but also because one would need to introduce the new field ω and consider beta functions which are different from the original ones.

- Tokareva:

Could you comment on the gravitational conformal anomaly, which, to my knowledge, is very hard to cancel?

- 't Hooft:

Here I depart considerably from standard gravity, since I introduce a Weyl squared action which takes over at small distances/high energies to renormalize the theory. Weyl gravity, which has $1/k^4$ in the propagator, also generates anomalies. These anomalies are of a different kind than those looked at so far, therefore one has to be careful in comparing possibly different situations. The kind of anomalies that you would encounter, and that could jeopardize renormalization, are associated to scaling.

- Tokareva:

One more question: is the ω field a ghost?

- 't Hooft:

The ω field is treated as a physical field, however there is an added gauge invariance. Since the field itself is not gauge invariant, it can be used to fix the gauge and then it disappears altogether. An alternative gauge fixing could give a field $\omega$and a ghost, where the latter cancels out the first. So yes, I would call ω a ghost.

- Tokareva:

So the ω field disappears. Does the Fadeev-Popov ghost remain?

- 't Hooft:

Whenever one fixes the gauge in a more subtle way so as to preserve the renormalizability of the theory, one has to adopt the full Fadeev-Popov procedure, which implies that Fadeed-Popov ghosts will indeed be present.

- *Kovalenko:*

Do you think that, if it exists, the algebra which satisfies the condition of vanishing beta functions, is a unique one? Or would this give rise to a landscape of solutions for the Standard Model couplings and masses?

- *'t Hooft:*

Very interesting question, to which I am afraid I have no answer. Since the algorithms for a beta function are quite complex, we do not have models for which beta functions vanish automatically already at one loop. In principle it can be done: it is relatively simple for gauge couplings, and there are some extremal principles for the Yukawa couplings in order to find the fixed point. However there is an extra scalar whose algebra is rather arbitrary, and it remains very difficult to judge which algebra would be most suited. One possibility would be to take an algebra with large N: in particular one would want to compute the fixed point of the gauge, Yukawa and scalar-scalar couplings, in this order (since it works at one loop level). This procedure could yield a converging coupling expansion at large N, and this could give a way of generating algebras which can be handled perturbatively.

- *Grad:*

Which variables should become operators when one tries to quantize this theory?

- *'t Hooft:*

The theory is easiest to understand when one is near the perturbative domains, when the coupling constants are small and masses are all comparable with the Planck mass; in this case the fields themselves are the only operators needed. If, however, the coupling constants become large because the fixed point is not close to the origin, depending on the algebra, then one has to ask how fields transform as operators.

- *Grad:*

Usually QFT field operators should commute at points of space time which are separated by a space-like interval, outside the light cone, thus making a metric an operator one faces problems in defining causality.

- *'t Hooft:*

You are quite right. This is a very good reason to remain in the perturbative domain. If the Weyl action has a large coefficient the propagator becomes very small, thus leading you close to the perturbative regime. This is in analogy with the Yang-Mills case. If one is close to the perturbative regime the problem you mentioned formally exists but it doesn't show up at finite orders of perturbation expansion, so one is not yet in trouble. Problems arise if one tries to sum the perturbation theory because the sum does not exist. My theory of QM gives me some suspicions about how to handle this situation, but this is a completely different chapter of physics.

- *Grad:*

Do you consider black hole thermodynamics in your theory?

- *'t Hooft:*

Black holes were the important indication that something of this sort had to be happening; in particular if one considers Hawking radiation and the information paradox, two observers on opposite sides of a Hawking radiation curtain disagree about seeing radiation, about the ω function and therefore about the curvature which in turn is described by the metric. So in fact one can describe Hawking radiation in such a way that it only affects the scalar part of gravity.

- *Laszlo:*

What would happen if you used a Palatini type choice of your field variables, i.e. that the covariant derivative is independent from your metric, and the compatibility condition comes as a field equation? Wouldn't this solve your k^4 problem?

- *'t Hooft:*

No, I do not think so. The difference lies in understanding what is merely a field redefinition, and what consists in an actual change of the physics, particularly in perturbative gravity. Perturbatively, a transition from $g_{\mu\nu}$ to the vierbein field amounts to a transformation which leaves all physics unchanged. In fact, gravity is very robust against such field redefinitions in general.

- *Anastasiou:*

How would the fermionic interaction terms have to be restricted in order to cancel all scaling anomalies?

- *'t Hooft:*

A renormalizable theory requires all chiral anomalies to cancel as well. In the present theory all anomalies need to cancel, including scaling and chiral ones. For the chiral one, we have to make sure that the structure constants d_{abc} of the gauge group for the left and the right fermions satisfy the same algebra. The scale anomalies are more involved, and they would cause restrictions for all Yukawa couplings.

September 26, 2015

LOW–ℓ CMB FROM STRING–SCALE SUSY BREAKING?

A. SAGNOTTI

Scuola Normale Superiore and INFN
Piazza dei Cavalieri 7
56126 Pisa ITALY

ABSTRACT

Models of inflation are instructive playgrounds for supersymmetry breaking in Supergravity and String Theory. In particular, combinations of branes and orientifolds that are not mutually BPS can lead to *brane supersymmetry breaking*, a phenomenon where non–linear realizations are accompanied, in tachyon–free vacua, by the emergence of steep exponential potentials. When combined with milder terms, these exponentials can lead to slow–roll after a fast ascent and a turning point. This leaves behind distinctive patterns of scalar perturbations, where pre–inflationary peaks lie well apart from an almost scale invariant profile. I review recent attempts to connect these power spectra to the low–ℓ CMB, and a corresponding one–parameter extension of ΛCDM with a low–frequency cut Δ. A detailed likelihood analysis led to $\Delta = (0.351 \pm 0.114) \times 10^{-3}\,\mathrm{Mpc}^{-1}$, at 99.4% confidence level, in an extended Galactic mask with $f_{sky} = 39\%$, to be compared with a nearby value at 88.5% in the standard $f_{sky} = 94\%$ mask. In these scenarios one would be confronted, in the CMB, with relics of an epoch of deceleration that preceded the onset of slow–roll.

Based on the presentations at the International School for Subnuclear Physics, 53rd Course, "The Future of our Physics Including New Frontiers", Erice, June 24 – July 3 2015, and at "Physics on the Riviera 2015", Sestri Levante, September 16–18 2015

1 INTRODUCTION

It is often stated that Physics thrives on crises, and Cosmology [1] is somehow coming to the rescue these days, with a wealth of data which add new motivations to explore the possible impact, in the Early Universe, of ideas that emerged in Particle Physics and are still awaiting an experimental vindication.

The study of CMB anisotropies is a notable example in this respect. It has already contributed significantly to granting Cosmology an unprecedented accuracy that rests, to a large extent, on the underlying theory of cosmological perturbations. Rather involved to begin with, this theory has the virtue of resting largely on linear phenomena and brings to the forefront, within the theory of inflation [2], measurable aspects of the merging of Quantum Mechanics and General Relativity. The highlight of all this is a slight tilt of the CMB primordial power spectrum [3],

$$\mathcal{P}_R(k) \sim k^{n_s-1} , \tag{1.1}$$

for which PLANCK recently obtained the result [4]

$$n_s = 0.968 \pm 0.006 , \tag{1.2}$$

so that an additional, concrete window opens up on inflation, one of the most enticing ideas that have emerged in Cosmology. This early phase of accelerated expansion, proposed as a natural way to account for the flatness of the Universe and for the apparent paucity of its inhomogeneities, could have been realized if a scalar field, the inflaton, underwent a slow–roll motion in the very early Universe, thus simulating the repulsive effect of a positive cosmological constant before releasing its energy into the vacuum.

Once a slow–roll phase finds its place in Cosmology, it becomes natural to inquire how it started. This type of investigation suffered, over the years, from the lack of compelling scenarios to address it that can be blamed, to some extent at least, on our incomplete grasp of String Theory [5]. At any rate, as a departure from de Sitter space, inflation brings along naturally high–scale supersymmetry breaking in scenarios suggested by String Theory and Supergravity [6]. The models that will guide the ensuing discussion were inspired by *brane supersymmetry breaking* [7], a peculiar mechanism that shows up in String Theory, in orientifold constructions [8] involving collections of branes and orientifolds that are not mutually BPS. In its simplest manifestation, the ten–dimensional Sugimoto model of [7], the introduction of $O9_+$ orientifold planes requires the presence of anti D9-branes, and this induces the breaking of supersymmetry at the string scale and the emergence of non–linear realizations [9] in vacua that are classically stable, in that they are free of tachyon excitations. In the following I shall concentrate on some amusing scenarios that these constructions suggest for the Early Universe, trying to draw from them some lessons for the low–ℓ CMB. There is a lot to learn, however, from systems of this type in many respects, since unbalanced forces exist among their extended objects [1], and non–BPS systems of BPS branes seem to provide an ideal entry point into the dynamics of String Theory with broken supersymmetry.

We made an early encounter with this problem in the early 1990's, with M. Bianchi and G. Pradisi, which is reported in [11]. We had efficient world–sheet rules to deal with

[1]See [10] for a recent review with stimulating critical assessments.

supersymmetric systems, but occasionally they seemed to run into trouble and to require what, in the current geometric language of [12], would amount to negative numbers of branes. O_+-planes had already made an indirect appearance in the toroidal compactifications of [8], and when their nature was elucidated in [13] these types of systems were reconsidered in connection with the issue of supersymmetry breaking. The O_+-planes are BPS objects, characterized by equal values for their tension $T > 0$ and charge $Q > 0$, and it is the need to cancel the latter that brings about anti D-branes, anti-BPS objects with $T > 0$ and $Q < 0$. The unbalanced tension reflects itself, in the string frame, in the emergence of an exponential potential for the dilaton ϕ,

$$V \sim T e^{-\phi} , \tag{1.3}$$

whose exponent is determined by the Polyakov expansion.

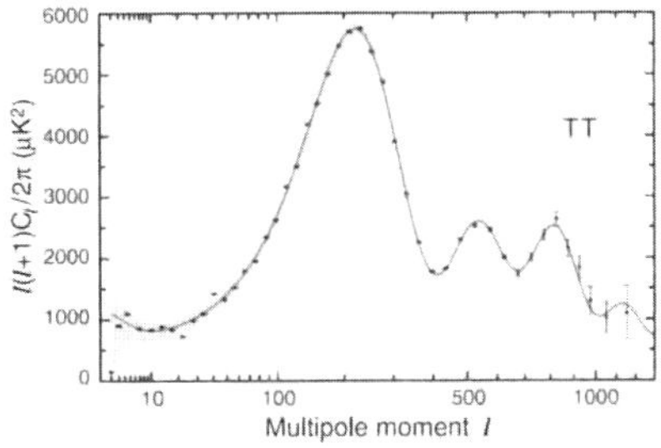

Figure 1: Deviations from the concordance model ΛCDM show up in the low–ℓ end of this nice WMAP9 plot of the CMB angular power spectrum. Most of them are foreshadowed by Cosmic variance, but for a low quadrupole and a dip around $\ell = 20$.

Since the potential (1.3) has no critical point, resummation procedures are to be invoked to connect these systems to flat backgrounds [14], but Cosmology and inflation provide a more natural setting. Mild exponential potentials supporting inflation were indeed studied since the 1980's [15, 16], but remarkably the exponent in eq. (2.8) sits precisely at a (steep) *critical value* where it starts to become impossible, for ϕ, to emerge from an initial singularity while descending the potential [17] (at least in the two–derivative approximation [18]). Hence, the *climbing scalar* generically reaches up to a point where it reverts its motion and starts to descend. If a critical exponential is accompanied by other contributions capable of supporting an inflationary phase, a picture builds up for the onset of inflation with a burst of pre–inflationary slow-roll around the turning point. This can introduce, in the power spectrum of scalar perturbations, a pre–inflationary peak well apart from the almost scalar invariant profile (1.1) that is eventually attained [19], in contrast with what happens if the scalar attains slow-roll directly, in which case the peak is contiguous to the almost scale invariant profile [20]. The study of a decelerating inflaton in connection with quadrupole depression, in different types of scenarios, attracted some attention during the last decade or so, and an (incomplete) list of references to earlier related works can be found in [20].

The key issue that I shall address here is whether the types of feature emerging from brane supersymmetry breaking, or at least signs of a decelerating inflaton, could be hiding behind the anomalies that have long emerged in the low–ℓ tail of the observed CMB angular power spectrum (APS). To this end, let me begin with fig. 1, which is relatively simple to read since it displays, in the region of interest, only a few binned data without error bars. Leaving aside momentarily the shadowed region, the reader will spot a very low quadrupole, followed by a large oscillation around $\ell = 5$, a milder peak and a sizable dip around $\ell = 20$. A closer look at the first multipoles displayed in fig. 2 reveals that matters are more murky, but a low quadrupole, a sizable peak following it around $\ell = 5$, a smaller one around $\ell = 15$ and a dip around $\ell = 20$ appear common to all these results.

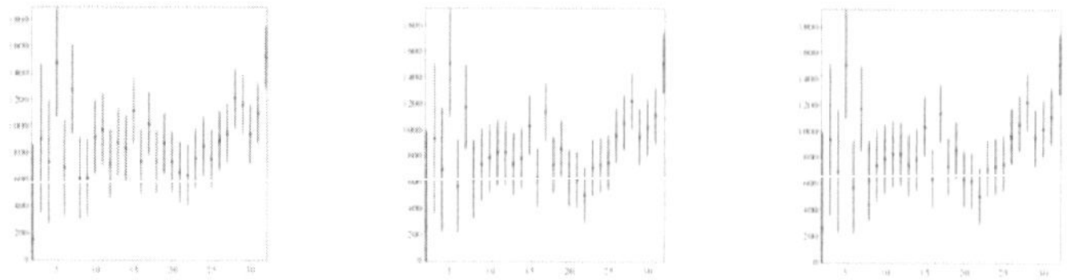

Figure 2: A closer look at the low–ℓ end ($\ell \leq 32$) reveals complicated patterns, although a low quadrupole, a peak around $\ell = 5$, a smaller one around $\ell = 15$ and a dip around $\ell = 20$ recur visibly in WMAP9 (left), PLANCK 2013 (center) and PLANCK2015 data (right).

Two facts make this type of analysis fascinating. To begin with, the theory of inflation associates the low–ℓ range to the earliest epochs that are accessible via the CMB. Moreover, the lower end of the CMB anisotropy spectrum contains, in principle, pristine information about primordial power spectra of scalar perturbations. The Universe behaves somehow as an optical filter for these frequencies, so that [25]

$$A_\ell \sim \ell(\ell+1) \int \frac{dk}{k} P_R(k)\, j_\ell(k\Delta\eta)^2 \simeq P_R\left(k = \frac{\ell}{\Delta\eta}\right), \tag{1.4}$$

where the last step is justified since the spherical Bessel functions build up a narrow kernel. There is, however, a big complication: Cosmic variance, an overall incertitude on the interpretation of low–ℓ data that reflects the limited information that we can collect on them. It is the reason for the shadowed region in the left end of fig. 1. We can indeed detect a single realization of the anisotropy pattern from our special vantage point in the Universe while, for any ℓ, the information at our disposal rests on $2\ell + 1$ values of the "magnetic quantum number" $m \in [-\ell, \ell]$, with expected fluctuations that behave proportionally to $\frac{1}{\sqrt{2\ell+1}}$. Caution is more than legitimate, although in scenarios that originate from higher dimensions, which admittedly I am unable to advocate at present in a convincing form, one would be dealing, in principle, with many more data. For instance, 55 independent data would be contributing to the quadrupole in $D = 10$! Keeping cosmic variance well in mind, I shall thus continue to examine the low–ℓ anomalies, drawing some inspiration from String Theory and Supergravity.

Let me stress, however, that the preceding considerations, and in particular the models of high–scale supersymmetry breaking, do not force pre–inflationary dynamics to affect the low–ℓ end of the CMB anisotropy. Rather, the possibility of accessing, via the CMB, some information related to the onset of inflation would represent a sort of cosmic coincidence, as the recent onset of an accelerated phase. This type of scenario is in principle compatible with a short inflation, and if it were correct the CMB would be encoding information on how inflation started, not only on its steady state, thus making the whole picture even more concrete. Future experiments might grant further support to these ideas, or perhaps rule them out altogether. An important datum, in this respect, is that in scenarios of this type the tensor–to–scalar ratio r tends to exceed, at low frequencies, the value that it attains in the slow–roll phase [24].

In order to minimize the overlap with some preceding reviews, I shall begin in Section 2 with a quick discussion of the climbing phenomenon, illustrating it also with reference to a simple analytic model introduced in [26]. In Section 3 I shall describe some simple inflationary models combining these features and corresponding χ^2 tests related to first 30 multipoles. The highlight will be that it is simple, and even natural, to capture a low quadrupole, a large peak around $\ell = 5$ reflecting the turning point of the inflation and even a dip around $\ell = 20$, via a climbing scalar. However, all these models share the unpleasing feature of yielding power spectra where the eventual almost scale invariant profile is approached too slowly. In Section 4 I shall begin by reviewing some analytic Coulomb–like models of primordial power spectra that were proposed in [24]. These well-motivated departures from eq. (1.1) at low k define two–parameter extensions of ΛCDM that can account for a decelerating inflaton. Their most important new parameter, a scale Δ, characterizes typical energy scales at the epoch where slow–roll inflaton would have started. Its determination will bring about a crucial new ingredient, the choice of Galactic mask.

2 Climbing scalars, in brief

Exponential potentials result in interesting cosmological solutions that have been explored since the 1980's [15, 16], but a key feature lies well beyond slow–roll profiles and went unnoticed for a while. This is the *climbing behavior* of a minimally coupled scalar in the presence of *(over)critical* exponentials [17]. In brief, a scalar field ϕ minimally coupled to gravity in D dimensions, which evolves in a spatially flat FLRW Universe cannot emerge from an initial singularity while descending an exponential potential

$$V = V_0 \, e^{2\gamma\varphi} \,, \tag{2.1}$$

if $\gamma \geq 1$. This simple description of the result applies to all D, albeit in terms of the non–standard field φ, which is related to a canonically normalized field ϕ according to

$$\varphi = \sqrt{\frac{D-1}{D-2}} \, \phi \tag{2.2}$$

For example, in terms of a canonically normalized scalar field ϕ, in four dimensions the *critical* exponential potential would read $e^{\sqrt{6}\,\phi}$.

A field of this type will be called a *climbing scalar*. The nature of the transition at $\gamma = 1$ and the analogy with a Newtonian particle in a viscous medium were reviewed in [26, 27], and more recently in [28]. Here I shall only mention, in passing, some intriguing facts, drawn largely from [17]:

- the Polyakov expansion implies that the ten–dimensional Sugimoto model of [7], the simplest realization of brane supersymmetry breaking, has precisely $\gamma = 1$;

- once the scalar is forbidden to emerge while descending the potential, the string coupling is bounded from above, rendering this behavior naturally protected against string loops (although not against α' string corrections [18]);
- the dynamics of φ is fully determined up to a single constant, φ_0, which gauges the impact with the steep exponential and has important effects on the primordial power spectrum of scalar perturbations;
- the KKLT uplift [29] can be naturally ascribed to brane supersymmetry breaking, and the climbing behavior continues to occur also in presence of a KKLT axion. Moreover, the transition at $\gamma = 1$ continues to occur if the potential is dominated asymptotically by a steep exponential;
- in lower dimensions, the dilaton mixes with the breathing mode and one of the resulting combinations, Φ_t, retains a critical exponential potential for all $D < 10$. If the other, Φ_s, is somehow stabilized, the climbing behavior persists for $D < 10$ [26, 27];
- if Φ_s is somehow stabilized, a space-time filling p-brane that couples to the dilaton, in the string frame, proportionally to $e^{-\alpha\phi}$, contributes in four dimensions an exponential potential [26, 27]
$$V \sim e^{2\gamma\varphi} , \tag{2.3}$$
where
$$\gamma = \frac{1}{12}(p + 9 - 6\alpha) . \tag{2.4}$$

The simplest models to describe the onset of inflation via a climbing phase rest on the two–exponential potentials

$$V(\varphi) = V_0 \left(e^{2\varphi} + e^{2\gamma\varphi}\right) , \tag{2.5}$$

which lead to a spectral index $\left(\gamma < \frac{1}{\sqrt{3}}\right)$ [2]

$$n_s = \frac{1 - 9\gamma^2}{1 - 3\gamma^2} . \tag{2.6}$$

There is some entertaining numerology: one could somehow associate $\gamma = \frac{1}{12}$ to an NS-fivebrane, and this value would result in a spectral index $n_s \simeq 0.96$.

In the next section I shall compare the toy models of eq. (2.5) to the low–ℓ end of the CMB angular power spectrum. However, in [19] we also described a more realistic modification of the Starobinsky potential of [2],

$$V(\varphi) = V_0 \left[e^{2\varphi} + \frac{1}{2}\, e^{2\gamma\varphi} + \left(1 - e^{-\frac{2}{3}\varphi}\right)^2\right] , \tag{2.7}$$

which can grant, with a proper choice of φ_0, about 60 e–folds of inflation, with a spectral index $n_s \simeq 0.96$ and a tensor to scalar ratio $r \simeq 0.15$ for a large fraction of it. The results

[2] The first model of this type was considered in [17], and involved the non–BPS D3-brane found in [30] following [31].

that I shall present will be very similar in the two cases, since they rest on the behavior near the turning point, and I shall also explain how, adding to these potentials a small gaussian bump close to the steep exponential,

$$\Delta V \;=\; V_0\, a_1\, e^{\,-\,a_2(\varphi+a_3)^2}\;, \tag{2.8}$$

with a_1 of a few percent and a_3 of order one, one can introduce other interesting features in power spectra, making them amusingly close to fig. 2 [19].

I would like to conclude this section with a brief illustration of an exact solution found in [26]. It corresponds to the class of two–exponential potentials

$$V(\varphi) \;=\; V_0\left(e^{\frac{2}{\gamma}\,\varphi}\;+\;e^{2\gamma\varphi}\right)\;, \tag{2.9}$$

where $0 < \gamma < 1$, which are much steeper than those of eq. (2.5). However, their solutions are qualitatively similar, can be obtained analytically and exhibit clearly the origin, in this case, of the climbing phenomenon. Letting

$$ds^2 \;=\; e^{\,2\,\mathcal{A}(t)}dt^2 \;-\; e^{\,\frac{2\,\mathcal{A}(t)}{D-1}}\,d\mathbf{x}\cdot d\mathbf{x}\;, \tag{2.10}$$

which entails a peculiar choice for the time variable, a minimally coupled Einstein–scalar system leads to the reduced action principle

$$\mathcal{L} \;=\; \frac{1}{2}\left(\dot\varphi^2\;-\;\dot{\mathcal{A}}^2\right)-\mathcal{V}_0\left(e^{\,2\,(\mathcal{A}+\gamma\,\varphi)}\;+\;e^{\,\frac{2}{\gamma}\,(\gamma\,\mathcal{A}+\varphi)}\right)\;, \tag{2.11}$$

which can be turned into the manifestly separable one

$$\mathcal{L} \;=\; \frac{1}{2}\left(\dot{\widehat{\varphi}}^2\;-\;\dot{\widehat{\mathcal{A}}}^2\right)\;-\;\mathcal{V}_0\left(e^{\,2\,\sqrt{1-\gamma^2}\,\widehat{\mathcal{A}}}\;+\;e^{\,\frac{2}{\gamma}\,\sqrt{1-\gamma^2}\,\widehat{\varphi}}\right) \tag{2.12}$$

via the "Lorentz boost"

$$\begin{aligned}\widehat{\mathcal{A}} \;&=\; \frac{1}{\sqrt{1-\gamma^2}}\,\left(\mathcal{A}\;+\;\gamma\,\varphi\right)\,,\\ \widehat{\varphi} \;&=\; \frac{1}{\sqrt{1-\gamma^2}}\,\left(\varphi\;+\;\gamma\,\mathcal{A}\right)\,.\end{aligned} \tag{2.13}$$

The key feature underlying the climbing behavior is therefore, in this case, an impenetrable barrier on which $\widehat{\varphi}$ can only impinge to undergo a reflection. The exact solution,

$$e^{\varphi} \;=\; e^{\varphi_0}\left[\frac{\sinh(\omega\gamma\tau)}{\cosh\omega(\tau\;-\;\tau_0)}\right]^{\frac{\gamma}{1-\gamma^2}}\;,\qquad e^{\mathcal{A}} \;=\; e^{\mathcal{A}_0}\left[\frac{\cosh^{\gamma^2}\omega(\tau\;-\;\tau_0)}{\sinh(\omega\gamma\tau)}\right]^{\frac{1}{1-\gamma^2}}\;, \tag{2.14}$$

where

$$\omega^2 \;=\; \frac{\lambda}{\gamma}\,\sqrt{1-\gamma^2}\;e^{\,2\,\mathcal{A}_0\,\sqrt{1-\gamma^2}}\;, \tag{2.15}$$

displays, as advertised, the typical hard–bounce behavior described at length in [19].

3 χ^2 TESTS OF THE FIRST FEW MULTIPOLES ($\ell \leq 32$)

Arriving at the Mukhanov–Sasaki (MS) equation was a major achievement, which took a number of years and a proper assessment of the actual observables. Many details can be found in [1], but the end result that will concern us is simply

$$\left(\frac{d^2}{d\eta^2} + k^2 - W_S(\eta) \right) v_k(\eta) = 0 \,. \tag{3.1}$$

Although here one is solving an *initial–value* problem, this minimal form should ring a bell to anyone familiar with one–dimensional scattering in Quantum Mechanics, especially since the final aim of the analysis is to compute, in the $\epsilon \to 0^+$ limit, the power spectrum

$$P(k) = \frac{k^3}{2\,\pi^2} \left| \frac{v_k(-\epsilon)}{z(-\epsilon)} \right|^2 \,. \tag{3.2}$$

The key difference with respect to one–dimensional scattering is transparent in the WKB limit, since the solutions are not tuned to decay after a reflecting wall. Starting, in a region where $W_s \simeq 0$, from the oscillatory Bunch–Davies solutions

$$v_k(\eta) \sim e^{-ik\eta} \,, \tag{3.3}$$

W_s builds up a Coulomb–like barrier after several e–folds, where the growing mode dominates, so that the familiar tunneling formula leaves way to

$$v_k(-\epsilon) \underset{\epsilon\to 0+}{\sim} \frac{1}{\sqrt[4]{|W_S(-\epsilon) - k^2|}} \exp\left(\int_{-\eta^\star}^{-\epsilon} \sqrt{|W_s(y) - k^2|}\, dy \right) \,. \tag{3.4}$$

Consequently, larger W_S build up larger $P(k)$.

I can now complete this brief discussion adding some more details on the nature of W_S and of the corresponding potential, W_T, which determines tensor perturbations. To begin with, one can prove the two limiting behaviors,

$$W_{S,T} \underset{\eta\to-\eta_s}{\sim} -\frac{1}{4}\,\frac{1}{(\eta+\eta_s)^2} \,, \qquad W_{S,T} \underset{\eta\to-0^-}{\sim} \frac{\nu^2 - \frac{1}{4}}{\eta^2} \qquad \left(\nu = \frac{3}{2}\,\frac{1-\gamma^2}{1-3\gamma^2} \right) \tag{3.5}$$

which manifest themselves close to the initial singularity and after several e–folds of slow–roll. These results follow from the expressions

$$W_S = \frac{1}{z}\,\frac{d^2 z}{d\eta^2} \,, \quad \text{with} \quad z = e^{\frac{\mathcal{A}}{3}}\,\frac{d\varphi}{d\mathcal{A}} \,, \tag{3.6}$$

and

$$W_T = \frac{1}{\mathcal{A}}\,\frac{d^2\mathcal{A}}{d\eta^2} \,. \tag{3.7}$$

A key lesson to be drawn from eqs. (3.6) and (3.7) is that scalar perturbations are highly affected, in general, by the scalar dynamics, in contrast with tensor perturbations. A neat result derived in [24], which applies in the region where the mild exponential dominates,

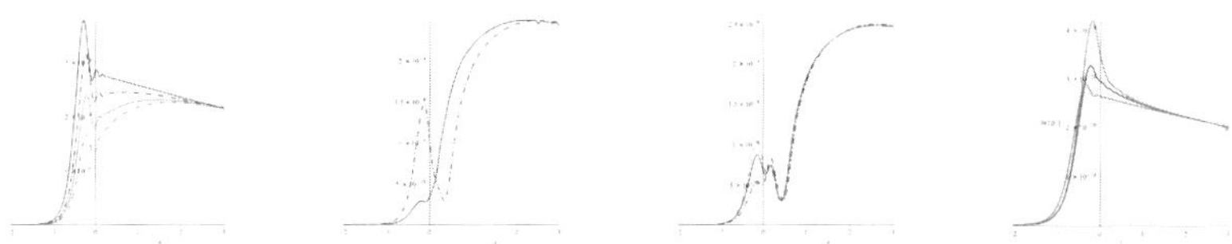

Figure 3: *First three:* power spectra of scalar perturbations for the double–exponential potential of eq. (2.7) with a small gaussian bump as in eq. (2.8). The horizontal axis displays x, where $k = 10^x$, and these power spectra are normalized, in an arbitrary but convenient fashion, so that they all meet as they approach the almost scale invariant profile of eq. (1.1). *Last:* some corresponding power spectra of tensor perturbations. Their only feature is the overshoot discussed in [24], which manifests itself also in the presence of the gaussian bump of eq. (2.8).

is that *the ratio W_S/W_T is less than one when the scalar is about to attain slow roll.* As a result, one can argue from the WKB expression of eq. (3.4) that, in these models, the tensor–to–scalar ratio r is larger, at low k, than its limiting value in the slow–roll region. The argument reflects the deceleration of the scalar field, since the slow–roll parameter ϵ is larger before the scalar attains slow–roll.

I can now describe the power spectra in fig. 3, all of which approach for large k the almost scale invariant profile of eq. (1.1) [19]. Let me begin from the first group of plots, which obtain for values of φ_0 that are near the low end of the range of interest. Initially, the scalar hardly meets the gaussian bump and behaves as in the presence of a single mild exponential. The power spectrum presents the typical pre–inflationary peak studied in [20, 21], which falls directly on the almost scale invariant profile of eq. (1.1). However, as φ_0 is increased, it begins to experience the gaussian bump, which reflects it, before it reaches its inflection, like a steep exponential would. Consequently, the pre–inflationary peak moves away from the almost scale invariant profile [19], since the scalar slows down momentarily close to the turning point but reverts faster and faster as φ_0 is increased. Increasing φ_0 even further we thus come to the second group of plots: the peak becomes smaller and smaller, until a featureless growth is attained, along the lines of what was found in [24] and [19] for the two–exponential case. As the scalar reaches beyond the inflection point, however, it almost comes to rest near the top and the pre–inflationary peak grows dramatically. Then, increasing φ_0 further, the scalar reaches beyond the top, so that it can finally approach the steep exponential and undergo a reflection nearby. Now it slows down twice, in climbing the gaussian bump and then near the steep exponential, and as a result *two pre–inflationary peaks* build up, followed by *deep trough* which reflects the fast exit from the gaussian bump. This last type of feature is well represented in the third group of plots. As we shall see shortly, it appears favoured by the low–ℓ CMB, with the deep trough giving rise to a dip around $\ell = 20$. Finally, some typical spectra of tensor perturbations are collected in the last entry in fig. 3. They undergo indeed similar, but less dramatic changes as φ_0 is varied, as I had anticipated.

In order to compare these power spectra with the low–ℓ CMB, in [19] we followed a simple

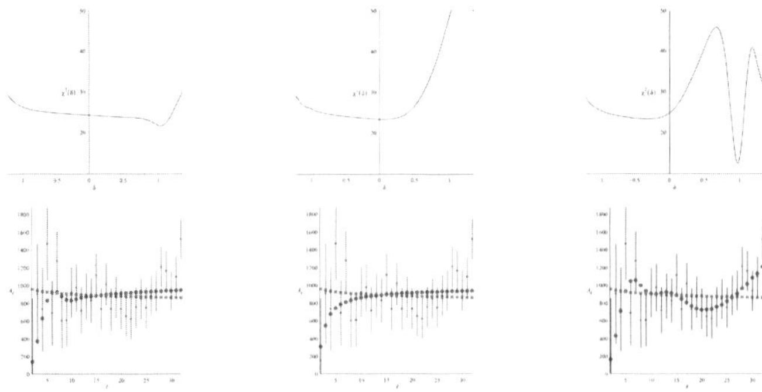

Figure 4: The dependence of $\chi^2(\mathcal{M}, \delta)$, minimized with respect to M, on δ, for three significant cases corresponding to the first three collections of power spectra in fig. 3 (first row), and the corresponding optimal angular power spectra (second row).

procedure guided by the order parameter

$$\chi^2(\varphi_0, \mathcal{M}, \delta) \; = \; \sum_{\ell=2}^{32} \frac{\left(A_\ell(\varphi_0, \mathcal{M}, \delta) \; - \; A_\ell^{\rm WMAP9}\right)^2}{\left(\Delta A_\ell^{\rm WMAP9}\right)^2} \; , \tag{3.8}$$

where the $\mathcal{A}_\ell$ were computed resorting to eq. (1.4) of the Introduction [25],

$$A_\ell\,(\varphi_0, \mathcal{M}, \delta) \; = \; \mathcal{M}\; \ell(\ell+1) \int_0^\infty \frac{dk}{k}\; \mathcal{P}_R\big(k, \varphi_0\big)\, j_\ell{}^2\big(k\, 10^\delta\big) \; , \tag{3.9}$$

in terms of the normalization $\mathcal{M}$ and of a second parameter, δ. $\mathcal{M}$ fixes essentially the energy scale of the process, and was determined, in all cases that we examined, minimizing analytically $\chi^2(\varphi_0, \mathcal{M}, \delta)$. The second parameter is more interesting: it effects an horizontal displacement, in semi–logarithmic scale, of the power spectra of fig. 3, making it possible to optimize their correlation with features present in the angular power spectra of fig. 2.

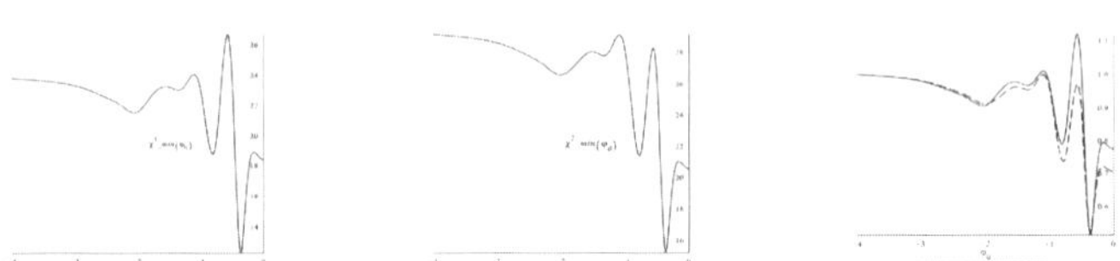

Figure 5: The optimal values of the minimum values of $\chi^2(\varphi_0, \mathcal{M}, \delta)$ as a function of φ_0 for WMAP9 data (left), PLANCK 2013 data (center) and their comparison (right) when both are normalized with respect to their values for $\varphi_0 = -4$, which are respectively 23.82 and 29.14.

The simple order parameter δ displays interesting behaviors, and the reader will not fail to recognize amusing analogies with Landau–Ginzburg theory. The left portion of fig. 4 corresponds to an intermediate case in the left portion of fig. 3, where the pre–inflationary peak has barely separated from the almost scale invariant profile. The shallow but narrow minimum met when varying δ identifies a horizontal displacement that hooks the pre–inflationary peak to the features present around $\ell = 5$. The central portion corresponds to the central portion of fig. 3, and more precisely to the featureless spectrum displayed there. Here the shallow and wide minimum met when varying δ is driven by the quadrupole depression. Finally, the right portion corresponds to the right portion of fig. 3, and now a deep, narrow minimum signals a stronger correlation between the features around $\ell = 5$ and the dip around $\ell = 20$. The results obtained with WMAP9 and PLANCK 2013 are nicely consistent, up to an overall normalization, as can be seen in fig. 5, whose entries display the dependence on φ_0 of the minimum value attained by $\chi^2(\mathcal{M}, \delta)$ as one varies $\mathcal{M}$ and δ. Although the ratio χ^2/DOF between this minimum χ^2 and the actual number of unconstrained parameters can drop below 50% of its value for eq. (1.1), all these models suffer from a slow approach to an almost scale invariant power spectrum that makes them, at most, interesting toy examples. On the other hand, confining the attention to quadrupole depression and paying some attention to the Galactic mask, as I can now explain, one can come closer to a more realistic, if still partial, description of low–ℓ anomalies.

4 Full spectrum analysis and the Galactic mask

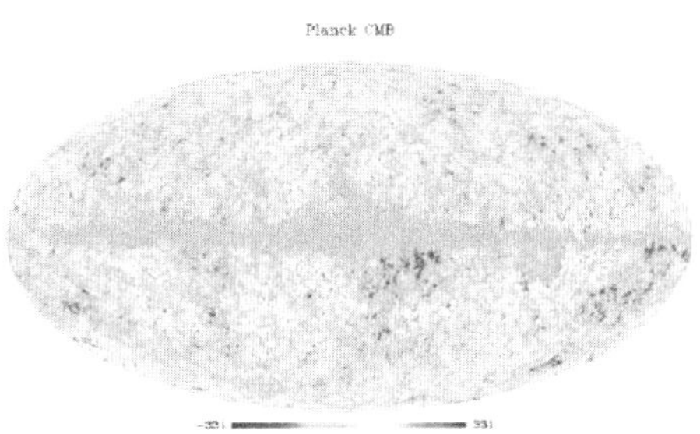

Figure 6: All studies of the CMB rest on a choice of Galactic mask, aimed at reducing contaminations while leaving access to the largest possible fraction of the sky. The mask used in PLANCK 2015 corresponds to $f_{sky} = 94\%$, *i.e.* it grants access to 94% of the sky.

In this section, I would like to summarize some recent steps leading to an extension of ΛCDM where a new parameter, a scale Δ, accounts for the apparent lack of power at low ℓ. Adding one more parameter in full-fledged likelihood tests is a delicate operation, which could easily destabilize the celebrated determination of the others or lead to inconclusive results. Eqs. (3.1) suggest a convenient way to proceed, since W_S must cross the η–axis as a system that started out in a kinetically dominated phase approaches slow roll.

A simple way to model this effect is to solve the MS equation for the class of Coulomb–like potentials

$$W_S = \frac{\nu^2 - \frac{1}{4}}{\eta^2} \left[c \left(1 + \frac{\eta}{\eta_0} \right) + (1-c) \left(1 + \frac{\eta}{\eta_0} \right)^2 \right] , \tag{4.1}$$

which approach for $\eta \to 0^-$ the attractor behavior of eq. (3.5) with

$$n_s = 4 - 2\nu , \tag{4.2}$$

and change sign at $\eta = -\eta_0$ with a slope determined by $c - 2$. The corresponding power spectra can be computed exactly [24]:

$$\mathcal{P}_R(k) \sim \frac{(k\,\eta_0)^3 \exp\left(\frac{\pi\left(\frac{c}{2}-1\right)\left(\nu^2-\frac{1}{4}\right)}{\sqrt{(k\,\eta_0)^2 + (c-1)\left(\nu^2-\frac{1}{4}\right)}}\right)}{\left|\Gamma\left(\nu + \frac{1}{2} + \frac{i\left(\frac{c}{2}-1\right)\left(\nu^2-\frac{1}{4}\right)}{\sqrt{(k\,\eta_0)^2 + (c-1)\left(\nu^2-\frac{1}{4}\right)}}\right)\right|^2 \left[(k\,\eta_0)^2 + (c-1)\left(\nu^2-\frac{1}{4}\right)\right]^{\nu}} . \tag{4.3}$$

In particular, for $c = 2$ this complicated expression reduces to a simple modification of eq. (1.1), which I would like to present in the form

$$\mathcal{P}_R(k) = A\,(k/k_0)^{n_s-1} \to \frac{A\,(k/k_0)^3}{\left[(k/k_0)^2 + (\Delta/k_0)^2\right]^{\nu}} , \tag{4.4}$$

where k_0 is a pivot scale. For $c > 2$ the power spectra in eq. (4.3) embody caricature pre–inflationary peaks, reminiscent of those present in the first and fourth groups of plots of fig. 3.

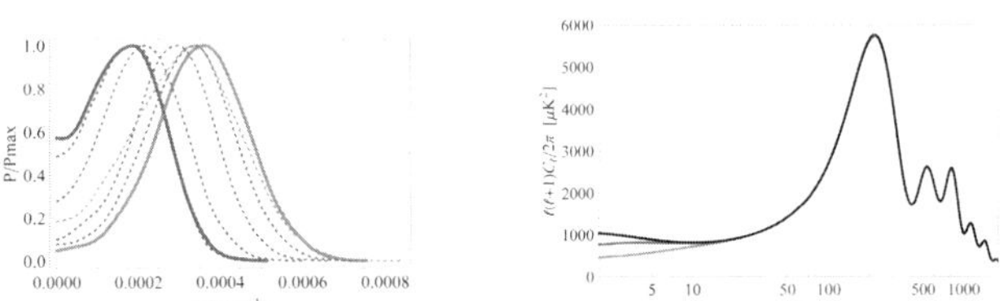

Figure 7: *Left:* posteriors for the parameter Δ in the power law (4.4), for different Galactic mask: solid blue for the 94% mask, thick red for a +30° extension, dotted blue for the intermediate masks +6°, +12°, +18°, +24°, and dotted red for +36°. *Right:* best fit APS models for ΛCDM with a standard mask (black), for ΛCDM with a +30° extension (gray, barely visible behind the previous case), for ΛCDM+Δ with a standard mask (blue), and for ΛCDM+Δ with a +30° extension (red). Only the $\ell \lesssim 15$ range is affected.

I can now summarize a few relevant facts [32]:

- if only Δ is introduced, the determination of the standard cosmological and nuisance parameters is not sensibly affected, even if polarization data are taken into account. The detection of Δ improves with wider masks obtained extending blindly the standard PLANCK mask that leaves out a fraction $f_{sky} = 94\%$, up to $f_{sky} = 39\%$, and then starts to deteriorate. At the same time, its mean value nearly doubles. On the other hand, if Δ is not introduced, widening the mask does affect the determinations to some extent. In [32] we thus obtained, with $f_{sky} = 39\%$, the optimal determination

$$\Delta = (0.351 \pm 0.114) \times 10^{-3}\,\text{Mpc}^{-1} , \tag{4.5}$$

which corresponds to about 99% CL and to the red curves in fig. 7. As we stressed in [32], this result should be compared to a nearby value obtained, at 88% CL, in the standard mask with $f_{sky} = 94\%$;

- with also c in eq. (4.3), the joint determination of both parameters becomes less effective, while the preferred c is compatible with a mild pre–inflationary peak;
- the idea that a short inflation might have left some imprints in the CMB APS was already pursued, in interesting directions, in a number of papers including those listed in [20], and more recently it was also considered in [21]. We came to it from a different perspective in [17, 19, 24], and the work in [32] differs from previous or other present approaches in three respects: the well–motivated parametrization in eqs. (4.3) and (4.4), the use of the best available data from PLANCK 2013 and from PLANCK 2015, including polarization effects, and above all the use of different Galactic masks. The end result is the relatively good evidence for the scale Δ in eq. (4.5). Clearly, the significance of the wider masks deserves further investigations.

I would like to conclude with some unpublished material that I presented in preliminary form at Erice and more fully at Sestri Levante. It is aimed at justifying, within a simple χ^2 analysis as in Section 3, the peculiar effect of the Galactic mask on the determination of Δ. Referring to fig. 8, one should keep in mind that the primordial power spectra of fig. 3 are in semi–logarithmic scale. Widening the masks cuts out oscillations in the first multipoles, as can be seen comparing figs. 2 and 8, leaving behind a slower growth. This increases Δ, shifting to the right $\mathcal{P}_R(k)$, in semi-logarithmic scale, via the parameter δ. However, this *dilates* its profile in the linear scale of eq. (1.4), washing out pre–inflationary peaks. As a result, very different power spectra yield APS that are essentially indistinguishable, for $\ell \leq 32$, from those resting on the analytic expression in eq. (4.4).

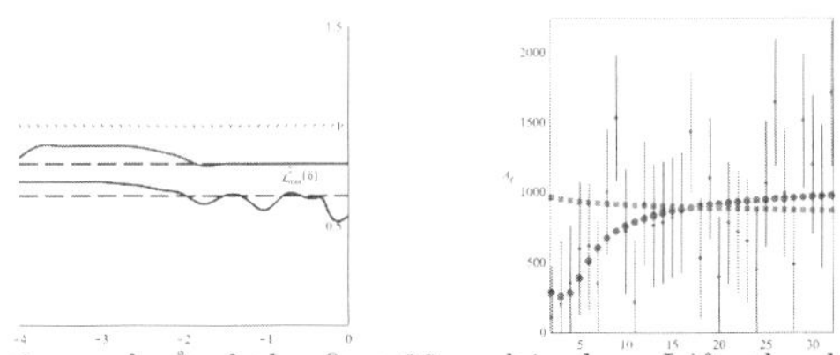

Figure 8: Larger galactic masks and the first 30 multipoles. *Left:* the dependence of $\chi^2_{min}(\mathcal{M}, \delta)$ on φ_0 essentially disappears with PLANCK 2013 data and a +30° mask (red, continuous) and with PLANCK 2015 data and a +30° mask (blue, continuous). *Right:* the optimal low–ℓ APS corresponding to fig. 4 stretches out and becomes indistinguishable from the APS determined by the analytic profile of eq. (4.4).

Acknowledgements

I would like to thank Prof. A. Zichichi for kindly inviting me to lecture at Erice. I am also grateful to the CPhT–E. Polytechnique and to the CERN Ph-Th Department for their kind hospitality during my sabbatical year, and to E. Dudas, A. Gruppuso, A. Kehagias, N. Kitazawa, N. Mandolesi, P. Natoli, M. Porrati, R. Stora, A. Yeranyan, and especially to S. Ferrara, for very stimulating collaborations. This work was supported in part by Scuola Normale Superiore and by INFN-CSN4 (IS Stefi).

References

[1] For reviews see: V. Mukhanov, "Physical foundations of cosmology," Cambridge, UK: Univ. Pr. (2005); S. Weinberg, "Cosmology," Oxford, UK: Oxford Univ. Pr. (2008); D. H. Lyth and A. R. Liddle, "The primordial density perturbation: Cosmology, inflation and the origin of structure," Cambridge, UK: Cambridge Univ. Pr. (2009); D. S. Gorbunov and V. A. Rubakov, "Introduction to the theory of the early Universe: Cosmological perturbations and inflationary theory," Hackensack, USA: World Scientific (2011).

[2] A. A. Starobinsky, Phys. Lett. **B 91** (1980) 99; D. Kazanas, Astrophys. J. **241** (1980) L59; K. Sato, Phys. Lett. B **99** (1981) 66; A. H. Guth, Phys. Rev. **D 23** (1981) 347; A. D. Linde, Phys. Lett. **B 108** (1982) 389; A. Albrecht and P. J. Steinhardt, Phys. Rev. Lett. **48** (1982) 1220; A. D. Linde, Phys. Lett. **B129** (1983) 177; For a recent reviews see: J. Martin, C. Ringeval and V. Vennin, Phys. Dark Univ. **5-6** (2014) 75 [arXiv:1303.3787 [astro-ph.CO]].

[3] V. F. Mukhanov and G. V. Chibisov, JETP Lett. **33** (1981) 532 [Pisma Zh. Eksp. Teor. Fiz. **33** (1981) 549].

[4] R. Adam *et al.* [Planck Collaboration], arXiv:1502.01582 [astro-ph.CO]; P. A. R. Ade *et al.* [Planck Collaboration], arXiv:1502.01589 [astro-ph.CO]; N. Aghanim *et al.* [Planck Collaboration], [arXiv:1507.02704 [astro-ph.CO]].

[5] For reviews see: M. B. Green, J. H. Schwarz and E. Witten, "Superstring Theory", 2 vols., Cambridge, UK: Cambridge Univ. Press (1987); J. Polchinski, "String theory", 2 vols. Cambridge, UK: Cambridge Univ. Press (1998); C. V. Johnson, "D-branes," USA: Cambridge Univ. Press (2003) 548 p; B. Zwiebach, "A first course in string theory" Cambridge, UK: Cambridge Univ. Press (2004); K. Becker, M. Becker and J. H. Schwarz, "String theory and M-theory: A modern introduction" Cambridge, UK: Cambridge Univ. Press (2007); E. Kiritsis, "String theory in a nutshell", Princeton, NJ: Princeton Univ. Press (2007).

[6] D. Z. Freedman, P. van Nieuwenhuizen and S. Ferrara, Phys. Rev. **D 13** (1976) 3214; S. Deser and B. Zumino, Phys. Lett. **B 62** (1976) 335. For a recent review see: D. Z. Freedman and A. Van Proeyen, Cambridge, UK: Cambridge Univ. Pr. (2012) 607 p.

[7] S. Sugimoto, Prog. Theor. Phys. **102** (1999) 685 [arXiv:hep-th/9905159]; I. Antoniadis, E. Dudas and A. Sagnotti, Phys. Lett. **B 464** (1999) 38 [arXiv:hep-th/9908023]; C. Angelantonj, Nucl. Phys. **B 566** (2000) 126 [arXiv:hep-th/9908064]; G. Aldazabal and

A. M. Uranga, JHEP **9910** (1999) 024 [arXiv:hep-th/9908072]; C. Angelantonj, I. Antoniadis, G. D'Appollonio, E. Dudas and A. Sagnotti, Nucl. Phys. **B 572** (2000) 36 [arXiv:hep-th/9911081].

[8] A. Sagnotti, in Cargese '87, "Non-Perturbative Quantum Field Theory", eds. G. Mack et al (Pergamon Press, 1988), p. 521, arXiv:hep-th/0208020; G. Pradisi and A. Sagnotti, Phys. Lett. **B 216** (1989) 59; P. Horava, Nucl. Phys. **B 327** (1989) 461, Phys. Lett. **B 231** (1989) 251; M. Bianchi and A. Sagnotti, Phys. Lett. **B 247** (1990) 517, Nucl. Phys. **B 361** (1991) 519; M. Bianchi, G. Pradisi and A. Sagnotti, Nucl. Phys. **B 376** (1992) 365; A. Sagnotti, Phys. Lett. **B 294** (1992) 196 [arXiv:hep-th/9210127] . For reviews see: E. Dudas, Class. Quant. Grav. **17** (2000) R41 [arXiv:hep-ph/0006190]; C. Angelantonj and A. Sagnotti, Phys. Rept. **371** (2002) 1 [Erratum-ibid. **376** (2003) 339] [arXiv:hep-th/0204089].

[9] E. Dudas and J. Mourad, Phys. Lett. B **514** (2001) 173 [hep-th/0012071]; G. Pradisi and F. Riccioni, Nucl. Phys. B **615** (2001) 33 [hep-th/0107090].

[10] J. Polchinski, arXiv:1509.05710 [hep-th].

[11] A. Sagnotti, In *Erice 1992, Proceedings, From superstrings to supergravity* 116-125. and Rome II Univ. - ROM2F-93-005 (93/02,rec.Mar.) 10 p. C [hep-th/9302099].

[12] J. Polchinski, Phys. Rev. Lett. **75** (1995) 4724 [hep-th/9510017].

[13] E. Witten, JHEP **9802** (1998) 006 [hep-th/9712028].

[14] E. Dudas, G. Pradisi, M. Nicolosi and A. Sagnotti, Nucl. Phys. B **708** (2005) 3 [hep-th/0410101].

[15] F. Lucchin and S. Matarrese, Phys. Rev. **B 32** (1985) 1316.

[16] J. J. Halliwell, Phys. Lett. **B 185** (1987) 341; L. F. Abbott and M. B. Wise, Nucl. Phys. **B 244** (1984) 541; D. H. Lyth and E. D. Stewart, Phys. Lett. **B 274** (1992) 168; I. P. C. Heard and D. Wands, Class. Quant. Grav. **19** (2002) 5435 [arXiv:gr-qc/0206085]; N. Ohta, Phys. Rev. Lett. **91** (2003) 061303 [arXiv:hep-th/0303238]; S. Roy, Phys. Lett. **B 567** (2003) 322 [arXiv:hep-th/0304084]; P. K. Townsend and M. N. R. Wohlfarth, Phys. Rev. Lett. **91** (2003) 061302 [arXiv:hep-th/0303097]; Class. Quant. Grav. **21** (2004) 5375 [arXiv:hep-th/0404241]; R. Emparan and J. Garriga, JHEP **0305** (2003) 028 [arXiv:hep-th/0304124]; E. Bergshoeff, A. Collinucci, U. Gran, M. Nielsen and D. Roest, Class. Quant. Grav. **21** (2004) 1947 [hep-th/0312102]; J. G. Russo, Phys. Lett. **B 600** (2004) 185 [arXiv:hep-th/0403010]; A. A. Andrianov, F. Cannata and A. Y. Kamenshchik, JCAP **1110** (2011) 004 [arXiv:1105.4515 [gr-qc]].

[17] E. Dudas, N. Kitazawa and A. Sagnotti, Phys. Lett. **B 694** (2010) 80 [arXiv:1009.0874 [hep-th]].

[18] C. Condeescu and E. Dudas, JCAP **1308** (2013) 013 [arXiv:1306.0911 [hep-th]].

[19] N. Kitazawa and A. Sagnotti, JCAP **1404** (2014) 017 [arXiv:1402.1418 [hep-th]], EPJ Web Conf. **95** (2015) 03031 [arXiv:1411.6396 [hep-th]], Mod. Phys. Lett. A **30** (2015) 28, 1550137 [arXiv:1503.04483 [hep-th]].

[20] A. D. Linde, Phys. Rev. D **59** (1999) 023503 [hep-ph/9807493]; C. R. Contaldi, M. Peloso, L. Kofman and A. D. Linde, JCAP **0307** (2003) 002 [astro-ph/0303636]; Y. -S. Piao, B. Feng and X. -m. Zhang, Phys. Rev. D **69** (2004) 103520 [hep-th/0310206]; Y. -S. Piao, Phys. Rev. D **71** (2005) 087301 [astro-ph/0502343]; D. Boyanovsky, H. J. de Vega and N. G. Sanchez, Phys. Rev. D **74** (2006) 123007 [astro-ph/0607487]; C. Destri, H. J. de Vega and N. G. Sanchez, Phys. Rev. D **78** (2008) 023013 [arXiv:0804.2387 [astro-ph]]; F. J. Cao, H. J. de Vega and N. G. Sanchez, Phys. Rev. D **78** (2008) 083508 [arXiv:0809.0623 [astro-ph]]; R. K. Jain, P. Chingangbam, J. -O. Gong, L. Sriramkumar and T. Souradeep, JCAP **0901** (2009) 009 [arXiv:0809.3915 [astro-ph]]; E. Ramirez and D. J. Schwarz, Phys. Rev. D **80** (2009) 023525 [arXiv:0903.3543 [astro-ph.CO]], Phys. Rev. D **85** (2012) 103516 [arXiv:1111.7131 [astro-ph.CO]]; R. K. Jain, P. Chingangbam, L. Sriramkumar and T. Souradeep, Phys. Rev. D **82** (2010) 023509 [arXiv:0904.2518 [astro-ph.CO]]; C. Destri, H. J. de Vega and N. G. Sanchez, Phys. Rev. **D 81** (2010) 063520 [arXiv:0912.2994 [astro-ph.CO]]; E. Ramirez, Phys. Rev. D **85** (2012) 103517 [arXiv:1202.0698 [astro-ph.CO]].

[21] Z.-G. Liu, Z.-K. Guo and Y.-S. Piao, Phys. Rev. **D 88** (2013) 063539 [arXiv:1304.6527 [astro-ph.CO]]; M. Cicoli, S. Downes and B. Dutta, JCAP **1312** (2013) 007 [arXiv:1309.3412 [hep-th], arXiv:1309.3412]; F. G. Pedro and A. Westphal, arXiv:1309.3413 [hep-th]; R. Bousso, D. Harlow and L. Senatore, arXiv:1309.4060 [hep-th]; Z. -G. Liu, Z. -K. Guo and Y. -S. Piao, arXiv:1311.1599 [astro-ph.CO]; Y. F. Cai, E. G. M. Ferreira, B. Hu and J. Quintin, arXiv:1507.05619 [astro-ph.CO].

[22] C. L. Bennett *et al.* [WMAP Collaboration], Astrophys. J. Suppl. **208** (2013) 20 [arXiv:1212.5225 [astro-ph.CO]]; G. Hinshaw *et al.* [WMAP Collaboration], Astrophys. J. Suppl. **208** (2013) 19 [arXiv:1212.5226 [astro-ph.CO]].

[23] P. A. R. Ade *et al.* [Planck Collaboration], Astron. Astrophys. **571** (2014) A1 [arXiv:1303.5062 [astro-ph.CO]], Astron. Astrophys. **571** (2014) A16 [arXiv:1303.5076 [astro-ph.CO]], Astron. Astrophys. **571** (2014) A15 [arXiv:1303.5075 [astro-ph.CO]].

[24] E. Dudas, N. Kitazawa, S. P. Patil and A. Sagnotti, JCAP **1205** (2012) 012 [arXiv:1202.6630 [hep-th]].

[25] V. F. Mukhanov, Int. J. Theor. Phys. **43** (2004) 623 [astro-ph/0303072].

[26] P. Fré, A. Sagnotti and A. S. Sorin, Nucl. Phys. B **877** (2013) 1028 [arXiv:1307.1910 [hep-th]].

[27] A. Sagnotti, Phys. Part. Nucl. Lett. **11** (2014) 836 [arXiv:1303.6685 [hep-th]].

[28] S. Ferrara and A. Sagnotti, arXiv:1509.01500 [hep-th].

[29] S. Kachru, R. Kallosh, A. D. Linde and S. P. Trivedi, Phys. Rev. D **68** (2003) 046005 [hep-th/0301240].

[30] E. Dudas, J. Mourad and A. Sagnotti, Nucl. Phys. B **620** (2002) 109 [hep-th/0107081].

[31] A. Sen, JHEP **9806** (1998) 007 [hep-th/9803194], JHEP **9808** (1998) 010 [hep-th/9805019], JHEP **9808** (1998) 012 [hep-th/9805170].

[32] A. Gruppuso and A. Sagnotti, arXiv:1506.08093 [astro-ph.CO], to appear in IJMPA; A. Gruppuso, N. Kitazawa, N. Mandolesi, P. Natoli and A. Sagnotti, arXiv:1508.00411 [astro-ph.CO].

LECTURER: A. SAGNOTTI

Low-ℓ CMB from String-Scale SUSY Breaking?

Scientific Secretaries: A.Avdoshkin, N.Astrakhantsev

DISCUSSION

- *Tokareva:*

My question is the following. If inflation lasted much longer than 60 efolds, we could not say anything about how inflation started from observations. Don't you think that the case in which we can observe something entails a lot of fine-tuning? Why the duration should be close to 60 efolds, rather than 100 efolds or more?

- *Sagnotti:*

Well, there is no reason a priori. Apart from the fact that, typically, at the level of the effective field theories that one constructs, there are stability issues to keep the potential flat for very long. The "eta" problem is an example. So, basically, one can try with this kind of situation, although one brings to a limit a lot of other problems. Because, for example, all the issues of diluting inhomogeneities and the like, with a short inflation, become marginal. So, you might say: what do you put it for? If you make it short and do not dilute things a lot. Actually, if you look at things a little more carefully, you can realize that there is a compensating effect. Namely, the first multipoles suffer the most from this problem, because they are the ones that exited the horizon earlier. They do come back with this kind of pollution, if you will, but at the same time they give you a blurred image of the sky, so there is a compensation effect. Surely, when you insert these power spectra modified in the infrared, you insert a bias in your data analysis because you insert some prejudice that what you see is this kind of effect. On the other hand, this is natural, as I argued, since the Mukhanov-Sasaki potential changes sign as scalar slows down. What you can do is to see whether the resulting scale plays any kind of role in the analysis, and to get really more convinced you might need additional information related to other observables that point in the same direction. For instance, I did not talk about tensor perturbations in this kind of models, but there are mild effects that are generic for this kind of potentials. Roughly speaking, the tensor spectrum of a scalar that slows down tends to increase in the low-k region. Moreover, other anomalies that I learned about from my cosmologist friends may point toward a short inflation. There are apparent unjustified alignments among the first multipoles. So, with the quadrupole pointing in a certain direction, the octupole appears strangely close to the same direction. Which could be a sign of a sky that was not completely smoothed. More than this, I cannot say, of course. One is pushing things to a limit and seeing whether it makes sense. These kinds of scenarios are motivated by the fact that, in String Theory, when you break supersymmetry the vacuum is modified and the resulting potential is generically runaway. Runaway potentials have a natural meaning in this kind of context, whereas if you try to do Physics with them in flat space, they are a disaster. You have to do resummations long before they make sense. To make sense of this type of models, one is led to try and see whether they have these kinds of effects. The second step of seeing whether in the CMB we do see this kind of effect is not logically connected. You could have the effect hidden there, in the past of many folds of inflation, and then it all becomes merely a theoretical observation. If a scalar moves fast and then slows down, you have some sort of depression of the power spectrum before attaining the

Chibisov-Mukhanov tilt, and you have also some sort of scalar bremsstrahlung, with typically some oscillations, but then you stop there. I agree: expecting to see effects of this sort in the CMB is not logically motivated, but why not? If the phenomenon is there, let us try to see, and what is the best scenario? The best scenario is that the phenomenon is sitting there for us, and as I have argued, it is not obviously contradicting to the idea of homogeneity. This is why people push inflation down to 60 folds, also for other reasons.

- *Teresi:*

Just a curiosity. Maybe you said that this morning, but I missed this detail. When you look at the acoustic oscillations from Plank data there is a point in the low multiple region that is very well outside the cosmic variance bound. Did you manage to fit that as well?

- *Sagnotti:*

Let me tell how the story went. The story went in two steps. The first step I call a Landau-Ginzburg-like semi-analytic test. This was done with my colleague Kitazawa, and we played this game precisely since one can do it semi-analytically. And it works up to l=20-30, this is a typical bound, with some types of models that are, so to speak, motivated by String Theory to some extent.

- *Teresi:*

Sorry, I am talking about that point.

- *Sagnotti:*

I know what you are talking about and I can tell you what I know, how close we went to something that is nice-looking in that respect. The critical exponential forces the scalar to hit and bounce, and in bouncing the scalar slows down. And the slowing down produces typically a pre-inflationary peak. Intuitively, the scalar slows down and the power spectrum tries to go up as long as scalar is in slow roll, while when it speeds up again it goes down. A double peak arose after adding the single small perturbation that one can add without disturbing inflation. You have to be very mild, otherwise the scalar stops, and then you don't do anything. Then you get this kind of phenomenon, a double peak followed by a very steep rise. If you try to compare with the first few multiples, you get this type of situation. So you see that you get this dip around $l = 20$ and peaks around $l = 5$, very naturally. You cannot move them. You can play with this kind of thing. This kind of figure is in my slides. The reason why I was not really bringing it to your attention is that models of this kind have a very slow rise. And from the CMB you know that you have to attain the spectral index behavior within a little bit more than one decade. Now, this led naturally to another way of looking at the problem, a model independent analytic method. The Mukhanov-Sasaki equation is a one-dimensional Schroedinger equation that one can solve exactly for certain forms of the potential, an in particular for a Coloumb-like potential. One can compare with data the resulting power spectra, since they attain quickly the attractor behavior. These systems lend themselves to a full cosmological analysis, which is what I hinted to at the end. This cosmological analysis is good in some respects and rough in others. Because here you cannot understand this peculiar point. You can understand the average and you can give a parametrization, if you want, to this decrease of the spectrum. This parametrization determines in a reasonable way, modulo the subtlety with the galactic masking, to 99.4% confidence level a scale of about 2800 Mpc, to be compared with the 3000 Mpc of the observable patch. The effect is close to the size of the observable patch, of course, since one is modifying only the first few multiples. But in this way of doing things one is not sensitive anymore to small details. As I showed you at some point, this analytic model can also account for a very rough peak, but adding another parameter gives a worse result for the first parameter. Instead of 99% confidence level, the determination decreases to 97.5%

confidence level. In general, the more parameters one adds the worse the determination becomes. In fact, when I was starting to communicate with my cosmologist friends, building up the analysis of string-inspired spectra, we were not getting anywhere since we had too many parameters.

The LAA impact on technology R&D: from past to future.

Horst Wenninger, CERN

LAA was the first special program at CERN[1] dedicated to prepare the future beyond the ongoing CERN Large Electron-Positron Collider (LEP) construction program 1980 to 1989.Unlike all previous CERN projects (PS, BOOSTER, BEBC, ISR, SPS)LEP approval was obtained under the condition to stay within the CERN budget level. No extra project budget was assigned for this major European scientific infrastructure project! This required to set priorities: CERN facilities and many experiments had to be closed to free resources for LEP construction and for the ongoing proton-antiproton SPS collider experiments - one of the few European science programs leading to a Nobel Prize!

The idea behind LAA was to develop technologies and detectors for experiments in view of a high intensity proton-proton collider in the LEP tunnel, discussed as early as 1984. LAA funds were used to hire dedicated staff (physicist, engineers, technicians) and to form collaborations - supported by LAA- in order to prepare the future beyond LEP. About forty LAA staff and eighty unpaid scientist worked together for six years. LAA activities are published in over 350 papers and journals.

Examples of detector developments within the LAA collaborations were the spaghetti electromagnetic calorimeter, multi-wire drift chambers, scintillation fibre trackers, micro-strip detectors - to list a few examples – and most important the advanced precision tracking and read-out electronics, IPSA tube and GaAs crystals, silicon pixels detectors, CMOS chips and ASIC/VLSI chips, and last but not least, resistive plate chambers (RPCs).Great impact of LAA on LHC came through people -engineers, physicists, technicians- recruited for the LAA activities, who helped to setup the Large Hadron Collider (LHC) experiments and now participating in the experiments. Another important impact of LAA on LHC was through the timely push on new technologies, in particular: the built-up at CERN of competences in micro-electronics, CMOS chips, ASIC/VLSI chips, micro-strip-, silicon pixel development and related technics [2].

Two decades of microelectronics at CERN were enabled by the LAA project.

In 1988, the AMPLEX multiplexed read-out chip was used in UA2. Hybrid pixel devices, with a read-out chip "bump bonded" to the detector, were used in WA97 in the mid-1990s. By 2002, CERN had developed a bump-bonded 8000-channel pixel for the ALICE silicon-pixel detector at the LHC.A more recent development in micro electronics based on LAA know-how is the successful NINO chip, an ultra fast and low-power front-end amplifier / discriminator ASIC, designed for the multi-gap resistive plate chamber and many applications elsewhere [3].

There is also a spectacular spin-off project related to LAA:“ **Medipix** “(1. Phase).

CERN has often been the incubator for the development of innovative technologies.
Most people know meanwhile about the World Wide Web, originating at CERN in the 1990th, but very few people know about the capacitive touch screens, invented for the consoles of the SPS Control Room in 1973 together with the tracker ball and the computer-programmable knob. Only after the progress in micro-electronics, however, the touch screen on computers and the i-phone became a commercial success.
Following LEP start of operations the LAA program was complemented by the CERN DRDC program, which is still today supporting some of the development projects:

PLAFOND	Platform for Developing Neutrino Detectors
RD-18	CRYSTAL CLEAR R&D on scintillation materials for novel ionizing radiation detectors for HEP, medical imaging and industrial applications
RD39	Cryogenic Tracking Detectors
RD42	Development of Diamond Tracking Detectors for the LHC High Luminosity Experiments
RD50	Development of Radiation Hard Semiconductor Devices for Very High Luminosity Colliders
RD51	Development of Micro-Pattern Gas Detectors
RD52	Dual-Readout Calorimetry for Energy Measurements
RD53	Development of pixel readout integrated circuits for extreme rate and radiation

RD51: the collaboration involves about 450 authors, 75 Universities and Research Laboratories from 25 countries in Europe, America, Asia and Africa.
RD51 collaboration studies two types of micro-pattern gas detectors:

(i) MICROMEGAS with a narrow gap of (50-100 μm) PPC with thin cathode mesh insulating gap-restoring wires or pillars(4)
(ii) GEM made of metal-coated polymer foils with 70 μm holes at 140 mm pitch(5)

More recently other micro pattern detector schemes, offers the potential to develop new gaseous detectors with unprecedented spatial resolution, high rate capability, large sensitive area, operational stability and radiation hardness

RD52: the collaboration studies the idea of dual (triple) readout methods for energy measurements. This idea was started by R. Wigmans, a former LAA collaborator, following the basic principle: measure components of electromagnetic showers separately from hadron shower components (dual readout) and also separately from slow neutron components (triple readout).

We understand that research and development (R&D) for experiments was required to prepare for LHC, but why is the title of the present talk:

"LAA impact on technology R&D: from past to future?"

What is the importance of a special programme and R&D fund provided by Italy to CERN and implemented by the CERN Council in 1986 at CERN under the Project Leader Prof. A. Zichichi? Why should this have an impact not only on past but even on present & future activities?

The original idea of the LAA project to perform technology R&D as an independent research program with its own, independent funding was the most effective support in the past to prepare new challenging projects. This idea has meanwhile been adopted for present and future R&D initiatives.
For accelerators, this topic has been addressed at the 2014 IPAC conference (Dresden, Germany) by R. Brinkmann, head of the DESY XFEL machine group (his statement at the conference):
" Efforts and progress to perform accelerator R&D as an independent research program with its own, independent funding are described for the example of the Helmholtz ARD program in Germany. Links to efforts in other countries are discussed and an outlook to future accelerator research is given."

The schedule of LHC outlined below indicates a period of R&D prior to prototyping and fabrication and construction for every phase of the program. During the construction of LEP within the fixed CERN budget, however, no resources were free to cover the R&D for the next step envisaged: the LHC in the LEP tunnel.
It was then when LAA – just at the right time – allowed to support this development activities in parallel with LEP installation.

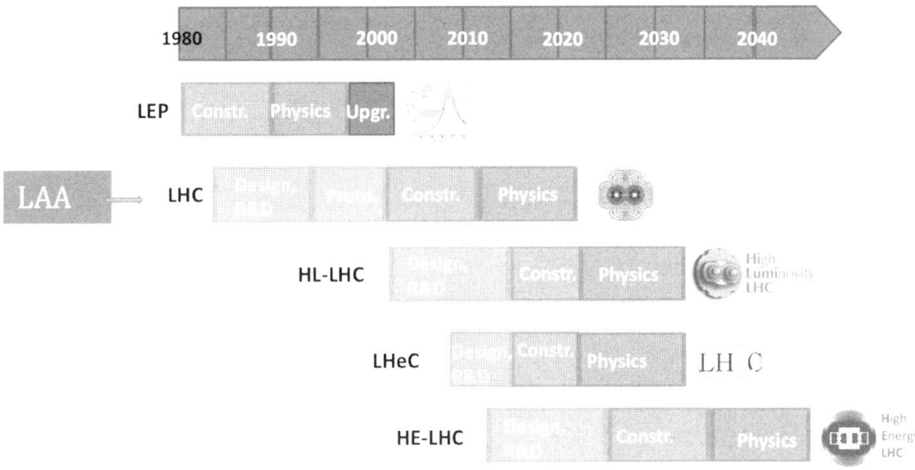

In Europe, the EU Framework programs /in future Horizon 2020 / are also providing funds for the joint R&D of science - industry – society. Today, accelerator and detector development projects are often organized in large worldwide collaborations. This creates overheads and inefficiencies but also competition and challenges.

LAA and DRDC programs were CERN specific collaborations, peer-reviewed by CERN committees, and implemented with minimal bureaucracy.

The outline of the EU program Horizon 2020 shows that emphasis is given on coupling research to innovation and to societal challenges. Horizon 2020 is the biggest EU Research and Innovation program ever with nearly € 80 billion of funding available over 7 years (2014 to 2020) in addition to the private investment that this money will attract. It promises more breakthroughs, discoveries and world-firsts by taking great ideas from the lab to the market.

Resource sharing: Science: 24 441 M€ - Industries: 17 016 M€Society:29 679 M€

In the spirit of the EU programs, COST-BENEFIT ANALYSIS will be mandatory for any EU Horizon 2020 grant above 50 M€. Such an analysis has been performed for the LHC scientific project and resulted in[(7)]:

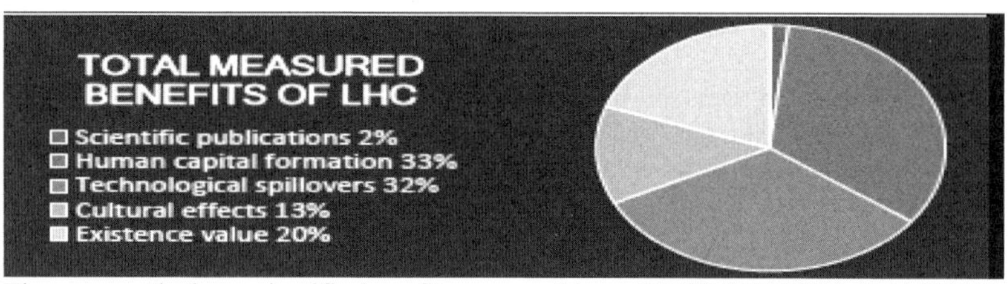

The apparently low scientific benefit measured by a COST-BENEFIT ANALYSIS of this nature reflects simply the fact of the limited number of scientists in the world and not the importance of science!

We need to find resources avoiding unreasonable bureaucratic overheads for the scientific community preparing their experiments. In the past we had the chance to get dedicated R&D resources during LEP construction at CERN while preparing in parallel the LHC in the LEP tunnel with funds provided by the special program LAA-
No conditions were attached to the LAA project, thus avoiding overhead costs.

Of course, there is a need to evaluate the relevance for society of major new infrastructure project funding. R&D funding, however, should be assigned with a minimum of overheads.

Back to LHC and its scientific program: The long-term planning for the LHC experimental program at CERN and in particular plans for LHC Detector R&D from Run1 to Run 3 are indicated in the graph (courtesy: S. Kühn, University Freiburg):

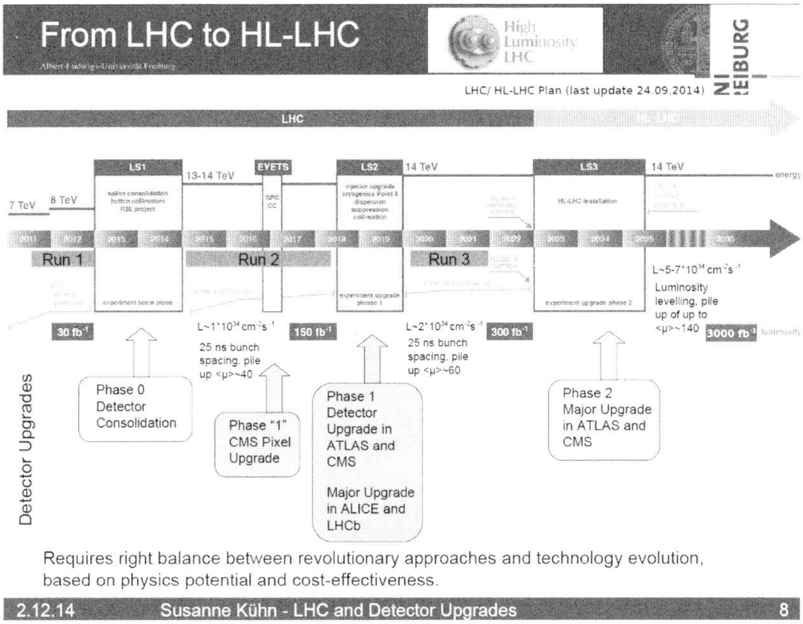

For the accelerator the table of consolidation and R&D works performed during the shut-down after Run 1are outlined by Mike Lamont, head of the LHC operations group, as follows:

- Measure all splices and repair the defective ones
- Consolidate interconnects with new design (clamp, shunt)
- Finish installation of pressure release valves (DN200)
- Magnet consolidation - exchange of weak cryo-magnets
- Consolidation of the DFBAs
- Measures to further reduce SEE (R2E): relocation, redesign, shielding…
- Install collimators with integrated button BPMs (tertiary collimators and a few secondary collimators)
- Experiments consolidation/upgrades

Figure 1: consolidation work in the LHC tunnel after Run 1 during LS1 shut-down

After start-up in 2015 the challenges of the increased energy from 7 TeV to 13 TeV and bunch spacing of 25ns with a ß* = 0.5 m or lower are: *Less margin to critical surfaces (risk of quenches), less particle losses allowed and hence set-up beam limits reduced, dipole magnets run into saturation with effects on the field quality and all hardware closer to limits (power converters, beam dump, cryogenics)*. This is outlined in the table of the graph below:

	50 ns	25 ns
GOOD	• Lower total beam current • Higher bunch intensity • Lower emittance	• Lower pile-up
BAD	• High pile-up • Need to level • Pile-up stays high • High bunch intensity – instabilities...	• More long range collisions: larger crossing angle; higher beta* • Higher emittance • Electron cloud: need for scrubbing; emittance blow-up; • Higher UFO rate • Higher injected bunch train intensity • Higher total beam current

Due to advantage of lower pile-up runs are scheduled for 25 ns bunch spacing.
Reference 6: CERN experiment web pages outline upgrade works to cope with higher energy and later also higher intensity. A short content list is summarized below:

LHC Season 2: Major work at the experiments for Run 2

ALICE: A LARGE ION COLLIDER EXPERIMENT

LHCB: LARGE HADRON COLLIDER BEAUTY

LONG SHUTDOWN 1 - ALICE
Among the consolidation and improvements to its 19 sub-detectors, they installed a new calorimeter (DCAL) extending the range covered by the electromagnetic calorimeter. This EMCAL now covers a wider angle, allowing measurements of the energy of the photons and electrons over a larger area. The TRD (Transition Radiation Detector) that detects particle tracks and identifies electrons has also been completed with the addition of 5 more modules.

LONG SHUTDOWN 1 - LHCb
A new forward shower detector, the RICH detector and the Trigger system upgrade to allow for offline analysis at 12.5 kHz were main upgrade items. Important is the new **HeRSCheL** detector which was installed to distinguish rare processes in which particles are observed in the detector but not along the beam-pipe. In addition a section of the beryllium beam pipe replaced.

LONG SHUTDOWN 1 – ATLAS
New chambers were added in the muon spectrometer and the calorimeter readout consolidated. The forward detectors were upgraded to provide a better measurement of the LHC luminosity A new aluminium beam pipe to reduce the background.
The whole detector readout system was improved to be able to run at 100 KHz and all data acquisition software and monitoring applications were re-engineered.
The trigger system was redesigned, going from 3 levels to 2 while implementing smarter and faster selection algorithms. A very ambitious upgrade of simulation, reconstruction and analysis software was completed, and a new generation of data management tools on the GRID was implemented

LONG SHUTDOWN 1 – CMS
Priority was to mitigate the effects of radiation on the performance of the Tracker, by equipping it to operate at low temperatures (down to -20°C).This required changes to the cooling plant, and extensive work on the environment control of detector and cooling distribution to prevent condensation or icing.

A fourth measuring station was added to each muon end-cap to maintain discrimination between low-momentum muons and background as the LHC beam intensities increase. Complementary to

this installation at each end of the detector of a 125-tonne composite shielding wall to reduce neutron backgrounds.
A luminosity-measuring device, the Pixel Luminosity Telescope, was installed on either side of the collision point around the beam-pipe.
Other major activities included replacing photo-detectors in the hadron calorimeter with better-performing designs, moving the muon readout to more accessible locations for maintenance, installation of the first stage of a new hardware triggering system and consolidation of the solenoid's magnet cryogenic system and of the power distribution.
The software and computing systems underwent a significant overhaul during the shutdown to reduce the time needed to produce analysis datasets.

ATLAS: A TOROIDAL LHC APPARATUS / CMS: COMPACT MUON SOLENOID

WHAT IS ATLAS? WHAT IS CMS?
The ATLAS and CMS detectors have a broad physics programme including investigating the recently discovered Higgs Boson, searching for extra dimensions and particles that could make up dark matter, as well as continuing systematic studies of the Standard Model.

The LHC roadmap oulines that- following run 1, with its successful discovery of the Higgs Boson and the shut-down LS 1 – theexperimental program will continue with increased collision anergy

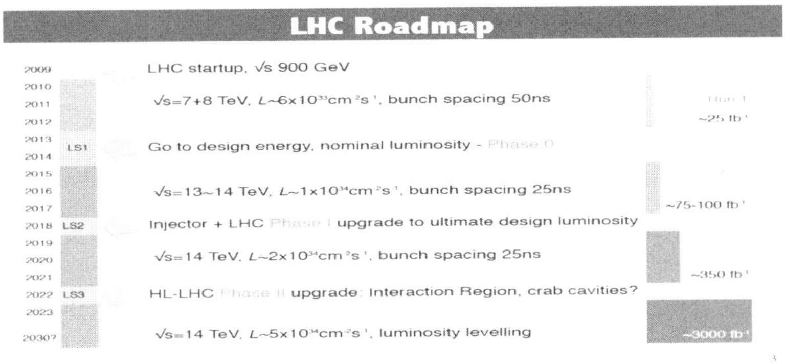

and indeed first record energy collisions have been recorded in June 2015

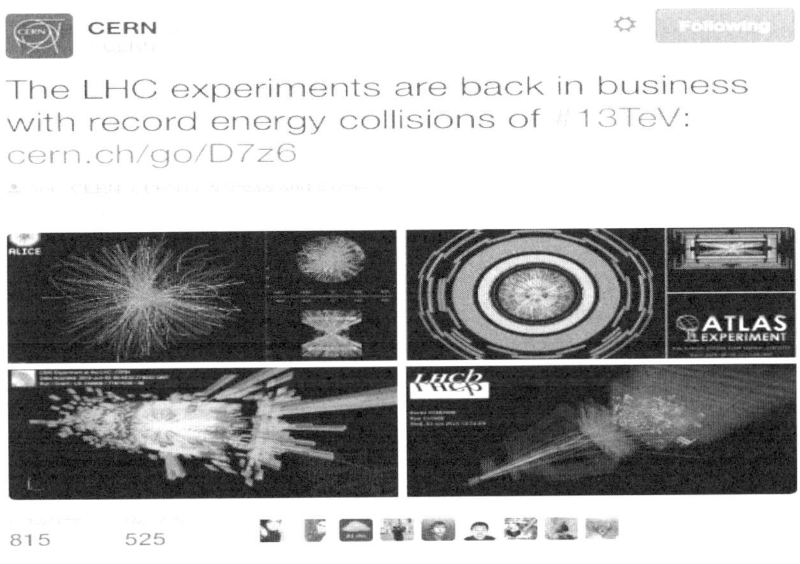

Figure 2: First collisions in LHC experiments taken during preparation of LHC Run 2

Run 2 may take longer than indicated above to accumulate enough data while preparing for the next step: the high luminosity run HL-LHC:

The main thrusts of this step to HL-LHC are the installation of

(i) wide aperture Nb_3Sn triplet qua m for collimation in region dispersion suppression

(iii) large aperture NbTi separator magnets/twin aperture magnets at interaction region

(iv) Crab cavities to reduce the effect of the crossing angle (see Figure 3)
(v) enhanced collimation for 500 MJ beams

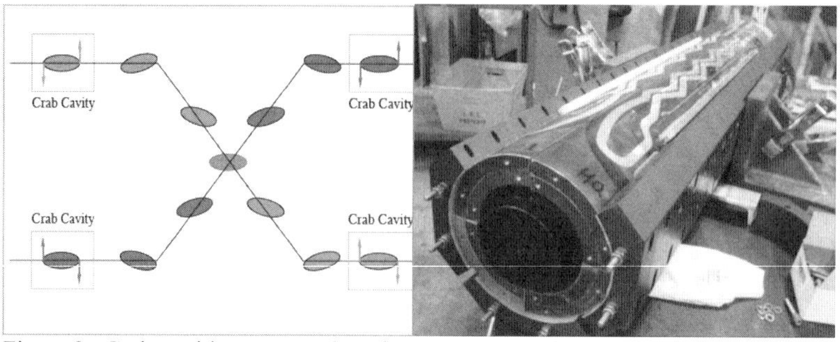

Figure 3: Crab-cavities to rotate bunches prototype of crab-cavity

The conclusion of the 2014 ECFA HL-LHC meeting can be summarized as follows:

It is anticipated that 3 - 4 years of R&D and prototyping, followed by 5 - 6 years of construction, are needed to complete the largest upgrades.
Funding Agencies will need to invest now in targeted R&D to develop cost effective technical solutions, which CERN can help facilitate, and this needs to proceed rapidly and with adequate resources.

At the ultimate luminosity of LHC there will beproblems such as pile-up, particle density, radiation damage resulting in performance loss of pixel tracker, trigger, end-caps, electronics,...

In addition, to prepare the far future, the FCC Workshop in May 2015brought together 62 participants to constitute governance bodies, commit to the project plan and align the organisation, structures and processes of 16 institutions from 10 countries. The goal of the project is to conceive a post-LHC research infrastructure around a 100 km circular energy-frontier hadron collider capable of reaching 100 TeV collisions. The project officially started on 1 June and will run for four years. The total estimated budget of 11.2 million Euros includes a 2.99 million Euro contribution from the Horizon 2020 programme on developing new world-class research infrastructures. EuroCirCol is a building block in the globally coordinated strategy of the FCC study to produce a global design for a global machine. Main areas for the design study are indicated below:

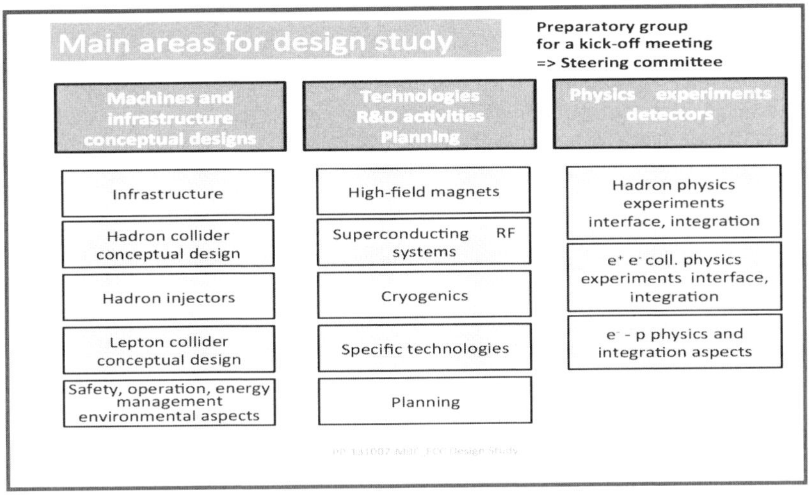

For the experiments of the past and future, the silicon pixel detectors and integrated readout and trigger electronics play an important role.
The CERN LAA project allowed CERN to acquire and develop the necessary competence and helped the development of the technologies. The evolution over time of the surface area of silicon detector systems in particle physics and in recent, space-based experiments - shown in the attached graph:

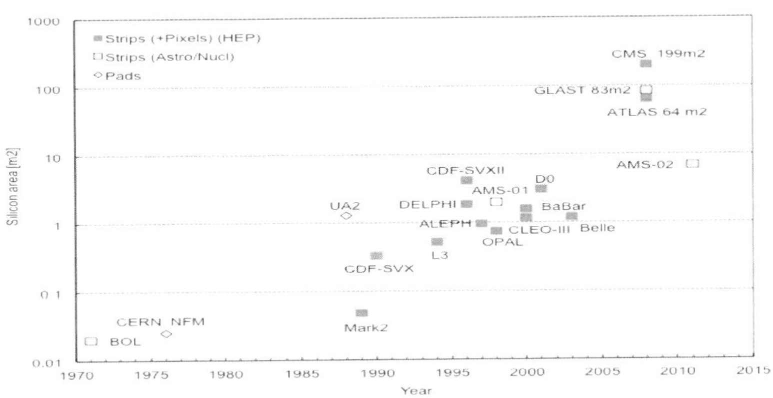

Evolution over time of the surface area of silicon detector systems in particle physics and in recent, space-based experiments. **Source Yoshinobu Unno, KEK.**

In conclusion:

We need to find resources avoiding unreasonable bureaucratic overheads for the scientific community preparing their experiments. In the past we had the chance to get dedicated R&D resources during LEP construction at CERN while preparing in parallel the LHC in the LEP tunnel with funds provided by the special program called LAA, implemented at CERN by Italy under the project leader, Professor A. Zichichi. This allowed CERN to build up competences for instance in micro-electronics, silicon-strip and -pixel detector technology.
No conditions were attached to the LAA project, thus avoiding overhead costs. Cost & Value to society analyses are expected today when funding requests are put forward to the EU Horizon 2020 programs. Of course, there is a need to evaluate the relevance for society of major new infrastructure project funding. R&D funding, however, should be assigned with a minimum of overheads.
We need to find enough support for the experiment upgrades to continue and profit from LHC, which is now running up in energy and is supposed to run up in luminosity. As an example the development of radiation hard silicon detectors including increased data reduction at the detector level with trigger logistics integrated in the readout electronics that has to be prepared to cope with the immense data flow in future experimental runs. This is what LAA type R&D funding could provide.

References:
(1) CERN EP 87-122, July 1987, The LAA Project, A. Zichichi

(2) CERN EF 89-14, CERN LAA SD 89-11, LAA Progress Report 1988-1989
Development of Integrated CMOS Circuits and Silicon Pixel Detectors in the CERN-LAA Project,
F. Anghinolfi et al.

(3) NINO chip, an ultra fast and low-power front-end amplifier / discriminator ASIC designed for the multi-gap resistive plate chamber , F. Anginolphi et al.NIMA 533 (2004) 183-187

(4) Y. Giomataris et al, Nucl. Instr. and Meth. A376(1996)239 ;
(5) F. Sauli, Nucl. Instr. and Methods A386(1997)531

(6) Web pages with information on LHC detector upgrades

(7) Cost-Benefit Analysis of the LHC to 2025 and beyond / Was it worth it ?
Massimo Florio (Milano) - CERN colloquium

Detailed work on LHC experiments for Run 2 are outlined in related web pages

http://press.web.cern.ch/sites/press.web.cern.ch/files/file/backgrounder/2015/05/ls1_workexperiments.pdf

for experiment ATLAS
https://cds.cern.ch/journal/CERNBulletin/2014/18/News%20Articles/1696879?ln=en

for experiment CMSCMS
http://cds.cern.ch/journal/CERNBulletin/2014/18/News%20Articles/1696880

for experiment ALICE
https://cds.cern.ch/journal/CERNBulletin/2014/16/News%20Articles/1694094?ln=en

for experiment LHCb
https://cds.cern.ch/record/1694095?ln=en

LECTURER: H. WENNINGER

The LAA impact on technology R&D: from past to future.

Scientific Secretaries: T. Drozhzhova, D.T. Larsen

DISCUSSION

- *Wenninger:*

Research and Development (R&D) for experiments is required to continue during the coming years to make full use of LHC, running at the highest energies and with unprecedented luminosities. We need to find resources avoiding unreasonable bureaucratic overheads for the scientific community preparing their experiments. In the past we had the chance to get dedicated R&D resources during LEP construction at CERN while preparing in parallel the LHC in the LEP tunnel with funds provided by the special program called LAA, implemented at CERN by Italy under the project leader, Professor A. Zichichi. This allowed CERN to build up competences for instance in micro-electronics, silicon-strip and pixel detector technology. No conditions were attached to the LAA project, thus avoiding overhead costs. Cost & Value to society analyses are expected today when funding requests are put forward to the EU Horizon 2020 programs. Of course, there is a need to evaluate the relevance for society of major new infrastructure project funding. R&D funding, however, should be assigned with a minimum of overheads. We need to find enough support for the experiment upgrades to continue and profit from LHC which is now running up in energy and is supposed to run up in luminosity. As an example I mention the development of radiation hard silicon detectors including increased data reduction at the detector level with trigger logistics integrated in the readout electronics that has to be prepared to cope with the immense data flow.

This is my main message. This is what LAA type R&D funding could provide.

NEUTRINOS. MEASURING THE UNEXPECTED

ALESSANDRO BETTINI

University of Padua. Department of Physics; INFN. Padua division (Italy)

Abstract

The history of neutrino physics is an instructive example of how the physics knowledge advances in an interplay between theoretical developments and progress of the experimental art. Neutrinos, in particular, have always surprised us, behaving differently from the expectations.
In the first part of the lecture I shall summarise the principal steps of neutrino history.
I then shall give a flash on the status of our knowledge and an outlook on the perspectives for the next years.

1. Introduction

Neutrinos are the most difficult to study of the known elementary particles. The reason is not only their very small cross section but also their extremely small masses joined to the *V–A* character of their coupling to the *W*.
In the 85 years since the Pauli desperate remedy, neutrinos have shown to behave very often differently from expectations giving us a series of surprises. On the other hand, neutrinos have been often kind with us. In several occasions, some unknown parameter, when finally measured after many efforts, resulted to have a value large enough to make our life less difficult in the subsequent steps. Examples are the Large Mixing Angle (LMA) solution of the solar neutrino puzzle and the pretty large value of the smallest of the mixing angles, θ_{13}.
Of the known particles, neutrinos are the only ones that do not behave as assumed in the Standard Model (SM). The SM assumes the flavour eigenstates to be stationary and, as such, to have a definite mass. Experiments say that this is false, neutrinos born with a flavour may be detected with another flavour. The SM assumes the masses of neutrinos to be null, experiments show they are not. The SM assumes neutrinos to be helicity eigenstates with negative eigenvalue, antineutrinos with positive. Again this is false. SM assumes neutrinos and antineutrinos to be distinguished by the lepton number, which is not experimentally proven. SM assumes three generations, of neutrinos. Ignoring some so-called anomalies, which are effects comparable or smaller than the statistical and systematic uncertainties, experiments agree on that, up to now (LEP experiments have given $N_\nu = 2.984 \pm 0.008$ and cosmology $N_{\nu-eff} = 3.39^{+0.54}_{-0.51}$).
Neutrino properties have been and are studied with a combination of complementary experiments using different types of detectors and different sources. The latter are both natural (cosmic rays collisions in the atmosphere (ν_e and ν_μ at production) the sun [(ν_e at the origin), the supernova explosions (all flavours)] and artificial [proton accelerators (mainly ν_μ and $\bar{\nu}_\mu$ at production) and power reactors $\bar{\nu}_e$ at production)]
Looking back at this history teaches a lot on the how physics knowledge develops. It should also be noticed that, in the case of neutrinos, several historic developments are ignored or are even found distorted in the literature.
In this lecture I shall briefly summarise the basic historical steps and then focus on our present knowledge and on the perspectives for the future.

2. The first period, from theory to reality. 1929-1956

In the 1920s' three elementary particles were known, one negative, the electron, one positive, the proton and one neutral, the photon. The models of atomic nuclei were built out of these particles. A nucleus of atomic number A and charge Z was made of A protons and $A-Z$ electrons. For example the ^{14}N nucleus would consist of 14 protons and 7 electrons. Consequently, its spin should be semi-integer (rather than integer as it is) and it should obey Fermi statistics (1926).

1928-1929. F. Rasetti[1] performed a series of beautiful experiments measuring the Raman rotation spectra of a number of diatomic gases, in particular homo-nuclear ones, like N_2 and O_2. The rotational bands showed lines of alternate intensity. The lines correspond to different nuclear spin states. It is easy to see that the ratio between the numbers of symmetric and antisymmetric spin states, for a given spin I, is $(I+1)/I$. From the spectra one can then learn the statistics obeyed by the nucleus. It resulted to be wrong, in the current nuclear model. For example ^{14}N obeys Bose, rather than Fermi statistics. For a very clear explanation see ref. [2].

Independently on that another important problem was the continuous spectrum, rather than a mono-energetic line, of the beta decays of the nuclei.

4 December 1930. W. Pauli assumes the existence of a new neutral particle, which he calls neutron, but is what we know as neutrino, a name given a couple of years later by Fermi. On the occasion of a nuclear physics meeting a Zürich, to which he cannot attend, he writes a famous letter beginning with (in English translation): *Dear radioactive Ladies and Gentlemen...unfortunately I cannot appear in Tübingen since I am indispensable here in Zurich because of a ball on the night of 6-7 December."* He continues with:

"*Considering the 'wrong' statistics of ^{14}N and ^{6}Li nuclei, as well as the continuous β spectrum, I have hit upon a desperate remedy to save the "exchange theorem" of statistics and the energy theorem".*

There should be one more type of particle in the nucleon, the "neutron", which is neutral, has a small mass, <0.01 proton mass and is emitted with the electron in the beta decay.

1932. J. Chadwick[3] discovers the neutron, a neutral particle of mass similar to proton constituent of the nuclei. The problem of the wrong statistics is solved.

1933. E. Fermi proposes his theory of beta decay[4],[5]. The abstract says[4]: "*Theory of β rays emission of radioactive substances, built on the hypothesis that the electron emitted by the nuclei do not exist before the decay. On the contrary they are created together with a neutrino, similarly to the creation of a light quantum that goes with a quantum jump of an atom. Comparison of theory and experiment.*" Notice that: a. this is the creation of a new fundamental force, b. the electron and the neutrino are created in the moment of the decay, they do not pre-exist. The idea will lead to the concept of creation and destruction operators.

Notice that "neutrino" is still one. There is not yet any difference between neutrino and antineutrino. By the way, we still do not know if such difference exists.

1934. H. Bethe and R. Peirls[6]. If neutrinos are produced in β decay, the inverse process, $\nu + {}^{Z}_{A}N \rightarrow e^- + {}^{Z+1}_{A}N$, must happen. They estimate the cross section as $< 10^{-44}$ cm^2, corresponding, they notice, to one interaction every10^{16} km of solid matter (1000 light years), concluding: "*it is therefore absolutely impossible to observe processes of this kind with the neutrinos created in nuclear transformations*".

And Pauli is told to have said[7]: "*I did a horrible thing, which no theorist should do, I postulated a particle that cannot be detected.*"

Their estimate of the cross section is close to the today known value, however, fortunately, their conclusion was premature. Indeed, two factors will help, the Avogadro number size and the coming in operation of nuclear reactors of the GW scale in the 1950s'.

The interaction rate per unit target nucleon at 10 m distance from 1 GW reactor is $R = N\sigma\Phi = 6\times10^{23}\times10^{-48}\left(m^2\right)10^{17}\left(m^{-2}s^{-1}\right) = 6\times10^{-8}mol^{-1}s^{-1}$. Consequently, the expected rate in 1 t detector with nuclei of, say A=100, is R=2/h. This is not so small. The real problem to control, as we shall see, is the background.

1939. H. Crane[8]. The radiochemical method. The salt bag experiment. The radiochemical method had been used since decades with neutrons. The target material under study was exposed to neutrons for several weeks and then the induced radioactivity was measured. The idea of Crane is to try doing the same to detect (the effect of) neutrinos. Indeed consider the inverse beta reaction (no difference between neutrino and antineutrino yet)

$$\nu + {}^{35}Cl \rightarrow {}^{35}S + e^{+} \qquad (1)$$

in which the final isotope is radioactive. Exposing for some time a quantity of Cal to neutrinos one might be able to look if any, certainly not many, ^{35}S atoms have been produced, by detecting their decays.

Crane used a 1 mc capsule of radioactive material in a three-pound bag of common salt (Nail) wit an exposure of 90 days. He then extracted the S atoms, if any, with classical chemical methods and counted what he had extracted in an ionization counter. No activity was detected, corresponding to the □<10^{-30} cm^2. Crane was well aware that his limit was some 12 orders of magnitude larger than the expectations. However, he had made the first step towards the radiochemical neutrino detection. The problems were still enormous, including finding a method to separate the produced radioactive species at the level of one in 10^{30}.

The next steps are due to B. Pontecorvo in 1945-46. He was then working in the nuclear British programme at the Chalk River facility in Canada, where a heavy water nuclear reactor was under development. Indeed, an atomic bomb cannot be made without a nuclear reactor. The programme was called, as a coverage, "Tube alloys". Obviously, the internal reports were classified. Two of these, PD-141 and PD-205, are on neutrino detection with the radiochemical method. While the latter is well know, the former came to the attention of the community only recently. Let us see.

1945. B. Pontecorvo[9]. PD-141 Report. The report was classified till 1964 (sic!).

B. P. states here that "*the detection via inverse beta is not out of question*". He thanks J. Guèron for the suggestion of using as a target Cl in C Cl_4, an inexpensive liquid that can be assembled in large masses. The reaction proposed is (1) as it was for Crane.

1946. B. Pontecorvo[10]. PD-205, classified till 1948. This is the Cl-Ar method proposal, namely to exploit the reaction

$$\nu + {}^{37}Cl \rightarrow {}^{37}Ar + e^{-} \qquad (2)$$

Notice that now the Cl isotope is ^{37}Cl rather than ^{35}Cl, in order to produce ^{37}Ar that is radioactive with a sufficiently long half life. Notice also that, we know now but was unknown then, that Cl-Ar reaction can be induced by neutrinos, not by antineutrinos. The argument for having Ar to extract is that it is a volatile and inert gas and as such should be separable at 1/10^{30} level. Time will show that Pontecorvo was right. However, his estimation of the cross section was two orders of magnitude too large and, in addition, the problem of background was not attacked.

1949. L. Alvarez. 1949. Report UCRL-238. Not classified. Alvarez heard of a claim by Fireman to have found evidence for the neutrino-less double beta decay, 0ν2βdecay a process, that, as we shall see, can happen only if neutrino is equal to antineutrino. It was not true, but the news stimulated Alvarez to study in all the details an experiment based on the Cl-Ar reaction with a nuclear reactor as the source and a detector of 40 t of CCl_4. He correctly evaluated the cross section in 2 10^{-45} cm^2 and addressed the background problem. Indeed there are other processes, the most important one being the (p,n) reactions induced by cosmic rays, able to transform Cl into Ar. He found out that the background was to high on the surface, proposing to perform the experiment several metres underground.

1949-50. The liquid scintillator. Solid scintillator detectors had been known since long times. The crystal, when crossed by a charged particle, like an electron, emits a flash of light that is amplified and converted to a charge pulse by a photomultiplier. However, such detectors had limited sizes due, mainly, to the lack of transparency. In 1949-50, three different groups[12], [13], [14] discovered that suitable ternary liquid mixtures scintillate too, but with very long attenuation lengths. The technique allows the detection of the electrons hit by neutrinos. It gave fundamental contributions to neutrino physics, including neutrino discovery. Detectors of tonne and kiloton size are built also now.

1956. F. Reines and F. Cowan[15] discover the neutrino, to be precise the electron antineutrino, in the well known experiment at the Savannah River reactor (0.8.GW), the most powerful of the time. The reaction was

$$\bar{\nu} + p \rightarrow n + e^{+} \tag{3}$$

We just recall that the target was Cd doped water in two tanks sandwiched between liquid scintillator detectors. On 14.7.1956 a telegram was sent by Cowan and Reines to Pauli: "We are happy to inform you that we have definitely detected neutrinos from fission fragments by observing inverse beta decay of protons. Observed cross section agrees well with expected six time ten to the minus forty four square centimetres". The reply of Pauli was: "Thanks for message. Everything comes to him who knows how to wait".

1958. R. Davis[16]. If neutrino and antineutrino are the same particle, the Cl-Ar reaction should be induced by reactor neutrinos

$$\bar{\nu}_e + {}^{37}\mathrm{Cl} \rightarrow {}^{37}\mathrm{Ar} + e^{-} \tag{4}$$

With the high power reactor of Savannah River operational, R. Davis set up an experiment with a target of 11 400 l (18.5 t) of C_2Cl_4. The sensitivity was 20 times larger than the minimum needed for detection if neutrino = antineutrino. No event was detected.

We can conclude that the neutral particle produced in β^{+} decay does not induce the Cl-Ar reaction. The further conclusion, assumed in the SM, that neutrino and antineutrino are different particles, is however arbitrary. The Cl-Ar reaction might not happen for a different reason, namely the *V-A* structure of the interaction and the smallness of neutrino mass.

3. Is neutrino a completely neutral particle? The problem is posed

1937. E. Majorana[17] is not happy of the Dirac electron theory due to the asymmetry, in its derivation, between particle and antiparticle. He writes a fundamental paper titled: "Symmetric positron electron theory". The abstract reads: "*We show that it is possible to achieve complete formal symmetrisation in the electron and positron quantum theory by means of a new quantization process. The meaning of Dirac equations is somewhat modified and it is no more necessary to speak of negative-energy states; nor to assume, for any other type of particles, especially neutral ones, the existence of antiparticles, corresponding to the "holes" of negative energy*". But, much more important, the new approach allows "*also to build a substantially novel theory for the particles deprived of electric charge (... hypothetical neutrinos)*".

The Dirac solution of the Dirac equation is a bi-spinor ψ, with four degrees of freedom, corresponding to the two helicity states of electron and positron. Its components are complex.

It is well known that only the anticommutators, the Clifford algebra, of the γ matrices are defined. Majorana question is: can the γ matrices be chosen to have all the components of ψ real? The answer is positive, namely

$$\tilde{\gamma}_0 = \sigma_2 \otimes \sigma_1, \quad \tilde{\gamma}_1 = i\sigma_1 \otimes 1, \quad \tilde{\gamma}_2 = i\sigma_3 \otimes 1, \quad \tilde{\gamma}_3 = i\sigma_2 \otimes \sigma_2 \tag{5}$$

satisfy the Clifford algebra and are pure imaginary. Consequently, being all the $i\tilde{\gamma}_\mu$ real, the Dirac equation

$$\left(i\tilde{\gamma}^\mu \partial_\mu - m\right)\tilde{\psi}(x) = 0 \tag{6}$$

can govern a $\tilde{\psi}$ with all real components. The Majorana solution has only two degrees of freedom corresponding to the two helicity states of a completely neutral particle of spin ½. Completely neutral means that it has no charge of any type, including lepton number. We know completely neutral bosons, the photon, the *Z*, two of the gluons. Notice that a completely neutral boson can be mass less or massive. A completely neutral fermion must have a mass. If the mass term would be zero, Eq. (6) would not distinguish between Dirac and Majorana solutions.

In his article Majorana did not afford the question on how to experimentally distinguish between the two solutions. He simply stated that adding that it was probably "*not yet possible to ask to the experience to decide between this new theory and the simple extension of the Dirac equations to the neutral particles*".

After three quarters of a century and an impressive experimental progress, we do not have the answer yet.

1935. M. Goeppert-Mayer[18], theoretically proposes the existence that the double beta disintegration accompanied by two neutrinos (2ν2β)

$$ {}^{Z}_{A}N \rightarrow {}^{Z+2}_{A}N + 2e^- + 2\bar{\nu}_e \tag{7}$$

should exist when single beta decay is theoretically forbidden. She evaluates, with the Fermi theory, half life in excess of 10^{17} yr.

1937. G. Racah[19], then working in Florence, soon after the Majorana paper, gave the first proposal for possible detection. If neutrinos obey Dirac equation, neutrinos emitted in a β^- decay can induce only a β^+ process and vice versa. If they obey Majorana equation, any neutrino can produce both electrons and positrons. This was a very nice idea but unfortunately cannot work, for reasons that Racah did not know, *V–A* and the smallness of neutrino mass, as we shall soon see.

1939. W. H. Furry[20]. *In the older theory double-β disintegration involves emission of four particles, two electrons (or positrons) and two antineutrinos (or neutrinos), and the probability is extremely small. In the Majorana theory only two particles – the electrons or positrons – have to be emitted, and transition probability is much larger.* (0ν2β).

Once more, he did not know that neutrino coupling to *W* is *V–A* and that neutrino masses are very small $m \ll E$. Unfortunately "the transition probability" of 0ν2β is not "much larger", but much smaller than that of 2ν2β. We now see why.

Electron neutrino, ν_e, is by definition the neutral particle produced with a positron in the β^+ decay. It has two helicity components. The negative one is largely dominant (and is the only one assumed to exist in the SM), the positive one has m/E amplitude (where m is the neutrino mass and E its energy). It is extremely small at the usual energies. Similarly, the antineutrino, $\bar{\nu}_e$, is the neutral particle produced with an electron in the β^- decay. It has a dominant positive helicity component and a small, of m/E amplitude, of negative helicity.

If neutrinos are Dirac particles, distinguished by the lepton number, when hitting a target both helicity components of ν_e produce electrons, both of $\bar{\nu}_e$ produce positrons.

If neutrinos are Majorana particles, the dominant negative helicity component of ν_e produce electrons, exactly as Dirac, but the positive helicity component produces positrons. Vice versa for the $\bar{\nu}_e$.

In conclusion, the difference between Dirac and Majorana is hidden in the small *m/E* component. In the 0ν2β the typical neutrino energy is 10 MeV, while neutrino masses are 100 meV (or less). The disintegration probability has to pay the factor $(m/E)^2 \sim 10^{-14}$. Furry was right, without this factor the 0ν2β would have been much faster than 2ν2β. Calculations show that the two would have been comparable for neutrino masses around 10 eV. But now we know that they are much smaller.

4. Three families

1947-50. Several physicists, including G. Bernardini, E. Hincks, B. Pontecorvo, G. Puppi, J. Steinberger, were looking at the question: "are the neutral particles produced in the μ decay and in β different or equal?" Even names as nu-prime and neutretto were proposed[21].

1958. G. Feinberg, **1960**, Lee and Yang showed to be essentially impossible to avoid the decay $\mu \rightarrow e\gamma$ if the two neutrinos are equal. The decay had not been observed and an upper limit $<10^{-8}$ had been experimentally established.

The problem will be solved with neutrino beams from proton accelerators of energies of the order of 10 GeV or more. Such accelerator became operational at the end of the 1950s, the 10 GeV synchro-phasatron at the Lebedev Institute in Moscow in April 1957, the 28 GeV Proton Synchrotron (PS) at CERN in the Autumn 1959 and the 33 GeV Alternate Gradient Synchrotron (AGS) at BNL in July 1960. However, the energy of the available proton beam is a necessary but not sufficient condition. In addition, a proper neutrino beam and a proper detector must be designed. In fact, the winner was the last starting, BNL. But let us see the main events.

1957. M. A. Markov charges one of his students, D. Fakirov, of a thesis "On the Possibility to Investigate the Interaction of High Energy Neutrinos with Matter at Accelerators''. The thesis, completed in 1958, contained the calculation of neutrino beams produced by the decays in flight of pions, produced by protons accelerated at energies between 6 an 100 GeV[22]. However, the two vs. one neutrino question was not mentioned.

1959. B. Pontecorvo[23], then working in Dubna, suggested to look at e^+ in interactions of neutrinos from decays of pions at rest in a target hit by the proton beam. The π are captured by nuclei and do not decay. If $\nu_\mu \neq \nu_e$, the decay chain of π^+ is

$$\pi^+ \rightarrow \mu^+ + \nu_\mu;\ \mu^+ \rightarrow e^+ + \nu_e + \bar{\nu}_\mu \tag{8}$$

Notice that there are no $\bar{\nu}_e$. The produced neutrinos would interact in a separate target, where the observation of any positron would prove that $\bar{\nu}_\mu = \bar{\nu}_e$.

Observe however that muon neutrinos produced by pions at rest have energy below the threshold for

$$\nu_\mu + N \rightarrow \mu + N' \tag{9}$$

and cannot be detected. Consequently, if $\nu_\mu \neq \nu_e$ nothing is observed. But a non observation cannot prove something positive. In addition, neutrinos from the decay at rest go in all the directions uniformly filling the 4π solid angle. The fraction reaching the detector would be very small, of the order of 10^{-3} or less at 10 m distance. This is certainly a price that cannot be paid in neutrino experiment.

As a matter of fact, no neutrino beam was built in the Soviet Union in that period.

1959. M. Schwartz, unaware of the activity in Russia, started at Columbia University the study of a possible neutrino beam (discussing with T. D. Lee)[24], including preliminary measurements of the pion yield, needed for a proper design of the proposal.
1962. M. Schwartz, L. Lederman, J. Steinberger et al. The second neutrino discovery[25]. The experiment is well known to the reader. I just remark that the main reasons of its success were: a. the location of the internal target in a straight section of adequate length, 3 m; b. the use of a novel detection technique, the spark chamber invented by S. Fukui and S. Myamoto in 1959[26].
Following the Schwartz ideas, a proposal for the two-neutrino experiment was presented also at CERN in 1961 using an internal target in the PS. The pions should emerge from the target in a short, 1.5 m long, straight section of the PS. However, Guy von Dardel computed that, because of neglect of the defocusing effect of the fringing field of the PS magnets, the neutrino flux had been overestimated by an order of magnitude. The event rate in the rather modest mass "Ramm" bubble chamber (1/2 t) would be only 3 or 4 events after several months of running. CERN management decided to cancel the experiment.
1963. A. Zichichi started at CERN, on the basis of pure experimental intuition, the search for the 3rd sequential lepton family, a replica of the first two. He named them "Heavy Lepton and its neutrino". The basic idea was to search for acoplanar lepton pairs of opposite charges, looking for the reaction

$$\begin{aligned} &p\bar{p} \to HL^+ + HL^- \\ &HL^+ \to e^+\nu_e\bar{\nu}_{HL}; HL^- \to \mu^-\bar{\nu}_\mu\nu_{HL} \quad \text{or} \quad HL^- \to e^-\bar{\nu}_e\nu_{HL}; HL^+ \to \mu^+\nu_\mu\bar{\nu}_{HL} \end{aligned} \tag{10}$$

The presence of an $e\mu$ pair associated with non coplanarity with the beam, due to the presence of neutrino is already a strong signature to extract the signal from the enormous hadronic background. Not sufficient however without proper experimental techniques for the electron and muon identification. With the techniques developed by Zichichi in the large acceptance double arm spectrometer rejection factors of several thousands were reached[27]. The Heavy Lepton was not found.
1967. A new experiment on the same line was proposed and performed by Zichichi and collaborator at ADONE at Frascati[28]. It was the highest energy e^+e^- collider at the time, $\sqrt{s}$=3 GeV, unfortunately not enough to produce the *HL*. It was discovered, and called τ, by M. Perl and collaborators at SLAC in 1975[29], using similar ideas, when a collider of sufficient energy became available.
2001. K. Niwa[30] and collaborators discover the third neutrino with the emulsion technique on a neutrino beam of the Fermilab.

5. The great surprise. Physics beyond the SM. 1960s

1957. B. Pontecorvo[31]. We have seen how a claim, which later was found to be wrong, of the discovery of $0\nu2\beta$ decay, stimulated L. Alvarez to transform the Pontecorvo idea of the Cl-Ar reaction in a fully developed experimental proposal. A similar episode happened when R. Davis was running his experiment, based to that technique, to try to detect reactor antineutrinos. In that period rumours of a positive evidence reached Pontecorvo in Dubna. They were wrong, but stimulated the idea of neutrino oscillations.
Pontecorvo wrote: "*Recently there came to our attention a paper of Davis* (still in press), *who investigated the production of* ^{37}Ar *from* ^{37}Cl *under bombardment of neutral leptons produced by a powerful reactor. Davis result - a measurable probability of the investigated process -if confirmed, definitely indicates that neutrino charge is not strictly conserved.* If the process were observed, it might be due to antineutrinos transforming into neutrinos. The mechanism might be neutrino antineutrino oscillation as, by analogy, that of the neutral kaons.

"...*neutrino and antineutrino are.... symmetrical and antisymmetrical combinations of two truly neutral Majorana particles* ν_1 *and* ν_2

$$\nu_1 = \frac{1}{\sqrt{2}}(\nu + \bar{\nu}); \quad \nu_2 = \frac{1}{\sqrt{2}}(\nu - \bar{\nu}) \tag{11}$$

It follows that neutrinos in vacuum can transform themselves in antineutrinos and viceversa....So, a beam of neutral leptons from a reactor which first consists mainly of antineutrinos will change in composition and at a certain distance...will be composed of neutrinos and antineutrinos in equal quantities.

He considered the sun as a possible source, but concluded that, unfortunately, the solar physics is not enough under control.

A posteriori, the rumours were wrong, neutrino-antineutrino oscillations do not exist, but oscillations between neutrino flavours do (notice that in 1957 only one flavour was known).

1964. J. Bahcall[32] and R. Davis[33] publish two face to face articles. The theoretical one, by Bahcall, contains the full development of the solar model and the prediction of the electron neutrino energy spectrum. The experimental one, by Davis, contained the design of an experiment to detect them, using the Cl-Ar reaction

$$\nu_e + {}^{37}\mathrm{Cl} \rightarrow {}^{37}\mathrm{Ar} + e^- \tag{12}$$

The flux is much smaller than at a reactor and the cosmic ray background much more relevant. The experiment needs to be done deep underground. It was hosted in the Homestake gold mine in the USA, 1480 m deep. The cosmic ray flux is reduced to ~ 4 μ/(d m^2), six orders of magnitude less that on the surface. The big detector consisted of 380 m^3 of perchlorethylene (C_2Cl_4), containing 133 t di ^{37}Cl. R. Davis was able to develop the chemical procedures capable to extract, with an efficiency close to 100%, the few Ar atoms per week (one every two days) from the huge tank.

On 11 August 1967. R. Davis writes to W. Fowler: *"Dear Willy, I do have a preliminary result from our first good run. The sample was taken June 22nd and counting has continued until today. I am now removing the sample and will rerun background. So we do have a result and during the last few weeks I have told a few people that are interested."*.

The measured solar ν_e flux was about 1/3 of the Bahcall predictions. It was the origin of the "solar neutrino puzzle". In the same years in which the SM was being developed, experimental evidence was growing for physics beyond it: neutrinos do change flavour.

6. Present status

Two mechanisms observed changing neutrino flavour: oscillation in vacuum, which corresponds to the kinetic part of the Hamiltonian, and transformation in matter (Mikehev-Smirnov-Wolfestein effect[34]), which is a dynamical phenomenon due to interaction potential between ν_e and electrons.

The mentioned phenomena may happen when neutrinos of definite flavour are not stationary states (mass eigenstates). The latter, ν_1, ν_2 and ν_3, have definite masses, m_1, m_2 and m_3. The two basis are linked by an orthogonal transformation that can be expressed in terms of three rotations, through angles that we shall call θ_{12}, θ_{23} and θ_{13}, and of phase factors. If neutrinos are Dirac particles, as assumed in the Standard Model, all but one of the phase factors can be absorbed, as in the case of quarks, in the wave functions of the states. However, neutrinos might be completely neutral, namely 'Majorana particles'. In this case two more phases, which we shall call Majorana phases ϕ_2 and ϕ_3, are physically observable.

In conclusion, writing c_{ij}=cosθ_{ij}, and s_{ij}=sinθ_{ij}, the unitary transformation is

$$\begin{pmatrix} \nu_e \\ \nu_\mu \\ \nu_\tau \end{pmatrix} = \begin{pmatrix} 1 & 0 & 0 \\ 0 & c_{23} & s_{23} \\ 0 & -s_{23} & c_{23} \end{pmatrix} \begin{pmatrix} c_{13} & 0 & s_{13}e^{-i\delta} \\ 0 & 1 & 0 \\ -s_{13}e^{i\delta} & 0 & c_{13} \end{pmatrix} \begin{pmatrix} c_{12} & -s_{12} & 0 \\ s_{12} & c_{12} & 0 \\ 0 & 0 & 1 \end{pmatrix} \begin{pmatrix} 1 & 0 & 0 \\ 0 & e^{i\phi_2} & 0 \\ 0 & 0 & e^{i\phi_3} \end{pmatrix} \begin{pmatrix} \nu_1 \\ \nu_2 \\ \nu_3 \end{pmatrix} \tag{13}$$

Nine are the quantities to measure, three masses, three mixing angles and three phases (causing CP violation). We now know the three mixing angles. As for the masses, we know the values

$$\delta m^2 = 75.4 \pm^{2.6}_{2.2} \text{ meV}^2 \qquad \left|\Delta m^2\right| = 2430^{+70}_{-90} \text{ meV}^2 \tag{14}$$

Indeed, the oscillation frequency is proportional to the absolute differences between two square masses. Consequently, the oscillation experiments, like those on atmospheric neutrinos and accelerator neutrinos are not sensitive to the absolute values of the masses and, at the first order, to the sign of the difference.
The matter MSW effect, that happens in the sun at the smaller mass difference, is sensitive to the sign of that difference too.
Due to the smallness of the square mass differences, neutrino oscillations happen, at energy of the order of GeV on flight lengths between source and detector of the order of hundreds to thousands kilometres. Consequently, many years of research with dedicated experiments on neutrino beams in Europe and in the USA did not discover oscillations. The development of fully equipped underground laboratories in Japan, Italy and Canada and of a series of experiments finally led to the discovery in 1989 of the oscillation of atmospheric neutrinos and the matter conversion of the solar neutrinos.
Presently, three are the major problems on the table, namely the measurement of: 1. The sign of Δm^2, 2. The "Dirac" phase δ producing CP violation (if different fro 0 and π), 3. The Majorana or Dirac nature of neutrinos, and the measurement, if they turn out to be the former, of the two "Majorana" phases.
Several experiments and projects are being developed for sign of $|\Delta m^2|$.
At the next to the leading order in $\alpha = \left|\delta m^2\right| / \left|\Delta m^2\right|$; 0.03 interference effects between the shorter and longer period oscillations appear in the oscillation probability. The JUNO experiment being built in China aims at detecting that very small effect. The source will be a set of six, 1.9 GW each, nuclear power stations, the detector a 22 kt liquid scintillator detector at 50 km distance. An energy resolution of 3% must be achieved.
Another subdominant effect, in the atmospheric neutrinos, is the following. For neutrinos crossing the earth $P(\nu_\mu \to \nu_\mu) \neq P(\bar{\nu}_\mu \to \bar{\nu}_\mu)$ due to opposite matter potential. The effect depends on the sign of Δm^2. In practice, the momentum spectra of up going and down going muons, from neutrino collisions, are compared in a Fe tracking magnetized calorimeter. This is the INO project funded and under construction in India.
Experiments on neutrino beams from accelerators are sensitive to sign of Δm^2 through matter effects too. The distance travelled by neutrinos in the crust, namely the base line, needs to be around 1000 km at least. Projects at such a very long base lines are being developed for a not very near future. The NOνA experiment that started to take data recently, at 810 km from Fermilab, might start to give some information.

Two types of experiments are needed to measure the Dirac phase. With a ν_μ produced at an accelerator one looks for ν_e appearance by detecting the neutrino through its scattering on electrons. The oscillation probability depends on θ_{13}, on δ and on sign of Δm^2. Consequently, θ_{13} must be known with great precision, independently on the other parameters. Indeed, the electron antineutrino disappearance probability depends on θ_{13} but is independent of δ. Two accelerator experiments are presently running, T2K in Japan and the already mentioned NOνA in the USA. Three reactor experiments are running too, DayaBay in China, RENO in

Korea and Double Chooz in France. They have already provided a tantalising hint that δ might be large, around $(3/2)\pi$. May be neutrinos will be kind with us again. Time will tell.

7. Double beta decay and the low energy frontier

If neutrinos are Majorana particles the extremely rare $0\nu2\beta$ decay may happen. The Feynman diagram at the quark level is shown in Fig. 1

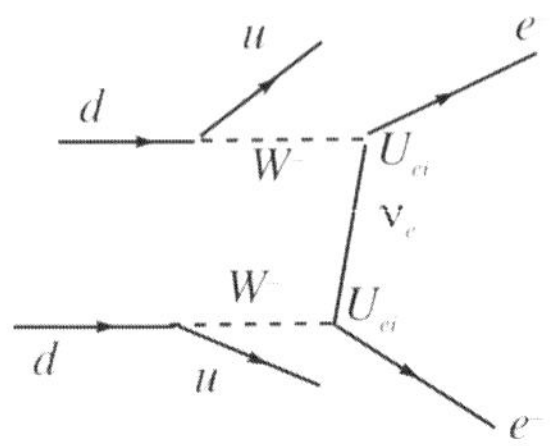

Fig. 1. Feynman diagram at the quark level of the $0\nu2\beta$

The Majorana neutrino, which is a completely neutral particle, is exchanged in the propagator joining the two decay vertices. There is no "charge" propagating between on vertex and the other, consequently there is no arrow on the propagator (notice that often this is not the case in the literature). At each vertex, Majorana neutrino couples exactly as a Dirac neutrino, namely with its left (namely $1-\gamma_5$) projection. The SM violation is in the propagator. The states created at one vertex has definite flavour, hence is a superposition of mass eigenstates.

$$\nu_e = U_{e1}\nu_1 + U_{e2}\nu_2 + U_{e3}\nu_3 \qquad (15)$$

In addition, the mass eigenstates are superposition of the two helicity states. The latter behave respectively as Dirac neutrinos and antineutrinos.

$$\nu_i \approx \frac{m_i}{E}\nu_{iL}^{+} + \nu_{iL}^{-} \qquad (16)$$

where the superscripts are the helicity values.

At one vertex the positive helicity, Dirac-antineutrino-like component matters, the negative helicity, Dirac-neutrino-like component, at the other vertex. The corresponding transition amplitude is (17)

$$M_{ee} = \left|\sum_i U_{ei}^2 m_i\right| = \left|c_{13}^2 c_{12}^2 m_1 + c_{13}^2 s_{12}^2 m_2 e^{i\phi_2} + s_{13}^2 m_3 e^{i\phi_3}\right| \approx \left|0.83m_1 + 0.55m_2 e^{i\phi_2} + 0.15m_3 e^{i\phi_3}\right|$$

To have the corresponding half life we must consider that the decay takes place in a nucleus, including a nuclear matrix element term M_{nucl} and, obviously, the phase space factor G.

$$T_{1/2} = G\left|M_{nucl}\right|^2 \left|M_{ee}\right|^2 \qquad (18)$$

The calculation of the nuclear matrix elements is complicated. A number of approximation techniques has been developed leading to results that agree one another within a factor of two. No evidence has been experimentally found so far. The most stringent limits are for $T_{1/2}\left({}^{136}\text{Xe}\right) > 2.1\times10^{25}$ yr for ^{76}Ge of the GERDA experiment[36] and, for ^{136}Xe, $T_{1/2}\left({}^{136}\text{Xe}\right) > 1.6\times10^{25}$ yr of EXO-200[36] and $T_{1/2}\left({}^{136}\text{Xe}\right) > 1.9\times10^{25}$ yr of KamLAND-Zen[37]. The running experiments as well as new ones coming in line aim to reach sensitivities of $T_{1/2}=10^{26}$ yr in the present decennium, corresponding to $|M_{ee}|$ between 50 and 150 meV, depending on the nuclear matrix elements.

If we are lucky enough, the present generation of experiments might find the signal. The measurement of the half life joined to a better knowledge of the nuclear matrix elements will then allow to extract a value of M_{ee} in Eq. (18). However, even if by that time all the neutrino masses will be known from other experiments, we shall still faced with the problem of extracting two unknowns, the Majorana phases, from one number. Fig. 2 is a cartoon representing the problem.

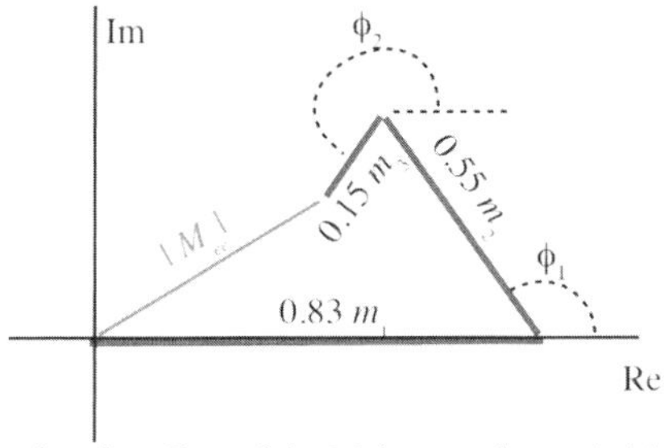

Fig. 2. Cartoon showing the effect of the Majorana phases (arbitrary values) on M_{ee}

The main problem in the Majorana neutrino physics is that the effects are proportional to $(m/E)^2$, a factor that is extremely small for typical nuclear energies of several MeV, because neutrino masses are not larger than, say, 100 meV. One should try to work with eV or sub eV energy scales, namely with atoms or molecules. However, at those energies, neutrino interaction cross sections become extremely small. One could however try to exploit coherent processes, where effects are proportional to the squares of the number of atoms Avogadro squared), and to transitions forced by an external source, rather than spontaneous.

Yoshimura and coll.[38], [39] are working on the Radiative Emission of Neutrino Pairs (RENP) process, shown in Fig. 3.

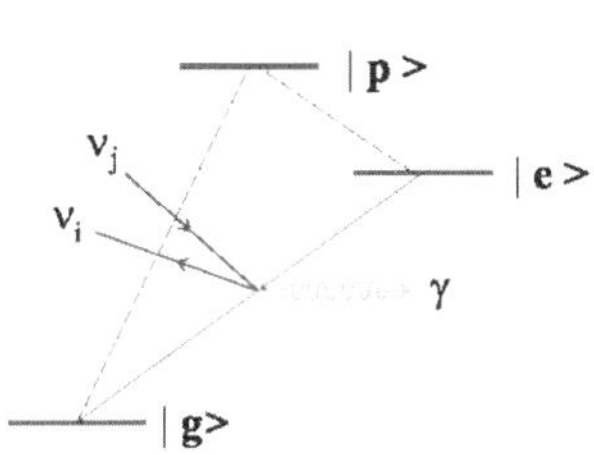

Fig. 3. Principles of the RENP

The atoms of the gas under study is contained in a transparent cell that is irradiated by two counter-propagating trigger LASER beams tuned on 1/2 E_{eg}. The latter is the energy difference between the excited state e and the ground state g in Fig. 3. These states are chosen to have the e to g dipole forbidden transition. The trigger field induces macroscopic coherent polarisation of the atoms. The rate is proportional to n^2. The decay

$$|e\rangle \rightarrow |g\rangle + \gamma + \nu_i + \nu_j \tag{19}$$

occurs via E1×M1 coupling to the level p.

Calculations show that detectable rates might be obtained with $n \approx O(N_A)$ and volumes of $O(10^2\ \mathrm{cm}^3)$.

Six thresholds corresponding to six different possible neutrino pairs

$$\omega_{ij} = \frac{E_{eg}}{2} - \frac{(m_i + m_j)^2}{2E_{eg}} \tag{20}$$

give redundant information on the absolute values of neutrino masses. The spectrum shape near the thresholds is sensitive to the sign of Δm^2 and to the mixing angles. $E_{eg} \approx 2$ eV looks to be adequate.

In this process, the difference between Majorana and Dirac, is in a few interference terms, which are small, but grow with decreasing energy. To detect them, we should work with sub-electron volt energy level differences.

The search is in its first steps. It is certainly very difficult, but the very low energy route may well result leading to physics beyond the Standard Model.

8. Conclusions

Neutrino physics history has been rich of surprises, Nature behaves often differently from what we think.
Neutrinos have shown us and continue to show physics beyond the Standard Model, since when the SM building started in the late 1960s.
Progress has been due mainly to the development of novel detector technologies, the creation of underground laboratories (were discoveries happened), reactor experiments and accelerator experiments.
Neutrino masses are seven orders of magnitude smaller than the lightest charged particle, the electron. These values, through the sea saw mechanism, point to physics at very high energy scales, close to the grand unification scale.
On the experimental side, progress will require development of detectors exploiting eV and sub-eV (atomic and molecular) physics. These are also the energy scales needed for the axion search.
Several experiments that are starting or being built now will give answers to the open problems in the next few years.
•the sign of Δm^2 from natural sources
•the Dirac phase δ, if we are lucky, with running experiments at accelerators
•the absolute mass scale, indirectly, from cosmology
•and, presumably, more surprises
More progress will require novel thinking and creativity in the experimental art.

References

1. F. Rasetti. *Nuovo Cimento* 6 (1929) 356; F. Rasetti. *Phys. Rev.* 34 (1929) 367; F. Rasetti. *Z. Physik* 61 (1930) 598
2. E. Fermi. *Molecole e cristalli*, Zanichelli Bologna, 1934. English translation: *Molecules Crystals and Quantum Statistics*, W.A.Benjamin, inc. New York 1966.
3. J. Chadwick, F. R. S. *Proc. Royal Soc. of London* A 136 (1932) 692, *Nature* 192 (1932) 312
4. E. Fermi. La Ricerca Scientifica, Vol II, n. 12 (1933)
5. E. Fermi. *Nuovo Cimento* 11 (1934) 1; English transl. Fred L. Wilson Am. J. Phys. 36 (1968) 1150
6. H. Bethe & R. Peierls, Nature 133, 532 (1934); H. Bethe & R. Becher, Rev. Mod. Phys. 8, 82 (1936)
7. Quoted by Reines, in the preface to *Spaceship Neutrino* by C. Sutton (1992), p. XI
8. H. R. Crane Phys. Rev. **55** (1934) 501; H.R. Crane Rev. Mod. Phys. **20** (1948) 278
9. B. Pontecorvo. *On a method of detection of free neutrinos. Report PD-141*. Chalk River (1946). Canada
10. B. Pontecorvo. *On a method of detection of free neutrinos. Report PD-205*. Chalk River (1946). Canada
11. L. Alvarez. Report UCRL-238
12. M. Ageno et al. Atti Acc. Naz. Lincei **6** (1949) 626; Phys Rev **79** (1950) 720
13. H. Kallmann Phys. Rev. **78** (1950) 621
14. G. I. Reynolds, F.B. Harrison, G. Salvini Phys Rev. **78** (1950) 488
15. C. L. Cowan et al. *Science* **124** (1956) 124
16. R. Davis, Unesco Conference 1958. Paris. Vol. 1 p. 728
17. E. Majorana. *Nuovo Cimento* **14** (1937) 171
18. M. Goeppert-Mayer "Double beta disintegration" Phys. Rev. **48**
19. G. Racah. *Nuovo Cimento* **14** (1937) 323

20. W. H. Furry "On the transition probabilities in double beta-decay" Phys Rev **56** 1184
21. See for example: B. Pontecorvo *The infancy and youth of neutrino physics. Some recollections.* Jour. De Phys. 1982. n. 12, Vol 43, p C8-221
22. I am grateful to S. Petcov for information on the Fakirov thesis (in Bulgarian)
23. B. Pontecorvo, 1960. Electron and Muon neutrinos. Sov. J. Phys. JETP 10: 1236-1240
24. M. Schwartz. Nobel Lecture 8.12.1988
25. G. Danby et al. *Phys. Rev. Lett.* **9** (1962);36
26. S. Fukui and S. Myiamoto. *Nuovo Cimento* **11** (1959) 113
27. M. Bernardini et al. INFN/AE-67/3, 20 March 1967
28. A. Zichichi et al. CERN-63-26. N P Div/ 27:6:1963; T. Massam et al. *Nuovo Cimento* **39** (1965) 464
29. M. L. Perl et al. *Phys. Rev. Lett.* **35** (1975) 1489
30. K. Kodama et al. *Phys. Lett.* **B504** (2001) 218
31. B. M. Pontecorvo. JINR Preprint P-95, Dubna, 1957
32. J. N. Bahcall; *Phys. Rev. Lett.* **12** (1964) 300
33. R. Davis Jr. *Phys. Rev. Lett.* **12** (1964) 303
34. S. P. Mikheyev and A. Smirnov; Yad. Fiz. 42, 1441 [Sov. J. Nucl. Phys. 42, (1985) 913]
35. M. Agostini *et al.* (GERDA) Phys. Rev. Lett. **111** (2013) 122503
36. M. Auger et al., (EXO-200). *Phys. Rev. Lett.* **109**, (2012) 032505
37. A. Gando et al., (KamLANDZen) *Phys. Rev. Lett.* **110** (2013) 062502
38. Y. Yoshimura, Phys Rev. **D** 75 (2007) 113007
39. D. N. Dinh et al. Phys Lett B **719** (2013) 154

LECTURER: A. BETTINI

Neutrinos. Measuring the unexpected

Scientific Secretaries: A. Salakh, W. Wang

DISCUSSION

- *Klopt:*

Could you elaborate more on the types of experiments suitable for measuring neutrino mass?

- *Bettini:*

The determination of the sign of the large square mass difference (called neutrino mass hierarchy) may be done in several ways. The first way is to produce a neutrino beam with a proton accelerator. These beams are at the origin mainly composed of ν_μ. One then looks for the appearance of ν_e with a detector at a distance. The sign of Δm^2 is determined via matter effect, in the earth crust. Consequently the baseline should be at least several hundred kilometres. The NOvA experiment in the USA, which is starting now, may become, in several years, sensitive enough. A second way is using power reactors as (anti) neutrino sources, and looking for $\bar{\nu}_e$ disappearance at about 50 km. In this case the sign of Δm^2 is determined through the interference of the longer and shorter period oscillations. Such an experiment is Juno in China, which is building a 20 kt liquid scintillator detector. The main difficulty is reaching the necessary energy resolution, which should be better than 3%. The target looks to be reachable, starting from the experience of BOREXINO at LNGS that has already a 5% energy resolution (even of on a much smaller sensitive mass). A third possibility is the detection of μs produced by atmospheric $\nu\mu$, measuring their charge and momentum. This was the idea of the MONOLITH proposal at LNGS, which unfortunately did not materialize, but is now planned in the India-based Neutrino Observatory, which is now funded and is being built. A fourth possibility is to be lucky, and have a supernova explosion in our galaxy. There are several detectors ready to detect the neutrino burst. These data would certainly help to disentangle the hierarchy.

- *'t Hooft:*

The title of the talk given to you by Prof. Zichichi was "neutrinos; measuring the unexpected", while you only talked about things that were unexpected in the past. How do you prepare for unexpected observations in the future?

- *Bettini:*

A famous physicist remarked that predicting things is difficult, particularly things in the future. Maybe there are sterile neutrinos or very massive neutrinos, which might be seen in the future…

- *'t Hooft:*

That's a good example.

- *Bettini:*

As a matter of fact there is no sound indication for sterile neutrinos. The so-called anomalies have a marginal statistical evidence (not more than three standard deviations) and, more important, they are smaller than the systematic uncertainties. Anyway, in the LNGS programme an exposure of BOREXINO at an artificial neutrino source of several MCi has been financed by the Italian government. The experiment will be sensitive enough. May be it will show surprises.

- *Gnesi:*

Do you have some idea of any effort going on to build beta spectrometers which can measure the neutrino masses?

- *Bettini:*

The electron neutrino effective mass can be measured, or limited, with a spectrometer, as you mention. KATRIN at Karlsruhe is coming in operation. Other techniques are based on bolometers..

- *Benedek:*

May I add a comment on this topic? You made a reference to the use of bolometers. Such experiments need to be sensitive to electron neutrino effective mass values below 100 meV. For that a very low energy end point is needed. In addition, the total mass of the detector must be large. Unfortunately, in the solid state materials there are oscillations in the energy spectrum near the end point. The effect covers the looked for behaviour due to neutrino mass. However, methods recently developed in surface physics, namely the electron-energy-loss spectroscopy and the single electron spectroscopy, might make the distortion of the beta spectrum due to masses below 100 meV observable, regardless the mentioned oscillations.

Highlights from INFN

Fernando Ferroni*
Sapienza Università di Roma & INFN
P.za A Moro 2, Roma, I-00184 –Italy

1) A very interesting period in HEP

High Energy Physics is at the border between the glorious past of Standard Model and the uncertain future of New Physics. The extraordinary success of LHC in finding the long awaited Higgs boson exhausts the roads, that although impervious, were somewhat marked, Indeed we are in front of a list of questions, none of which has a guaranteed answers from the experiments we are doing, the accelerators we have, the technology we exploit.

A non exhaustive shopping list for the researchers of near future comprehends:

- DarkMatter exists, but what is made of ?
- Dark Energy is a total mystery
- Our existance in the broader context of Flavour Physics is a puzzle
- Neutrino even after 80 years of presence in Science is still little understood

2) INFN today

INFN is a research institute diffused all over the Italian country, living mostly in a blessed symbiotic relation with 20 physics departments of universities. In addition it comprises 4 national labs, 3 specialised centres (computer centre- applied physics- Ph.D. school), an infrastructure for Gravitational Waves search, 1 lab for Cultural Heritage and 1 for Superconductivity . A couple of thousand people (staff and university associates) are active in its research groups thanks to a budget in the 300 MEuro range.

**email fernando.ferroni@roma1.infn.it*

Fig. 1 : Distribution of INFN structures on the italian map.

2.1 INFN ballistic program

There are a lot of projects where substantial human and capital resources have been invested. The continuation of these projects is what we call baseline (or ballistic) program . The main items are:

- LHC at CERN
- Dark Matter searches at LNGS
- Neutrino Physics at LNGS
- Gravitational Wave search at EGO (Cascina)
- Accelerator Physics at LNF, LNL, LNS
- Cosmic rays in space, on ground, underground and underwater

Here some picture best illustrating these fields.

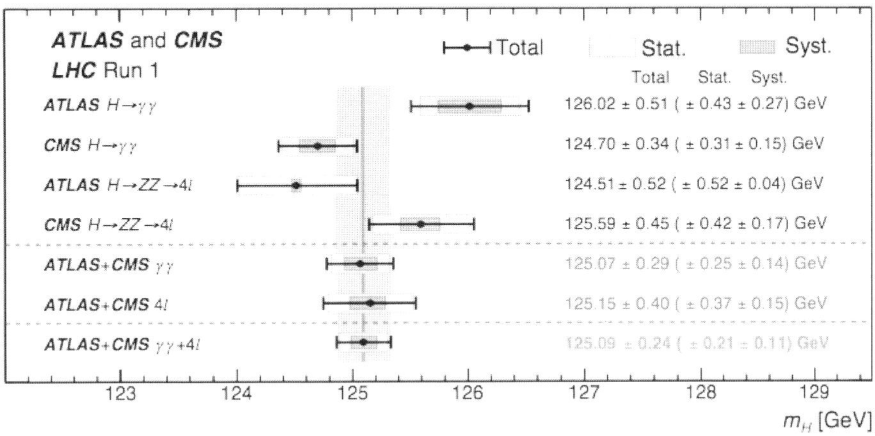

Fig.2 Higgs boson mass as measured [1] at LHC

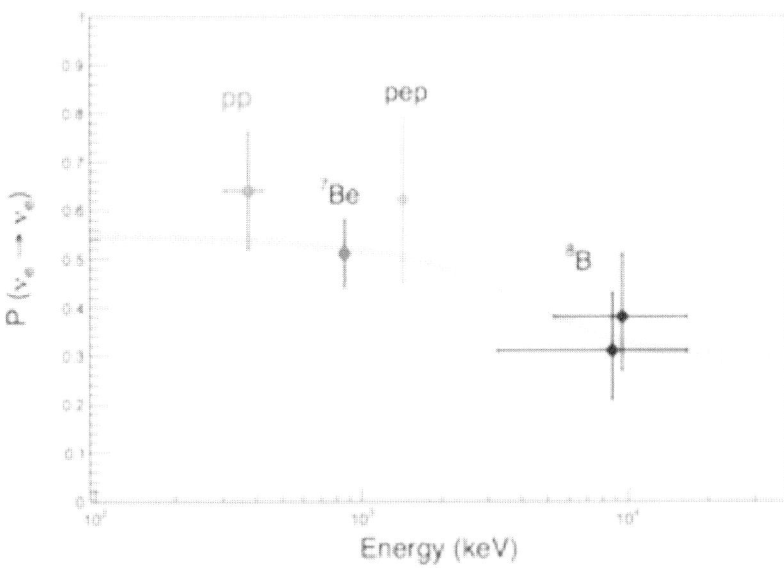

Fig.3 Solar neutrinos as measured [2] by Borexino [3] experiment at LNGS.

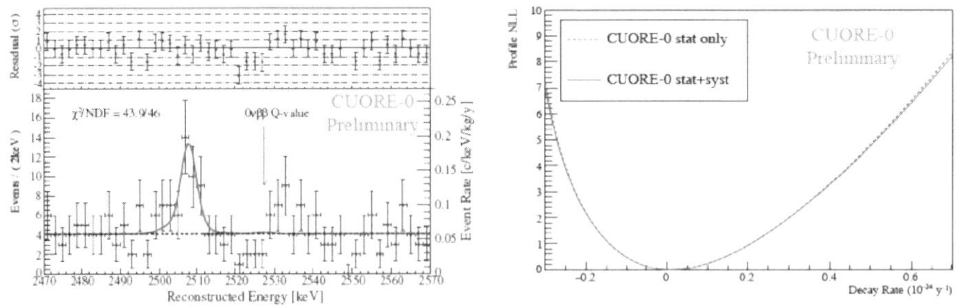

Fig.4 CUORE-0 [4] results on neutrinoless double beta decay at LNGS.

Neutrinos are specially dear to INFN. Indeed the name neutrino was coined by Enrico Fermi as a word play on neutrone, the Italian name of the neutron. Moreover neutrino is the only particle known to us that could be a Majorana fermion [5].

Neutrino poses us a lot of challenges for the future as:

- How much does a neutrino weigh ?
- What is the mass ordering (hierarchy)
- Is neutrino a Majorana or Dirac particle
- Do more (sterile) neutrinos exist ?
- Do neutrinos violate CP ?
- Can we observe the CNB, a picture of the universe 1 second old ?

For the observation of very energetic neutrinos coming from phenomena generated by the so called violent universe, INFN is leading the construction of an underwater observatory (KM3Net [6]) off the Sicilian coast, 100 km far from Capo Passero.

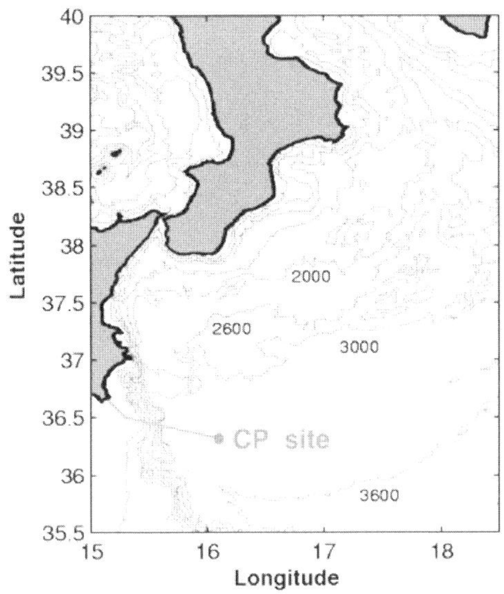

Fig.5 Site for the KM3Net experiment.

Dark matter is the most vibrant field of research today. The elusive nature of this unknown component of our Universe calls for a wide range of unbiased searches. Indeed experiments are performed at LHC hoping to produce particles of it, in space hoping to observe the products of its annihilation and most important in the cosmic silence of underground laboratories through its elastic scattering with nuclei of ordinary matter.

INFN is in a privileged position as it possesses, thanks to the vision of Professor A. Zichichi, the largest and most accessible underground laboratory in the world: the Gran Sasso lab.

Here many experiments are carried out and some are in preparation. One of the trigger is the unexplained result fo the DAMA/LIBRA [8] collaboration that ha s observed a signal that looks very much alike to what you would expect from Dark Matter although not confirmed as yet by other experiments.

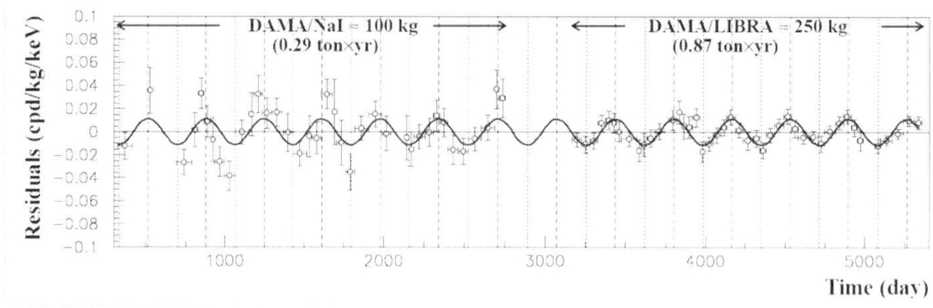

Fig. 6 Annual modulation [9] observed in DAMA/LIBRA experiment.

Looking for a signal we are hosting at LNGS XENON1T [10], the evolution of two previous smaller scale detectors. One which (XENON100 [11]) contends to LUX [12] the best limit on the non observation of Dark Matter.

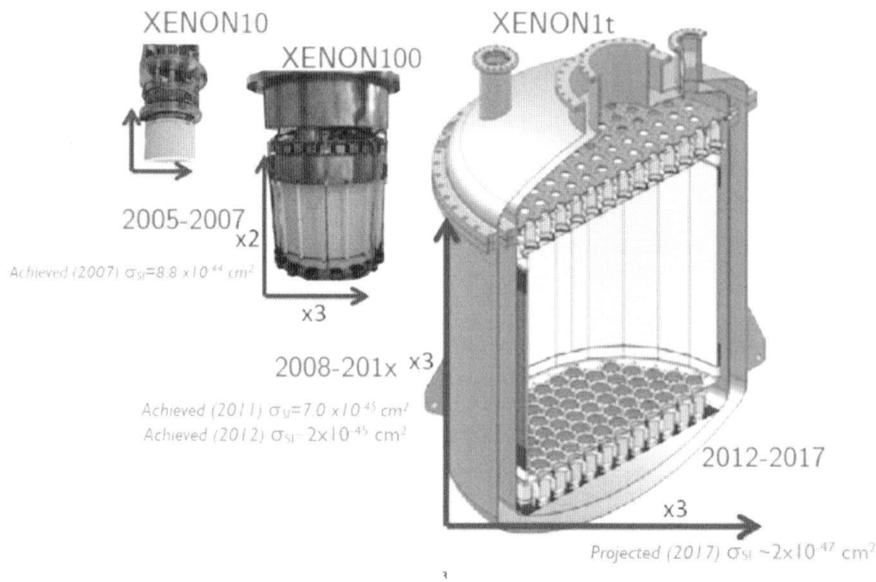

Fig. 7 : The XENON detector family

Since the name of the game for this type of detectors based on noble liquids is having zero background we are also preparing an experiment based on Argon (DarkSide [13]).

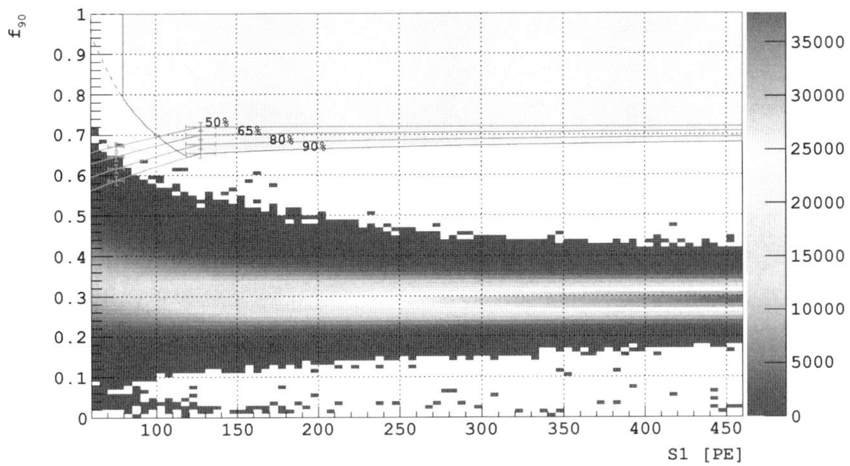

Fig. 8 : Results from DarkSide-50 prototype [14] showing that zero background is indeed achievable.

This experiment involves a complex collaboration with industries in order to provide Argon-40 free from the radioactive isotope 39, a facility able to produce square meters of SiPM and a very large chamber for e-beam welding.

The next big challenge of INFN is in the search and discovery of gravitational waves. In 2016 the Virgo interferometer [15] located in Cascina, near Pisa, shall restart its operation after animportant upgrade with an improved sensitivity that will allow it to explore a much bigger portion of the universe.

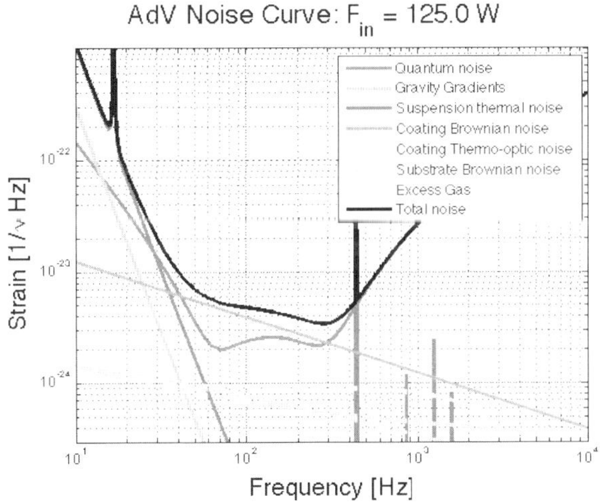

Fig. 9 : Expected sensitivity of Advanced Virgo.

2.2 What next?

Integral part of the INFN mission is to maintain a focus on physics at accelerators and on novel technologies for accelerating particles. In this respect all the national labs will play a role.

LNL will operate SPES [16], a 70 MeV cyclotron capable of delivering 500 μA of current on two different lines, one for producing radioactive ion beams and the other dedicated to rare medical isotopes.

LNS has a project for substantially refurbishing its superconducting cyclotron making it capable of achieving currents that allows to study [17] a series of double charge exchange processes which will provide an invaluable input to the calculation of nuclear matrix elements for neutrinoless double beta decay beside a rich physics program.

LNGS is acquiring a few MeV accelerator for the experiment LUNA [18] whose scope is to describe the entire evolution of a star by precisely measuring all the nuclear cross sections involved.

LNF is concentrating its effort on SPARC [19] , a FEL where new techniques for accelerating particles will be tested. In particular by creating a plasma either by a 300TW laser or by injecting very short and very close electron bunches and then using it for accelerating a bunch of electrons that will surf on the plasma waves.

As for new lines of research INFN has started an activity in cosmology, both for CMB studies in the experiment LSPE [20] which will fly a balloon over the North pole for measuring the photon polarization and by starting a collaboration with EUCLID [21] on dark energy studies.

Feasibility studies on axion searches are also on going as many other ideas are at level of R&D.

3. Conclusions

INFN has a well balanced portfolio of initiatives. The ballistic physics program is rich, bringing fruits and with perspectives of being productive for many years. The very basic competence on accelerator physics is kept alive through initiatives both at home and by participating in the construction of facilities abroad. The reflection on what next and/or what else has started and it is already producing interesting idea. The real challenge is in securing enough money to keep the momentum and in recruiting the best young researchers.

References

[1] G. Aad et al., Phys. Rev. Lett. 114 (20152) 191803

[2] O.Smirnov et al., Phys. Part. Nucl. 46 (2015) no.2, 166

[3] G. Alimonti et al., Nucl. Instrum. Meth. A600 (2009) 568

[4] K. Alfonso et al., Phys. Rev. Lett. 115 (2015) no. 10, 102502

[5] E. Majorana, Il Nuovo Cimento, 14 (1937)171

[6] M. de Jong, AIP Conf. Proc. 1666 (2015) 040004

[7] G. Alimonti et al., Astropart. Phys. 16 (2002) 205

[8] P. Belli et al., Phys. Rept. 307 (1998) 269

[9] R. Bernabei at al., Eur. Phys. J. C73 (2013) 2648

[10] E. Aprile, arXiv:1206.6288 (2012)

[11] E. Aprile et al., Phys. Rev. Lett. 109 (2012) 181301

[12] D. S. Akerib et al., Nucl. Instrum. Meth. A704 (2013) 111

[13] C.E. Aalseth et al., Adv. High Energy Phys. 2015 (2015) 541362

[14] P.D. Meyers et al., Phys. Procedia 61 (2015) 124

[15] F. Acernese et al., Class. Quant. Grav. 32 (2015) no.2, 024001

[16]F. Gramegna et al., Acta Phys. Polon. B46 (2015) no.3, 591

[17] C. Agodi et al., AIP Conf. Proc. 1686 (2015) 020001

[18] A. Guglielmetti, EPJ Web Conf. 78 (2014) 07001

[19]M. Ferrario et al., Nucl. Instrum. Meth. B309 (2013) 183

[20] S. Aiola et al., arXiv:1208.0281 (2012)

[21] http://sci.esa.int/euclid/57039-euclid-dark-universe-mission-ready-to-take-shape/

LECTURER: F. FERRONI

Highlights from INFN

Scientific Secretaries: I .Rudenok, I. Ivanchenko

DISCUSSION

- *Yeletskikh:*

Can you comment on the idea of observing the relic neutrinos?

- *Ferroni:*

Nowadays the idea of observing the relic neutrinos looks like scientific fiction. These neutrinos have about the same temperature as the photons of the CMB. The only way to look for them is when they hit a nucleus that does not have a threshold for the emission of one electron. Tritium for example. However it does not help, because you have the normal processes that will hyde them. So you have to invent a method to improve the sensitivity of the experiments. The only idea so far is to have the targets for the relic neutrinos that will be placed on the surface of some graphene-made folded object. The total surface probably would be like the one of the detectors on LHC although compacted in a small space. The entire surface shall be 'painted' with Tritium. And then you will have a very sophisticated calorimeter with a resolution of a fraction of eV (project PTOLHEMY at Princeton). It is like a science fiction, but nevertheless I believe fantasies like those should be encouraged. And one day we will measure them.

- *Savitskyi:*

What is the status of cosmological neutrino measurements regarding mass hierarchy? Isn't it an easier goal to achieve than the direct measurement of the neutrino mass? Is it possible to perform such measurement?

- *Ferroni:*

As far as I know, it is possible to obtain the total mass of the neutrinos via cosmological fit. Today the limit, depending on the model you use and the assumptions you make, is in the region (100, 200, 300 meV). The mass hierarchy is a hard thing to measure. But we know how to do this. For example, the mass hierarchy can be extracted from neutrino oscillations. So such experiments as Nova or Juno or Orca will probably succeed sometime after 2020.The direct value of the mass can be obtained in KATRIN experiment in Germany, where the are electrons that come from Tritium decay through a sophisticated spectrometer. You measure the end points of the spectrum which shall deviate from the one you expect if the neutrino was massless. Another method we are testing in my country is to measure the decay in a bolometric pixel detector (made from Holmium). You might be able to get the energy solution. So in short the answer is: the mass hierarchy determination will possibly be achieved within 10 years. But for the exact mass it's a hard thing to predict. I can't promise a time scale.

- *Teresi:*

Mine is actually a comment. We already have evidence for neutrino mass scale from cosmological observations. The problem is how much you trust your understanding of the related astrophysics.

- *Ferroni:*

I agree, these results come from a fit depending on one's theoretical prejudice. Not from direct measurement.

Highlights from ALICE

Paolo Giubellino
INFN Torino and CERN Geneve
E-mail: `paolo.giubellino@CERN.ch`

The LHC has delivered for the first time collisions of Nuclei in November 2010, at an energy of 2.76 TeV per nucleon pair, which represents a jump of more than an order of magnitude over the highest energy nuclear collisions ever studied before. The high energy, the quality of the state-of-the art detectors, and the readiness of the experimental collaborations at the LHC have allowed a rich harvest of important scientific results. In fact, many of the most cited papers published about LHC results from the various experimental collaboration deal with nuclear collisions. The overall interest in the program has grown constantly form those first collisions, and now all 4 major LHC experiments include the study of Nuclear Collisions in their scientific program. In this lecture a short overview will be given of how the results from the LHC, and in particular from the ALICE experiment, have provided new insight on the properties of matter under extreme conditions of temperature and pressure, analogous to the conditions present in the early phases of the evolution of the Universe. Results from both the nucleus-nucleus and the proton-nucleus runs will be presented. Finally, a very short outlook to the future of nuclear Collisions at the LHC will be given.

1.Why high-energy Heavy-Ion Collisions

Collisions between very high-energy nuclei, normally referred to as Heavy-Ions since they are atoms stripped of their electrons to allow acceleration, produce each a huge number of particles, over ten thousand in a single event at the LHC. A typical event is shown in Fig. 1. The study of such extremely complex events poses a formidable challenge both to the detections systems and to the software and analysis methods to be used to treat and eventually understand the data. So, it is legitimate to ask first of all the question of why experimenters are willing to face such extreme problems, what are the underlying questions which we aim to answer.

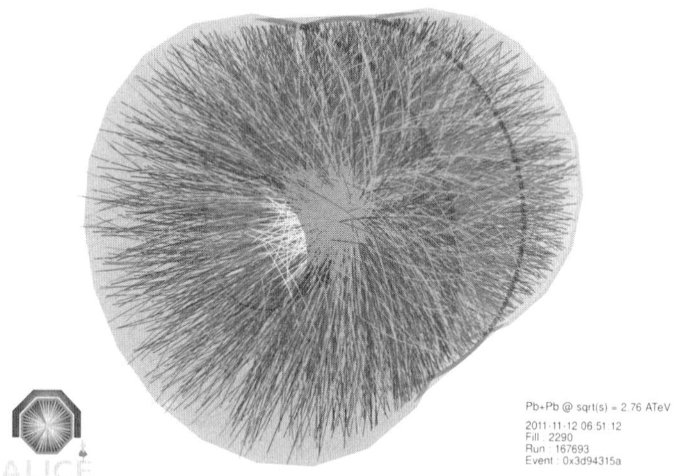

Fig. 1 A collision of two Lead Nuclei as seen by the ALICE detector at the LHC. (CERN-EX-1111290 04)

The strong force, which holds together the nucleons, protons and neutrons, in nuclei and the quarks within the nucleons, is described to great precision by a very successful theory, called Quantum Chromo Dynamics (QCD), in which the quarks possess a color charge and interact exchanging force carriers, the gluons, which are also colored. A fundamental feature of the interaction is the fact that it grows with the distance between the quarks, much like a spring connecting two balls. Therefore, any attempt to isolate a quark are in vain, since the energy stored in the spring will eventually be more than the mass of a quart-antiquark pair: the spring will break generating a pair, and the quarks will not be isolated. This property is called "confinement" and has major consequences. In particular, most of the mass of the hadrons is to be attributed to this feature: more technically, one can say that the mass is acquired dynamically because of the interaction between the quarks, whose Higgs mass would account for just a small fraction, about 1%, of the mass of the light hadrons as shown in Fig. 2. Therefore, since the mass of the matter around us is essentially the mass of the nuclei of the atoms, it can be said that, while the Higgs mechanism is at the base of the masses of the quarks, strong interaction is at the origin of most of the mass of ordinary matter.

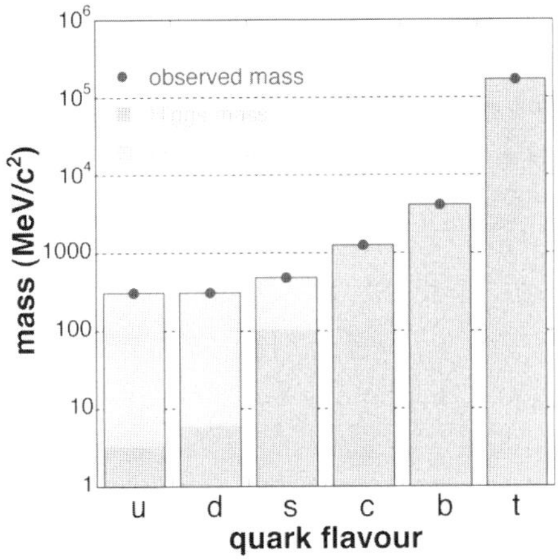

Fig. 2 QCD and Higgs mass components for the light and heay quarks from Ref [1]

The methodof colliding high-energy nuclei to heat and compress hadronic matter dates from the early eighties, but the path had been indicated already several years before. It was in 1975 that T.D. Lee pronounced the famous sentence "it would be interesting to explore new phenomena by distributing a high amount of energy or high nuclear density over a relatively large volume", and the idea of deconfinement had been around for some time. Already in 1965 Hagedorn had observed that the mass spectrum of hadronic states led naturally to the concept of a "limiting" or "critical" temperature beyond which the concept of hadron would cease to make sense. In 1973 the development of QCD as the theory describing strong interactions introduced the concept of asymptotic freedom[3][4]: at small distances quarks and gluons would be loosely interacting, or quasi-free. This fundamental result earned D.J.Gross and F.Wilczek, H.D. Politzer the Nobel Prize for Physics in 2004. The motivation reads:"for the discovery of asymptotic freedom in the theory of the strong interaction".

In 1975, N. Cabibbo and G. Parisi[5]proposed for the first time the existence of a ``different phase of the vacuum in which quarks are not confined", and drew the first schematic diagram of hadronic matter with temperature andbaryonic density as axes. Two regions were defined: at low-T and low baryonic density, ordinary hadrons, at high-T, high-μ_B a deconfined state: a new field of research was born!In the same year, J.C. Collins and M.J. Perry[6]developedthe idea that matter under extreme density conditions, as in neutron star cores and in the early universe, is made of quarks instead of hadrons, which overlap and lose their identity. Their argument was based on the factthat quarks interact weakly when they are close together, and therefore at high densities could bedescribed as a gas of free massless quarks. Finally, in 1980 E. Shuryak introduced [7]the term quark gluon plasma (QGP) to describe the state of nuclear matter in which quarks are deconfined, stressing the analogy between a classical plasma made by ionized atoms and the state made of colored strongly interacting objects.

In a nucleus-nucleus collision the energy of the incoming projectiles is dissipated in the relatively large volume defined by the overlap region of the two colliding nuclei,thus creating the conditions for a phase transition to deconfined quark matter by heating.

So, by colliding nuclei at high energies we compress and heat a system with a large number of quarks, to form a "bubble" of deconfined strongly-interacting matter, which in a very short time, of the order of 10^{-23}s, expands and cools until ordinary hadrons reconstitute, just as they did in the evolution of primordial Universe, some 10 millionth of a second after the Big Bang.

Over the last three decades, a vigorous experimental program, carried both in Europe and in the United States, has dramatically increased our understanding of nuclear matter under extreme conditions of temperature and density, allowing a real exploration of the phase diagram of strongly interacting matter. Our current understanding is graphically summarized in Fig. 3, with the trajectories allowed by the most important accelerators in operation or under construction: the two colliders LHC (in Europe) and RHIC (in the USA) and the fixed-target high-luminosity facility FAIR under construction in Germany. In addition, a new collider for Heavy Ions exploring the same energy range as FAIR is under construction at the JINR international laboratory in Dubna, near Moskow.

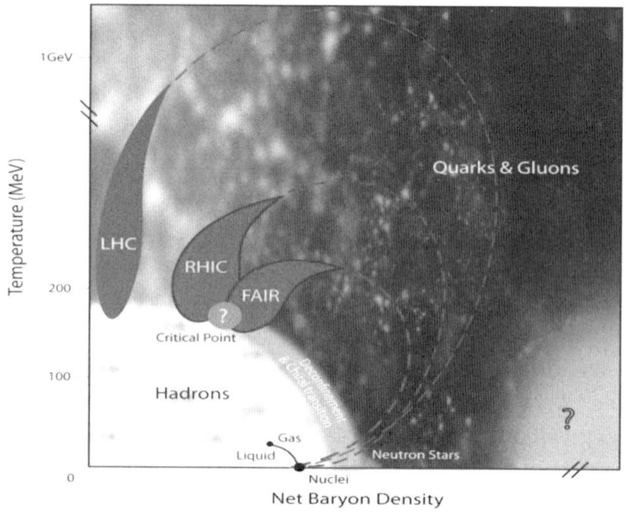

Fig.3 The phase diagram of nuclear matter, from Ref[1]

Together, these programs aim at providing answers to questions such as:

What are the properties of matter at the highest temperaturesand densities?

What is the QCD equation of state? How can we test it?

What are the dominant microscopic mechanisms ofQCD non-equilibrium dynamics and thermalization?

How does hadronization proceed dynamically? How is it changed in dense QCD matter?

How can the QCD phase diagram be efficiently explored?

Clearly, a comparison of the experimental results with theory is not a trivial task. Since the transition to the deconfined phase is outside the limits of validity of the perturbative QCD, a number of different approaches has been developed to derive the equation of state and the transition temperature. Only at the LHC some aspects of the interactions will be accessible with perturbative methods. Effective lagrangians have been extensively used, and also potential models, and even calculations based on the AdS/CFT correspondence have been providing a fresh look and new insight. Still, the most important and widely used are QCD calculations on a space--time lattice. They generally predict a phase transition to occur, and are able to determine its temperature. Current results indicate a critical temperature around $T_c \sim 170$ MeV, or about 10^{12} K, 100000 times higher than at the center of the Sun, corresponding to a critical energy density of $\varepsilon_c \sim 700$ MeV/fm^3

In the following, I will give a brief overview of the main experimental results obtained so far at the LHC, just little over one year after the very first collision of Lead Nuclei was recorded in the experiments. Three experiments joined in this pioneering work: ALICE[9], a dedicated experiment specifically designed to study nuclear collisions, and both ATLAS [10]and CMS[11], the two general-purpose experiments of the LHC, primarily designed for the study of high-p_T processes in pp collisions. I will focus mostly on the ALICE results.

While I will not describe the experimental apparata, I would like to remind here that ALICE has a very different optimization compared to the other LHC experiments. For the understanding of nuclear collisions it is essential that as many of the produced particle as possible are measured and identified, for the widest possible range of transverse momenta (typically from 0.1 to 100 GeV/*c*) and for all particle types: electrons, muons, photons and hadrons. So, the two guiding principles in the ALICE design have been robust tracking, able to handle the very large number of tracks which characterize Heavy Ion Collisions, and Particle Identification capability. The first is provided by very high granularity tracking detectors providing 3-Dimensional space points (several hundred millions of points) operating in a moderate magnetic field and with very low material. To identify leptons, photons and hadrons, including short-lived particles, such as hyperons, D and B mesons, ALICE features essentially all known techniques: dE/dx measurement, Cherenkov & Transition radiation detectors, Time Of Flight, calorimeters, muon filter and topology (secondary vertices, kinks).

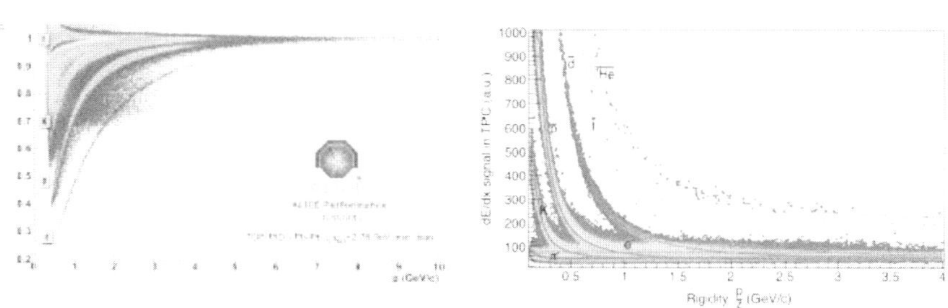

Fig. 4: PID in ALICE with TOF (left panel) and TPC (right panel) from ref.[12]

As an example of the ALICE PID capability, Fig. 4 shows the of the performance of the Time Of Flight (TOF) and of the Time Projection Chamber (TPC)'s energy loss measurement. Both are performing better than originally specified, a remarkable result in detector systems of such complexity observing events with thousands of tracks! The TOF has provided a time resolution below 100 picoseconds, and the TPC an energy loss measurement with a precision better than 6%.

In figure 5 (left panel) the momentum coverage of the various ALICE PID detectorsis shown, demonstrating the extent of the coverage achieved. In addition, short-lived particles are identified through their decay topologies: in the right panel of figure 5 the impact parameter resolution expected in ALICEis shown, which is adequate for the measurement of D and B meson decays. K and Λ decays can be efficiently reconstructed for transverse momenta up to more than 10 GeV/*c*. The very fine granularity of the tracking detectors, providing a large number of space points per track, provides high efficiency tracking for charged particles even at the highest particle multiplicities foreseen at the LHCwith Pb beams, with a lower limit in momentum lower than 200 MeV/*c*. The very long track length measured allows ALICE to achieve an excellent momentum resolution ($\delta p/p < 5\%$ at 100 GeV/*c*). Several smaller detectors (ZDC, PMD, FMD, T0 andV0) used for global event characterization and triggering are located at small angles. The combination of the central and forward detectors provides coverage for the measurement of charged multiplicity which extends from -3.4 to +5 in pseudorapidity. An Electromagnetic Calorimeter, EMCal, covers about a third of the azimuth and provides improved energy resolution and triggering for jets. A high-resolution crystal photon spectrometer, PHOS, is placed at a very large radial distance from the interaction point (4.6 m). The muon spectrometer covers the forward rapidities, between-2.5 and -4, and features a complex arrangement of absorbers, starting very close to the interaction point (90cm) to minimize muon background from weak hadronic decays, and fourteen planes of tracking and triggering chambers, using 2-D pad readout to handle the high multiplicities.

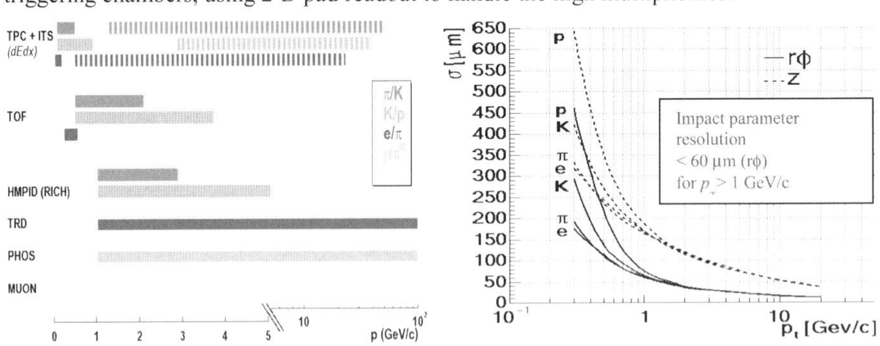

Figure 5: Left: particle identification in ALICE: momentum range and detectors used. Topological identification, not included in this plot, can reach over 10 GeV/*c* for K and Λ. Right: ALICE impact parameter resolution as a function of transverse momentum.

The hardware trigger in ALICE combines the input from detectors with fast trigger capability (T0, V0, ZDC, SPD, TOF, PHOS, EMCal, Muons, ACORDE). It operates at several levels (pretrigger, L0, L1 and L2) to satisfy the individual timing requirements of the different detectors, as the ALICE electronics is in general not pipelined. In addition, a software based High-Level Trigger (HLT) using a farm of up to 1000 multiprocessor PCs performs a preprocessing of the data allowing for a compression of the data volume by a factor of 3.5. It also allows real timedata quality monitoring.

2. ALICE results

The first LHC heavy-ion results were already published during the 2010 lead–lead run, lasting for a month. In fact, there were three papers submitted almost simultaneously: the multiplicity measurement [13] and the measurement of v_2 coefficient of the azimuthal anisotropy [14] by ALICE, and by the ATLAS collaboration a paper on jet-energy imbalance [15]; these appeared together in one issue of Phys. Rev. Lett. One of the most spectacular findings at RHIC was that the matter generated in heavy-ion collisions flows like a liquid with very low viscosity, almost at the limit of what is allowed for any material in nature. This tells us that the constituents of this QGP are quite different from freely interacting quarks and gluons, and led to the definition of the matter under study as sQGP, for strongly interacting Quark Gluon Plasma, a liquid very different from the expected gas of quasi-free quarks and gluons. The first LHC azimuthal anisotropy measurement [14] confirms the RHIC results: elliptic flow of particles with the same transverse momenta is almost identical at the two energies. The nearly-perfect fluid has been found to be opaque to even the most energetic partons (quarks and gluons), which appear as jets of particles from the collisions, an effect known as jet quenching. This is an interpretation of the reported strong jet-energy imbalance [15]. However, the physical mechanisms underlying these phenomena are not well understood. Another manifestation of jet quenching manifests is a reduced yield of high-transverse-momentum-particle in central collisions compared to that expected from the measurements in proton–proton reactions. To express such reduction a properly normalized ratio of yields in heavy-ion and in proton–proton interactions – the nuclear modification factor R_{AA} – is used. Since proton-proton collisions are rescaled according to the number of binary collisions, R_{AA} should be well below one at low transverse momentum, where particle production is rather proportional to the number of participant nucleons, and grow toward one for increasing transverse momenta. The suppression pattern observed by ALICE [16] gives factor about 7 ($= 1/R_{AA}$) lower charged-particle production in lead–lead collisions at transverse momentum around 6GeV/c, a striking evidence of the energy loss in the extreme density matter.

The size of the pion-emitting source in central lead-ion collisions is deduced from the shape of the Bose-Einstein peak in the two-pion correlation functions [17]. The collective flow makes the size of the system appear smaller with increasing momentum of the pair. This behaviour is clearly visible for the radii measured in the ALICE experiment. The results for measurements of the radius of the pion source in three dimensions indicate a short duration for the emission, hence an "explosive" emission. Time when the emission reaches its maximum is 10–11fm/c, significantly longer than it is at RHIC, see fig. 6 (left side). Moreover, the product of the three components at low pair-momentum, an estimate of the homogeneity volume of the system at decoupling, is twice as large as at RHIC, see fig. 6 (right side).

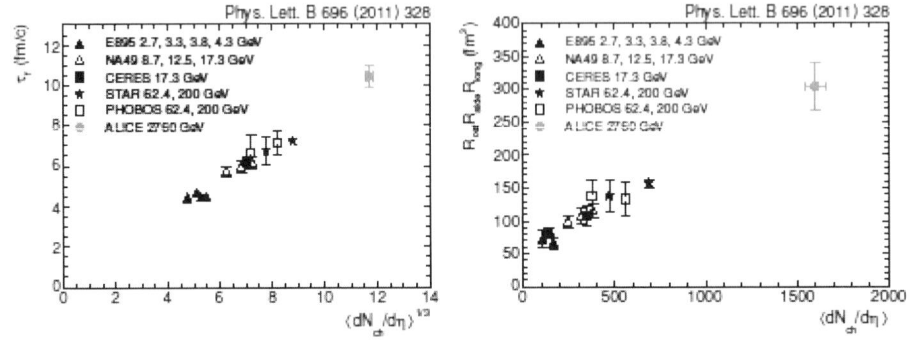

Figure 6. Bose–Einstein pion-interferometry results from different experiments [17]. Estimate of lifetime of emitting source as a function of cube root of particle density (left side). Volume of homogeneity region as a function of particle density (right side).

One of the crucial measurements for the characterization of the fireball produced in heavy-ion collisions are the spectra of identified hadrons, which encode the collective expansion velocity developed in the QGP and in hadronic stages. Moreover, the overall abundances of identified hadrons are believed to be fixed at hadronization, thus they indicate the temperature when chemical composition was established. The measured spectra [18] show an increase of about 10% in the radial-flow velocity when compared to RHIC results. At present, however, the yield ratios observed by ALICE seem to challenge both previous experiments and theory. While the K/π, Ξ/π and Ω/π ratios are compatible with the expectations from the thermal model with a temperature for the "chemical freeze-out" of about 165MeV, as in previous observations, the p/π ratio points to a significantly lower temperature. On the experimental side, there are indications of a similar effect at lower energies, which call for further investigations. On the theoretical side, a number of different possibilities are being investigated, none of them conclusive at the moment.

As mentioned above, the first results on elliptic flow were published [14] during the initial Pb–Pb run. Flow is a fundamental observable since it carries information on the equation of state and the transport properties of matter created in a heavy-ion collision. This observable relates final state anisotropies with features of the initial one, thus allowing study of the medium response and characteristics. The azimuthal anisotropy in particle production is a clear experimental signature of collective flow. It is caused by multiple interactions between the constituents of the created matter and the initial asymmetries in the spatial geometry of a non-central collision. The amount of elliptic flow measured by the v_2 coefficient, integrated over transverse momentum, increases by 30% compared to RHIC energies. However, this increase is entirely a consequence of the growth in transverse momenta. The overall azimuthal distribution of particles emitted in a heavy-ion collision can be described with a Fourier expansion, whose coefficients relate to different aspects of the initial state. ALICE measured also the higher harmonic coefficients of azimuthal distribution [19]; they exhibit much shallower centrality dependence than v_2, and the symmetry planes of v_2 and v_3 coefficients are uncorrelated (see fig. 7, left side). These observations point to fluctuations in the initial geometry as the origin of higher-order azimuthal asymmetries. The flow coefficients can also be studied with a different approach, using two-particle correlations in azimuthal angle [20]. A Fourier decomposition of the correlation function gives squares of different v coefficients, and describes well the structures seen in azimuthal correlations, see fig. 7 (right side), such as the "long-range ridge" (a small angle correlation between particles distanced in longitudinal direction) and the "Mach cone" (a correlation on two sides of back-to-back direction). The collective response to initial spatial anisotropy that causes elliptic flow economically explains these puzzling features, once event-by-event initial-state density fluctuations are considered. The ALICE experiment has analyzed azimuthal asymmetry also separately for different particle species, which constrains models incorporating a realistic initial state and hydrodynamic evolution.

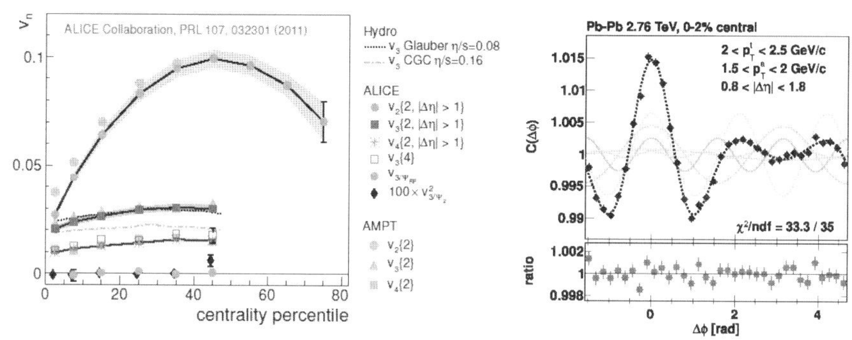

Figure 7. Measurements of azimuthal anisotropy in lead–lead collisions at LHC. Dependence of different v_n coefficients (see legend) on collision centrality [19] (left). Fourier decomposition of two-particle azimuthal correlation function in 2% most central collisions [20] (right).

The two-particle azimuthal correlations at higher transverse momenta measure particle-yield modifications in jet-like structures [21]. When compared to the expectation from proton–proton interactions, the observed yield around high-transverse-momentum particle is slightly higher, while the yield on the opposite side is reduced. This measurement assists the understanding of jet-quenching phenomena. The ALICE collaboration continues to contribute to this topic by studying reconstructed jets, underlying event fluctuations, modifications of jet-fragmentation function, jet-particle composition and providing detailed studies of the dependence of the energy loss on the flavor and mass of theparton traversing the QGP.

Heavy-flavour particles are particularly effective probes of a very dense and hot medium formed in nucleus–nucleus collisions, sensitive to its energy density, through the mechanism of in-medium energy loss. At LHC energies, heavy-flavour particles are copiously produced and thus provide ideal tools for QGP studies. The nuclear modification factor R_{AA} is well established as a sensitive observable for the study of the interaction of hard partons with the medium. Parton energy-loss is caused by the strong interaction, hence the amount of energy loss depends on the colour charge of parton. Therefore, quarks are predicted to lose less energy than gluons.

In addition, heavy quarks (up to some high momentum), which are slower than light, cannot emit gluons within the so-called "dead-cone" around their trajectory; this effect is expected to reduce the energy loss of heavy quarks with respect to light ones. Thus, a pattern of gradually decreasing suppression (i.e. increasing R_{AA}) should emerge when going from the light-flavour hadrons (e.g. pions), which mainly come from gluons, to the heavier D and B mesons: $R_{AA}(\pi) < R_{AA}(D) < R_{AA}(B)$. The measurement and comparison of these different probes provides, therefore, a test of the colour-charge and mass dependence of parton energy-loss. The ALICE collaboration has measured the production of the charmed mesons D^0 and D^+, detecting their hadronic decays in lead–lead collisions [22]. In central collisions a large suppression with respect to expectations at large transverse momentum was found, indicating that charm quarks undergo a strong energy loss in the hot and dense state of strongly-interacting matter formed at the LHC. This is the first time that D meson suppression has been measured directly in central nucleus–nucleus collisions. The results show a suppression by a factor 4–5, almost as large as for charged pions, above 5 GeV/*c* (see fig. 8). At lower momenta, there is an indication of smaller suppression for D than for π mesons.

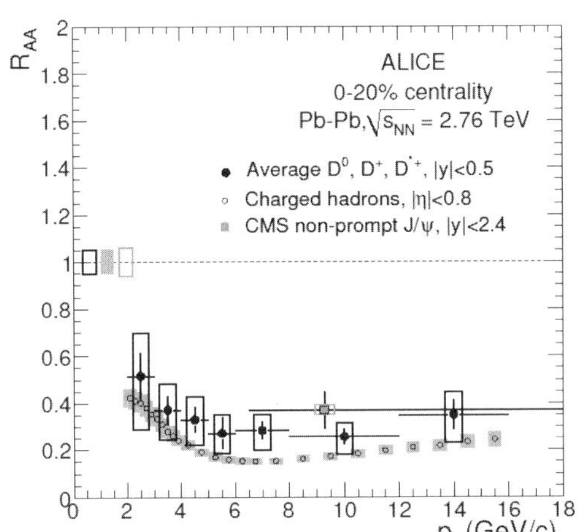

Figure 8. Charmed-meson nuclear modification factor as a function of transverse momentum compared to that of charged hadrons and of non-prompt (mostly from B-decays) J/ψ mesons (as measured by CMS).

Suppression of charmonium production was for a long time considered as one of the main probes for a deconfined medium. At large enough temperatures bound charm–anticharm states are supposed to be dissolved due to Debye screening. However, at LHC energies, new mechanisms of charmonium production in the QGP could occur because of a large number of charm quarks. Around one hundred charm–anticharm pairs are expected to be produced in a central lead–lead collision. Several dynamical transport models predict that charm and anticharm quarks could combine at later stages of the interaction, leading to an enhancement of charmonium production in the most central collisions. ALICE detects charmonium down to very low transverse momentum in two different regions: the central barrel in the dielectron

channel and the forward muon arm in the dimuon channel. The detection at low transverse momentum is crucial because the recombination of the charm and anticharm quarks is expected to be the main production mechanism for charmonium below 3GeV/*c*. The different regions allow for the study of QGP with different charm densities. ALICE has studied the nuclear modification factor R_{AA} for J/ψ mesons as a function of collision centrality [23], [26]. The results indicate that the J/ψ nuclear modification factor shows little dependence on centrality (see fig. 9, left), a trend that is different from that observed at lower energies. For central and mid-central collisions the J/ψ R_{AA} is larger at the LHC than that measured at RHIC. In a complementary study, the CMS collaborations at the LHC have measured a smaller value for the J/ψ nuclear modification factor at transverse momenta above 6.5 GeV/*c*. The nuclear modification factor exhibits a marked dependence on transverse momentum (see fig. 9, right), with a strong increase for values below 3GeV/*c*.
These observations hint at the recombination of charm and anticharm quarks in the QGP as the main mechanism for J/ψ production in central lead–lead collisions at LHC energies. As shown in fig. 9, the effect of suppression followed by recombination accounts for both the shape and the magnitude of the effect.

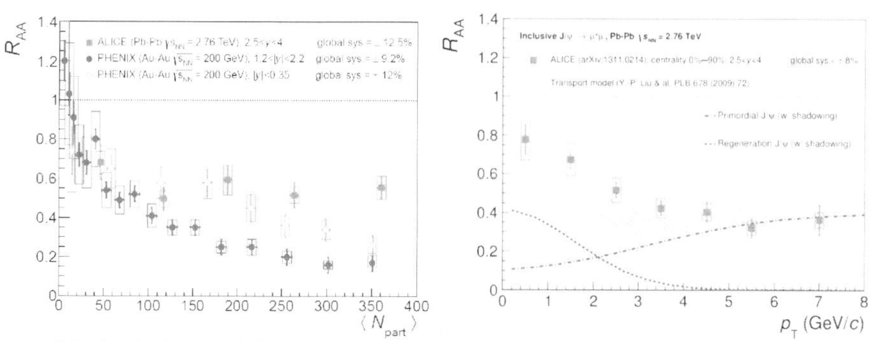

Figure 9. Left: Nuclear modification factor for J/ψ as a function of collision centrality, expressed as number of participants. ALICE data are compared to those from the PHENIX experiment at RHIC. Right: Dependence of the J/ψ nuclear modification factor on transverse momentum, showing the contributions of regenerated and primordial J/ψ (from [26])

It is clear from these results that a coherent picture of the characteristics of the extreme density matter of which our Universe was made of in the first microseconds of its existence. Even more, the tool chosen to study it, heavy-ion collisions at the world's most powerful accelerator, has proven to be the right one, allowing for a continuous improvement of our understanding. The results shown here are but a small subset of the wealth of results obtained so far, and many more continue to flow: some papers have already been published using the 2015 data at 5.02 TeV! It is indeed a very exciting time for the physics of the Quark-Gluon Plasma.

3. Results from the pp and pA runs

A wealth of results has been produced by ALICE using pp collision data, including for example the first measurement of the polarization of the J/ψ particle or the first measurement of the zero-net baryon number in the central rapidity region. Here we will mention only that at the very high energies of the LHC even pp collisions might exhibit features linked to the properties of strongly interacting matter at extreme densities, and that this study of "collectivity in small systems" has become one of the hot new research subjects at the LHC. Indeed, very suggestive structures have been observed in pp events by the CMS Collaboration, featuring long-range angular correlations among the produced particles, known as "the Ridge" [27]. Recently, the ALICE Collaboration has observed an enhanced production of strange and multi-strange hadrons in high-multiplicity proton-proton (pp) interactions, one of the earliest proposed indicators for the formation of a Quark-Gluon Plasma (QGP) state.

When studying the difference between pp and PbPb collisions it is essential to understand the initial wave function of the colliding nuclei, and thus disentangle 'initial state effects" from those due to the hot and dense matter created in PbPb collisions. Understanding the structure of the initial state is very interesting in its own right, because at the LHC energy, experiments probe the structure of the nucleus in a novel and unexplored QCD regime of very low values of the longitudinal parton-momentum fraction (x <10–3). In this kinematic region, the extremely high gluon density is expected to saturate by means of strongly non-linear coherent processes, leading theorists to predict the existence of yet another pre-collision state of matter – the so-called colour glass condensate (CGC).Moreover, pPb collisions at the LHC open a new window for the study of high density strongly interacting matter, allowing a study as a function of the volume of the system.

The results from the basic control measurements, the charged-particle multiplicity density and transverse-momentum spectrum, were in agreement with expectations from theoretical models that incorporate present knowledge of the function describing the longitudinal distribution of partons inside the nucleus [24]. In addition, as shown in the left panel of fig 10, when compared with the results obtained for pp collisions at 2.76 and 7 TeV, on one hand the charged-particle multiplicity in pPb collisions was found to scale roughly with the mean number of nucleons participating in the collision, as expected for particle production through soft processes. On the other hand, the particle spectrum at large transverse momenta – beyond 3–4 GeV/c – was found to scale with the mean number of binary nucleon–nucleon collisions, as expected for particle production through hard processes. Also charmed mesons showed no significant deviation from binary nucleon–nucleon scaling, as shown in Fig. 10. The above observations confirm that the suppression of high transverse-momenta hadrons and jets observed in central PbPb collisions can be attributed to a final-state effect, namely, the energy loss experienced by partons traversing the medium created in these collisions.

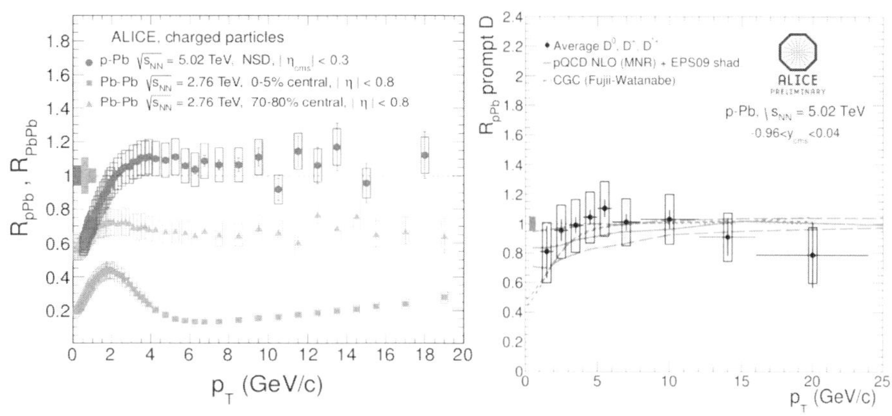

Figure 10. Left (from ref [24]): modification factor for charged particles. Right: Ratio of charmed mesons production in pPb collisions to the one measured in pp scaled by the number of collisions R_{pPb}. $R_{pPb}$is plotted as a function of pT for prompt D mesons (average of D0, D+ and D*+) in p-Pb collisions at √sNN=5.02 TeV. Results are compared to model calculations of initial-state effects.

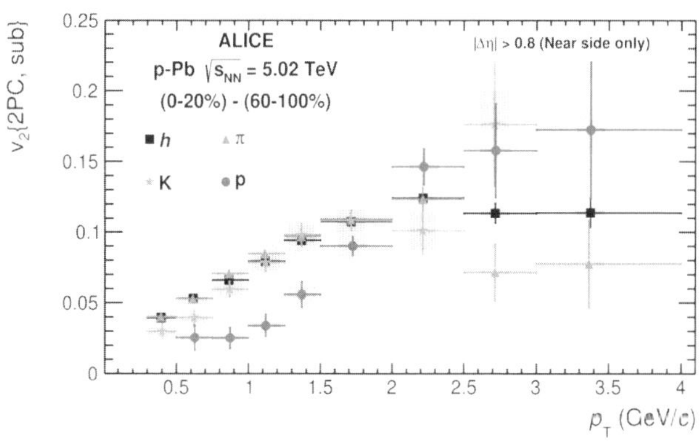

Figure 11.v_2 measured in p-Pb collisions at √sNN=5.02 TeVfor identified particles, from [28]

The study of two-particle correlations in high-multiplicity pPb events led instead to a surprising result: a near-side, long-range (elongated in pseudorapidity) correlation, forming a ridge-like structure observed in high-multiplicity pp collisions, accompanied by an essentially symmetrical away-side ridge, opposite in azimuth ,was observed [25]. This double ridge was revealed after the short-range correlations arising from jet fragmentation and resonance decays were suppressed by subtracting the correlation distribution measured for low-multiplicity events from the one for high-multiplicity events. Similar long-range structures in heavy-ion collisions have been attributed to the collective flow of particles emitted from a thermalized system undergoing a collective hydrodynamic expansion. The origin and nature of this structure are currently subject of debate and research. In the subsequent months, and more evidence of features which in heavy-ion collisions have been attributed to collective behavior have emerged, in particular a characteristic mass scaling in the measurement of v_2 harmonic coefficients of the azimuthal anisotropy [28], shown in Fig 11, and in the baryon-to-meson ratios.

So the questions of how collectivity and even hydrodynamic behaviour could take place in such small systems, and of the compatibility with the energy loss results are currently hotly debated. The whole community interested in the study of strongly interacting matter at extreme conditions is eagerly waiting for the further developments of these intriguing findings.

4. The future of ALICE

The LHC operates in a quite different mode than previous CERN accelerators. During running, only a very short stop of operation takes place around the end of the year, while most of the time is devoted to quasi-continuous running. After O(3) years of operation, a long shutdown of more than a year allows for major maintenance and improvements of both the accelerator and the detectors.

Data taking for ALICE in the so-called RUN1 has included, apart from data taking with pp collisions, three periods of data taking with nuclear beams. In 2010 the ALICE detector collected with a minimum-bias trigger about 30 millions of lead–lead collisions at centre-of-mass energy of 2.76TeV per nucleon pair. In 2011 there was a short period of proton–proton running at the same energy at the start of the LHC operation, to gather comparison data; these were used for normalization in practically all measurements described above. The rest of the proton running proceeded at (a standard) energy of 7TeV. At the end of the year, the LHC switched again to the heavy-ion mode. This time the instant luminosity grew to above $10^{26}cm^{-2}s^{-1}$, and was above the design value for energy 2.76TeV per nucleon pair. ALICE used triggers for different types of lead–lead collisions: central, semi-central, dimuons, photons, jets, ultraperipheral, and others, and collected about 100 million events, inspecting over $0.1nb^{-1}$ of integrated luminosity. At the beginning of 2013, the Heavy-Ionrun has been dedicated to proton–lead collisions, to improve the baseline comparison, taking into account modifications of structure functions in nuclei. ALICE collected $0.03nb^{-1}$ of luminosity in the four weeks of pA running, half of which in pA mode and half in Ap mode, to take into account the asymmetry of the apparatus.

Since then the LHC operation has been paused for the upgrade necessary to increase the collision energy. The Long Shutdown (LS1) lasted until early 2015. During this period, ALICE has carried a very intense program of detector consolidation, installation of new detectors (DCAL, a second arm of the ALICE electromagnetic calorimeter, a set of forward scintillator counters to improve the coverage at large rapidities, the last five supermodules of TRD and a fourth supermodule of PHOS).

During RUN2, after the restart in 2015, the ALICE collaboration aims to complete its originalprogramme, collecting $1nb^{-1}$ of heavy-ion collisions at the higher collision energy (5.5TeV per nucleon pair in the center of mass, being the design value). This will provide a 10-fold increase in statistics, on top of a better quality of data and the higher energy. In addition, a second proton-nucleus run is planned, in order to further explore the very exciting features discovered in RUN1. This agenda will be completed before the second long shutdown of the

LHC, planned to start in2019. The first year of running has started very successfully, and ALICE is well on track to fulfill the RUN2 goals.
A majorupgrade of the ALICE detector allowing for high-luminosity heavy-ion running after this period, to reach a tenfold increase in integrated luminosity in RUN3 and RUN4 (extending to the late 2020's) has been proposed by the ALICE Collaboration. The extended physics program justifying the LHC operation in heavy-ion mode beyond 2020, which would imply collecting over $10nb^{-1}$ of data, has been included among the top priority goals of the European Strategy for High Energy Physics. The Letter of Intent [29], defining the scientific goals of the Upgrade, has been approved in late 2012, and the five Technical Design Reports discussing the technical implementation of the upgrade have been approved successively in 2013, 2014 and 2015.
After having confirmed the main discoveries obtained at RHIC, ALICE has entered an exciting phase of new measurements, allowing a much broader and deeper study of the QGP. At the same time, ALICE is already preparing, on the basis of what has been learnt so far, a next step in more detailed characterization of the extreme state of matter produced at LHC; new, unexpected discoveries may come!

References

[1] NUPECC Long Range Plan 2010: Perspectives of Nuclear Physics in Europe, published by the European Science Foundation
[2] R. Hagedorn, NuovoCim. Suppl. 3 (1965) 147-186
[3] D.J. Gross and F. Wilczek Phys. Rev. D 8(1973) 3633—3652.
[4] H. D. Politzer Phys. Rev. Lett. 30 (1973) 1346-1349.
[5] N. Cabibbo and G. Parisi Phys. Lett. B 59 (1975) 67-69
[6] J.C. Collins and M.C. Perry~M.C. Phys. Rev. Lett.34(1975) 1352—1356.
[7] E.V. Shuryak Phys. Rept. 61 (1980) 71—158
[8] B. Muller and J. Nagle, Ann. Rev. Nucl. and Part. Phys. **56** (2006) 93
[9] ALICE Collaboration, Aamodt K, et al., JINST 3:S08002 (2008).
[10] ATLAS Collaboration, Aad G, et al., JINST 3:S08003 (2008).
[11] CMS Collaboration, Adolphi R, et al., JINST 3:S08004 (2008).
[12] A. Kalweit for the ALICE Collaboration, J.Phys.G G38 (2011) 124073.
[13] K. Aamodt et al. (ALICE collaboration), Phys. Rev. Lett. **105** 252301 (2010).
[14] K. Aamodt et al. (ALICE collaboration), Phys. Rev. Lett. **105** 252302 (2010).
[15] G. Aad et al. (ATLAS collaboration), Phys. Rev. Lett. **105** 252303 (2010).
[16] K. Aamodt et al. (ALICE collaboration), Phys. Lett. **B696** 30 (2011).
[17] K. Aamodt et al. (ALICE collaboration), Phys. Lett. **B696** 328 (2011).
[18] M. Florisfor the ALICE collaboration, J. Phys. **G38** 124025 (2011).
[19] K. Aamodt et al. (ALICE collaboration), Phys. Rev. Lett. **107** 032301 (2011).
[20] K. Aamodt et al. (ALICE collaboration), Phys. Lett. **B708** 249 (2012).

[21] K. Aamodt et al. (ALICE collaboration), Phys. Rev. Lett. **108** 092301 (2011).

[22] B. Abelev et al. (ALICE collaboration), JHEP **1209** 112 **(2012)**

[23] B. Abelev et al. (ALICE collaboration), Phys.Rev.Lett. **109** 072301(2012)

[24] B. Abelev et al. (ALICE collaboration), Phys.Rev.Lett. **110**082302 (2013)

[25] B. Abelev et al. (ALICE collaboration), Phys. Lett. **B 719** 29(2013)

[26] B. Abelev et al. (ALICE collaboration), Phys. Lett. **B 743** (2014) 314-327

[27] S. Chatrchyan et al. (CMS Collaboration) Phys. Lett. B 718 (2013) 795

[28] B. Abelev et al. (ALICE Collaboration) Phys. Lett. B 726 (2013) 164-177

[29] B. Abelev et al. (ALICE collaboration), J. Phys. G 41 (2014) 087001

The High Luminosity LHC Project

Lucio Rossi

CERN, Accelerator & Technology Sector

University of Milan, Physics Department

Abstract: The High Luminosity LHC is one of the major scientific project of the next decade. It aims at increasing the luminosity reach of LHC by a factor five for peak luminosity and a factor ten in integrated luminosity. The project, now fully approved and funded, will be finished in ten years and will prolong the life of LHC until 2035-2040. It implies deep modifications of the LHC for about 1.2 km around the high luminosity insertions of ATLAS and CMS and relies on new cutting edge technologies. We are developing new advanced superconducting magnets capable of reaching 12 T field; superconducting RF crab cavities capable to rotate the beams with great accuracy; 100 kA and hundred meter long superconducting links for removing the power converter out of the tunnel; new collimator concepts, etc... Beside the important physics goals, the High Luminosity LHC project is an ideal test bed for new technologies for the next hadron collider for the post-LHC era.

Keywords: Hadron Colliders, Luminosity, LHC, High Field Magnets

1. Introduction

The Large Hadron Collider (LHC) was successfully commissioned in 2010 for proton proton collisions with a 7 TeV centre-of-mass energy and delivered 8 TeV centre-of-mass proton collisions from April 2012 to the end of 2013. The announcement given by CERN on 4 July 2012 about the discovery of a new boson at about 125 GeV, the long-awaited Higgs particle, is the first fundamental discovery, hopefully the first of a series that LHC can deliver. Beyond this, the LHC has the potential to go on and help answer some of the key questions of the age: the existence, or not, of supersymmetry; the nature of dark matter; the existence of extra dimensions. It is also important to continue to study the properties of the Higgs – here the LHC is well placed to do this in exquisite detail. The LHC will remain the most powerful accelerator in the world for at least the next two decades. Its full exploitation is the highest priority of the European Strategy for particle physics [1] as well as for the Particle Physics Strategy of the US [2] and, to a certain extent, Japan. To extend its discovery potential, the LHC will need a major upgrade in the 2020s to increase its luminosity (and thus collision rate) by a factor of five beyond its design value. The integrated luminosity design goal is an increase by a factor of ten. The necessary developments will require still about 10 years to prototype, test, build and install. The novel machine configuration, the High Luminosity LHC (HL-LHC) [3], will rely on a number of key innovative technologies representing exceptional technological challenges.

HL-LHC federates the efforts and R&D of a large international community towards the ambitious HL-LHC objectives. This participation will be required for the execution of the construction phase as a global project. In particular, the US LHC Accelerator R&D Program (LARP) [4] has developed some of the key technologies for the HL-LHC, such as the large-aperture niobium–tin (Nb_3Sn) quadrupoles and the crab cavities. The proposed governance model is tailored accordingly and should pave the way for the organization of the construction phase.

2. Luminosity and HL-LHC goal

The LHC baseline programme until 2025 is shown schematically in Figure 1. After approaching nominal energy at 13 TeV centre-of-mass energy in 2015, it is expected that the LHC will reach the design luminosity of 1 × 1034 cm−2 s−1. This peak value should give a total integrated luminosity of about 40 fb−1 per full operational year. In the period 2015–2024 the LHC will hopefully further increase the peak luminosity. Margins in the design of the nominal LHC are expected to allow, in principle, about two times the nominal design performance. The baseline programme for the next ten years is depicted in Figure 1, where the peak luminosity double than nominal and the expected value for integrated luminosity are indicated.

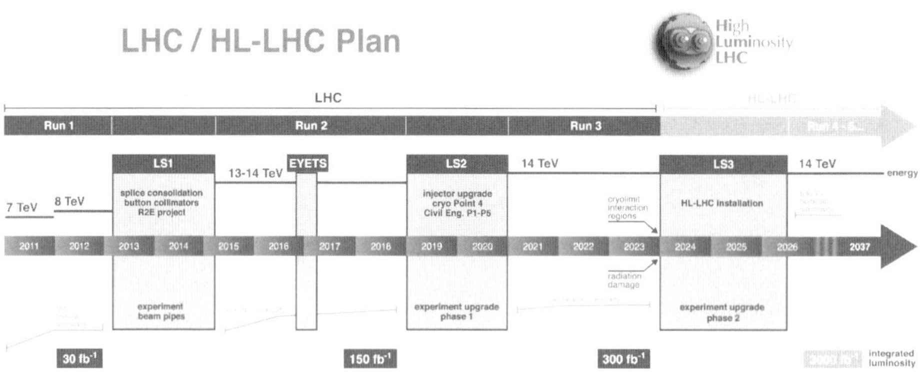

Fig. 1. Schematic plan of the LHC machine and HL-LHC, with indicated the trend for peak luminosity and the integral luminosity at each long shutdown (LS)

After 2024 the statistical gain in running the accelerator without a significant luminosity increase beyond its design value will become marginal. Therefore, to maintain scientific progress and to explore its full capacity, the LHC will need to have a decisive increase of its luminosity. This is why, the priority to fully exploit the potential of the LHC has been confirmed as the first priority among the 'High priority large-scale scientific activities' in the new European Strategy for particle physics – Update 2013 [1]. The importance of the LHC luminosity upgrade for the future of high energy physics has been also recently re-affirmed by the May 2014 recommendation by the Particle Physics Project Prioritization Panel (P5) to the High Energy Physics Advisory Panel (HEPAP), which in turn advises the US Department of Energy (DOE) [2]. In Japan, the updated KEK roadmap in 2013 states that 'The main agenda at LHC/ATLAS is to continually participate in the experiment and to take a proactive initiative in upgrade programmes within the international collaboration at both the accelerator and detector facilities.'

In this context, at the end of 2010 CERN created the High Luminosity LHC (HL-LHC) project [3]. Started as a design study, and after the approval of the CERN Council of 30 May 2013 and the insertion of the budget in the CERN Medium Term Plan approved by Council in June 2014, the HL-LHC has become CERN's major construction project for the next decade.

The main objective of the High Luminosity LHC design study is to determine a set of beam parameters and the hardware configuration that will enable the LHC to reach the following targets:

1) **peak luminosity of 5×10^{34} cm^{-2} s^{-1} with levelling**, allowing

2) an **integrated luminosity of 250 fb^{-1} per year with the goal of 3000 fb^{-1}** in about a dozen years after the upgrade. This integrated luminosity is about ten times the expected luminosity reach of the first twelve years of the LHC lifetime.

The main goals are the installation of the main hardware for the HL-LHC and the commissioning of the new machine configuration during LS3 (long shutdown n.3), scheduled for 2024–2026, while taking all actions to assure a high efficiency in operation until 2035.

Since all equipment is being designed with a margin of 50%, regarding reaching the required luminosity, we are defining the concept of ultimate parameters. By using these margins we should be able to push our machine to about 7–7.5 × 10^{34} cm^{-2} s^{-1} of peak, levelling luminosity, therefore of course increasing the total pile-up in the detectors up to 200. This luminosity level should enable the collect of up to 300–350 fb^{-1}/year. Also, in terms of total integrated luminosity, we think we can define an ultimate value of about 4000 fb^{-1}. It must be said that, while at first examination there is no showstopper for these performances, the ultimate parameters are not yet consolidated as the nominal parameters.

All of the hadron colliders in the world before the LHC have produced a combined total integrated luminosity of about 10 fb^{-1}. The LHC delivered nearly 30 fb^{-1} by the end of 2012 and should reach 300 fb^{-1} in its first 13–15 years of operation. The High Luminosity LHC is a major, extremely challenging, upgrade. For its successful realization, a number of key novel technologies have to be developed, validated, and integrated. The work was initiated quite early: ideas were circulating at the beginning of LHC construction [5] and this continued throughout construction [6]. From 2003, LARP (see Section 1.2.2) has been the main and continuous motor for technological development devoted to the LHC upgrade [4]. After a period during which the upgrade was conceived in two phases [7], all studies were unified in 2010 under the newly formed High Luminosity Project[3],[8]. The first step consisted in launching a Design Study under the auspices of EC-FP7 with the nickname *HiLumi LHC*, which, following approval by the EC in 2011 [9], has been instrumental in initiating a new global collaboration for the LHC matching the spirit of the worldwide user community of the LHC experiments. The High Luminosity LHC project is working in close collaboration with the CERN project for the LHC Injector complex Upgrade (LIU) [10], the companion ATLAS and CMS upgrade projects of 2019–2020 and 2024–2026 and the upgrade foreseen in 2019–2020 for both LHCb and Alice.

2.1 Luminosity

The (instantaneous) luminosity L can be expressed as:

$$L = \gamma \frac{n_b N^2 f_{rev}}{4\pi\, \beta^*\, \varepsilon_n} R; \quad R = 1 \Big/ \sqrt{1 + \frac{\theta_c\, \sigma_z}{2\sigma}}$$

where γ is the proton beam energy in unit of rest mass; n_b is the number of bunches per beam: 2808 (nominal LHC value) for 25 ns bunch spacing; N is the bunch population. $N_{\text{nominal 25 ns}}$: 1.15×10^{11} p ($\Rightarrow$0.58 A of beam current at 2808 bunches); f_{rev} is the revolution frequency (11.2 kHz); β^* is the beam beta function (focal length) at the collision point (nominal design 0.55 m); ε_n is the transverse normalized emittance (nominal design: 3.75 μm). R is a luminosity geometrical reduction factor (0.85 at a β^* *of* 0.55 m of, down to 0.5 at 0.25 m); θ_c is the full crossing angle between colliding beam (285 μrad as nominal design); and σ, σ_z are the transverse and longitudinal r.m.s. sizes, respectively (nominally 16.7 μm and 7.55 cm, respectively)

With the nominal parameter values shown above, a luminosity of 1×10^{34} cm^{-2} s^{-1} is obtained, with an average pile-up (number of events in the same bunch crossing) of $\mu = 27$ (although $\mu = 19$ was the original forecast at LHC approval due to uncertainties in the total proton cross-section at higher energies).

2.2 Limitations of LHC luminosity

Before discussing the new configuration it is useful to recall the hardware limitation that are likeable to make luminosity to saturate, either because they become vulnerable to breakdown and accelerated aging, or because they may become a bottleneck for operation in a higher radiation environment.

a. Inner triplet magnets. After about 300 fb^{-1} some components of the inner triplet quadrupoles and their corrector magnets will have received a dose of 30 MGy, entering into the region of possible radiation damage. The replacement of the triplet magnets must be envisaged before damage occurs because changing the low-beta triplet is a long intervention, requiring a one- to two-year shutdown and must be coupled with major detector upgrades.

b. Cryogenics. To increase intervention flexibility and machine availability it is planned to upgrade the cryogenics plant to make possible a separation between superconducting RF (SCRF) and magnet cooling in Point4. This would avoid the need to warm-up an entire arc in the case of triplet region intervention.

c. Collimation. The collimation system will need an upgrade that takes into account the needs for the lower impedance required for the planned increased beam intensities. A new configuration will also be required to protect the new triplets in IR1 and IR5. Also requiring special attention are the dispersion suppressor (DS) regions of Point7, where a leakage of off-momentum particles into the first and second main superconducting dipoles has been already identified as a possible LHC performance limitation. The most promising concept is to substitute an LHC main dipole with dipoles of equal bending strength (~120 T·m) obtained by a higher field (11 T) and shorter length (11 m) than those of the LHC dipoles (8.3 T and 14.2 m). The room gained is sufficient for the installation of special collimators.

d. Radiation to electronics (R2E) and superconducting (SC) links for the remote powering of cold circuits. New radiation resistance electronics need to be implemented to cope with increased beam intensity of HL-LHC. However, the power converters of the new magnets and associated electrical feedboxes (DFBs), would be placed out of the tunnel, in an adjacent aside gallery. Displacement of power converters to distant locations without large consumption is possible only thanks to a novel technology: superconducting links (SCLs) made from HTS (YBCO or Bi-2223) or MgB_2 superconductors.

e. Quench Protection System (QPS), machine protection and remote manipulation. These systems will potentially become problematic, along with the aging of the machine and the radiation level that comes with higher performance levels of 40 fb^{-1} to 60 fb^{-1} per year. In addition, remote manipulation must be planned to allow the replacement of collimators, magnets, vacuum components, etc., according to the 'as low as reasonably achievable' (ALARA) principle. While full robotics is difficult to implement, given the conditions on the ground, remote manipulation, enhanced reality and supervision are the key to minimizing the radiation doses sustained during interventions.

2.3 Luminosity Levelling

Both the consideration of energy deposition by collision debris in the interaction region magnets, and the necessity to limit the peak pile-up in the experimental detector, impose an a priori limitation upon peak luminosity. The consequence is that HL-LHC operation will have to rely on luminosity levelling. As shown in Figure 2 left, the luminosity profile without levelling quickly decreases from the initial peak value due to 'luminosity burn' (protons consumed in the collisions). The collider is designed to operate with a constant luminosity at a value below the virtual maximum luminosity. The average luminosity achieved is almost the same as that without levelling, see Figure 2 right. The advantage, however, is that the maximum peak luminosity is lower.

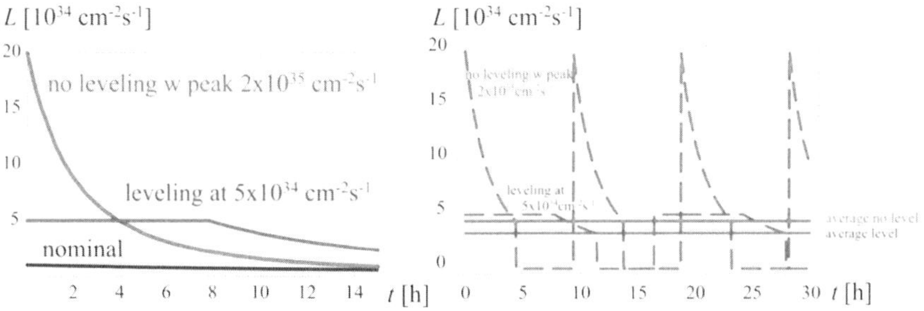

Fig. 2. Left: luminosity profile for a single long fill: starting at nominal peak luminosity (black line), with upgrade no levelling (red line), with levelling (blue line). Right: luminosity profile with optimized run time, without and with levelling (blue and red dashed lines), and average luminosity in both cases (solid lines).

Because of the levelled luminosity limit, to maximize the integrated luminosity one needs to maximize the fill length. This can be achieved by maximizing the injected beam current. Other key factors for maximizing the integrated luminosity and obtaining the required 3 fb^{-1}/day (see Figure 3) are a short average machine turnaround time, an average operational fill length that exceeds the luminosity levelling time, and good overall machine efficiency. The machine efficiency is essentially the available time for physics after downtime for fault recovery is taken into account. Closely related is the physics efficiency – the fraction of time per year spent actually providing collisions to the experiments. For integrated luminosity the efficiency counts almost as much as the virtual peak performance.

The HL-LHC with 160 days of physics operation a year needs a physics efficiency of about 40%. The overall LHC efficiency during the 2012 run, without luminosity levelling, was around 37%. The requirement of an efficiency higher than the one of the present LHC, with a (levelled) luminosity five times that of nominal, will be a real challenge. The project must foresee a vigorous consolidation for the high intensity and high luminosity regime: the High Luminosity LHC must also be a high availability LHC.

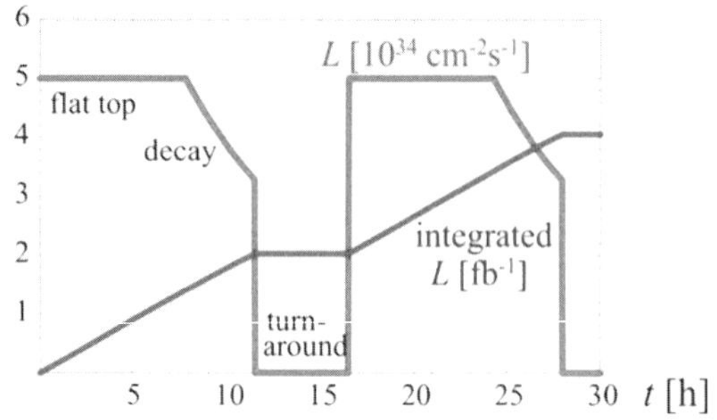

Fig. 3. Luminosity cycle for HL-LHC with levelling and a short decay (optimized for integrated luminosity).

2.4 High Luminosity LHC Beam Parameters

Table 1 lists the main parameters foreseen for high luminosity operation. The regular 25 ns bunch spacing is the baseline operation mode; however, a special bunch structure, where 8 regular bunches, 25 ns spaces, are followed by 4 empty bunches, also spaced 25 ns, is kept as a possible alternative in case the e-cloud or other unforeseen effects undermine regular 25 ns bunch spacing operation. A slightly different parameter set at 25 ns (batch compression and beam merging scheme (BCMS)) with very small transverse beam emittance is also shown and might be interesting for HL-LHC operation in case operation with high beam intensities results in unforeseen emittance blow-up.

For the HL-LHC there is a need to increase the beam brightness, a beam characteristic that must be maximized at the beginning of beam generation and then preserved throughout the entire injector chain and in LHC itself. The LIU project has as its primary objective increasing the number of protons per bunch by a factor of two above the nominal design value while keeping emittance at the present low value.

With the above parameters values, is possible to obtain the luminosity cycle illustrated in Figure 3, showing that in a perfect day we can obtain some 3 fb^{-1}/day.

2.5 Reduction Factor

A classical route for a luminosity upgrade is to reduce β^* by means of stronger and larger aperture low-β triplet quadrupoles. However a reduction in β^* values implies not only larger beam sizes in the triplet magnets but also an increase in crossing angle, which reduces the luminous region size and thus the gain in peak luminosity.

Stronger chromatic aberrations coming from the larger β-functions inside the triplet magnets may furthermore exceed the strength of the existing correction circuits. The peak β-function is also limited by the possibility to match the optics to the regular beta functions of the arcs. A previous study has shown that in the nominal LHC the practical limit for β^* is 30 cm to 40 cm versus the nominal 55 cm. However, a novel scheme called Achromatic Telescopic Squeeze (ATS) uses the adjacent arcs as enhanced matching sections. The increase of the beta-functions in these arcs can boost, at constant strength, the efficiency of the arc correction circuits. In this way a β^* value of 15 cm can be envisaged, and flat optics with a β^* as low as 5 cm in the plane perpendicular to the crossing plane could be realized. For such a β^* reduction the triplet quadrupoles need to double their

Table 1: list of LHC and HL-LHC beam parameters

HL-LHC Parameters V4.2 (Last updated 22-September-2015)

Parameter	Nominal LHC (design report)	HL-LHC 25ns (standard)	HL-LHC 25ns (BCMS) [9]	HL-LHC 8b+4e[12]
Beam energy in collision [TeV]	7	7	7	7
N_b	1.15E+11	2.2E+11	2.2E+11	2.3E+11
n_b	2808	2748	2604	1968
Number of collisions in IP1 and IP5 [1]	2808	2736	2592	1960
N_{tot}	3.2E+14	6.0E+14	5.7E+14	4.5E+14
beam current [A]	0.58	1.09	1.03	0.82
x-ing angle [μrad]	285	590	590	554 [10]
beam separation [σ] [11]	9.4	12.5	12.5	12.5 [10]
β^* [m]	0.55	0.15	0.15	0.15
ε_n [μm]	3.75	2.50	2.50	2.20
ε_L [eVs]	2.50	2.50	2.50	2.50
r.m.s. energy spread	1.13E-04	1.13E-04	1.13E-04	1.13E-04
r.m.s. bunch length [m]	7.55E-02	7.55E-02	7.55E-02	7.55E-02
IBS horizontal [h]	80 -> 106	18.5	18.5	13.1
IBS longitudinal [h]	61 -> 60	20.4	20.4	17.6
Piwinski parameter	0.65	3.14	3.14	3.14
Total loss factor R0 without crab-cavity	0.836	0.305	0.305	0.304
Total loss factor R1 with crab-cavity	(0.981)	0.829	0.829	0.828
beam-beam / IP without Crab Cavity	3.1E-03	3.3E-03	3.3E-03	3.9E-03
beam-beam / IP with Crab cavity	3.8E-03	1.1E-02	1.1E-02	1.3E-02
Peak Luminosity without crab-cavity [$cm^{-2}\,s^{-1}$]	1.00E+34	7.18E+34	6.80E+34	6.38E+34
Virtual Luminosity with crab-cavity: Lpeak*R1/R0 [$cm^{-2}\,s^{-1}$]	(1.18E+34)	19.54E+34	18.52E+34	17.40E+34
Events / crossing without levelling and without crab-cavity	27	198	198	246
Levelled Luminosity [$cm^{-2}\,s^{-1}$]	-	5.00E+34 [5]	5.00E+34	3.63E+34
Events / crossing (with leveling and crab-cavities for HL-LHC) [8]	27	138	146	140
Peak line density of pile up event [event/mm] (max over stable beams)	0.21	1.25	1.31	1.28
Leveling time [h] (assuming no emittance growth) [8]	-	8.3	7.6	9.5
Number of collisions in IP2/IP8	2808	2452/2524 [7]	2288/2396	1163/1868
N_b at LHC injection [2]	1.20E+11	2.30E+11	2.30E+11	2.40E+11
n_b / injection	288	288	288	224
N_{tot} / injection	3.46E+13	6.62E+13	6.62E+13	5.40E13
ε_n at SPS extraction [μm] [3]	3.40	2.00	< 2.00 [6]	1.70

[1] Assuming one less batch from the PS for machine protection (pilot injection, TL steering with 12 nominal bunches) and non-colliding bunches for experiments (background studies...). Note that due to RF beam loading the abort gap length must not exceed the 3μs design value

[2] An intensity loss of 5% distributed along the cycle is assumed from SPS extraction to collisions in the LHC

[3] A transverse emittance blow-up of 10 to 15% on the average H/V emittance in addition to the 15% to 20% expected from intra-beam scattering (IBS) is assumed (to reach the 2.5 μm/3.0 μm of emmitance in collision for 25ns/50ns operation

[4] As of 2012, ALICE collided main bunches against low intensity. satellite bunches (few per-mill of main bunch) produced during the generation of the 50ns beam in the injectors rather than two main bunches, hence the number of collisions is given as zero.

[5] For the design of the HL-LHC systems (collimators, triplet magnets,..), a design margin of 50% on the stated peak luminosity was agreed upon.

[6] For the BCMS scheme emittances down to 1.4 μm have already been achieved at LHC injection which might be used to mitigate excessive emittance blowup in the LHC during injection and ramp.

[7] The lower number of collissions in IR2/8 wrt to the general purpose detectors is a result of the agreed filling scheme, aiming as much as possible at a democratic sharing of collisions between the experiments.

[8] The total number of events/crossing is calculated with an inelastic cross-section of 85 mb (also for nominal), while 100 mb is still assumed for calculating the proton burn off and the resulting levelling time

[9] BCMS parameters are only considered for injection and as a backup parameter set in case one encounters larger than expected emittance growth in the HL-LHC during injection, ramp and squeeze

[10] The crossing angle for the 8b+4e alternative could be reduced down to about 400 μrad (9 σ) thanks to the lower number of long ranges.

[11] Minimum normalized long-range beam-beam separation at minimum β*

[12] The 8b+4e variant represents a back-up scenario for the basline 25ns operation in case of e-cloud limitations. The parameters are still evolving but are stated for the sake of performance reach comparison.

aperture, and require a peak field 50% above the present LHC. This implies the use of new, advanced, superconducting technology based on Nb_3Sn.

The drawback of very small β^* is that it requires a larger crossing angle. This causes a reduction of the geometrical luminosity reduction factor R. In Figure 4 the reduction factor is plotted vs. β^* values.

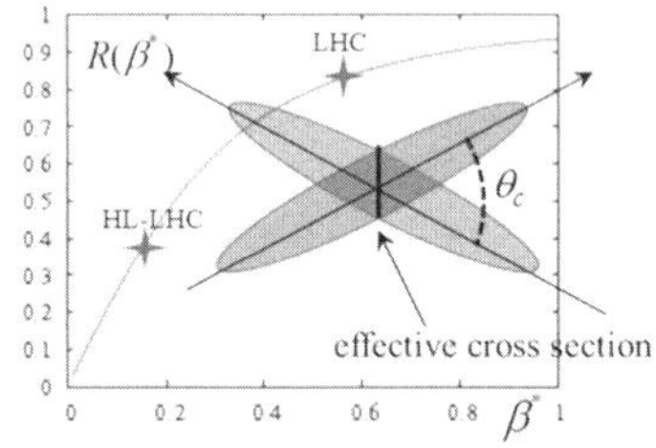

Fig. 4. Behaviour of geometrical luminosity reduction factor vs. β* for a constant normalized beam separation with the indication of two operational points: nominal LHC and HL-LHC. The sketch of bunch crossing shows the reduction mechanism.

Use of special superconducting RF crab cavities, capable of generating transverse electric fields that rotate each bunch longitudinally by $\theta_c/2$, such that they effectively collide head on, overlapping perfectly at the collision points and cancelling –almost– the reduction factor R, as illustrated in Figure 5. While the primary function of the crab cavities is to boost the virtual peak luminosity, they can also be used in combination with dynamic β^* variation during the fill. Finally, the crab cavities can be used to tilt the bunches in a direction perpendicular to the plane of crossing, providing pile-up and pile up density control and an additional handle for luminosity levelling through the so-called 'crab-kissing' scheme.

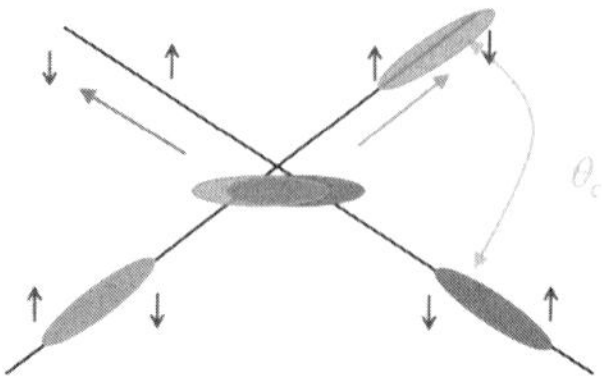

Fig.5. Effect of the crab cavity on the beam (small arrows indicate the torque on the bunch generated by the transverse RF field).

2.6 Luminosity Performance

The set of parameters indicated in the Table I would allow to reach 3000 fb^{-1} about twelve years after installation. A possible evolution of the luminosity performance is illustrated in Figure 6.

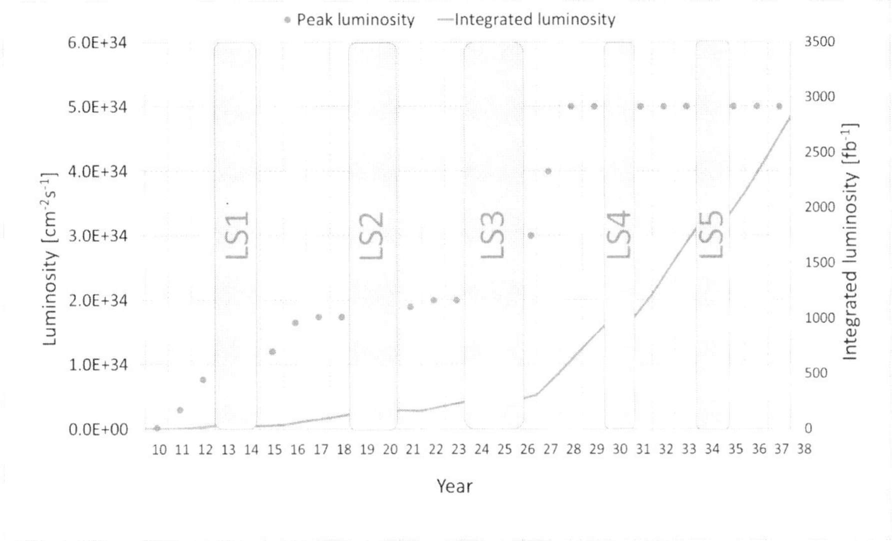

Fig. 6. Peak luminosity and integrated luminosity vs time for the LHC machine and HL-LHC evolution. (Courtesy of M. Lamont, CERN).

3. Hardware and Main Technology Advance

3.1 New Magnet Technology for HL-LHC

For the HL-LHC various new magnets will be needed. Some of them will be designed with the same technology as the LHC ones, with Nb-Ti superconductors. They are more difficult than the LHC magnets because of the larger apertures and then larger forces and stored energy. However, they can be considered an evolution of the LHC magnets. So here, we will focus on the new magnet technology that is being developed for HL-LHC, which makes use of new type of superconductor (Nb_3Sn).

The replacement of the present LHC inner triplet (IT) quadrupoles is the backbone of the luminosity upgrade. The new quadrupoles made use of the new more advanced superconducting material, Nb_3Sn. It is the same as the one used for the ITER project, however for HL-LHC we need a performance in critical current density $J_c > 2500$ A/mm^2 at a field of 12 T (temperature of 4.2 K), i.e. three times higher than the one required for the ITER superconducting magnets. Thanks to the Nb_3Sn the peak field that can be reached is about 12 T in operation, when working at 1.9 K temperature. To reach this operating goal, the superconductor will be designed for about 15 T maximum performance, while the magnet structure will be designed for about 13.5 T. The comparison between the IT quadrupole and the LHC main dipoles or IT quadrupole can be summarized as follow: LHC dipoles have 50 mm aperture and $B_{peak} \cong 8.4$ T, LHC IT quadrupoles have 70 mm aperture and $B_{peak} \cong 8$ T, while the HL-LHC IT quadrupoles (named MQXF) have 150 mm aperture and feature $B_{peak} \cong 12$ T.

In addition to the IT quadrupoles, another type of high field magnet will make use of Nb_3Sn technology: to make room for special collimation in the Dispersion Suppressor, a few LHC dipoles, 15 m long, will be replaced 11 T field dipoles about 11 m long. This would leave about 4 m of room to place a Cold-Warm-Cold bypass to lodge at room temperatures collimators to intercept beam losses in the Dispersion Suppression regions. The 11 T has an aperture of 60 mm, similar to the LHC dipole, however the 30% more field made use of Nb_3Sn also necessary. In addition to the high field, the fact that has to be powered in series with all the 154 dipoles in a sector (1/8 of the LHC ring) put significant challenges on this magnet. The Nb_3Sn has been developed mainly by the US program LARP. And in Figure 7 the evolution of the cross section from the present LHC triplet to the final MQXF is illustrated. More information on the Nb_3Sn magnet program for HL-LHC is in [11] and [12].

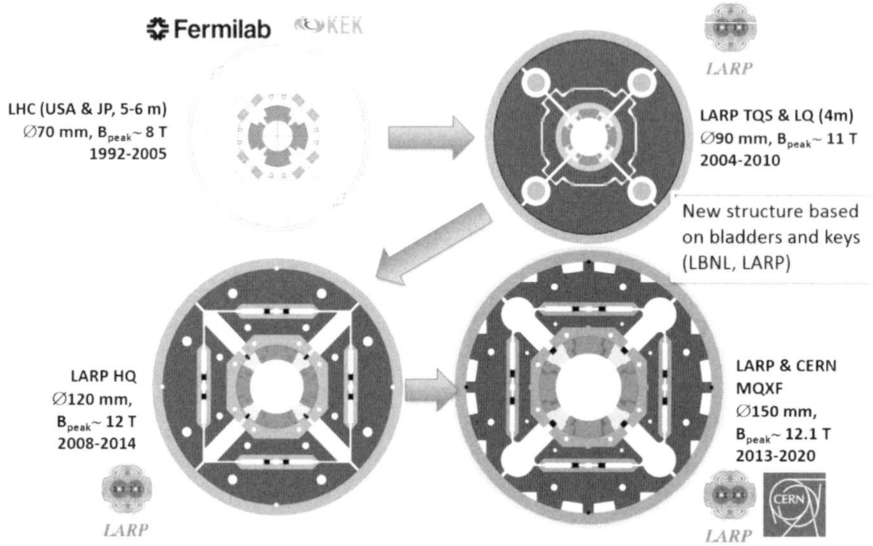

Fig. 7. Evolution of the cross section of the Inner Triplet Quadrupole, form LHC (top left) to HL-LHC (bottom right). Courtesy of G. Ambrosio (FNAL), P. Ferracin and E. Todesco (CERN).

The challenges of this technological breakthrough (it will the first time that Nb_3Sn magnets are used in accelerators) is illustrated by the plot of Figure 8 were the evolution along the years of high field for colliders is illustrated. Then plot clearly shows how much HL-LHC step is necessary on the way toward higher field for future accelerator. Indeed the 100 km project called FCC is at present based on dipole magnets to be operated at 16 T, which can be a credible goal only if the 12 T regime of HL-LHC would be secured. Even the FCC operation at 18-20 T (indicated as High Energy LHC in Figure 8) is based on great part on Nb_3Sn technology, besides using HTS technology, therefore the effort of HL-LHC would be essential also in this case.

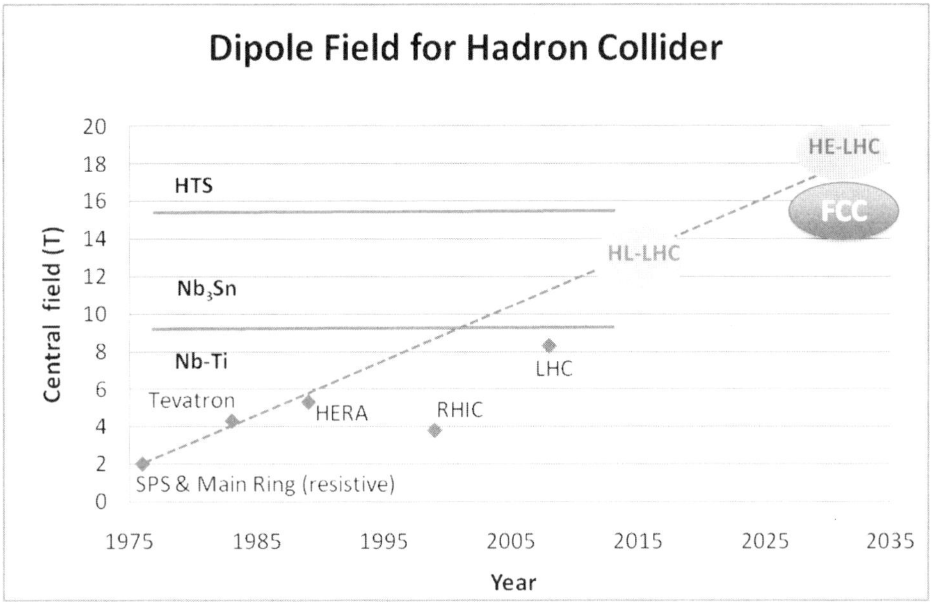

Fig. 8. Evolution of peak field in past accelerators and what is foreseen for near (HL_LHC) and far future projects.

3.2 Crab Cavity

Another technological novelty of the project is the use of the crab cavities [13]. Crab cavity concept has been tested in KEK-Belle on an electron-positron ring. What is new in HL-LHC is the use on proton ring (which requires a much lower level of noise and much greater accuracy) and the use of special shape compact crab cavity, in place of the usual elliptical shape typical for electron beams. The transverse size of HL-LHC compact crab cavity is less than half than would be with classical elliptical shape, allowing installation in the very compact LHC insertion regions around ATLAS and CMS experiment. Not only the cavity is of new conception: cavities placed at opposite side of the collision point must give exactly opposite rotation kick, with an accuracy that can be measure in 0.001° of RF phase, i.e. with synchronisation in time of fee tens of fs. A detailed description is not possible here, however in Figure 9 is reported a view of both types of crab cavities under development for HL-LHC: the RF Dipole (RFD) and the Double Quarter Wave (DQW) cavity. The former is more suited for defecting beam in the vertical plane while the second in the horizontal plane (ATLAS and CMS have vertical and horizontal beam crossing, respectively).

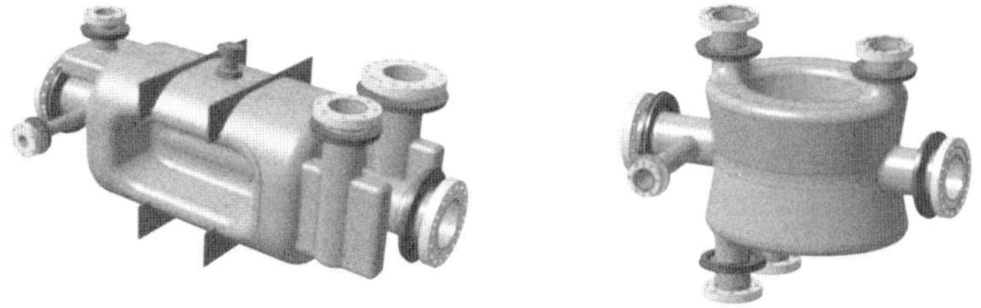

Fig. 9. Schematic view of the crab cavity (left: DQW & Right: RFD). Courtesy of R. Calaga (CERN).

3.4 Collimators

Many types of collimators are needed for HL-LHC [14]. The most advanced concept is making use of Molybdenum Carbide- Graphite (Mo-Gr) composites, possibly coated with pure Mo. Other lower Z refractory coatings are presently under study. The Mo coating considered now would reduce the surface resistivity by about a factor 10 to 20 compared to Mo-Gr and by more than a factor 100 compared to present Carbon-fibre Carbon composites (CFC). The benefit on the impedance budget of the collimation system would be significant: in the relevant frequency range, impedance would be reduced to 10 % of the one of the CFC jaws. In Figure 10 is shown a scheme of the new material collimator jaws.

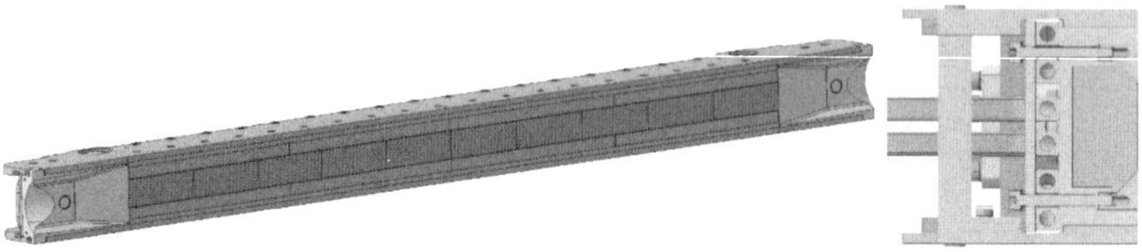

Fig. 10. Preliminary design of the TCSMP jaw. This features 10 MoGr blocks. Also note lengthened jaw taperings, futher reducing contribution to HOM RF instabilities in the geometrical transition zones. Courtesy of S. Redaelli (CERN).

3.5 Superconducting Links

To connect the far-away placed power converters to the superconducting magnets in the tunnel, about 100 kA would need to be transported for a length of about 100 m. We need such connections on each side of both high luminosity Points, IP1 and IP5. The only way to avoid several MW of dissipation in these four connections is using superconducting links [15]. To gain temperature margin and stability in operation we intend to make use of new superconductors: either MgB_2 operating in the 4-20 K range or HTS (high temperature superconductors) which can be operated up to 30-40 K temperature. The baseline is use of MgB_2, which is much cheaper than HTS and comes in form of round wire, suitable to easy cabling configuration. A possible scheme of such cable, the cabling and the first cable test are shown in Figure 11.

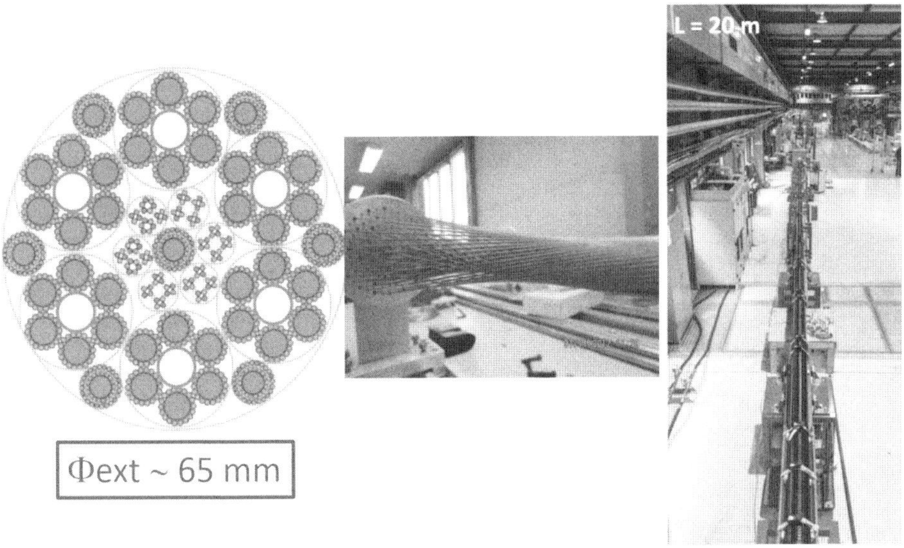

Fig. 11. Left: cross section of the MgB_2 cable; center: cabling MbB_2 round strands one element of the cable shown in left; right: first 20 m long prototype cable, containing part of the cable shown in left, in the flexible cryostat (black tube) under test at CERN.

4. Civil Engineer

The HL-LHC needs a considerable new construction, both in surface and underground. On each site of the high luminosity insertion, Point 1 and Point 5, about seven new buildings will be constructed, one for the 12 m large pit to descend the various equipment. In addition, a 300 m long gallery will be excavated and equipped aside of the LHC tunnel, to locate the magnet power converter, the powering of the Crab Cavity, protection equipment and many other equipment. At one side a larger gallery will host the new cryogenic plants we want to install at Point 1 and Point 5, capable of 18 kW 4.2 K power, to deliver about 3 kW of cooling at HeII. The civil engineer construction and its technical infrastructure is one of the major challenge of the HL-LHC project, in terms of cost and schedule, since recent measurement of beam disturbances induced by vibrations have ruled out the possibility of excavating during Run3, see Figure 1. Therefore main excavation works must be executed during LS2, see Figure 1. In Figure 12 is illustrated the underground works for Point 1. In total more than 30,000 m3 of volume must be excavated on both P1 and P2 at around 100 m underground.

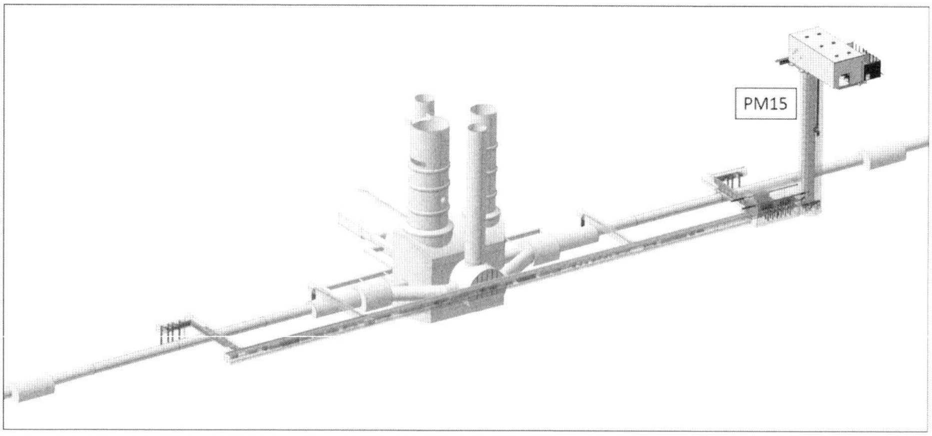

Fig. 12. The new aside gallery (tunnel in front view with equipment inside) running 300 m long aside LHC tunnel (in grey), with its own pit (PM15) and the four short connection tunnels. Courtesy of P. Fessia, I. Bejar Alonso and J. Osborne (CERN).

References

[1] The European Strategy for Particle Physics Update 2013, CERN-Council-S/106, adopted at a special session in the Brussels on 30 May 2013. http://cern.ch/council/en/EuropeanStrategy/ESParticlePhysics.html

[2] Building for Discovery: Strategic Plan for U.S. Particle Physics in the Global Context, in http://science.energy.gov/hep/hepap/reports/

[3] Lucio Rossi, "LHC Upgrade Plans: Options and Strategy", Proceedings of IPAC2011, San Sebastián, Spain, 01/09/2011, pp. 908-912

[4] LARP Proposal (2003), http://www.uslarp.org/

[5] O. Bruning, R. Cappi, R. Garoby, O. Grobner, W. Herr, T. Linnecar, R. Ostojic, K. Potter, L. Rossi, F. Ruggiero (editor), K. Schindl, G. Stevenson, L. Tavian, T. Taylor, E. Tsesmelis, E. Weisse, and F. Zimmermann, "LHC Luminosity and Energy Upgrade: A Feasibility Study", LHC Project Report n. 626, CER 1 December 2002

[6] J. Strait et al., "Towards a New LHC Interaction Region Design for a Luminosity Upgrade"http://accelconf.web.cern.ch/AccelConf/p03/PAPERS/MOPA006.PDF

[7] R. Ostojic, "Overview of the LHC IR Upgrade Phase-1" Proceedings of the Chamonix 2010 Workshop on LHC Performance, CERN-ATS-2010-026

[8] L. Rossi and O. Brüning, "High Luminosity Large Hardon Collider – A description for the European Strategy Preparatory Group", CERN-ATS-2012-236

[9] 7th Framework Programme, Theme INFRA-2011-2.1.1., "FP7 High Luminosity Large Hadron Colider Design Study", Grant Agreement n. 284404

[10] LHC Injector Upgrade – Technical Design Report – Volume I: Protons, J. Coupard et al (eds), CERN-ACC-2014-0337

[11] L. Bottura, G. de Rijk, L. Rossi, E. Todesco, "Advanced Accelerator Magnets for Upgrading the LHC", IEEE Transactions on Applied Superconductivity, June 2012, Vol. 22, No. 3.

[12] E. Todesco et al, " Superconducting Magnet Technology for the Upgrade", in *The High Luminosty Large Hadron Collider",* Oliver Brüning and Lucio Rossi eds, Advanced Series on Directions in High Energy Physics – Vol. 24, World Scientific, 2015.

[13] R. Calaga, E. Jensen, G. Burt and A. Ratti, "Crab Cavity Development", in *The High Luminosty Large Hadron Collider",* Oliver Brüning and Lucio Rossi eds, Advanced Series on Directions in High Energy Physics – Vol. 24, World Scientific, 2015.

[14] S. Redaelli et al., "Cleaning Insertions and Collimation Challenges", in *The High Luminosty Large Hadron Collider",* Oliver Brüning and Lucio Rossi eds, Advanced Series on Directions in High Energy Physics – Vol. 24, World Scientific, 2015.

[15] A. Ballarino and J.P. Burnet, " Powering of the High-Luminosity Triplets", in *The High Luminosity Large Hadron Collider",* Oliver Brüning and Lucio Rossi eds, Advanced Series on Directions in High Energy Physics – Vol. 24, World Scientific, 2015.

LECTURER: L. ROSSI

High Luminosity LHC Project

Scientific Secretaries: S.Mamedov, A.Rebesh

DISCUSSION

- *Simoniello:*

The last topic you have touched on slide 63, is a very nice reduction of the pile-up with the crab-kissing. Could you explain a bit more. How does it actually work? Always with the tilt of the beam?

- *Rossi:*

The idea is to reduce pile up density rather than total pile up. Take the beam: two bunches, i.e. the two cylinders collide with a crossing angle. Take for example the CMS where there is horizontal crossing (i.e. the two beams lay in the horizontal plane). The two beams that should cross with a reduced cross section because of the angle, they are kicked transversally and we do it in such a way that they cross exactly head-on, i.e. at the crossing the bunches are perfectly parallel, despite their trajectories have an angle. We can also use half of the crab cavity to go further: to tilt the two beams in the other plane (for CMS the vertical plane). This we call crab-kissing scheme. In this way the two beams cross in zero time. In a "standard" crossing even with crab gymnastic you will always have a little bit of angle and the real crossing takes time and it is short in length. If you apply crab-kissing the two bunches cross each other at the same time but the length where they cross is elongated, so the number of collisions is spread over a longer space, i.e. the pile up density is reduced. Suppose that we have four crab cavities per side. Two cavities we use to change the crossing angle and the other two for this crab-kissing scheme. It is quite complicated and it has not been proved yet. It was published by Stéphane Fartoukh who invented this. We have still to prove it, it will take long. The manipulation of crossing angle is the baseline of the project, however the crab-kissing is not yet the baseline of the project but we are working on it. We hope in a few years to have it as a baseline. We are also discussing whether to install a second harmonic system (RF harmonic system) in the accelerator that helps to stretch the bunch length. Because, basically, if you have one RF system, you have only a certain length of the beam, because you can only use a few degrees of the RF cycle. But with the second harmonic, basically, you can make the beam either shorter or longer.

- *Spiro:*

Just to complete the question. What is the concept of the crab waist cavities?

- *Rossi:*

It is different. The crab waist that was devised by Pantaleo Raimondi (previously in Frascati now in ESRF) is different. Basically it has to do with the fact that with some special sextupoles one can change the shape of the beam, better of the single bunch, at the collision point. You have to realize that when the focal length at collision β^*, is very small, i.e. comparable with the length of the bunch itself the size of the beam is not anymore constant. It is the hourglass effect.... The β^* is produced by the Inner Triplet (IT) quadrupole magnets, placed just before ATLAS, or CMS. If the variation length of the β* is shorter than the bunch, it means that the bunch itself will have a different β* inside the bunch. In other words, the bunch itself will not any more be a cylinder. It will be kind of an hourglass shape. This idea of Pantaleo counteracts this. In LHC we are at the limit, but we don't see a lot of the hourglass effect so we will not apply the crab waist scheme (which requires special sextupoles). This is typically used in an electron machine. I think they will apply this if for example they do the Chinese Lepton Collider.

- *Savitskyi:*

In 2013 during the 8 TeV LHC physics community got unexpected p-Lead Run after Lead-Lead programme. Is there any chance to get similar runs at High Luminosity LHC? Can there be any obstacle with these at HL-LHC?

- *Rossi:*

Of course LHC is in game for nuclei collisions, in particular for Lead-Lead and p-Lead runs even at HL-LHC. The time and period of actual run will depend from the needs of the experiments and from performance of the ALICE experiment in particular. Currently, it is scheduled to produce ions collisions till around 2030. Probably this will occur for one month (15% of time relatively to p-p program) per year. If expectations for Luminosity accumulation won't be achieved by ALICE, run can be extended. Are you satisfied with the answer?

- *Savitskyi:*

Honestly, I expected that there can be some obstacles with pile-up and so on during the high Luminosity. I mean that the high Luminosity set up of the collider or the detector set up can be incompatible with high track presence during the collision of different ions and so on.

- *Rossi:*

We do not touch the optics in point 2 where there is ALICE for example. When you run ions, it is compatible with the set up. For example, when you run ions, actually not only ALICE taking data but also ATLAS and CMS. CMS by the way is a very big community of heavy ions. So the luminosity in Alice depends on the interplay between the three experiments talking data during ion run. High Luminosity is perfectly compatible with the ions. It is not a problem.

- *Savitskyi:*

Actually I did not expect such answer. Usually people are concerned only with the proton-proton program and so on.

- *Rossi:*

No. The ions are fully part of the scene. It is fully compatible with the high lumi regime in ATLAS and CMS.The optics is compatible.

- *Tannenbaum:*

You mention, High Luminosity Pb+Pb runs in 2025-2026. Isn't Pb+Pb Luminosity limited by the reaction $Pb+Pb \rightarrow Pb+Pb+e^+e^-$ where the e^- is captured on a Pb nucleus which then is host. How high is that Luminosity limit?

- *Rossi:*

We may be limited by this effect which you say creates some species of ions with different magnetic rigidity that are going into the magnet. But for this we are working on improving the collimation. Actually, the first 11 Tesla dipole for the new collimation in the cold region to counteract this ion loss, is foreseen to go to ALICE. It's not sure that we really need it, but we are preparing for this actually for ALICE in order to avoid this possible limitation to the luminosity, which for the moment is limited by ALICE itself. I can tell you that we discovered to our surprise that who today is collecting more ion collision is CMS, it's not ALICE. CMS and ATLAS are working at double peak luminosity than ALICE, that's why ALICE wants to upgrade and to have much faster collection. Of course, ALICE is specialized, they see very well the time of flight and so on, but in terms of peak luminosity ATLAS and CMS are better than ALICE, so they are burning more ions than ALICE.

- *Belov:*

I have two questions. The first one is the following. When you speak about Luminosity, do you mean stored or delivered Luminosity? And how do you measure or calculate the delivered Luminosity? Just by the formula which you provide?

- *Rossi:*

First of all, how do we measure the Luminosity. We have a system but we discovered that the best one is just to rely on the Luminosity measurement by the experiment. It is the most accurate measure because they need to know it very well. The Luminosity depends on many parameters. Since we measure each single parameter with a certain accuracy we are able to determine what is the Luminosity at least at the beginning. And then we control the proton burning, the decrease of intensity of the beam and so on. But the best measurement is coming from the experiment itself. What is the difference in delivered and collected the Luminosity? We always say the Luminosity that we deliver to the experiment. In reality the experiment - maybe Peter can correct me - collects a Luminosity which is observed of 10-15% less. The example is that the delivered Luminosity is 25 fb^{-1} and the stored is 22 fb^{-1} (it is just a particular case).I think Peter can comment more.

- *Jenni:*

I think I can comment on that. I actually showed a slide. We are typically speaking about CMS and ATLAS. We recorded about 95% of the delivered Luminosity when the machine is declared stable and then come still the data preparation phase where one checks whether the quality of data is perfect and the end result is typically 90% of the stably delivered Luminosity which is very remarkable if one compare it to the previous experiment. Maybe I can also just comment that most precise determination of the Luminosity comes indeed from the experiments with quite a lot of work. It is typically in the order of 2 to 3% precision. And it uses actually a very old method to determine the overlap area of the beam by running at low Luminosity. The beams cross each other and determine essentially the effective overlap of the beams.

- *Belov:*

The second question. You emphasized that the helium temperature is 1.9 K. Could you please repeat why or why not 1 K or 3 K? Because the temperature of boiling is about 4 K.

- *Rossi:*

Most of the magnets in all previous accelerators (Tevatron, RHIC and so on) work at 4.2 K or 4.5 K which is the normal liquid helium boiling temperature at the normal pressure. So why we are using 1.9 K, 1.8 K? It's because niobium-titanium, the material that we are using, improves its superconductive properties going down in temperature. We have to understand that superconductivity is a short-cut of a property that, in principle, should appear at 0 K, It is basically a transition phase in a space of three parameters: temperature, as in any thermodynamic system, the intensive property is the magnetic field and the extensive property is the current. So the magnetic field plays the role of pressure and the current plays the role of volume of the ideal gas thermodynamic system. And you have to be below a certain critical surface to have superconductivity. That means, if you decrease the temperature, you have more room to have higher field and more current. The critical temperature of niobium is 9 K. But at 9 K it just becomes barely superconductive, you have no possibility to send a lot of current and to have a high field. Going down in the temperature, you transform really the margin of temperature in room for current and field, let's put it this way. So, niobium-titanium is very sensitive to this and, practically, going down in temperature from 4 K to 1.8–1.9 K, basically you can have 2 T more field in the magnets. Our magnet will be limited to probably 6 to 7 tesla at 4.2 K... Let's make an example: SSC, the superconductive machine of the United States, the Super Superconductive Collider, quoted by Peter Jenni. It was designed to work at 4.2 K and was pushing the magnet at the maximum and the magnet was set for 6.6 K. It was already very pushing. Maybe more pushing than our 8.3 K, in proportion. So it's 1.8 T difference. This means, if translated in energy, that we can have 1.2 TeV per beam:, it means we gain 2.5 TeV center of mass by going down to 1.9 K. It is not enormous, but it is something,: that is why we got to superfluid helium. Why don't we go down? Because most of the gain, when you are at 1.8 K, is there. And if you push more, of course, you are fighting against thermodynamics. The capability to take out entropy decreases, it becomes more and more difficult. You can go to 1.4 K with a tremendous effort, you have to spend a lot more, let's put it this way. There is also a third reason which is that the superfluid helium has zero mechanical friction. And actually the superfluid helium you can describe as a mixture of a real absolute superfluid component plus a resisting component, mechanical resistance in this case. It turned out that the capability to transfer heat of superfluid helium is maximum at around 1.8–1.9 K. So thermically it is even better to stay at 1.8-1.9 K respect 1.4 K.. So there are a lot of principles and a lot of technical reasons to stay around this temperature.

- *Campbell:*

I have two questions. And they are both about the collimation system. The first question is when you go through the primary collimator, you increase the shower. How do you prevent particles from the shower in the collimators from travelling down the beampipe?

- *Rossi:*

Particles are also going along the beam chamber. The probability that a particle is perfectly aligned is almost zero. That means, eventually, either it goes off and will be caught. Actually we don't have collimation only in one point of the machine; we have collimation in two points of the machine. Do not forget that the beam oscillates. Let me do only parentheses because about the β. What is the β? It is the fact that the beam oscillates because there are convergent lenses which are the quadrupole. And, like any lenses, the stronger the lens, the shorter is the focal length. The β is the focal length of the beam. This is why inside experiments, the β*, the β is very small because the beam is strongly focused. So, going back, the beam oscillates and sooner or later it will be caught by one element collimation system. When it goes in the halo of the beam there is no hope that it stays stable here. Eventually some particles go to the magnet. But it's only one out of ten thousand or one out of a hundred thousand.

- *Jenni:*

But mathematically these particles are at lower momenta so they are swept into the collimators which come afterwords.

- *Savitskiy:*

Can I comment here? For example in the CMS case, there was a CASTOR experiment which was using halo-muons, for example, for the calibration. These halo-muons appeared at the first collimator which is closest to the CMS. Actually, this also can happen in this case. Those halo-muons will move towards the beam direction.

- *Rossi:*

This is the part of the results. Some particles are identical to the parts of the background. In fact, we do a lot of calculation for this. Some can eventually arrive to detectors. This is proton 7 TeV energyThere can be also a much more dangerous thing. For example, we couldn't avoid the splash in ALICE at certain point, it seems to me, because there was the injection error. Despite the fact that collimation is the the main absorber, some particles can leak.

- *Campbell:*

My second question is about the absorber. What do you use to observe damage energy and what will work in your new high luminosity, higher energy plans?

- *Rossi:*

This is a good point. In fact, the primary collimators are never heavy metal. Because they will absorb too much and they will be destroyed. In fact, the primary collimator is basically graphite, light element. We want to spread, let's say, particles. And this one, the new for HiLumi, is a mixed, it's graphite and carbon. And now we want also, by the way, to put copper, because graphite has a lot of resistance, so the beam sees it as a resistor. And eventually the final absorber, you see here, is copper and tungsten. And in high luminosity we want to do it all in tungsten, the final absorber. We are also testing a new type of collimation based on special crystal that can collimate protons by channeling. If it works, it's very good. Of course, it's a little bit dangerous because in crystal collimation, the real key point is the absorber. Suppose that instead of the primary collimator you put something that takes all of the beam halo (and even a fraction of the core of the beam) and puts it aside via channeling. But then we need to add an absorber that is capable to absorb all particles in a short range, it is a real problem. So the absorber is a real point. Do not forget that the dump of the LHC, the one that is under the airport shown by Peter Jenni is really a huge system basically made of iron and concrete. And there are special kicker magnets that dilute the beam. Because otherwise the beam will continue to perforate and eventually will go up and I don't know where, in some regions, some nice valleys of Switzerland and Peter will not be happy. This to stress that absorbing the LHC beam is not trivial.

- *Savitskiy:*

Once I was asked a similar question. Where actually LHC dumps the beam and it was very interesting because usually you cannnot find such information easily. But what happens when the beam starts to heat the storage where the beam has dumped? What happens with the heat? It won't warm up the airport, right?

- *Rossi:*

No in fact, because we don't just have a simple kicker that takes out the beam. We have the kicker that takes out the beam, and we have special kickers that we call dilution kickers, they spread the beam in such a way that a bunch goes here, the other bunch goes there, the other bunch goes in another place and so on. The 1500 bunches (3000 bunches in the future) are spread over a wide volume or hit a wide surface exactly to avoid energy concentration. Without the dilution kicker we can not operate the LHC, for lack of safety in the beam dump. Nothing to be worry but it is true, it is a critical point. And, I repeat, there is a gallery where we lose the beam where we put a long dump and we play this game of painting the beam over a large surface.

Highlights from BNL and RHIC 2015

M. J. Tannenbaum *
Physics Department, 510c,
Brookhaven National Laboratory,
Upton, NY 11973-5000, USA
mjt@bnl.gov

1 Introduction

The Relativistic Heavy Ion Collider (RHIC) at Brookhaven National Laboratory (BNL) is one of the two remaining operating hadron colliders (the other being the LHC at CERN); and the first and only polarized proton collider. BNL is a multipurpose laboratory, quite different in scope from Fermilab and CERN, with many "cutting edge" major research facilities in addition to RHIC. BNL, which is owned by the U.S. Government but operated by a "management and operating (M&O) contractor" was founded by nine major northeastern universities in 1947 to promote basic research in the physical, chemical, biological and engineering aspects of the atomic sciences and for the purpose of the design, construction and operation of large scientific research facilities that individual institutions could not afford to develop on their own. In addition to RHIC, BNL is now home to the National Synchrotron Light-source-II (NSLS-II), the NASA Space Radiation Laboratory, a Tandem Van de Graaf, an Accelerator Test Facility, a Linac Isotope Producer, the Long Island Solar Farm, a Center for Functional Nanomaterials as well as Radiochemistry, Biological Imaging, Environmental & Climate Sciences groups and a Nonproliferation and National Security Department. BNL has a distinguished history in nuclear & particle physics and accelerator science [1] as well as many other discoveries [2]. So far, Nobel Prizes have been awarded to 12 scientists who were either BNL staff members or performed their Nobel work at BNL [3].

2 News from BNL since ISSP2014

The news this past year was very positive in many ways. The "management and operating (M&O) contract" for BNL, for which there was a solicitation last year [4], was awarded to the present management, BSA, which had run the lab for the past 15 years, so that there was a very smooth transition. The major event was the startup and dedication of the "World's brightest Synchrotoron Light Source", NSLS-II, by Secretary of Energy Ernest Moniz (who was also an important participant in the USA, France, Germany, UK, China, Russia negotiations with Iran which were ongoing during the ISSP2015 school and reached a

*Research supported by U. S. Department of Energy, DE-SC0012704.

historic agreement on a Joint Comprehensive Plan of Action on July 14, 2015 to limit Iran's Nuclear Energy activities to civilian purposes).

In addition to the successful RHIC run, several other activities took place which related to the future. The proposal for a new "Large-Acceptance Jet and Upsilon detector for RHIC" using hadron calorimetry outside a thin-coil superconducting solenoid (originally called sPHENIX) had two important milestones. The laboratory management called a general meeting on June 16, 2015 [5] "to move expeditiously to form a new detector collaboration to take advantage of the physics opoortunities" offered by this proposal. The result was the formation of an Institutional Board for the new collaboration, a committee to develop the ByLaws of the new collaboration and to identify spokesperson candidates, working groups to develop a Pre-Conceptual Design Report for a BNL Director's Cost and Schedule review, Nov 9-10, 2015, and a plan to hold a collaboration meeting December 10-12, 2015.

The second milestone was the mid-winter shipment and safe arrival of the (made in Italy) BABAR superconducting solenoid magnet from SLAC on January 16 to BNL on February 3, 2015, which went smoothly (Fig. 1). The solenoid arrived in mint condition as determined from detailed acceptance tests by the BNL Superconducting Magnet Division.

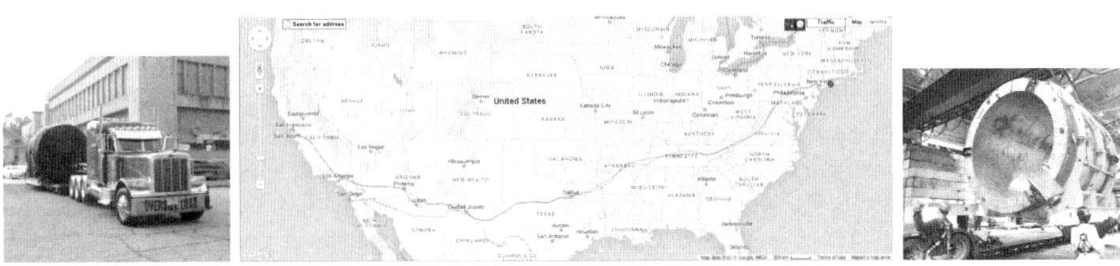

Figure 1: a)(left) BABAR solenoid leaves SLAC on Air Ride Trailer b) (center) Route across U.S.A. c) (right) Unloading at BNL.

Also important for the future of Nuclear Physics in the U.S. was the completion of the Long Range Planning exercise for the Nuclear Science Advisory Committee (NSAC) in April 2015, with the final version to be submitted to the Department of Energy (DOE) in October 2015 [6], which was unfortunately leaked to the press in May with an article in Nature [7] with the headline "Billion-dollar collider gets thumbs up". The collider in question is an electron-ion collider, which if located at BNL would be called eRHIC, for which BNL has proposed a highly advanced and energy efficient FFAG electron accelerator/storage ring based on an Energy Recovery Linac that I discussed here last year [4]. The Nature article made two points that I wish to discuss further.

The first involves cost, which a review by an NSAC expert subcommittee [8] put at \$1.5 Billion for either the JLAB or BNL version, which was equal to the JLAB proposal but \$0.5 Billion higher than BNL's proposed cost because of the committee's concern of the higher technical risk of the novel advanced accelerator design. Hopefully, ongoing R&D and similar work for the CERN LHeC project [9] will overcome the present technical risk and reduce the cost, because many pundits questioned the feasibility of the U.S. alone building an EIC at the \$1.5 Billion price. The second point made in the article was that "The machine should also solve a puzzle about the proton that has baffled physicists for nearly 30 years"—"strangely the spins of its three constituent quarks add up to only about 1/3 of its own spin." Evidently, Nature [7] did not know about the latest RHIC-spin results (Fig. 2) which show

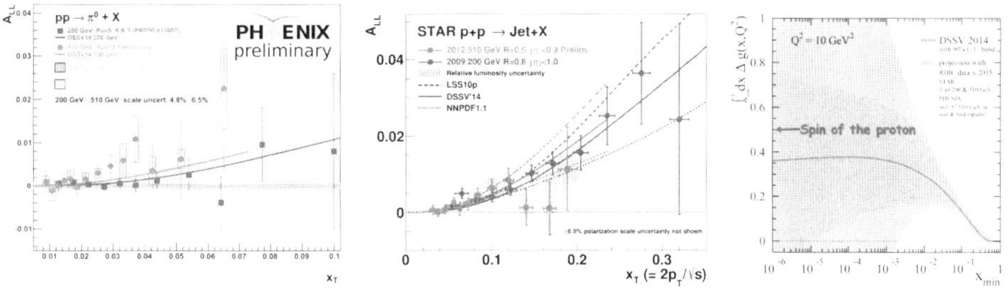

Figure 2: a)(left) PHENIX preliminary measurements of A_{LL} of π^0 [11] with DSSV14 fit[12] b) (center) STAR measurement of A_{LL} of jets [11] with DSSV14 fit[12] c)(right) Integral of the helicity asymmetry of gluon structure function $\Delta g(x, Q^2)$ from DSSV14 fit[12] as a function of $x_{\rm min}$ [11].

a significant non-zero gluon spin contribution to the proton from measurements of A_{LL}, the two-spin asymmetry for the scattering of two longitudinally polarized protons: [10]

$$A_{LL} = \frac{1}{P_1 P_2} \frac{(\sigma^{++} + \sigma^{--}) - (\sigma^{+-} + \sigma^{-+})}{(\sigma^{++} + \sigma^{--}) + (\sigma^{+-} + \sigma^{-+})} = \frac{1}{P_1 P_2} \frac{\sigma^{++} - \sigma^{+-}}{\sigma^{++} + \sigma^{+-}} \quad \text{if parity is conserved,} \qquad (1)$$

where $\sigma^{++} \equiv N^{++}/L^{++}$ is the measured cross section with both beams having '+' helicity, N^{++} is the measured number of events for an integrated luminosity L^{++}, with analogous notation for the other helicity combinations; and P_1 and P_2 are the polarizations of the two beams. The parton helicity asymmetries are related to the proton asymmetries by QCD [10].

For those not familiar with the standard nomenclature: the the helicity asymmetry of the structure function $a(x, Q^2)$ for a parton a (where a represents e.g. u-quark, $\bar{u}$-quark, gluon, etc.) is defined as $\Delta a(x, Q^2) \equiv a^+(x, Q^2) - a^-(x, Q^2)$ and the "+" and "-" refer to partons with the same or opposite helicity as the parent proton. There is a helicity sum rule for the spin 1/2 proton:

$$\frac{1}{2} = \frac{1}{2}\Delta\Sigma + \Delta G + L_q + L_g$$

where $\Delta\Sigma \equiv \int_0^1 dx\, [\Delta u(x, Q^2) + \Delta\bar{u}(x, Q^2) + \Delta d(x, Q^2) + \ldots]$ is the combined quark and anti-quark spin contribution, $\Delta G \equiv \int_0^1 dx\ \Delta g(x, Q^2)$ is the gluon contribution; and L_q and L_g are possible quark and gluon angular momentum contributions. The quark contribution, which has been measured in DIS [13] to be $\Delta\Sigma \approx 0.25$, thus accounts for only $\sim 1/4$ of the proton spin. "The inclusive DIS measurements have, however, very little sensitivity to gluons" [12]. The new results from Fig. 2 are the first evidence of a finite gluon polarization in the proton [12]. Notice that the best fit of DSSV14 in Fig. 2c gives $\Delta G \geq 0.36$ for $x_{\rm min} \leq 10^{-3}$. If improved measurements at $x = 10^{-3}$ by future RHIC runs keep the present best fit of DSSV14 with smaller errors, it is conceivable that $\Delta\Sigma/2 + \Delta G$ could add up to 1/2 with sufficiently small errors to solve the proton spin puzzle.

3 RHIC Operations in 2015 and accelerator future plans

Physics data taking for the 15th run at RHIC started on February 11, 2015 and ended on June 22, all at $\sqrt{s_{_{NN}}}$ =200 GeV. The principal objectives were: i) a comparison p+p

RHIC Run	Year	Species	Energy	Ldt
Run-1	2000	Au+Au	130 GeV	1 μb-1
Run-2	2001-2	Au+Au	200 GeV	24 μb-1
Run-2		Au+Au	19 GeV	0.4 μb-1
		p+p	200 Gev	150 nb-1
Run-3	2002/3	d+Au	200 GeV	2.74 nb-1
		p+p (L)	200 GeV	0.35 nb-1
Run-4	2003/4	Au+Au	200 GeV	241 μb-1
		Au+Au	62.4 GeV	9 μb-1
Run-5	2005	Cu+Cu	200 GeV	3 nb-1
		Cu+Cu	62.4 GeV	0.19 nb-1
		Cu+Cu	22.4 GeV	2.7 μb-1
		p+p (L)	200 GeV	3.8 pb-1
Run-6	2006	p+p (T+L)	200 GeV	10.7 pb-1
		p+p	62.4 GeV	100 nb-1
Run-7	2007	Au+Au	200 GeV	813 μb-1
Run-8	2007/2008	d+Au	200 GeV	80 nb-1
		p+p (T)	200 GeV	5.2 pb-1
		Au+Au	9.2 GeV	
Run-9	2009	p+p (L)	200 GeV	16 pb-1
		p+p (L)	500 GeV	14 pb-1
Run-10	2010	Au+Au	200 GeV	1.3 nb-1
		Au+Au	62.4 GeV	100 μb-1
		Au+Au	39 GeV	40 μb-1
		Au+Au	7.7 GeV	260 mb-1
Run-11	2011	p+p (L)	500 GeV	27 pb-1
		Au+Au	200 GeV	915 μb-1
		Au+Au	27 GeV	5.2 μb-1
		Au+Au	19.6 GeV	13.7 M events
Run-12	2012	p+p (L)	200 GeV	9.2 pb-1
		p+p (L)	510 GeV	30 pb-1
		U+U	193 GeV	171 μb-1
		Cu+Au	200 GeV	4.96 nb-1
Run-13	2013	p+p (L)	510 GeV	156 pb-1
Run-14	2014	Au+Au	15 GeV	44.2 μb-1
		Au+Au	200 GeV	2.56 nb-1
		He3+Au	200 GeV	134 nb-1
Run-15	2015	p+p (L)	200 GeV	59.9 pb-1
		p+Au (T)	200 GeV	206.2 nb-1
		p+Al (T)	200 GeV	690.8 nb-1

Years	Beam Species and	Science Goals	New Systems
2014	Au+Au at 15 GeV Au+Au at 200 GeV ^{3}He+Au at 200 GeV	Heavy flavor flow, energy loss, thermalization, etc. Quarkonium studies QCD critical point search	Electron lenses 56 MHz SRF STAR HFT STAR MTD
2015-16	$p^{\uparrow}+p^{\uparrow}$ at 200 GeV $p^{\uparrow}$+Au, $p^{\uparrow}$+Al at 200 GeV High statistics Au+Au Au+Au at 62 GeV ?	Extract η/s(T) + constrain initial quantum fluctuations Complete heavy flavor studies Sphaleron tests Parton saturation tests	PHENIX MPC-EX STAR FMS preshower Roman Pots Coherent e-cooling test
2017	$p^{\uparrow}+p^{\uparrow}$ at 510 GeV	Transverse spin physics Sign change in Sivers function	
2018	No Run		Low energy e-cooling install. STAR iTPC upgrade
2019-20	Au+Au at 5-20 GeV (BES-2)	Search for QCD critical point and onset of deconfinement	Low energy e-cooling
2021-22	Au+Au at 200 GeV $p^{\uparrow}+p^{\uparrow}$, $p^{\uparrow}$+Au at 200 GeV	Jet, di-jet, γ-jet probes of parton transport and energy loss mechanism Color screening for different quarkonia Forward spin & initial state physics	sPHENIX Forward upgrades ?
≥ 2023 ?	No Runs		Transition to eRHIC

Figure 3: a)(left) Year, species and proton polarization (**L**ongitudinal or **T**ransverse), $\sqrt{s_{NN}}$ and integrated luminosity of RHIC runs. b) (right) Future run schedule and new equipment.

measurement for the STAR heavy flavor tracker (HFT) which had its first (Au+Au) run in 2014[1] ; ii) comparison p+p measurements for the PHENIX central (VTX) and forward micro-vertex (FVTX) detectors and the new MPC-EX detector (a forward charged particle tracker and EM pre-shower detector to complement the forward EM Calorimeters, located at $3.1 \leq |\eta| \leq 3.7$ inside the Muon Pistons); iii) p+Au and p+Al runs for a better baseline of nuclear effects than the previous d+Au measurements. In all cases the protons were polarized: longitudinally for the p+p measurements and transversely for the p+A measurements. Fig. 3a gives the species, c.m. energies ($\sqrt{s_{NN}}$), polarizations and integrated luminosities obtained for all RHIC runs, with new equipment and future plans given in Fig. 3b. The luminosity performance is given in Fig. 4. Note that the p+p integrated luminosity at $\sqrt{s_{NN}}$ =200 GeV in 2015 exceeds the sum of all previous p+p integrated luminosity at this $\sqrt{s_{NN}}$.

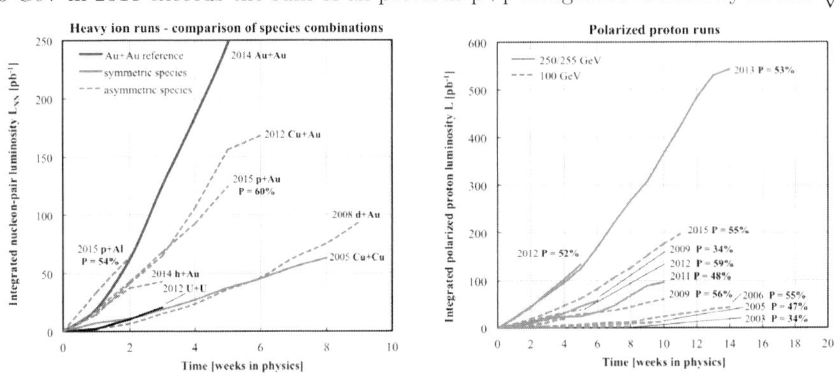

Figure 4: a)(left) A+B performance, where the nucleon-pair luminosity is defined as $L_{\rm NN} = A \times B \times L$, where L is the luminosity and A, B are the number of nucleons in the colliding species. b) (right) Polarized p+p performance. Note that the thin lines by the years are pointers. Courtesy Wolfram Fischer.

[1]The STAR and PHENIX detectors have been operating at RHIC all 15 years [14] with regular upgrades.

3.1 Nobody's perfect—the first asymmetric Z/A run

Because of the asymmetric p+A running, the final focussing DX magnets near the interaction regions had to be moved. Aperture restrictions were identified and fixed; masks for the protection of the detectors in case of a quench were installed. However since this was the first such run at RHIC, abort kicker prefires were not adequately shielded and the MPC detector was damaged by several prefires and removed from service on June 1 (Fig. 5). The VTX and FVTX detectors were also removed as a precaution. The good news is that only the p+Al measurement was affected by these problems.

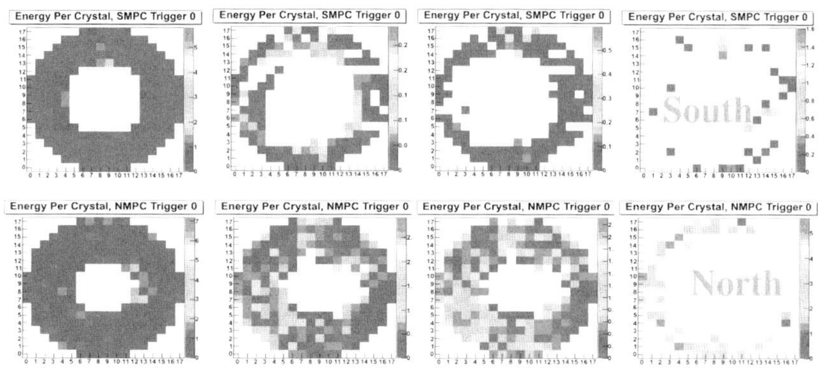

Figure 5: MPC damage from yellow-ring abort kicker prefires: top row South MPC, bottom row North MPC. From left to right: Start of run, Feb. 11; after May 11; after May 28; after June 1.

4 10th Anniversary Celebration of the Perfect Liquid

High energy nucleus-nucleus collisions provide the means of creating nuclear matter in conditions of extreme temperature and density where a phase transition is expected from a state of nucleons containing confined quarks and gluons to a state of quarks and gluons, deconfined from their original nucleons, covering a volume that is many units of the confinement length scale ($\sim$ 1 fm) in which the q and g, with their color charge fully exposed, freely traverse the medium composed of a large density of similarly exposed color charges. This state of nuclear matter was originally given the name Quark Gluon Plasma (QGP) [15], a plasma being an ionized gas. The QGP was discovered at RHIC, and announced on April 19, 2005. However the results at RHIC [16] indicated that instead of behaving like the anticipated gas of free quarks and gluons, the matter created in heavy ion collisions at nucleon-nucleon c.m. energy $\sqrt{s_{NN}} = 200$ GeV appears to be more like a *liquid*. This matter interacts much more strongly than originally expected, as elaborated in peer reviewed articles by the 4 RHIC experiments [17, 18, 19, 20], which inspired the theorists [21] to give it the new name "sQGP" (strongly interacting QGP). These properties were quite different from the "new state of matter" claimed in a press-conference [22] by the CERN fixed target heavy ion program on February 10, 2000, which was neither peer-reviewed nor published.

I had a few additional quips in my discussion of this discovery in previous ISSP talks and proceedings [23]. Perhaps I was too harsh because in an older previous publication [16] I noted that it was not unreasonable to expect surprises in the search for the QGP, i.e. "Many of the predicted properties will be found but will not be the QGP; the QGP will have a few unpredicted or unexpected properties; the search will uncover many unexpected backgrounds and new properties of p+p and A+A collisions, some of which may be very interesting phenomena in their own right." In fact this is what has happened[2] and is still happening this year.

At the time of the proposals for experiments at RHIC (also for the LHC [25]) in 1990-91, the "gold-plated" signature for deconfinement in the QGP was the suppression of J/Ψ production caused by the Debye screening of the QCD potential in the QGP so that the c and $\bar{c}$ quarks could not bind [26]. However, as predicted a year later, in 1987 [27], enhanced production of c and $\bar{c}$ quarks in A+A collisions and recombination of $c\bar{c}$ into J/Ψ might (and did [23]) "hinder" J/Ψ suppression as evidence for the QGP.

In 1997, an additional signal of QGP formation was proposed [28], Jet Quenching—energy loss from coherent LPM radiation of hard-scattered partons exiting the QGP, which would result in an attenuation of the jet energy and broadening (enhanced acoplanarity) of the di-jets formed by fragmentation of the outgoing partons. This has been a robust signature of QGP formation (although more complicated than the original proposal) and is one of the few remaining signals without an alternate explanation.

In my opinion, the key QGP discoveries at RHIC, so far, are the following:

- Suppression of high p_T hadrons from hard-scattering of initial state partons [29]; also modification of the away-side jet [30].
- Elliptic flow at the Hydrodynamic limit as a near ideal fluid [31] with shear viscosity/entropy density, η/s, at or near the quantum lower bound $\eta/s \approx \frac{1}{4\pi}$ [32].
- Elliptic flow of particles proportional to the number of constituent quarks [33, 34, 35].
- Mid-rapidity transverse energy, $dE_T/d\eta$ and charged particle multiplicity density $dN_{\rm ch}/d\eta$ proportional to the number of constituent quark participants N_{qp}. [36, 37].
- Higher order flow moments proportional to density fluctuations of the initial colliding nuclei [38, 39].
- Suppression and flow of heavy quarks roughly the same as that of light quarks [40].
- QCD hard prompt photons not suppressed [41], do not flow [42].
- Production [43] and flow [42] of thermal soft photons.

4.1 A few illustrations

4.1.1 $dE_T/d\eta$ and $dN_{\rm ch}/d\eta$ proportional to constituent quark participants N_{qp}

The closely related charged particle density $dN_{\rm ch}/d\eta$ and transverse energy density $dE_T/d\eta$ at mid-rapidity are composed of soft particles with low transverse momenta and a steeply falling exponential spectrum, $d\sigma/p_T dp_T \propto \exp -6p_T$ in p+p collisions. The first measurements [44] of the total charged multiplicity $N_{\rm ch}$ in p+A collisions at nucleon-nucleon c.m. energies $\sqrt{s_{NN}}$ =10-20 GeV showed that $N_{\rm ch}$ was proportional to the number of nucleons, $N_{\rm part}$, that participate (i.e are struck) in the collision of the nuclei. [3] $N_{\rm part}$ scaling was originally expected

[2]It may be interesting for some readers to look back at some of the early predictions for the QGP [24].

[3]$N_{\rm part}$ is an estimate of the centrality of the collision, the overlap of the two colliding nuclei [16].

to be valid at RHIC energies but was not observed at mid-rapidity (Fig. 6a). The correct scaling found at mid-rapidity (including $\sqrt{s_{NN}}$ =10-20 GeV) is the number of (massive) constituent quark participants N_{qp} [45] (Fig. 6b,c).

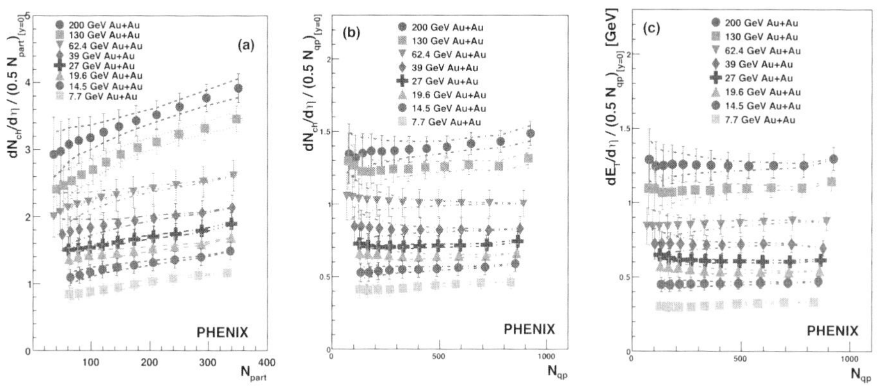

Figure 6: PHENIX[45] mid-rapidity (a) $dN_{\rm ch}/d\eta/(0.5N_{\rm part})$ as a function of $N_{\rm part}$, (b) $dN_{\rm ch}/d\eta/(0.5N_{qp})$ (c) $dE_T/d\eta/(0.5N_{qp})$ as a function of N_{qp}, in Au+Au collisions at the $\sqrt{s_{NN}}$ indicated.

4.1.2 Collective Flow—anisotropic production of particles

Collective flow (anisotropic production of particles) [46], or simply flow, is a collective effect which can not be obtained from a superposition of independent N+N collisions. "It does not assume necessarily the hydrodynamic flow, which in particular would require a thermalization of the system" [33] but this distinction is not generally discussed. The almond shaped overlap region of the A+A collision causes the particles to be emitted more favorably in the reaction plane (see Fig. 7a). The semi-inclusive single particle spectrum is modified by an expansion in harmonics of the azimuthal angle of the particle with respect to the reaction plane, $\phi-\Phi_R$:

$$\frac{Ed^3N}{dp^3} = \frac{d^3N}{p_T dp_T dy d\phi} = \frac{d^3N}{2\pi\, p_T dp_T dy}\left[1+\sum_n 2v_n \cos n(\phi-\Phi_R)\right]. \qquad (2)$$

The Fourier coefficient v_2, called elliptic flow, is predominant at mid-rapidity. Odd harmonics were thought to be forbidden by the symmetry $\phi \rightarrow \phi+\pi$ of the almond. Only in 2010 was it realized that the nuclear geometry fluctuated from event to event and did not respect the average symmetry [38].

Flow measurements contributed two of the most important results about the properties of the QGP: i) the scaling of v_2 of identified particles at mid-rapidity with the number of constituent-quarks n_q in the particle—v_2/n_q scales with the transverse kinetic energy per constituent-quark, KE_T/n_q, because particles have not formed at the time flow develops (Fig. 7b); ii) the persistence of flow for $p_T > 1$ GeV/c which implied that the viscosity is small [31], perhaps as small as a quantum viscosity bound from string theory [32], $\eta/s = 1/(4\pi)$, where η is the shear viscosity and s the entropy density per unit volume. This led to the description of the "sQGP" produced at RHIC as "the perfect fluid".

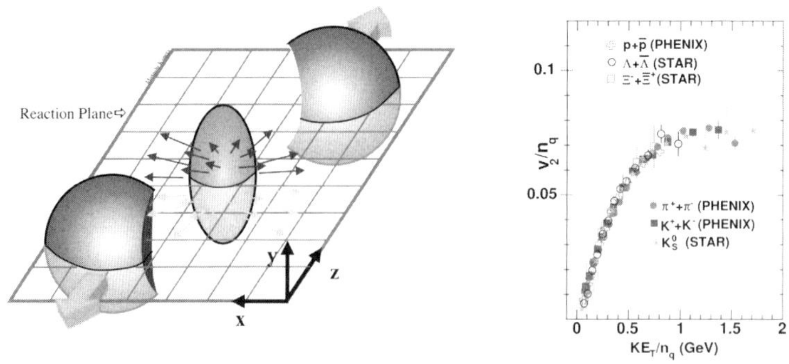

Figure 7: (a) (left) Almond shaped overlap zone generated just after an A+A collision where the incident nuclei are moving along the $\pm z$ axis. The reaction plane by definition contains the impact parameter vector (along the x axis) [47]. (b) (right) v_2/n_q vs. KE_T/n_q for baryons and mesons [48].

4.1.3 More surprises this year, flow in three small systems

Last year [4], I noted that results in 2013 observed what looked very much like collective flow in p+Pb at LHC and d+Au at RHIC. These systems were believed to be too small for collective effects which inspired the He^3+Au run at RHIC in 2014 to see whether triangular flow (v_3) would be more prominent with a 3 nucleon projectile (Fig. 8a). The He^3+Au result

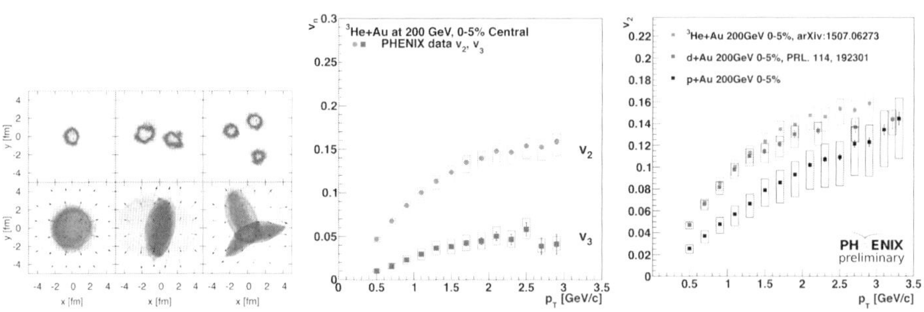

Figure 8: a) (left) (above) Simulated initial energy depositions for p, d and He^3 on a larger nucleus; (below) the QGP geometry resulting from hydrodynamic flow [6]. b) (center) PHENIX [49] v_2 and v_3 from He^3+Au with 0-5% centrality at $\sqrt{s_{NN}}$ =200 GeV. c)(right) v_2 from p,d,He^3+Au at $\sqrt{s_{NN}}$ =200 GeV[50].

(Fig. 8b) [49] did indeed show a strong v_3 as well a v_2 comparable to the d+Au (Fig. 8c). Perhaps no longer surprising, the p+Au data from this year's run [50] also showed a comparable v_2, but smaller, consistent with the more spherical initial geometry. Of special interest is how well the large v_3/v_2 ratio in He^3+Au corresponds to the ratio of the eccentricities in the initial geometry because translation of the geometrical anistropy into the flow anisotropy depends on the viscosity and dissipates with larger viscosities for larger v_n. The issue of whether this is really hydrodynamic flow is still one of the very hot topics in RHI physics.

4.1.4 Jet quenching at RHIC — Suppression of high p_T particles

The discovery at RHIC that π^0's produced at large transverse momenta are suppressed in central Au+Au collisions by a factor of ~ 5 compared to binary-collision scaling from p+p collisions (Fig. 9a) [29] is arguably *the* major discovery in Relativistic Heavy Ion Physics. At RHIC energies, the steeply falling exponential spectrum at low p_T becomes negligible relative to the power-law dependence of hard-scattering for $p_T \geq 3$ GeV/c [51]. The suppression in A+A collisions is presented as $R_{AA}(p_T)$, the ratio of the yield of e.g. π^0 per A+A collision of a given centrality to the binary-scaled p+p cross section at the same p_T, where $\langle T_{AA}\rangle$ is the average overlap integral of the nuclear thickness functions for that centrality

$$R_{AA}(p_T) = \frac{(1/N_{AA})\, d^2N^{\pi}_{AA}/dp_T dy}{\langle T_{AA}\rangle \quad d^2\sigma^{\pi}_{pp}/dp_T dy} \quad . \tag{3}$$

Fig. 9a illustrates another advantage of high p_T suppression as a QGP probe: at $\sqrt{s_{NN}}$ of CERN SpS fixed target and ISR A+A (i.e. $\lesssim$31 GeV), high p_T production is enhanced.

A related but possibly greater discovery, the opposite of what was predicted [52], is that heavy quarks observed using direct-single-$e^{\pm}$ from c and b-quark decay are suppressed comparably to light quarks (π^0) for $p_T \gtrsim 4$ GeV/c (Fig 9b) and also exhibit flow (Fig 9c) [40]. The important impact of this discovery was that it provided a demonstration that heavy quarks were strongly coupled to the medium with viscosity/entropy density $\eta/s \approx (1.3-2)/4\pi$, close to the quantum lower bound [32], reinforcing the 'perfect fluid', and stimulating a broad spectrum of possible explanations (e.g see Ref. [23]).

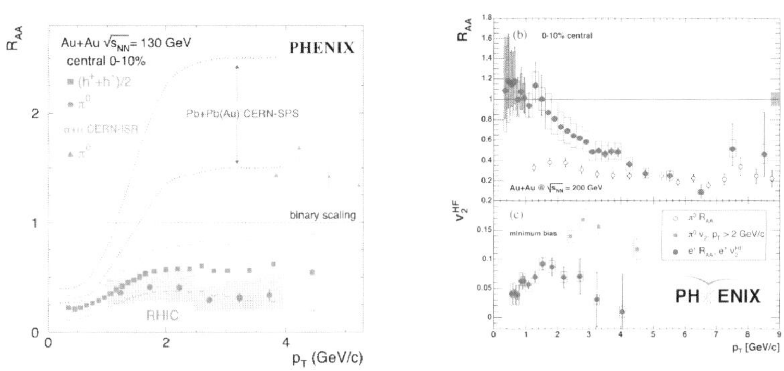

Figure 9: (a) (left) PHENIX [29] discovery of suppression ($R_{AA} \sim 0.3$) of charged hadrons and π^0 in Au+Au central collisions at $\sqrt{s_{NN}}$ =130 GeV compared to enhancement ($R_{AA} > 1.0$) at lower $\sqrt{s_{NN}}$. (right) PHENIX direct-single-$e^{\pm}$ in Au+Au at $\sqrt{s_{NN}}$ =200 GeV [40]: (b) R_{AA} (central) (c) v_2 (minimum bias) vs. p_T, compared to π^0.

5 RHIC Beam Energy Scan (BES) in search of a critical point—aided by Lattice QCD

This subject has remained one of my main interests since I got sandbagged by a press release from LBL during ISSP2011 [23] claiming that "By comparing theory with data from STAR, Berkeley Lab scientists and their colleagues map phase changes in the QGP" after I had criticized a different laboratory for physics by press release. It turned out that the LBL press release and publication were later found to be "not useful" [53] (i.e. "not correct" [23]). The good news is that this year my colleagues and I at PHENIX have actually made such a comparison and were able to use our measurements of net-charge fluctuations [54] together with a Lattice QCD calculation [55] to find both the freezeout temperature, T_f, and the Baryon Chemical Potential μ_B at several values of $\sqrt{s_{NN}}$ at RHIC, without particle identification!

To set the stage for the new measurement, Fig. 10a shows a proposed phase diagram for nuclear matter (which I emphasized in last year's proceedings [4] had an incorrect μ_B scale) as corrected by Frithjof Karsch; and Fig. 10b shows measurements of the total charge multiplicity $N_{\rm ch} = N_{\rm ch}{}^{+} + N_{\rm ch}{}^{-}$ as a function of centrality in Au+Au collisions at RHIC together with fits to Negative Binomial Distributions (NBD) [56], $\boxed{P(m) = \frac{(m+k-1)!}{m!(k-1)!}\frac{(\mu/k)^m}{(1+\mu/k)^{m+k}}}$, where $P(m)$ is normalized from $0 \leq m \leq \infty$ and $\mu \equiv \langle N_{\rm ch}\rangle = \langle m \rangle$.

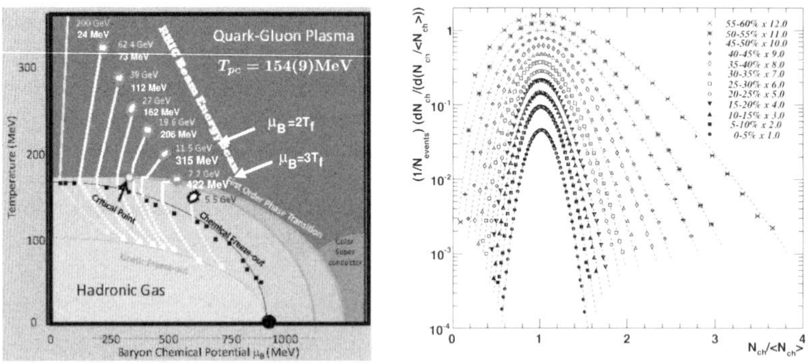

Figure 10: (a) Figure 20a from Ref. [4] corrected by F. Karsch. (b) PHENIX [56] fits of Negative Binomial Distributions to $N_{\rm ch}/\langle N_{\rm ch}\rangle$ measurements from $\sqrt{s_{NN}}$ =200 GeV Au+Au for the centralities indicated.

Figure 11 shows the new PHENIX $\Delta N_{\rm ch} = N_{\rm ch}{}^{+} - N_{\rm ch}{}^{-}$ distributions from Au+Au collisions, not yet corrected for the reconstruction efficiency $\epsilon \approx 0.69$ within the acceptance [54]. These distributions can not be compared directly to Lattice QCD predictions which use

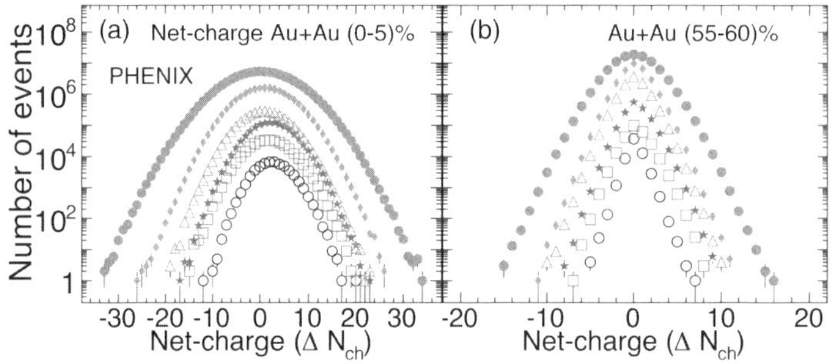

Figure 11: PHENIX [54] $\Delta N_{\rm ch} = N_{\rm ch}{}^{+} - N_{\rm ch}{}^{-}$ distributions uncorrected for the efficiency within the acceptance, $|\eta| < 0.35$, $\delta\phi = \pi$, $0.3 < p_T < 2.0$ GeV/c, in Au+Au at two centralities, for all $\sqrt{s_{NN}}$ measured.

cumulants from a Taylor expansion of the free energy $F = -T \ln Z$ around the critical temperature T_c, where Z is the partition function (or sum over states) which is of the form $Z \propto e^{-(E-\sum_i \mu_i Q_i)/kT}$ and μ_i are chemical potentials associated with conserved charges. The terms of the Taylor expansion, which are obtained by differentiation, are called susceptibilities, denoted χ. The connection of this method to the measurements is that the cumulant generating function in mathematical statistics is also a Taylor expansion of the ln of an exponential, where the n-th cumulant κ_n represents the n-th moment of a distribution with all n-fold combinations of the lower order moments subtracted. One of the most important properties of cumulants is that they are additive for the sum of statistically independent random variables, x and y where $\langle xy\rangle = \langle x\rangle \langle y\rangle$: $\kappa_n(x+y) = \kappa_n(x) + \kappa_n(y)$.

A recent important theorem from (of all places) Quantitative Finance [57] for "integer value Levy processes" (notably Poisson and Negative Binomial Distributions) has proved that the cumulants κ_j for the distribution $P(x-y)$ of the difference of samples from two such distributions, $P^+(x)$ and $P^-(y)$, with cumulants κ_j^+ and κ_j^-, respectively, are $\boxed{\kappa_j = \kappa_j^+ + (-1)^j \kappa_j^-}$ so long as the distributions are not 100% correlated. The individual ${N_{\rm ch}}^+$ and ${N_{\rm ch}}^-$ distributions are in fact NBD [54] so this theorem would allow a great simplification of the analysis but is presently used only as a check of the very complicated standard method which starts with the calculation of the cumulants of the $\Delta N_{\rm ch} = {N_{\rm ch}}^+ - {N_{\rm ch}}^-$ distributions from Fig. 11.

For each distribution the n-th order moment can be calculated: $\mu'_n \equiv \langle x^n\rangle = \sum_i x_i^n P(x_i)$, where $\mu'_1 \equiv \mu = \langle x\rangle$, $\mu_n \equiv \langle (x-\mu)^n\rangle$ and x_i represents a bin in the $\Delta N_{\rm ch}$ plot with a fraction $P(x_i)$ of the total number of events. Then one computes the cumulants κ_m from the moments, where presently the analyses compute only the first four cumulants:

$$\mu = \kappa_1 = \mu_1' \qquad \sigma^2 = \mu_2 \equiv \langle (x-\mu)^2\rangle = \kappa_2 = \mu_2' - {\mu_1'}^2 \qquad \mu_3 = \kappa_3 = \mu_3' - 3\mu_2'\mu_1' + 2{\mu_1'}^3$$
$$\mu_4 - 3\mu_2^2 = \kappa_4 = \mu_4' - 4\mu_3'\mu_1' - 3{\mu_2'}^2 + 12\mu_2'{\mu_1'}^2 - 6{\mu_1'}^4 \qquad (4)$$

The cumulants are then corrected for reconstruction efficiency. Ratios of the corrected cumulants are used to compare to the calculated ratios of susceptibilities because the volume dependences of the calculations cancel: $\mu/\sigma^2 = \kappa_1/\kappa_2 \approx \chi_1/\chi_2$; $S\sigma^3/\mu = \kappa_3/\kappa_1 \approx \chi_3/\chi_1$; $S\sigma = \kappa_3/\kappa_2 \approx \chi_3/\chi_2$; $\kappa\sigma^2 = \kappa_4/\kappa_2 \approx \chi_4/\chi_2$; where $S \equiv \kappa_3/\sigma^3$ and $\kappa \equiv \kappa_4/\kappa_2^2$ (Fig. 12).

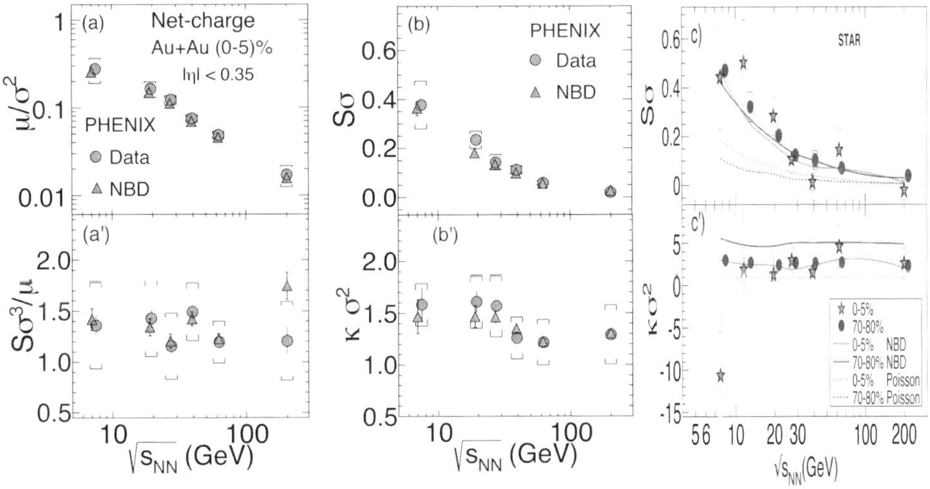

Figure 12: $\sqrt{s_{NN}}$ dependence of net-charge cumulant ratios in Au+Au central (0-5%) collisions. PHENIX [54]: (a) κ_1/κ_2, (a′) κ_3/κ_1, (b) κ_3/κ_2, (b′) κ_4/κ_2. STAR [58]: (c) κ_3/κ_2, (c′) κ_4/κ_2. The error bars are statistical and the caps are systematic uncertainties.

The new PHENIX net-charge results [54] for the corrected cumulant ratios shown in Fig. 12a,b are compared to the previous STAR net-charge measurements [58] for $S\sigma$ and $\kappa\sigma^2$ in Fig. 12c. There are two important observations from this comparison: i) for PHENIX the error on all the corrected cumulant ratios is 20-30%, while for STAR the error on $\kappa_1/\kappa_2 = \mu/\sigma^2$ is $< 1\%$ but for $\kappa_3/\kappa_2 = S\sigma$ is $\sim 40\%$ and for $\kappa_4/\kappa_2 = \kappa\sigma^2$ is $> 100\%$; ii) the PHENIX standard cumulant ratios and errors (circles) are checked by NBD fits to the $N_{\rm ch}{}^{+}$ and $N_{\rm ch}{}^{-}$ distributions plus the cumulant theorem [57] (triangles) and agree—which proves that the errors are correct and that the cumulant theorem works.

As illustrated in Fig. 13, the PHENIX (♦) errors allow both μ_B and T_f to be determined from the Lattice QCD calculations [55] while the large STAR (★) error on $R_{31} = \kappa_3/\kappa_1$ does not give a solution for T_f.

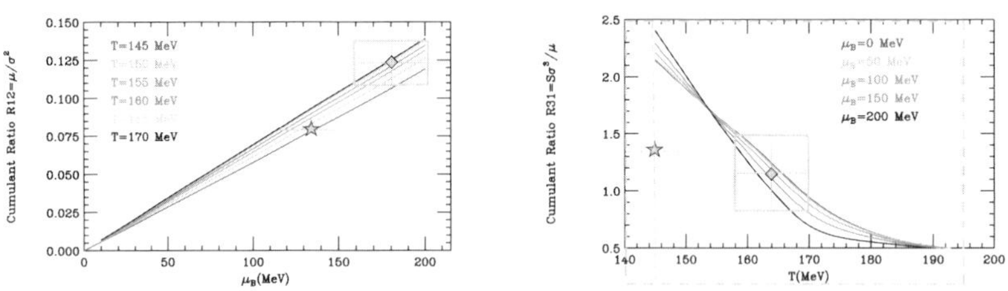

Figure 13: Lattice QCD calculations of cumulant ratios (lines) [55] (a) $R_{12} = \kappa_1/\kappa_2$ (b) $R_{31} = \kappa_3/\kappa_1$. The vertical lines in the boxes represent the error of the R_{12} and R_{31} measurements at $\sqrt{s_{NN}}$ =27 GeV and the horizontal lines the error of the best fit values of T_f and μ_B. The STAR measurement of R_{31} has a such a huge error that the data point (★) could go anywhere in the dashed box. The horizontal dashes and position of the ★ for R_{31} are an assumption [59] based on the STAR net-proton measurement [60].

Figure 14 shows that the PHENIX + Lattice results for T_f and μ_B from net-charge fluctuations, with no particle identification, are in excellent agreement with the best accepted analysis of T_f and μ_B from baryon/anti-baryon ratios. [61]. I believe that this is a first!

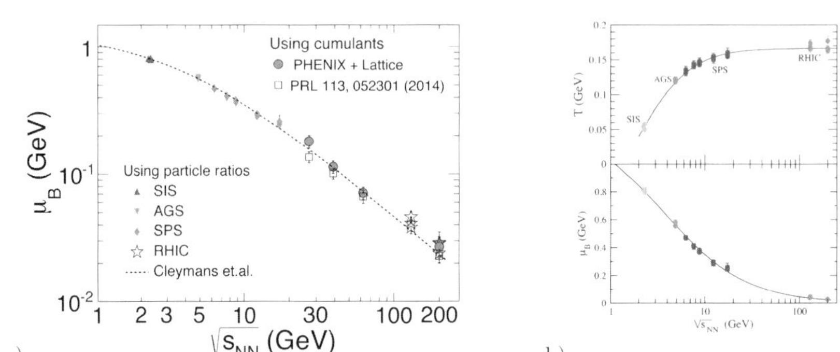

Figure 14: (a) $\sqrt{s_{NN}}$ dependence of μ_B from PHENIX+Lattice [54] net-charge results. Open squares are from STAR net-charge together with net-protons [59]. Dashed line and other data points are from (b), the best accepted analysis of T_f and μ_B vs $\sqrt{s_{NN}}$ from baryon/anti-baryon ratios [61].

5.1 Why are the STAR errors on R_{31} so large? Corrections?

A fraction of the tracks that fall on the detector acceptance are not reconstructed, a clearly random, thus Binomial, effect. This is further complicated if the efficiencies are different for ${N_{\rm ch}}^+$ (ϵ^+) and ${N_{\rm ch}}^-$ (ϵ^-). The standard Koch-Bzdak [62] Binomial efficiency correction, e.g. to correct κ_3 of an arbitrary net-charge distribution, is:

$$\kappa_3 = \kappa_1 + 2\kappa_1^3 - F_{03} - 3F_{02} + 3F_{12} + 3F_{20} - 3F_{21} + F_{30} - 3\kappa_1(N + F_{02} - 2F_{11} + F_{20}) \quad (5)$$

where the corrected values are $\langle N^+\rangle = \langle {N_{\rm ch}}^+\rangle/\epsilon^+$, $\langle N^-\rangle = \langle {N_{\rm ch}}^-\rangle/\epsilon^-$, $\kappa_1 = \langle N^+\rangle - \langle N^-\rangle$, $N \equiv \langle N^+\rangle + \langle N^-\rangle$. Also, a new quantity, the double Factorial Moment has been introduced: $\boxed{F_{ik} = \sum_{N_1=i}^{\infty}\sum_{N_2=k}^{\infty} P(N_1, N_2)\frac{N_1!}{(N_1-i)!}\frac{N_2!}{(N_2-k)!}}$, which looks simple, but is very difficult to compute because the joint probability $P(N_1, N_2)$, e.g. $P(13^+, 11^-)$ etc., is required. This means that to make the computation one must know not only the $\Delta N_{\rm ch}$ distribution but also both the N^+, the N^- distributions and their correlation. $P(N^+, N^-)$ can be obtained from the data by making a 3d lego plot with base axes N^+ and N^- and height equal to the normalized number of events with (N^+, N^-), but this costs a huge dilution of the statistical errors, so approximations or random-sampling of the lego plot are used. My guess is that this dilution plus the huge complication in e.g. the calculation of κ_3 in Eq. 5 are one reason for the huge STAR errors in $S\sigma = \kappa_3/\kappa_2$ and $\kappa\sigma^2 = \kappa_4/\kappa_2$ but I haven't verified it with them.

5.1.1 If you know the distribution, you know all the moments and cumulants

When I first saw net-charge distribution plots like Fig. 11, e.g. see Ref. [23], my thought was to fit them to NBD so that the title of this section would apply. However, they are not NBD, they are the difference between two NBD. This is why the cumulant theorem [57] is so important: once the cumulants, κ_j^+ and κ_j^-, of NBD fits to the individual ${N_{\rm ch}}^+$ and ${N_{\rm ch}}^-$ are known, the cumulants of the $\Delta N_{\rm ch}$ distribution are simply $\kappa_j = \kappa_j^+ + (-1)^j\kappa_j^-$. Perhaps equally important, the correction for efficiency to the measured ${N_{\rm ch}}^+$ and ${N_{\rm ch}}^-$ is a "binomial split" of an NBD [63]—if a population is NBD(μ_t, k) and divided randomly with probability p onto a smaller region, the distribution on the region is NBD($p\mu_t, k$), i.e. the mean changes but the k parameter remains constant. Thus if we measure $\mu^+ = \langle {N_{\rm ch}}^+\rangle$ in the detector with efficiency $p = \epsilon^+$, the efficiency corrected mean, μ_t^+ equals μ^+/ϵ^+ and e.g. the efficiency corrected κ_3/κ_1 is:

$$\frac{S\sigma^3}{\mu} = \frac{\kappa_3}{\kappa_1} = \frac{\kappa_3^+ - \kappa_3^-}{\kappa_1^+ - \kappa_1^-} = \frac{\mu_t^+[1 + 3(\frac{\mu_t^+}{k^+}) + 2(\frac{\mu_t^+}{k^+})^2] - \mu_t^-[1 + 3(\frac{\mu_t^-}{k^-}) + 2(\frac{\mu_t^-}{k^-})^2]}{\mu_t^+ - \mu_t^-} \quad , \quad (6)$$

which is a lot simpler than Eq. 5 and dramatically simpler than for $\kappa_4/\kappa_2 = \kappa\sigma^2$, which the reader can figure out from Table 1 compared to Ref. [62].

5.2 The End?

Of course this is still not the whole story. Since there can be no fluctuations of conserved quantities such as net-charge or net-baryon number in the full phase space, one must go to "locally conserved quantities" [64] in small rapidity intervals to detect a fraction which then fluctuates, i.e. varies from event to event. Then the question becomes what the required acceptance is for an experimental result to compare with the Lattice QCD calculations.

Table 1: Cumulants for Poisson, Negative Binomial Distributions
Measured with efficiency ϵ corrected to the true value

Measured Cumulant	Corrected Poisson	Corrected Negative Binomial
$\kappa_1 = \mu$	$\mu_t \equiv \mu/\epsilon$	$\mu_t \equiv \mu/\epsilon$
$\kappa_2 = \mu_2 = \sigma^2$	μ_t	$\mu_t(1 + \mu_t/k) \equiv \sigma_t^2$
$\kappa_3 = \mu_3$	μ_t	$\sigma_t^2(1 + 2\mu_t/k)$
$\kappa_4 = \mu_4 - 3\kappa_2^2$	μ_t	$\sigma_t^2(1 + 6\mu_t/k + 6\mu_t^2/k^2)$
$S \equiv \kappa_3/\sigma^3$	$1/\sqrt{\mu_t}$	$(1 + 2\mu_t/k)/\sigma_t$
$\kappa \equiv \kappa_4/\kappa_2^2$	$1/\mu_t$	$(1 + 6\mu_t/k + 6\mu_t^2/k^2)/\sigma_t^2$
$S\sigma = \kappa_3/\kappa_2$	1	$(1 + 2\mu_t/k)$
$\kappa\sigma^2 = \kappa_4/\kappa_2$	1	$(1 + 6\mu_t/k + 6\mu_t^2/k^2)$
$\mu/\sigma^2 = \kappa_1/\kappa_2$	1	$1/(1 + \mu_t/k)$
$S\sigma^3/\mu = \kappa_3/\kappa_1$	1	$(1 + 3\mu_t/k + 2\mu_t^2/k^2)$

Another issue is whether "net-protons" are equivalent to net-baryons since the neutrons are not measured. Further complicating this, the experiments measure in a fixed $\delta\eta$ interval but the rapidity distributions change (dramatically for protons because of stopping [65]) as the $\sqrt{s_{NN}}$ is reduced. Also Lattice QCD calculations are not now available for $\mu_B > 0.2$ GeV... There are many unresolved isues—stay tuned!!!

References

[1] See M. J. Tannenbaum, `http://www.pas.va/content/dam/accademia/pdf/sv119/sv119-tannenbaum.pdf` for some examples.

[2] See `https://www.bnl.gov/about/history/discoveries.php`

[3] See `https://www.bnl.gov/bnlweb/history/nobel/`

[4] M. J. Tannenbaum, Highlights from BNL and RHIC 2014, *International School of Subnuclear Physics, 52nd Course,* Erice, Sicily, Italy, 24 June–3 July 2014, arXiv:1504.02771v1.

[5] Workshop on *A Large-Acceptance Jet and Upsilon Detector for RHIC*, BNL, Upton, NY, USA, June 16, 2015. `https://www.bnl.gov/lajudr2015/`

[6] *2015 NSAC Long Range Plan*, `http://science.energy.gov/np/nsac/reports/`

[7] E. Cartlidge, Nature **521** (2015) 272.

[8] L. Edward Temple, Jr., NSAC Subcommittee Review of the EIC cost estimates, `http://science.energy.gov/~/media/np/nsac/pdf/20150403/Temple_NSAC_Sub-Committee_Review_of_the_EIC.pdf`

[9] *LHeC Workshop*, CERN and surroundings, 24-26 June 2015, `https://indico.cern.ch/event/356714/timetable/#20150625.detailed`

[10] G. Bunce, N. Saito, J. Soffer and W. Vogelsang, Ann. Rev. Nucl. Part. Sci. **50** (2000) 525.

[11] E-C. Aschenauer, *et al.*, The RHIC SPIN Program: Achievements and Future Opportunities, arXiv:1501.01220v2.

[12] D. de Florian, R. Sassot, M. Stratmann and W. Vogelsang, Phys. Rev. Lett. **113** (2014) 012001.

[13] C. A. Aidala, S. D. Bass, D. Hasch and G. K. Mallot, Rev. Mod. Phys. **85** (2013) 655.

[14] M. Harrison, T. Ludlam, S. Ozaki, eds., Nucl. Instrum. Methods A**499** (2003) 235–880.

[15] E. V. Shuryak, Phys. Repts. **61** (1980) 71–158.

[16] For more details and an extensive list of references, see M. J. Tannenbaum, Rep. Prog. Phys. **69** (2006) 2005–2059.

[17] BRAHMS Collab. (I. Arsene *et al.*), Nucl. Phys. A**757** (2005) 1–27.

[18] PHOBOS Collab. (B. B. Back *et al.*), Nucl. Phys. A**757** (2005) 28–101.

[19] STAR Collab. (J. Adams *et al.*), Nucl. Phys. A**757** (2005) 102–183.

[20] PHENIX Collab. (K. Adcox *et al.*), Nucl. Phys. A**757** (2005) 184–283.

[21] D. Rischke and G. Levin, eds., Nucl. Phys. A**750** (2005) 1–171.

[22] Press Conference, *A New State of Matter Created at CERN*, `http://newstate-matter.web.cern.ch/newstate-matter/Story.html`

[23] M. J. Tannenbaum, Highlights from BNL-RHIC 2011-2013, Int. J. Mod. Phys. A**29** (2014) 1430017.

[24] M. J. Tannenbaum, Hadron Production in Relativistic Heavy Ion Interactions and the search for the QGP, *Quark-Gluon Plasma*, eds. B. Sinha, S. Pal, S. Raha (Springer-Verlag, Berlin, 1990), `http://www.osti.gov/scitech/biblio/7163593`

[25] ALICE Collab. (K. Aamodt *et al.*), JINST **3** (2008) S08002.

[26] T. Matsui and H. Satz, Phys. Lett. B**178** (1986) 416.

[27] T. Matsui, On the Charm Production in Ultra-relativistic Heavy Ion Collisions, *Proc. 2nd Workshop on Experiments and detectors for RHIC*, Berkeley, CA, May 1987, eds. H. G. Ritter, A. Shor, LBL-24604. `http://www.osti.gov/scitech/biblio/5302565`

[28] See R. Baier *et al.*, Ann. Rev. Nucl. Part. Sci. **50** (2000) 37–69, and references therein.

[29] PHENIX Collab. (K. Adcox *et al.*), Phys. Rev. Lett. **88** (2002) 022301.

[30] STAR Collab. (D. Hardke *et al.*), Nucl. Phys. A**715** (2003) 272c–279c.

[31] D. Teaney, Phys. Rev. C**68** (2003) 034913.

[32] P. K. Kovtun, D. T. Son, and A. O. Starinets, Phys. Rev. Lett. **94** (2005) 111601.

[33] For the development of the idea see S. A. Voloshin, J. Phys: Conf. Series **9** (2005) 276–279.

[34] PHENIX Collab. (S. S. Adler *et al.*), Phys. Rev. Lett. **91** (2003) 182301.

[35] PHENIX Collab. (A. Adare *et al.*), Phys. Rev. Lett. **98** (2007) 162301.

[36] S. Eremin and S. Voloshin, Phys. Rev. C**67** (2003) 064905.

[37] PHENIX Collab. (S. S. Adler *et al.*), Phys. Rev. C**89** (2014) 044905.

[38] B. Alver and G. Roland, Phys. Rev. C**81** (2010) 054905, ibid. **82** (2010) 039903(E).

[39] PHENIX Collab. (A. Adare *et al.*), Phys. Rev. Lett. **107** (2011) 252301.

[40] PHENIX Collab. (A. Adare *et al.*), Phys. Rev. Lett. **98** (2007) 172301.

[41] PHENIX Collab. (S. S. Adler *et al.*), Phys. Rev. Lett. **94** (2005) 232301.

[42] PHENIX Collab. (A. Adare *et al.*), Phys. Rev. Lett. **109** (2012) 122302.

[43] PHENIX Collab. (A. Adare *et al.*), Phys. Rev. Lett. **104** (2010) 132301.

[44] C. Halliwell, *et al.*, Phys. Rev. Lett. **39** (1977) 1499–1502.

[45] PHENIX Collab. (A. Adare *et al.*), arXiv:1509.06727v1 [nucl-ex].

[46] J.-Y. Ollitrault, Phys. Rev. D**46** (1992) 229–245, Nucl. Phys. A**638** (1998) 195c–206c.

[47] PHENIX Collab. (M. Kaneta *et al.*), J. Phys. G**30** (2004) S1217–S1220.

[48] PHENIX Collab. (C. A. Aidala *et al.*), Phys. Rev. Lett. **98** (2007) 162301.

[49] PHENIX Collab. (A. Adare *et al.*), Phys. Rev. Lett. **115** (2015) 142301.

[50] PHENIX Collab. (I. Nakagawa *et al.*), Results on Collectivity tests in high-multiplicity p+p and p+Au collisions at $\sqrt{s_{NN}}$ =200 GeV, Quark Matter 2015, Kobe, Japan, `https://indico.cern.ch/event/355454/session/31/contribution/378`

[51] PHENIX Collab. (A. Adare *et al.*), Phys. Rev. D**76** (2007) 051106(R).

[52] Yu. L. Dokshitzer and D. E. Kharzeev, Phys. Lett. B**519** (2001) 199–206.

[53] F. Karsch and K. Redlich, Phys. Rev. D**84** (2011) 051504(R).

[54] PHENIX Collab. (A. Adare *et al.*), arXiv:1506.07834v1 [nucl-ex].

[55] A. Bazavov *et al.*), Phys. Rev. Lett. **109** (2012) 192302.

[56] PHENIX Collab. (A. Adare *et al.*), Phys. Rev. C**78** (2008) 044902.

[57] O. E. Barndorff-Nielsen, D. G. Pollard and N. Shephard, Quant. Finance **12** (2012) 587–605.

[58] STAR Collab. (L. Adamczyk *et al.*), Phys. Rev. Lett. **113** (2014) 092301.

[59] S. Borsanyi *et al.*, Phys. Rev. Lett. **113** (2014) 052301.

[60] STAR Collab. (L. Adamczyk *et al.*), Phys. Rev. Lett. **112** (2014) 032302.

[61] J. Cleymans, H. Oeschler, K. Redlich and S. Wheaton, Phys. Rev. C**73** (2006) 034905.

[62] A. Bzdak and V. Koch, Phys. Rev. C**86** (2012) 044904.

[63] P. Carruthers and C. C. Shih, Phys. Lett. B**165** (1985) 209–216.

[64] M. Asakawa, U. Heinz, and B. Müller, Phys. Rev. Lett. **85** (2000) 2072.

[65] BRAHMS Collab. (I. G. Bearden *et al.*), Phys. Rev. Lett. **93** (2004) 102301.

LECTURER: M.J. TANNENBAUM

Highlights from BNL and RHIC 2015

Scientific Secretaries: S. Campbell, I. Gnesi

DISCUSSION

- *Klopot:*

How do experimentalists extract baryon density in heavy ion collisions and is it possible to do it for a single collision event?

- *Tannenbaum:*

You can extract the baryon density by taking the particle-anti-particle ratio, but the most interesting one is taking pbar-to-p ratio. There is a nice formula for this ratio:

$$\frac{\bar{p}}{p} = \frac{e^{-(E-\mu_B)/T}}{e^{-(E-\mu_B)/T}} = e^{-(2\mu_B)/T}$$

You can also do this for the strangeness ratio. But for the baryon chemical potential, you measure the pbar-to-p ratio and compare it to the Boltzmann equation with the above formula. So experimentally you use the time of flight detectors and momentum information to do the particle identification. We measure the number of protons and the number of anti-protons in the whole p_T spectrum (high and low p_T) and take the ratio. It might be possible to do it for individual events.

- *Kopot:*

Can you give more detail on the spin program at RHIC and in particular how can it help to solve the proton spin puzzle?

- *Tannenbaum:*

An example: W is a special case because we measure a parity violating asymmetry. We can measure a W by measuring the forward single muons at high p_T in 510 GeV pp collisions. And then we can calculate x_1 and x_2, because if the muon is far enough forward, energy/momentum conservation fixes the neutrino direction from W decay. On slide 92 of the pdf file there are equations for the single spin asymmetry for the W+ and W-. These equations depend on u(x1), d(x1), u(x2), d(x2) and theta, the angle of the lepton direction with respect to the W spin in the W rest system. If $x_1 << x_2$, meaning that the W is moving very backward relative to the polarized proton, then A_L^{W-} directly measures Δubar(x1)/ubar(x1) and A_L^{W+} measures Δdbar/dbar, where Δu,Δd, Δū, Δdbar measure the number of quarks of each type polarized along the direction of the proton polarization minus the number of quarks of each type polarized in the opposite direction. So far we haven't succeeded measuring Δdbar(x)/dbar(x) Δubar(x)/ubar(x) directly but in principle that's how you do it. This gives you information on the sea quark polarization of identified quarks because the W is coupled to flavor, not color. A similar thing can be done for the W-->e+ν using central arm electron measurements, where the W is at rest, which gives a Jacobian peak---as Zichichi predicted in 1964. Then we know $x1=x2=M_W/\sqrt{s}=80.4/500=0.16$. In principle we can do it for direct photons also since the reaction is g+q—>gamma+q so we can construct x1 and x2 by measuring the gamma + quark jet (or fragment) and then directly measure $\Delta g(x)/x$ but we haven't yet succeeded in doing this yet. In practice we measure the single spin asymmetry or the double spin asymmetry and then hand it over to the theorists who then can extract the gluon or quark contributions to the spin parton distribution functions.

- *Varnhorst:*

How can you vary the volume for a finite scaling analysis in the experiment?

- *Tannenbaum:*

I know how to vary the volume. The answer is that you can vary the solid angle of the measurement. The question always is that theorists do lattice QCD calculations for all particles. We measure in limited regions of pseudo-rapidity and azimuth. PHENIX measures +/- 0.35 in eta that's about +/- 20 degrees on either side of the transverse plane and STAR measures +/- 0.5 in eta around mid-rapidity. So we only measure one quarter or less of the particle distributions because they extend out to higher rapidities, typically +/-2, than the detectors can measure. We could make measurements by reducing the η acceptance; but you have to be careful in varying η because there are correlations. We could also vary the azimuthal acceptance in PHENIX. The question is that what volume should be used in the theory to compare to the measurement since we don't measure the full particle distribution. Theorists and experimentalists must work together to get an empirical description of what is going on to get to this effect.

- *Kovalenko:*

When you compare the fluctuations and cumulants to lattice results, your measurements depend on acceptance, particularly the transverse momentum range.

- *Tannenbaum:*

In azimuth,φ, we have a two-arm spectrometer we measured the net charge cumulant ratios in one arm, and in both arms and got the same answer. In η, you can check to see whether μ/k for the NBD fits remains constant with increasing η acceptance. If this is true then the cumulant ratios won't have anη dependence (see slide 83). This is still an open question: do we have to measure all the particles that are produced? Or can we measure a sample? We don't know. We need to find this empirically. For momentum, particularly lower p_T, looking at the p_T distribution should help. We need to be careful in how values like $\kappa\sigma^2$ change over the p_T range. For example on slides 65-66 you can see STAR's net proton $\kappa\sigma^2$ measurement in the published version is measured over the p_T range from 0.4-0.8 GeV/c with result $\kappa\sigma^2$=0.7+\-0.1 at $\sqrt{s_{NN}}$=7.7 GeV while in the new preliminary version measured over the p_T range from 0.4-2GeV/c the result for $\kappa\sigma^2$is dramatically different, $\kappa\sigma^2$+3.5+\-1.0 at $\sqrt{s_{NN}}$=7.7 GeV. Without them showing the net proton (actually proton since the anti-protons are negligible) distributions it is impossible to know what is happening.

- *Puccio:*

Are there any other possible measurements that can confirm the critical point observation?

- *Tannenbaum:*

The conventional idea is that you should see a non-monotonic behavior of a signal (such as $S\sigma$, $\kappa\sigma^2$, etc) as a function of $\sqrt{s_{NN}}$ when it passes through the critical point. Because correlation lengths diverge at the critical point it would cause these values to jump up, and possibly become infinite.

- *Ivanytskyi:*

You told us about techniques to define the critical endpoint which is essentially a property of infinite systems. Critical end point is a property of an infinite system. How do you evaluate finite size effects?

- *Tannenbaum:*

This is a problem for theorists. The QGP made in experiment is of a finite size radius of less than 10fmand length less than 30 fm and lasts 10^{-23}s. The trick is to find some effect.

- *Saviskyi:*

What was the evidence for the observation of the effect of ripples from particle-separating wave in primordial plasma? Was there some periodical change in the dynamics of the QGP?

- *Tannenbaum:*

It is controversial. You measure two-particle correlations looking for Bose-Einstein correlations, slide 86 and 87. Then you extract from them the R_{out}, R_{side} and R_{long} by fitting these correlation functions. And this is done from 10GeV to 2.7 TeV. Looking at $R_{out}^2 - R_{side}^2$ (slides 86, 87) you can see non-monotonic behavior with a bump. This is the only non-monotonic effect as a function of $\sqrt{s_{NN}}$ that has been seen at RHIC and LHC.

- *Kaur:*

Is there any measurement at RHIC which gives an estimate of the contribution of the gluons to the proton spin?

- *Tannenbaum:*

We measure the π^0 at mid-rapidity from polarized proton collisions. On slide 26 we have the double spin asymmetry as a function of x_T. At low x it is completely gluon dominated, that is why the scale ends at 0.1. There is a small but measureable non-zero A_{LL} for the π^0. We also measure jets which extend to x_T~0.3and show a larger A_{LL}. Unfortunately we can't measure the gluon contribution directly. Theorists calculate the polarized parton distribution functions and extract the best fit to the gluon $\Delta g(x,Q^2)$. The integral of the gluon contribution is 0.2, the quark contribution is 0.17, so we haven't explained the spin ½ of the proton. Now theorists have predicted its behavior at low x and these predictions if true would solve the proton spin problem. We need to make more measurements at low x to reduce the uncertainty of the theory by a factor of two or three. If this is possible and the theorists are correct we could solve the proton spin problem.

- *Lee:*

Page 39. The errors are dominated by systematic error. If we remove them, the shape is not consistent with a flat line. Does this mean that the quark participant scaling doesn't work well?

- *Tannenbaum:*

We measure the multiplicity distribution. The quark participant scaling works at 62 GeV; and below 62 it fails. The issue with the error bars is related to how it is plotted because these errors are correlated errors. If we ignore the ALICE data you see that it works well for the $\sqrt{s_{NN}}$ 200-62 GeV data.

- *Minkowski:*

Looking at the cumulant ratio plots as a function of $\sqrt{s_{NN}}$, I don't see an effect of a critical point. Can you explain this? I predicted a critical point.

- *Tannenbaum:*

We have yet to measure a clear signature of the critical point. We are going to do a measurement at 49GeV. We have to see if it is there.

LHCb - selected highlights and future prospects

Guy Wilkinson, on behalf of the LHCb collaboration

University of Oxford, Denys Wilkinson Building, Keble Road, Oxford, OX1 3RH, United Kingdom; PH Dvision, CERN, CH-1211, Switzerland

The importance of flavour physics as a method to search for non-Standard Model effects is explained. The LHCb experiment is introduced, and selected recent highlights are presented. Prospects are given for the new LHC run and for the Upgrade phase of the experiment.

1 Introduction – the importance of flavour physics

There are many good reasons why flavour physics [a] is an important and vibrant frontier of experimental activity:

- Many of the open questions in the Standard Model (SM) are found in the flavour sector. Why are there three generations? What determines the extreme hierarchy of quark masses? What determines the elements and structure of the Cabibbo-Kobayashi-Maskawa matrix? What is the origin of CP violation?
- Progress in flavour physics may have consequences for our understanding of cosmology, as it is well known that the CP violation that can be accommodated in the SM is inadequate to explain the matter dominated universe that we observe.
- Flavour physics is a proven tool of discovery. The observed rate of kaon mixing and the suppressed decay $K^0_{\rm L} \to \mu^+\mu^-$ necessitated the construction of the GIM mechanism and hence predicted the existence of the charm quark. Explaining CP violation in the kaon system within the SM required invoking a third quark generation. The first indication that the top quark was very massive followed from the observation of B-mixing.

It is the last point in particular which is worth stressing in this, the first year of LHC run 2. The LHC has been constructed to search for evidence of new physics (NP) at the TeV energy scale. As well as looking for direct production of new particles associated with this NP, however, an equally valid approach is to make precise measurements of processes which are suppressed in the SM, with the hope of discovering deviations with respect to the SM predictions.

The LHCb experiment is a project dedicated to flavour-physics measurements at the LHC, which has yielded many world-leading results and has exciting potential for the years ahead. In the following a brief overview is given of the experiment, and then selected recent results are presented. In conclusion, an overview is given of future prospects during run 2 and the upgrade to the LHCb detector which will be performed in the long shutdown that follows.

[a] In this article flavour physics is interpreted as hadronic flavour physics, although many of the arguments advanced apply equally well to leptons.

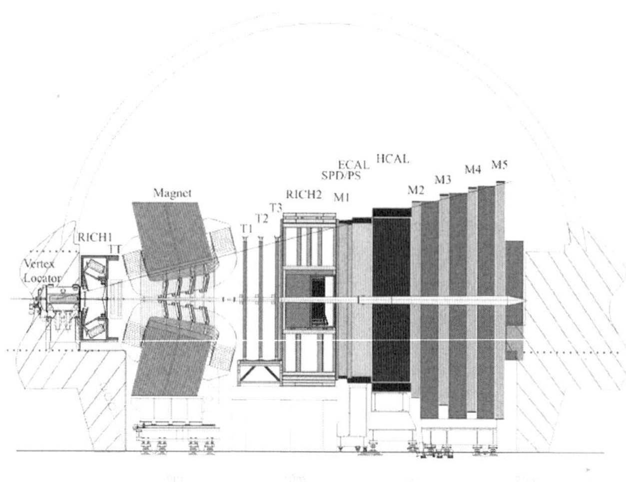

Figure 1: A schematic of the LHCb detector.

2 The LHCb detector

The LHCb detector, shown in Fig. 1, is situated at interaction point 8 of the LHC in the cavern previously used by the DELPHI experiment at LEP. It has a distinctive forward geometry which is chosen to take advantage of the forward- (and backward-) peaking production of b and c-quark pairs in pp collisions, and to give sensitivity down to low ($\sim 1\,\mathrm{GeV}/c$) transverse momentum.

A full description of the LHCb detector, and its operation, is beyond the scope of this article, but may be found in Refs. [1,2]. LHCb possesses three crucial attributes that distinguish it from the other experiments at the LHC and make it ideally equipped for flavour-physics studies.

1. Around the interaction point is a precise silicon strip Vertex Locator (VELO)[3]. The VELO is situated within a secondary vacuum, and the two sides of the detector are retracted during injection and then brought to within 8 mm of the beamline for data taking. The VELO allows the secondary vertices associated with b and c-hadron decays to be reconstructed with a longitudinal precision of around $250\,\mu m$ (with a dependence on decay multiplicity) and the proper lifetime of the hadrons to be measured to around 50 fs.

2. LHCb is equipped with a pair of Ring Imaging Cherenkov Counters (RICHes), one before the dipole magnet and one after[4]. These provide hadron identification, enabling excellent discrimination between *e.g.* pions and kaons from low momentum up to around 100 GeV/c. Such a capability is invaluable when attempting to separate different decay modes of heavy hadrons.

3. The experiment has a trigger system dedicated to flavour physics[5]. At the earliest stage a hardware trigger, the 'L0', uses simple signals mostly in the calorimeters and the muon system to reduce the rate of interesting events to 1 MHz. Thereafter, the detector information is read out into a farm of processors in which the 'High Level Trigger', an ensemble of software algorithms, is run. Selected events are output to storage at a rate of several kHz. As a result of this trigger system LHCb has good efficiency for final states involving hadrons, single leptons and photons, not only the simpler signature of dimuons.

In many flavour physics measurements it is necessary to measure the lifetime of the decaying

b or c-hadron, and therefore to know the correspondence between the secondary (decay) vertex and the primary (collision) vertex in which the hadron was born. This requirement motivates maintaining the rate of multiple interactions in each beam crossing at a lower level than is the case in ATLAS and CMS. Hence LHCb does not collect data at the maximum available luminosity of the collider. Rather the beams are kept slightly displaced, with the degree of displacement reduced during the course of a fill, as the intensity decays, in order to maintain constant luminosity throughout. During 2012 data were collected at a luminosity of $4 \times 10^{32}\mathrm{cm}^{-2}\mathrm{s}^{-1}$. A data set corresponding to an integrated luminosity of 3 fb^{-1} was accumulated during 2011 and 2012.

3 Selected physics results

At the time of the school LHCb had published over 260 papers. Here selected recent results are highlighted in three distinct areas: flavor-changing neutral currents and friends, CP violation and Unitarity Triangle tests, and hadron spectroscopy (progress in the latter topic is not expected to inform the search for NP, but is nonetheless very important for improving our understanding of how matter is constituted).

3.1 FCNC and other 'rare' decays

Flavour changing neutral currents cannot occur at tree level in the SM, and hence are mediated by amplitudes involving loop or box diagrams. For this reason they are suppressed and are sensitive to the possible contribution of new heavy particles which participate in the loop.

The observation of the ultra-rare decay $B_s^0 \to \mu^+\mu^-$

The most famous FCNC in b-physics is the ultra-rare decay $B_s^0 \to \mu^+\mu^-$. Not only an FCNC, this mode is also CKM and helicity suppressed, and predicted to have a branching fraction of a few 10^{-9} in the SM. Moreover this SM branching fraction can be calculated with $< 10\%$ uncertainty [6]. Many models of NP, in particular those with an extended scalar sector, such as SUSY, can give rise to deviations from the SM expectation. For this reason the decay mode has been searched for intensely at different colliders over the past 25 years with the hope of seeing it appear with an anomalous rate. First evidence was reported in 2013 by LHCb [7], followed soon after by the CMS collaboration [8]. Neither experiment, however, had sufficient data or sensitivity to claim an unambiguous observation, nor to measure the decay rate with interesting precision. For that reason it was decided to perform a joint analysis, fitting the samples of each experiment simultaneously with common assumptions concerning shared backgrounds *etc.*

The joint CMS-LHCb analysis [9] observes the decay $B_s^0 \to \mu^+\mu^-$ with a significance of 6.2 σ and measures a branching fraction of $(2.8^{+0.7}_{-0.6} \times 10^{-9})$, which is a little lower, but entirely compatible with, the SM expectation. The mass spectrum, after suitably weighting different subsamples in the analysis according to their signal sensitivity, is shown in the left-hand plot of Fig. 2. It is interesting to note that there is also a hint of the even rarer mode $B^0 \to \mu^+\mu^-$, at a rate a little higher, but again compatible with, the SM prediction. The numerical results are shown in the right-hand plot of Fig. 2 and are seen to lie at around 2 σ from the SM point. It will be fascinating to see how this picture evolves with the data of run 2.

Measuring observables in $b \to sl^+l^-$ processes

Decays of b-hadrons involving the quark-level process $b \to sl^+l^-$ are a rich laboratory to probe for NP effects. Although suppressed, the decays can be accumulated in relative abundance at LHCb. The most widely studied such channel is the mode $B^0 \to K^{*0}\mu^+\mu^-$. By analysing the decay kinematics of the final-state particles a host of observables may be measured, many of

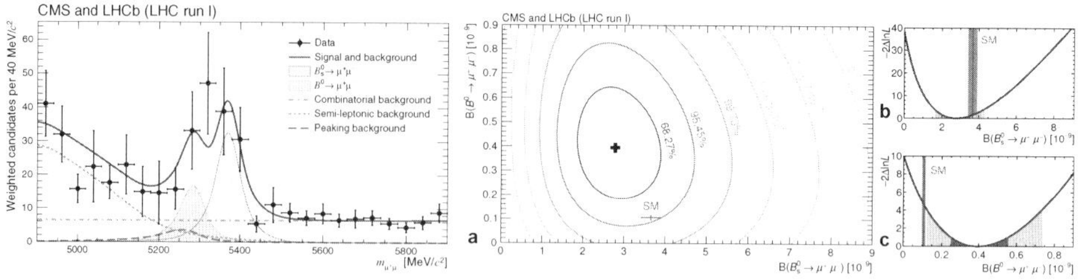

Figure 2: Results from the joint CMS-LHCb $B^0_{(s)} \to \mu^+\mu^-$ analysis. Left: mass spectrum. Right: constraints in the $B^0_s \to \mu^+\mu^-$ vs. $B^0 \to \mu^+\mu^-$ plane. From Ref. [9].

which are highly sensitive to the helicity structure of any contributing NP. One such observable is $A_{\rm FB}$, the forward-backward asymmetry of the dimuon system, measured in bins of q^2, the invariant mass of the dimuons. The left-hand plot in Fig. 3 shows preliminary results for the analysis of the full run-1 LHCb data set, compared with the predictions from theory [10], in bins where these predictions exist (note that the bins containing the resonant decays $B^0 \to K^{*0}J/\psi$ and $B^0 \to K^{*0}\psi'$ are excluded). It may be seen that the shape of the distribution in data broadly agrees with that expected from the SM, although the exact bin-to-bin agreement is not perfect.

Another observable of interest in this decay is the inscrutably named P_5'. This quantity does not have a simple physical correspondence, but is constructed from the amplitudes measured in the angular fit to the final-state particles, and is one of several observables that have been proposed as suitable for measurement on account of their limited sensitivity to form-factor uncertainties [13]. An LHCb analysis based on the first 1 fb^{-1} of run-1 data [14] had shown a significant discrepancy between data and SM expectation around $q^2 = 6$ GeV$^2/c^4$. It is very interesting to observe, therefore, that the updated preliminary measurement making use of the full data set, and shown in the right-hand plot of Fig. 3, still exhibits tension with respect to prediction in this region. Commentators have been intrigued by this behaviour and have sought to find a common pattern between this anomaly and other puzzles in the $b \to sl^+l^-$ sector, such as the tendency for the measured differential branching ratios to be slightly low (see *e.g.* Ref. [15] for a recent measurement with the decay $B^0_s \to \phi\mu^+\mu^-$, shown in Fig. 4.) or the 2.6σ violation of lepton universality found when comparing the decay rates of $B^+ \to K^+e^+e^-$ and

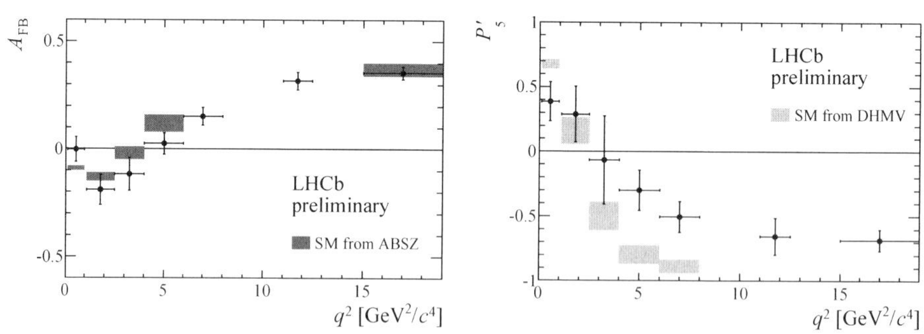

Figure 3: Preliminary measurements of observables in $B^0 \to K^{*0}\mu^+\mu^-$ by LHCb with 3 fb^{-1}. Left: the forward backward asymmetry $A_{\rm FB}$ vs. q^2, the invariant mass of the dimuon system. Right: the observable P_5' vs. q^2. Superimposed are SM predictions from Ref. [10] for $A_{\rm FB}$ and Ref. [12] for P_5'. From Ref. [11].

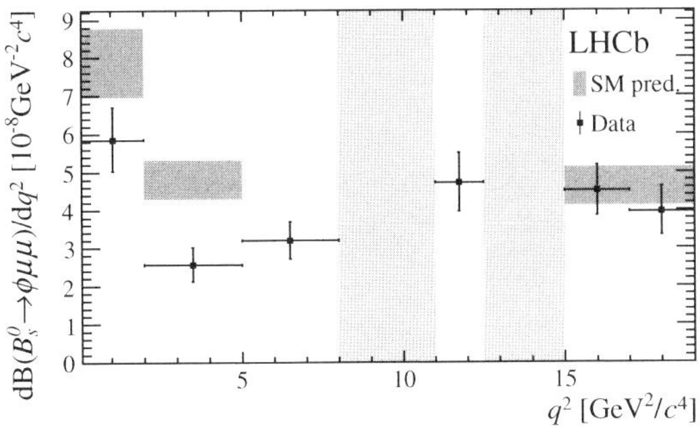

Figure 4: Measurement of the differential cross-section of the decay $B_s^0 \to \phi\mu^+\mu^-$ vs. q^2. Superimposed are the SM predictions from Refs. [10,18]. From Ref. [15].

$B^+ \to K^+\mu^+\mu^-$ [17]. An example of such a global analysis can be found in Ref. [16]. It is hoped that future measurements with run-2 data will clarify the picture.

The decay $B^0 \to D^{*+}\tau^-\bar{\nu}_\tau$

The decay $B^0 \to D^{*+}\tau^-\bar{\nu}_\tau$ is not a FCNC and proceeds by a tree-level amplitude. The observable $R(D^*)$ is measured by normalising the rate of this decay with that of $B^0 \to D^{*+}\mu^-\bar{\nu}_\mu$, and in the SM is calculated to be 0.252 ± 0.003 [19]. A measurement of $R(D^*)$, and the sister observable $R(D)$ involving $B \to D\tau^-\bar{\nu}_\tau$ decays, is of interest as a test of lepton universality, and one with sensitivity to the effects of a charged Higgs boson, or similar such particle, in mediating the decay. Studies at the e^+e^- B-factories, have drawn much attention to these observables, as they yield central values above the SM expectation. Particularly noteworthy is a measurement of $R(D^*)$ performed by the BaBar collaboration [20], which shows a 2.7 σ deviation.

It was not expected that LHCb could contribute in this area of study. The presence of neutrinos in the final state and the absence of any beam-energy constraint, which is a powerful tool in the B-factory analyses, make the isolation of $B^0 \to D^{*+}\tau^-\bar{\nu}_\tau$ decays extremely challenging at a hadron collider. Nonetheless this challenge has been overcome [21]. The analysis employs the decay signature $\tau^- \to \mu^-\nu_\tau\bar{\nu}_\mu$ and separates the signal mode from $B^0 \to D^{*+}\mu^-\bar{\nu}_\mu$ decays and other backgrounds through fits to the muon energy in the B rest frame and the missing invariant mass squared of the reconstructed particles, performed in bins of the squared four-momentum transfer to the lepton system. A result of $R(D^*) = 0.336 \pm 0.027\,(\text{stat.}) \pm 0.030\,(\text{syst.})$ is obtained, which is very similar to the BaBar central value and 2.1 σ above the SM expectation. The precision of the measurement is better than that obtained by the Belle collaboration [22].

The ensemble of $R(D^*)$ and $R(D)$ measurements present an intriguing picture, as shown in Fig. 5. An analysis, prepared by the Heavy Flavor Averaging Group, finds the average of the measurements to lie 3.9σ away from the SM predictions. A measurement of $R(D)$ from LHCb is eagerly anticipated.

3.2 CP violation and Unitarity Triangle tests

CP violation is accommodated within the SM by a single complex phase within the Cabibbo-Kobayaski-Maskawa (CKM) matrix. The matrix is unitarity and this property leads to a series of relations between the elements that may be depicted geometrically in the complex plane. One of these relations gives rise to the so-called Unitarity Triangle. A principal task of flavour physics is to make precise measurements of observables that allow the angles or sides of this triangle to

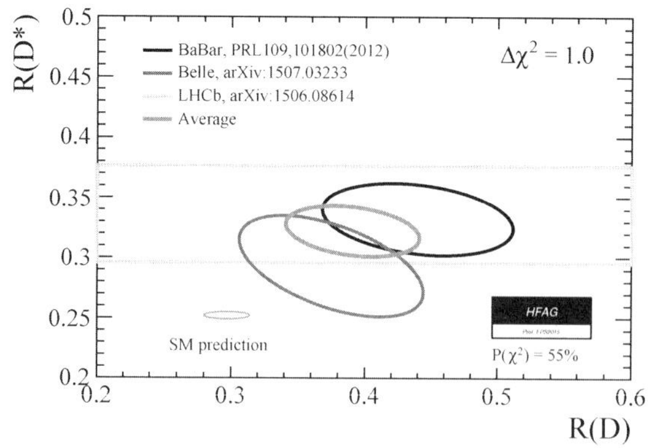

Figure 5: Results for $R(D^*)$ and $R(D)$ and the SM expectation. From Ref. [23].

be determined. Within the SM these measurements should all be consistent with the Unitarity Triangle/CKM paradigm, whereas the presence of NP may lead to internal inconsistencies. The Unitarity Triangle is shown in Fig. 6, with constraints superimposed from available relevant measurements in b-physics and kaon physics. The compatibility of these measurements has been analysed in a Bayesian (left-hand plot, UTfit collaboration [24]) and frequentist (right-hand plot, CKMfitter group [25]) framework. In both cases consistency is found, which shows that the CKM description of CP violation is adequate at leading order. More precise measurements are required to refine these tests and to probe for the contributions of NP. Examples are now given of how LHCb is contributing to this programme.

The angle γ

The angle γ, seen at the left-hand bottom side of the triangle in Fig. 6, has a special status in Unitarity Triangle studies since it can be measured in tree-level processes. It is the weak phase that exists between $b \to u$ and $b \to c$ transitions and can be accessed with minimal theoretical uncertainties by comparing observables between B^+ and B^- in $B^\pm \to DK^\pm$ decays, where D indicates a neutral charm meson reconstructed in a final state common to both D^0 and $\bar{D}^0$ mesons [26,27,28]. Since loop diagrams are not involved, the physics that results is expected to be

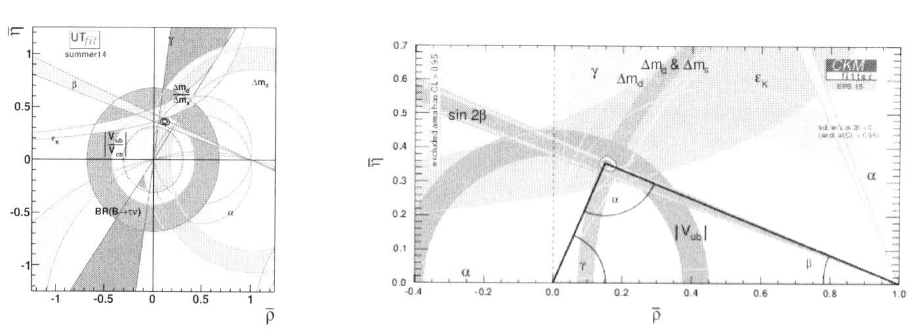

Figure 6: The Unitarity Triangle, showing constraints for the sides and angles coming from available measurements in flavour physics. Left: analysis by the UTfit collaboration [24]; right: analysis by the CKMfitter group [25].

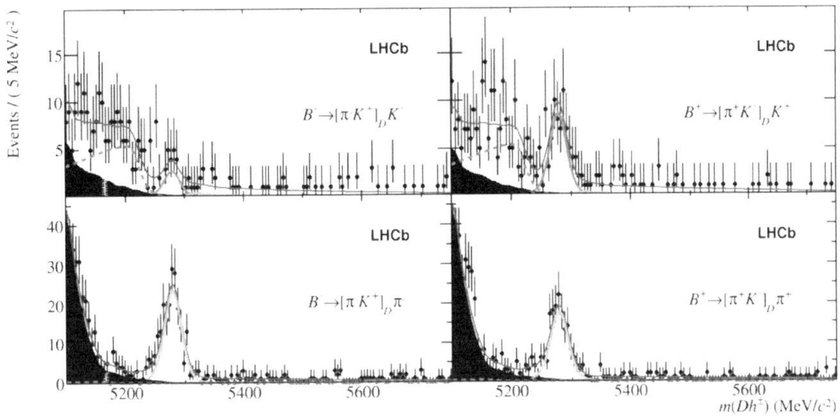

Figure 7: Plots of $B^\pm \to (\pi^\pm K^\mp)_D K^\pm$ and $B^\pm \to (\pi^\pm K^\mp)_D\pi^\pm$ with LHCb data corresponding to 1 fb^{-1} of integrated luminosity. The plots are divided between B^- (left) and B^+ (right). From Ref.[29].

wholly dominated by SM processes. Therefore a measurement of γ in this manner provides a SM benchmark against which other quantities, more sensitive to NP effects, can be compared.

Prior to LHCb, sensitivity to γ was poor due to the limited sample sizes avaliable. The situation is now changing. In Fig. 7 are shown the signal peaks accumulated in $B^\pm \to (\pi^\pm K^\mp)_D K^\pm$ decays with 1 fb^{-1} of data[29]. The visible branching fractions for these decay modes are $\sim 10^{-7}$, which was beyond the statistical reach of previous experiments. Clear signals are seen with the LHCb data, as well as a striking CP asymmetry, manifested in the different peak sizes for B^+ and B^-. The CP asymmetry carries clean information on γ. Also shown are the corresponding peaks for $B^\pm \to (\pi^\pm K^\mp)_D\pi^\pm$ decays, which are more abundant modes where a lower level of CP violation is both seen and expected. Other important measurements in this category of $B^\pm \to (\pi^\pm K^\mp)_D K^\pm$ decays include the case where the D meson is reconstructed in a multibody final state such as $K^0_S\pi^+\pi^-$. In this case the CP violation manifests itself as a difference in distributions across the Dalitz plot, which when interpreted brings information on γ. Examples of these Dalitz plots from the LHCb 3 fb^{-1} study can be seen in Fig. 8.

A global analysis[31] of all available LHCb results returns $\gamma = (72.9^{+9.2}_{-9.9})^\circ$. The precision is better than that obtained from the combined analyses of both the BaBar and Belle experiments. Further improvement is expected with the run-1 data, and an uncertainty of around 3° will be attainable with the data set expected in run 2.

The angle β

The angle β, seen at the right-hand bottom side of the triangle in Fig. 6, was well measured at the B-factories. Indeed, this achievement was the critical piece of evidence in confirming the broad validity of the CKM mechanism. Nonetheless improved precision is highly desirable. Information on β comes from observables which probe the interference between $B^0 - \bar{B}^0$ mixing and decay amplitudes. The box diagrams that mediate the mixing are promising entry points for NP contributions. The most sensitive approach for obtaining information on this angle is to measure $\sin 2\beta$, which is related to the CP asymmetry of B^0 and $\bar{B}^0$ mesons decaying into the CP eigenstate $J/\psi K^0_S$ as a function of decay time. The measurement therefore requires that the initial flavour of the b-hadron is known, which is performed through a procedure known as 'flavour tagging'. An example of flavour tagging is to use the charge of a muon from a semi-leptonic decay of the accompanying b-hadron in the event to infer the initial particle/antiparticle

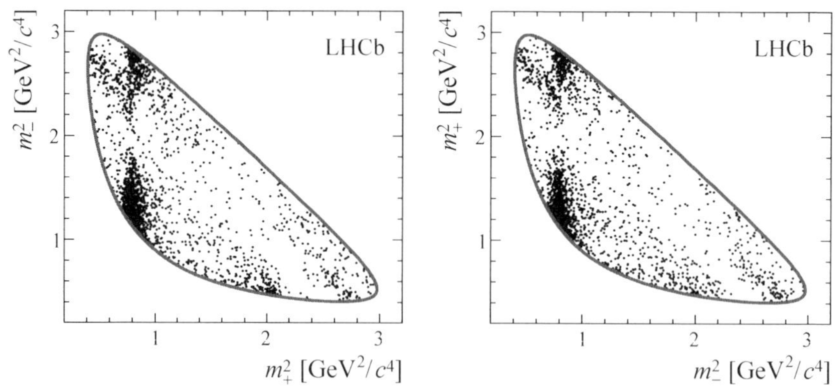

Figure 8: The Dalitz distribution of $D \to K_S^0\pi^+\pi^-$ decays from the process $B^\pm \to DK^\pm$ with LHCb data corresponding to 3 fb^{-1} of integrated luminosity. The plots are shown for B^- (left) and B^+ (right) decays. From Ref. [30].

nature of the b-quark in the hadron of interest.

A determination of β is more challenging at a hadron machine than at an $e^+e^- \to \Upsilon(4S) \to b\bar{b}$ facility. There are certain advantages that make the flavour-tagging intrinsically more reliable in the latter environment (the coherence of the $b\bar{b}$ state, the absence of fast-oscillating B_s^0 mesons *etc.*). Nonetheless the large number of b-hadrons collected by LHCb, and recent improvements to the flavour-tagging performance, have allowed for a measurement to be made with the full run-1 data set of $\sin 2\beta = 0.731 \pm 0.032\,(\text{stat.}) \pm 0.020\,(\text{syst.})$. In Fig. 9 is shown both the signal peak, containing over 40,000 decays, and the time-dependent CP asymmetry that is reconstructed from this sample. The result is of very similar precision to that achieved at BaBar and Belle and indicates that significantly improved knowledge of this critical parameter of the Unitarity Triangle can be expected with the run-2 data set.

V_{ub}/V_{cb}

The side of the Unitarity Triangle opposite to the angle β corresponds to the ratio of magnitudes of CKM elements V_{ub}/V_{cb}. This is not itself a CP violating observable, and is best measured through comparing the rate of $b \to u$ and $b \to c$ transitions, accompanied by a semileptonic

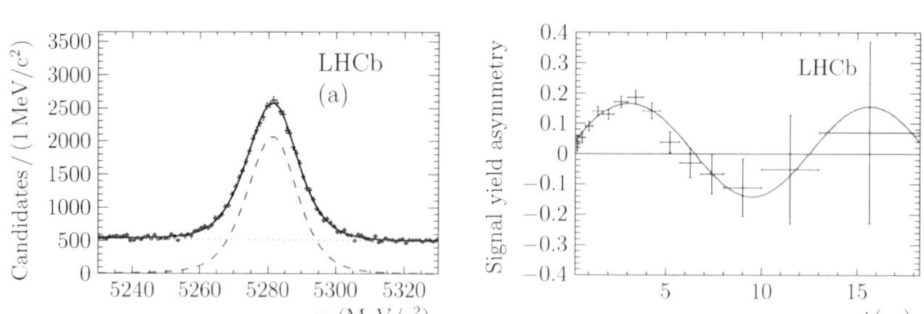

Figure 9: Left: mass peak of $B^0 \to J/\psi K_S^0$ decays. Right: time-dependent CP asymmetry measured with this sample. From Ref. [32].

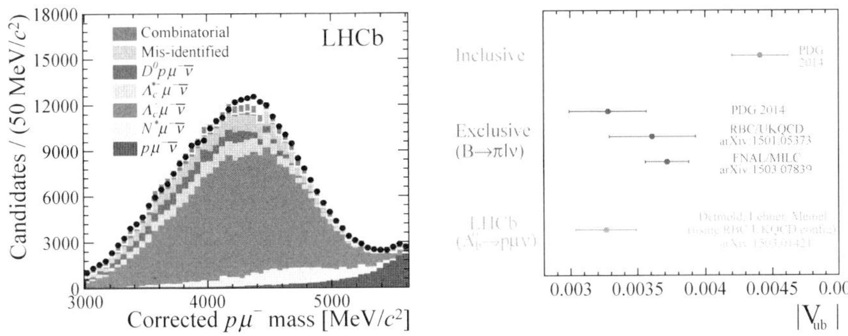

Figure 10: Left: discriminating variable constructed in LHCb V_{ub} analysis to isolate $\Lambda_b \to p\mu^-\bar{\nu}_\mu$ decays. Right: summary of V_{ub} measurements, showing the inclusive and exclusive results from the B-factories and the LHCb result. Also indicated are the LQCD inputs. From Ref. [33].

decay. As with the $B^0 \to D^{*+}\tau^-\bar{\nu}_\tau$ analysis, discussed above in Sect. 3.1, it was long assumed that measurements of V_{ub}/V_{cb} were the preserve of e^+e^- machines, because of the challenges posed by isolating a rare decay channel, *i.e.* one involving suppressed $b \to u$ transitions, with missing energy from the neutrino in the final state.

In fact these challenges have been overcome, and LHCb has recently published a precise measurement of $|V_{ub}|/|V_{cb}|$ [33]. To achieve this, advantage has been taken of the large number of b-baryons produced at the LHC and a first observation made of the suppressed decay mode $\Lambda_b \to p\mu^-\bar{\nu}_\mu$. The branching fraction for this decay is normalised to that of the favoured channel $\Lambda_b \to \Lambda_c^+\mu^-\bar{\nu}_\mu$. The branching fractions are measured in a restricted q^2 range, where q^2 is the invariant mass squared of the final-state leptons, of $q^2 > 15\,\mathrm{GeV}^2/c^4$ for the suppressed mode and $q^2 > 7\,\mathrm{GeV}^2/c^4$ for the favoured mode. The excellent vertex resolution and the particle identification capabilities of the experiment are key ingredients in allowing the signal mode to be reconstructed. The left-hand plot of Fig. 10 shows a clear enhancement from the signal at high values of the 'corrected $p\mu^-$ mass', which is a discriminant variable constructed in the analysis.

The ratio of branching fractions, which is found to be $\sim 1\%$, can be converted into a value of $|V_{ub}|/|V_{cb}|$ using input from lattice QCD (LQCD)[34]. A result of $|V_{ub}|/|V_{cb}| = 0.083 \pm 0.004 \pm 0.004$ is obtained, where the first uncertainty is experimental and the second is due to the uncertainty in the LQCD prediction. Finally, using the world average of $|V_{cb}|$ measured using exclusive decays [35], it is determined that $|V_{ub}| = (3.27 \pm 0.15 \pm 0.16 \pm 0.06) \times 10^{-3}$, where the first uncertainty is experimental, the second from LQCD and the third from $|V_{cb}|$.

The LHCb result is very valuable in itself and also indicates that further progress can now be expected in the measurement of CKM matrix elements, despite the challenges of the hadronic environment. Moreover, the result casts light on a longstanding anomaly seen at the B-factories concerned with the V_{ub} measurement. A significant difference exists between inclusive measurements of b-meson decays to charmless final states and those performed with an exclusive decay mode, *e.g.* $B \to \pi\mu\bar{\nu}_\mu$. The LHCb measurement favours the latter set of results, as is summarised in the right-hand plot of Fig. 10.

Studies in the B_s^0 sector

One of the most important tasks in flavour physics studies at the LHC is to search for CP violation in the B_s^0 sector. Prior to first collisions this system was largely unexplored, despite several pioneering measurements at the Tevatron and some time-integrated measurements per-

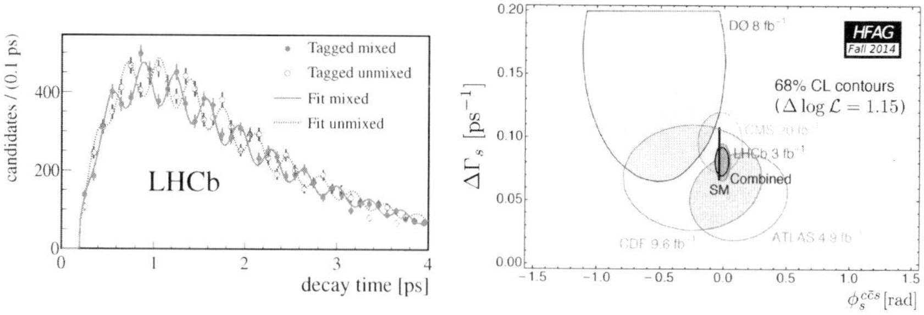

Figure 11: Left: flavour-tagged $B_s^0 \to D_s\pi$ decays in LHCb, showing $B_s^0 - \bar{B}_s^0$ oscillations[36]. Right: experimental constraints on $\Delta\Gamma_s$ vs. ϕ_s [23].

formed at the $\Upsilon(5S)$ by the Belle collaboration. The main goal is to gain sensitivity to the CP-violating phase ϕ_s, which quantifies the CP violation between $B_s^0 - \bar{B}_s^0$ mixing and decay (ϕ_s is not a parameter of the most commonly depicted Unitarity Triangle, which is shown in Fig. 6, but exists in another triangle that also arises from CKM unitarity relations). This phase is the B_s^0 analogue of the angle β, discussed above, but in the SM is expected to be very small. NP contributions in the $B_s^0 - \bar{B}_s^0$ mixing diagrams may lead to enhancements. The measurement is very challenging because of the need to resolve the very rapid $B_s^0 - \bar{B}_s^0$ oscillations, and the requirement to perform an angular analysis to separate out the CP-even and CP-odd contributions for the most sensitive channel, $B_s^0 \to J/\psi K^+K^-$.

LHCb has been optimised with the requirements of the ϕ_s measurement in mind. In particular, the excellent proper-time resolution of the experiment enables the $B_s^0 - \bar{B}_s^0$ oscillations to be clearly resolved, as is evident from the left-hand plot of Fig. 11. The run-1 precision attained on ϕ_s with $B_s^0 \to J/\psi K^+K^-$ decays alone[37] is around 20 times better than was achieved at the Tevatron, and additional sensitivity comes from the addition of the channel $B_s^0 \to J/\psi\pi^+\pi^-$ [38]. The combined result from both measurement is shown in the right-hand plot of Fig. 11, together with that for $\Delta\Gamma_s$, the width-splitting between the B_s^0 mass eigenstates, which is also measured in the analyses. Also displayed are the results of other experiments and the combined constraints[23]. Experiment is currently in good agreement with the SM expectation, but it is imperative to refine this test by increasing sensitivity up to the precision of the theory prediction. This will be possible with the LHCb Upgrade.

3.3 Spectroscopy and the search for exotic hadrons

In addition to providing an ideal environment to probe for NP, the copious supply of b and c-mesons and baryons produced at the LHC also affords excellent opportunities for observing and studying new hadronic states. Several such states have already been found with run-1 data, most of which can be accommodated within the minimal quark model, according to which mesons are constituted of a quark and antiquark, and baryons of three quarks. Notable examples include the discovery of the $\Xi_b^{\prime -}$ and Ξ_b^{*-} baryons [39] and the first observation of a heavy-flavour spin-3 meson [40].

Even more interesting, however, is the contribution that LHCb has made to our knowledge of 'exotics', hadrons that cannot easily be explained within the minimal quark model. One example is the determination of the quantum numbers of the $X(3872)$ through an angular analysis of a large sample of $B^+ \to X(3872)(\to J/\psi\pi^+\pi^-)K^+$ decays, to find $J^{PC} = 1^{++}$. Another, is the investigation of the $Z(4430)^-$, an enhancement found in the $\psi'\pi^-$ mass spectrum by the Belle collaboration when studying $B^- \to \psi'\pi^-K^+$ decays[42]. In the LHCb analysis[43] the same

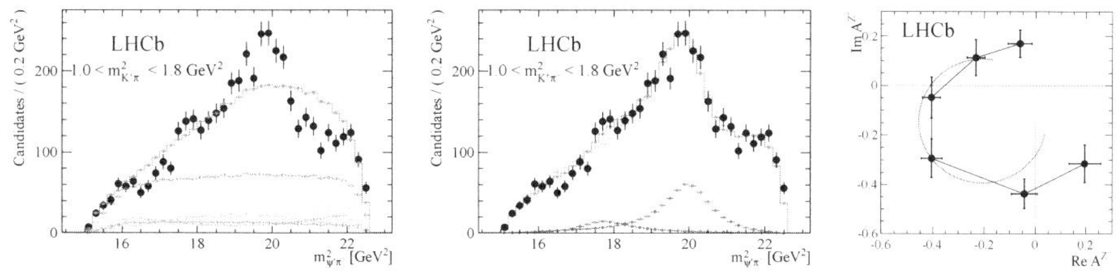

Figure 12: Plots from the LHCb $Z(4430)^-$ analysis [43]. Left: $m^2_{\psi'\pi^-}$ spectrum with no exotic contributions included in the fit model. The points are the data, the red histogram the model fit, and the other histograms the individual contributions. Middle: the same $m^2_{\psi'\pi^-}$ data spectrum with two exotic hadron contributions (shown in blue) included in the model. Right: argand diagram corresponding to the $Z(4430)^-$.

structure is seen, as is visible around $m^2_{\psi'\pi^-} = 19.5 \text{ GeV}^2/c^4$ in the data distributions that are shown in Fig. 12. Naively interpreted, this enhancement can be attributed to a four-quark state of content $c\bar{c}\bar{u}d$. From the mass spectrum alone, however, it cannot be excluded that the feature is an artifact arising from a conspiracy of interference effects between conventional K^{*0} resonances decaying into $K^+\pi^-$. The large size and high purity of the LHCb sample allows this possibility to be excluded. An amplitude model is constructed including all known minimal quark-model intermediate states that could contribute to the final state. The result is shown for the $\psi'\pi^-$ projection in the left-hand plot of Fig. 12; the model fails to describe the data – the p-value of the fit is $\sim 10^{-6}$. Satisfactory agreement between model and data is only achieved if an exotic contribution is added decaying directly to $\psi'\pi^-$, yielding a p-value of 12%. The best description is obtained, with p-value of 26%, if a second, wider, exotic resonance is added with mass around 4240 MeV/c^2. The fit in this scenario is shown in the middle plot of Fig. 12. A final piece of evidence comes from the Argand diagram shown in the right-hand plot of Fig. 12. Here no assumption is made about the shape of the narrow exotic resonance (and for simplicity, the broader resonance is discarded); rather its influence is represented by a complex amplitude, which is allowed to vary in bins of invariant mass across the region of interest. The Argand diagram representing the fitted amplitude shows a rapid variation of phase, which is exactly what is expected if the feature in the mass spectrum is a genuine resonance. From these studies it is concluded that the $Z(4430)^-$ is indeed a genuine particle with four quark composition. This is the first unambiguous observation of such a state.

Since the school a new pair of states have been found in the $J/\psi p$ mass spectrum of $\Lambda_b \to J/\psi K^- p$ decays [44]. A study has been performed following the same procedure as that developed in the $Z(4430)^-$ analysis, and again it is concluded that these resonances are genuine exotic particles – in this case pentaquark states.

4 The future

Run 2 of the LHC is now underway and the LHCb experiment is functioning well. There are several innovations in operation compared with run 1, which will lead to larger sample sizes and improve the analysis procedure. For example, data are now buffered on disk prior to running the last stage of the trigger, and full alignment and calibration are performed. The final trigger stage is only run when the results of this alignment and calibration are satisfactory, which enables 'offline-like' data quality to be achieved in the trigger. One consequence of this development is that many offline analyses can now be performed directly on the data output from the trigger, with no need for further offline processing. This is a revolutionary approach to data processing that marks a paradigm shift for High Energy Physics experiments. It is expected that the integrated luminosity collected during run 2 will be around twice that of run 1. This larger data set, coupled with the increase in production cross-section and the improvements in data processing, will lead to a significant improvement in statistical precision compared to run-1 measurements.

The LHCb experiment will be upgraded during the long shutdown (LS2) that will follow run 2. The key element of this Upgrade is to increase the readout frequency of all sub-detectors from 1 MHz to 40 MHz. This will allow the L0 hardware trigger to be discarded, and full event information to be considered and processed for every collision by the software trigger. This new scheme will lead to a significant increase in trigger efficiency, particularly for hadronic final states. Moreover, the operational luminosity of the experiment will be raised to $2 \times 10^{33}\,\mathrm{cm}^{-2}\mathrm{s}^{-1}$. All sub-detectors will be modified or replaced to comply with these new requirements. The Upgrade will enable the precision of many run-1 and run-2 measurements to be improved by an order of magnitude and will bring new physics goals within reach. More information can be found in Refs. [45,46,47,48,49,50].

5 Conclusions

The LHCb experiment is producing world-leading results in flavour physics and spectroscopy, as well as other areas which are not discussed in this report. It is impressive to see that some of these results are in topics where it was not expected that LHCb would contribute. All measurements are consistent with the expectations of the SM, although several anomalies do exist, which require more data to elucidate. The prospects for attaining significantly improved sensitivity with the data from run 2 and the LHCb Upgrade are excellent.

Acknowledgements

I wish to thank the organisers for their outstanding hospitality, and for arranging a very stimulating school in a delightful environment.

References

1. A.A. Alves Jr. *et al.* (LHCb collaboration), *The LHCb detector at the LHC*, JINST **3** (2008) S08005.
2. R. Aaij *et al.* (LHCb collaboration), *LHCb detector performance*, Int. J. Mod. Phys. **A 30** (2015) 1530022, arXiv:1412.6352.
3. R. Aaij *et al.*, *Performance of the LHCb Vertex Locator*, JINST **9** (2014) P09007, arXiv:1405.7808.
4. M. Adinolfi *et al.* *Performance of the LHCb RICH detector at the LHC*, Eur. Phs. J. **C 73** (2013) 2431, arXiv:1211.6759.
5. R. Aaij *et al.*, *The LHCb and its performance in 2011*, JINST **8** (2013) P04022, arXiv:1211.3055.
6. C. Bobeth *et al.*, *$B_{s,d} \to l^+l^-$ in the Standard Model with reduced theoretical uncertainty*, Phys. Rev. Lett **112** (2014) 101801, arXiv:1311.0903.
7. R. Aaij *et al.* (LHCb collaboration), *First evidence for the decay $B_s^0 \to \mu^+\mu^-$*, Phys. Rev. Lett. **110** (2013) 021801, arXiv:1211.2674.
8. S. Chatrchyan (CMS collaboration), *Measurement of the $B_s^0 \to \mu^+\mu^-$ branching fraction and search for $B^0 \to \mu^+\mu^-$ with the CMS experiment*, Phys. Rev. Lett. **111** (2013) 101804, arXiv:1307.5025.
9. V. Khachatryan *et al.* (CMS and LHCb collaborations), *Observation of the rare $B_s^0 \to \mu^+\mu^-$ decay from the combined analysis of CMS and LHCb data*, Nature **522** (2015) 68,

arXiv:1411.4413.
10. A. Bharucha, D.M. Straub and R. Zwicky, *$B \to Vl^+l^-$ in the Standard Model from light-cone sum rules*, arXiv:1503.05534.
11. LHCb collaboration, *Angular analysis of the $B^0 \to K^{*0}\mu^+\mu^-$ decay*, LHCb-CONF-2015-002.
12. S. Descotes-Genon, L. Hofer, J. Matias and J. Virto, *On the impact of power corrections in the prediction of $B \to K^*\mu^+\mu^-$ observables*, JHEP **12** (2014) 125, arXiv:1407.8526.
13. S. Descotes-Genon, J. Matias, M. Ramon and J. Virto, *Implications from clean observables for the binned analysis of $B \to K^*\mu^+\mu^-$ at large recoil*, JHEP **01** (2013) 048, arXiv:1207.2753.
14. R. Aaij *et al.* (LHCb collaboration), *Differential branching fraction and angular analysis of the decay $B^0 \to K^{*0}\mu^+\mu^-$*, JHEP **08** (2013) 131, arXiv:1304.6325.
15. R. Aaij *et al.* (LHCb collaboration), *Angular analysis and differential branching fraction of the decay $B^0_s \to \phi\mu^+\mu^-$*, arXiv:1506.0877.
16. W. Altmannshofer and D.M. Straub, *Implications of $b \to s$ measurements*, arXiv:1503.06199.
17. R. Aaij *et al.* (LHCb collaboration), *Test of lepton universality using $B^+ \to K^+l^+l^-$ decays*, Phys. Rev. Lett. **113** (2014) 151601, arXiv:1406.6482.
18. W. Altmannshofer and D.M. Straub, *State of new physics in $b \to s$ transitions*, arXiv:1411.3161.
19. S. Fajfer, J.F. Kamenik and I. Nisandzic,*$B \to D^*\tau\bar{\nu}_\tau$ sensitivity to new physics*, Phys. Rev. **D 85** (2012) 094025, arXiv:1203.2654.
20. J.P. Lees *et al.*, *Measurement of an excess of $\bar{B} \to D^{(*)}\tau^-\bar{\nu}_\tau$ decays and implications for charged Higgs bosons*, Phys. Rev. **D 88** (2013) 072012, arXiv:1303.0571.
21. R. Aaij *et al.*, *Measurement of the ratio of branching fractions $\mathcal{B}(\bar{B}^0 \to D^{*+}\tau^-\bar{\nu}_\tau)/\mathcal{B}(\bar{B}^0 \to D^{*+}\mu^-\bar{\nu}_\mu)$*, Phys. Rev. Lett. **115** (2015) 112001, arXiv:1506.08614.
22. M. Huschle *et al.*, *Measurement of the branching ratio of $\bar{B} \to D^{(*)}\tau^-\bar{\nu}_\tau$ relative to $\bar{B} \to D^{(*)}l^-\bar{\nu}_l$ with hadronic tagging at Belle*, arXiv:1507.03233.
23. Y. Amhis *et al.* (Heavy Flavor Averaging Group (HFAG)), *Averages of b-hadron, c-hadron and τ-lepton properties as of summer 2014*, arXiv:1412.7515 and online update at http://www.slac.stanford.edu/xorg/hfag
24. M. Bona *et al.* (UTfit collaboration), *The Unitarity Triangle fit in the Standard Model and hadronic parameters from lattice QCD: a reappraisal after the measurements of Δm_s and $BR(B \to \tau\nu)$*, JHEP **10** (2006) 081, hep-ph/0606167, updated results and plots available at http://www.utfit.org/UTfit/WebHome.
25. J. Charles *et al.* (CKMfitter group), *CP violation and the CKM matrix: assessing the impact of the asymmetric B factories*, Eur. Phys. J. **C 41** (2005) 1, hep-ph/0406184, updated results and plots available at http://ckmfitter.in2p3.fr.
26. M. Gronau and D. London, *How to determine all the angles of the unitarity triangle from $B^0_d \to DK_S$ and $B^0_s \to D\phi$*, Phys. Lett. **B 253** (1991) 483; M. Gronau and D. Wyler, *On determining a weak phase from CP asymmetries in charged B decays*, Phys. Lett. **B 265** (1991) 172.
27. D. Atwood, I. Dunietz and A. Soni, *Enhanced CP Violation with $B \to KD^0(\bar{D}^0)$ Modes and Extraction of the Cabibbo-Kobayashi-Maskawa Angle γ*, Phys. Rev. Lett **78** (1997) 3257, hep-ph/9612433; *Improved methods for observing CP violation in $B^+ \to KD$ and measuring the CKM phase γ*, Phys. Rev **D 63** (2001) 036005, hep-ph/0008090.
28. A. Giri, Y. Grossman, A. Soffer and J. Zupan, *Determining γ using $B^\pm \to DK^\pm$ with multibody D decays*, Phys. Rev. **D 68** (2003) 054018, arXiv:hep-ph/0303187; A. Bondar, *Proceedings of BINP special analysts meeting on Dalitz analysis, 24-26 Sep. 2002, unpublished.*

29. R. Aaij *et al.* (LHCb collaboraton), *Observation of CP violation in* $B^{\pm} \to DK^{\pm}$ *decays*, Phys. Lett. **B 712** (2012) 203, arXiv:1203.3662.
30. R. Aaij *et al.* (LHCb collaboraton), *Measurement of the CKM angle* γ *using* $B^{\pm} \to DK^{\pm}$ *with* $D \to K_S^0\pi^+\pi^-$, $K_S^0K^+K^-$ *decays*, JHEP **10** (2014) 097, arXiv:1408.2748.
31. R. Aaij *et al.* (LHCb collaboraton), *Measurement of the CKM angle* γ *from a combination of* $B^{\pm} \to Dh^{\pm}$ *analyses*, Phys. Lett. **B 726** (2013) 151, arXiv:1305.2050; LHCb collaboration, *Improved constraints on* γ*: CKM2014 update*, LHCb-CONF-2014-004.
32. R. Aaij *et al.* (LHCb collaboraton), *Measurement of CP violation in* $B^0 \to J/\psi K_S^0$ *decays*, Phys. Rev. Lett. **155** (2015) 031601, arXiv:1503.070809.
33. R. Aaij *et al.* (LHCb collaboraton), *Determination of the quark coupling strength* $|V_{ub}|$ *using baryonic decays*, Nature Physics **10** (2015) 1038, arXiv:1504.01568.
34. W. Detmold, C. Lehner and S. Meinel, $\Lambda_b \to pl^-\bar{\nu}_l$ and $\Lambda_b \to \Lambda_c l^-\bar{\nu}_l$ form factors from lattice QCD with relativistic heavy quarks, Phys. Rev. **D 92** (2015) 034503, arXiv:1503.01421.
35. K.A. Olive *et al.* (Particle Data Grobup), *Review of particle properties*, Chin. Phys. **C 38** (2014) 090001.
36. R. Aaij *et al.* (LHCb collaboraton), *Precision measurement of the* $B_s^0 - \bar{B}_s^0$ *oscillation frequency with the decay* $B_s^0 \to D_s^-\pi^+$, New J. Phys. **15** (2013) 053021, arXiv:1304.4741.
37. R. Aaij *et al.* (LHCb collaboraton), *Precision measurement of CP violation in* $B_s^0 \to J/\psi K^+K^-$ *decays*, Phys. Rev. Lett. **114** (2015) 041801, arXiv:1411.3104.
38. R. Aaij *et al.* (LHCb collaboraton), *Measurement of the CP-violating phase* ϕ_s *in* $\bar{B}_s^0 \to J/\psi\pi^+\pi^-$ *decays*, Phys. Lett, **B 736** (2014) 186, arXiv:1405.4140.
39. R. Aaij *et al.* (LHCb collaboraton), *Observation of two new* Ξ_b^- *baryon resonances*, Phys. Rev. Lett. **114** (2015) 062004, arXiv:1411.4849.
40. R. Aaij *et al.* (LHCb collaboration), *Observation of overlapping spin-1 and spin-3* $\bar{D}^0K^-$ *resonances at mass 2.86* GeV/c^2, Phys. Rev. Lett. **113** (2014) 162001, arXiv:1407.7574; R. Aaij *et al.* (LHCb collaboraton), *Dalitz plot analysis of* $B_s^0 \to \bar{D}^0K^-\pi^+$ *decays*, Phys. Rev. **D 90** (2014) 072003, arXiv:1407.7712.
41. R. Aaij *et al.* (LHCb collaboration), *Determination of the* $X(3872)$ *meson quantum numbers*, Phys. Rev. Lett. **110** (2013) 222001, arXiv:1302.6269; R. Aaij *et al.* (LHCb collaboration), *Quantum numbers of the* $X(3872)$ *state and orbital angular momentum in its* $\rho^0 J/\psi$ *decay*, Phys. Rev. **D 92** (2015) 011102, arXiv:1504.06339.
42. S.-K. Choi *et al.* (Belle collaboration), *Observation of a resonance-like structure in the* $\pi^{\pm}\psi'$ *mass distribution in exclusive* $B \to K\pi^{\pm}\pi'$ *decays*, Phys. Rev. Lett. **100** (2008) 142001; arXiv:0708.1790.
43. R. Aaij *et al.* (LHCb collaboration), *Observation of the resonant character of the* $Z(4430)^-$ *state*, Phys. Rev. Lett. **112** (2014) 222002, arXiv:1404.1903.
44. R. Aaij *et al.* (LHCb collaboration), *Observation of* J/ψ *resonances consistent with pentaquark states in* $\Lambda_b^0 \to J/\psi K^-p$ *decays*, Phys. Rev. Lett. **115** (2015) 072001, arXiv:1507.03414.
45. LHCb collaboration, *Letter of intent for the LHCb upgrade*, CERN-LHCC-2011-001.
46. LHCb collaboration, *Framework TDR for the LHCb upgrade*, CERN-LHCC-2012-007.
47. LHCb collaboration, *LHCb VELO upgrade technical design report*, CERN-LHCC-2013-021.
48. LHCb collaboration, *LHCb PID upgrade technical design report*, CERN-LHCC-2013-022.
49. LHCb collaboration, *LHCb tracker upgrade technical design report*, CERN-LHCC-2014-001.
50. LHCb collaboration, *LHCb trigger and online upgrade technical design report*, CERN-LHCC-2014-016.

LECTURER: G. WILKINSON

LHCb – Selected Highlights and Future Prospects

Scientific Secretaries: S. Wehle, C. Wiegand

DISCUSSION

- *Campbell:*

1st question: Will LHCb be able to take data during heavy ion running?
2nd question: Will you be able to take central events with the new tracker upgrade?

- *Wilkinson:*

1st question:Yes! The issue of ion running is a completely separate question from the upgrade. We intend to become involved in the ion run this year. Previously we did not participate because in high centrality collisions the occupancy is too high. This problem will remain, but we are excited by our possibilities in studying charm and beauty production in peripheral collisions. We participated in the proton ion collisions in run 1 with great success. And one other thing which can be done is fixed target collisions, by injecting a gas jet into the beam for luminosity measurements. For instance ions or protons on fixed target neon can be performed.

2nd question: We have not yet looked into our heavy-ion capabilities with the Upgraded detector.

- *Ke Li:*

1st question: The Belle experiment observed another state which can decay into pi+ phi' and the LHCb data seems to display an enhancement at the mass of this state (m=4.05GeV). Has this been considered?

2nd question: BaBar has not observed the Z(4430) state and points out that this might just be some K* related Background. Do you consider it?

- *Wilkinson:*

1st question: Yes. LHCb performed a detailed amplitude analysis and this state is considered, you can find the description in the paper.

2nd question: Yes. The paper has detailed discussion about this background. A full amplitude analysis considering many known K* states could not describe the data well. The addition of an exotic Z was required.

- *Albrecht:*

Concerning the partial wave decomposition for the Z(4430)-analysis: Is the Z(4430) described with a simple Breit-Wigner function? The other components of the fit (several K*-resonances) could be broad overlapping states. Are these described with Breit-Wigner functions, or with e.g. a K-matrix approach?

- *Wilkinson:*

Z(4430) is described with a Breit-Wigner shape. Other line shapes are used to describe the broader resonances, for example the LASS parameterization for the $K\pi$ S-wave.

- *Hogan:*

1st question: How to interpret amplitude plot on page 43?

2nd question: How do high q^2 points contribute to P5' discrepancy from the SM?

- *Wilkinson:*

1st question: The points are bins in the mass range – a resonance is shown by the circular behaviour of the points.

2nd question: The method used to calculate the SM prediction at low q^2 doesn't work at the above charm-resonance, therefore these points are omitted. The high q^2 measurement isn't part of the discrepancy.

- *Saviskyi:*

Usually in HEP, in such 4pi-geometry experiments like CMS, D0 and ATLAS, people are not interested in separating light flavours, but there is usage of b-taggers and c-taggers. How does the flavour separation work at experiments like LHCb and Belle?

- *Wilkinson:*

Basically in all the measurements shown there is a need of full reconstruction of B-mesons. Which means that jet information is irrelevant in the measurements. However jet tagging is possible in LHCb, as demonstrated by several recent publications. We have a c-tagger that has very high performance thanks to the fact that the vertex detector is so close to the beamline.

- *Belllantuono:*

Are there some hints about the existence of other exotic states like hybrids and glueballs?

- *Wilkinson:*

Glueballs should be lighter than the energy range considered and therefore hard to see in proton-proton inelastic collisions, as they cannot decay with the distinctive J/Ψ signature that we benefit from the study of heavier exotics. However glueballs and hybrids could be searched for via diffractive production in low-multiplicity processes involving double pomeron exchange.

Cosmology and Fundamental Physics

P. FERREIRA

The lecture notes are not included in this volume

LECTURER: P. FERREIRA

Cosmology and Fundamental Physics

Scientific Secretaries: L. Bellantuono, S. Stopyra

DISCUSSION I

- Cerchiai:

I have the following question. How is it possible to get better data using supernovae when there are so few of them? How do the statistics work?

- Ferreira:

Well, we will hit a limit, but we haven't hit the limit of supernovae, right? I'll probably get this wrong, but we've seen of order 10^3. LSST, which will be coming online in the next five years, had as its original prediction that we would get something like 10^5 supernovae. So, we're nowhere near the statistical limits i.e. the number of supernovae we might work with. Now there's a fundamental problem, which is that we don't understand supernovae well enough. As you probably know there are the following problems; we don't understand their evolution very well and we don't understand their environment very well, and so there are systematic errors which have already been mentioned today which we still don't have under control. But statistically it's not a problem.

- Klopot:

My question is about axions, so it is known to be a hypothetical particle which is one of the candidates for dark matter and as I understand it, it acquires mass due to some non-trivial QCD effect, so, QCD vacuum effect, so it is some kind of goldstone boson. Do you have any cosmological constraints on the mass of the axion and is the mass large enough to explain dark matter?

- Ferreira:

Yes. The QCD axion is a very heavy dark matter particle. It will behave non-relativistically through all the history which is of interest to us, for the history that we'll probe. And so it falls squarely into the regime of Cold Dark Matter. Unfortunately we won't be able to put constraints on the mass of the QCD axion, but it is quite possible, and there are a lot of people who believe that the QCD axion is the cold dark matter. Now, there's a possibility that there are other axions in the universe that don't come from QCD, but are from the same type of mechanism and this is something that has been advocated by people working on String Theory. They've said that it's quite conceivable that we should have many different types of axions, and the distribution of masses in axions should be of order one per decade in mass scale. Now what that means is that we should have very light axions; ultra-light axions. And the constraints that I showed you, at the end of my lecture, were exactly constraints on these types of axions, ultra-light axions. And the interesting thing about those axions is that if they're too light, they behave like dark energy and if they're too heavy they behave like dark matter. And we see evidence for dark energy and evidence for dark matter, so in some sense this is a kind of probability plot of how many of these axions that cosmological data allows, and if they're very light, they could be like dark energy, and we see that there is dark energy. But there is an intermediate regime where their behaviour is a little bit funny – it affects the power spectrum in such a way that we can constrain them and show that they're not very probable.

- Ivanytskyi:

My question is about the parameter omega which relates pressure and energy density. As you say, it's negative. However, I can't understand how positive energy density can lead to negative pressure. Moreover such a system seems not to be stable dynamically.

- Ferreira:

So, first of all, its w, not omega and you're talking about the equation of state. And you're absolutely right: if you look at the stuff we know, at the particles we know, the relativistic particles like photons, they have an equation of state where this w is equal to a third, or non-relativistic where this w is equal to zero, it doesn't make any sense to have a negative w. But you can do the following calculation. You can sit down and work out, what is the energy density of vacuum fluctuations. Sum up the zero point fluctuations in a field, and what you find is that the resulting energy and pressure, or the resulting energy and momentum is such that your pressure is minus the energy density. You have a very specific case there where it does actually behave like w is equal to minus one. I don't think I understand your question about it not being stable: I think you might be confusing the fact that it has negative pressure with the fact that it has negative energy. That's not what this is, it has negative pressure.

- *Bellantuono:*

My question is about the comparison between the quark gluon plasma that formed during the early evolution of the universe and the one produced in relativistic heavy ion collisions. So, are there differences, and what in the study of the two plasmas gives information on the other one?

- *Ferreira*:

That's really a good question, and let me say at once that I'm not sure how to answer that. The fact is that people who work on heavy ion collisions often say that they're looking at what the universe looks like at very early times and they say that they're looking at something at very high energies. I don't completely understand this argument, because basically in the very early universe you have this thermal system at very high temperatures and that's what you're looking at here, here you're smashing stuff together in completely different configurations, so my naive statement is: I don't completely understand the comparison apart from the fact that it's equating very high energies and very high temperatures.

LECTURER: P. FERREIRA

Cosmology and Fundamental Physics

Scientific Secretaries: A. Leithes, L. Hernandez

DISCUSSION II

- *Mayer:*

I have a question about slide 39 where you show the plot of the structure growth rate. You mentioned the huge error bars but still managed to fit a curve which corresponds to this logarithmic derivative to it. I could think of a constant which could fit to this data, so how do you justify fitting this curve to the data?

- *Ferreira*:

That curve is not a fit to that data. With that curve you take the Planck data, assume Lambda CDM, which is GR with a cosmological constant, and you predict that curve. The curve is not a fit to that data; you take the curve and overlay the data and this is what you find. I completely agree that if we had not used Lambda CDM the data is such that you can fit anything to it. Bear in mind there are 10% error bars on the data, so not just any curve would fit, it would be a curve which is close to that. But that curve is not a fit to that data, it is consistent with it.

- *Popov:*

Can you explain why cosmologists study theories only on classical level without quantum corrections?

- *Ferreira*:

It depends on which regime you are studying, if you are looking at the very early universe, indeed you have to worry about these quantum corrections you are looking at high energies. But look at the energy scale I am looking at and, as I said, I am looking at the late times, I am studying the system at energy scales of 10^{-3} eV, on very large scales, so even from an effective field theory point of view it does not make any sense to worry about quantum corrections on these macroscopic scales.

- *Zhidkova:*

A black hole in a universe without Lambda is described by the Schwarzschild metric. How do we describe the metric of a black hole in an expanding universe?

- *Ferreira*:

I think there are two questions there. The first is a black hole in an asymptotically flat universe, without Lambda, which is described by the Schwarzschild metric. If you add a cosmological constant you obtain a solution which is the Schwarzschild-De Sitter metric. In the Schwarzschild metric it is 1 – GM/c^2 r then you have to add in a + Lambda r/c^2. The g_00 of the metric and the g_rr you must correct that way. The second question is how can you embed these black holes in an expanding universe. You can do this using something called the Einstein-Straus model, where you can embed the solution for a black hole and you can asymptotically embed it into a Friedmann universe. There are very few situations in which you can do that but in this case you can.

- *Zhidkova:*

How does the gravitational radius of a black hole change in an expanding universe?

- *Ferreira*:

I do not think it does change. You are in a strong gravity regime which is unaffected by the expansion of the universe.

- *Savitskyi:*

What is the origin of primordial gravitational waves? Basically, why haven't they yet been confirmed?

- *Ferreira*:

We have not seen them yet. Unfortunately the problem is, as I said yesterday, you are trying to look at gravitational waves sitting within the galaxy; the polarization of the galaxy is a factor of a hundred to a thousand, to ten thousand greater than the polarization that you get from primordial gravity waves, so we have to find out a clever mechanism or a clever technique for analysing the data in order to extract the primordial gravitational waves; so that is the first answer. You asked me where these primordial gravitational waves come from. It is exactly the same mechanism that generates the density perturbations, where you have quantum fluctuation on small scale, in this case it is spin two fluctuations in space-time, the universe expands very roughly, it pushes them out exponentially, it stretches the wave lines of these gravitational fluctuations, these gravitons, to super horizon scales, to very large scales. At the end of the inflation these fluctuations are imprinted on space-time and as the universe expands the horizon gets bigger and these gravitational waves coming to the horizon, is the mechanism by which you produce it, it is quantum noise in the gravitational field which stretches the super horizon scales and then come back in.

- *Sola:*

As you said, this curve, f times sigma 8, is a prediction of the standard Lambda CDM model, not a fit to this data. There are other models which could potentially fit this data better. Sigma 8 is model dependent as it depends on the density contrast, transfer function etc. Dynamical vacuum energy models can improve the fit substantially.

- *Ferreira*:

My answer to this is twofold. Yes, I am aware that you can choose other models and you can vary parameters and get better fits. One of my students wrote a paper a few years ago showing precisely this. Unfortunately, I really do not take those results very seriously. The reason for this is if you look at the size of the error bars. Compare those with the scatter in the mean. This shows that people clearly have not yet understood the systematics in the calculation of these data points. There is a number of systematics; there is bias, there is non-linear corrections – which is really important. So, before I do any statistical analysis using likelihood curves, using Akaike information theory, I think we need to have better data. I do not think we are in a situation where we can say one model is better than another with this data. In fact if you look at the recent Planck paper, if you add the RSD data, this does indeed push it away from Lambda CDM, so there is clearly a problem, but this analysis does not include systematic effects. So I do not take it seriously.

- *Sola:*

I have another brief comment. I agree the error bars are big. But the fit I was talking about was not only on the linear growth data, it involves all kinds of data, SN data, BAO, H(z), CMB shift parameter, a lot of data points. When all these are included the result is the one I have mentioned.

- *Ferreira*:

There is a whole industry of people fitting their favorite models to the data. I think the big problem is that GR plus Lambda CDM unfortunately fits the data so well. What can we do?

Planck Satellite: Latest Data and Proof of Inflation

A. RIAZUELO

The lecture notes are not included in this volume

LECTURER: A. RIAZUELO

Planck Satellite: Latest Data and Proof of Inflation

Scientific Secretaries: V. Kovalenko, A. Baguet

DISCUSSION I

- *Kovalenko:*

You mention that with the Planck data one can extract information about electrons and baryons fraction as a function of the red shift. Could you please explain how they are related?

- *Riazuelo:*

When the Universe is hot, it is fully ionized. Then, it starts cooling down. Because the binding energy of Helium (singly or fully ionized) is higher than the one of the Hydrogen nuclei, there is a first transition, from He^{++} to He^{+}, which occurs at a redshift of ~ 6000. Then, there is a recombination of the second electron of helium followed by the recombination of hydrogen. The latter happens at thermodynamical equilibrium, while the first is not, and it depends on the expansion rate. Now, the cosmic fluid contains baryons and photons, which are coupled due to Thomson scattering. This scattering depends on the number of free electrons. The coupling between baryons and photons therefore decays with time. The rate of the decay depends on the respective He and hydrogen fractions and the number of free electrons, a quantity that decreases as the Universe is cooling down. Consequently, for a pure hydrogen medium, the electron fraction does not decrease with time prior to $z = 1500$ since hydrogen is fully ionized then. But in the real Universe, there is also helium, and the electron fraction varies at that epoch as helium recombination proceeds, therefore, there is a small imprint of the helium to hydrogen ratio in the CMB. As we can obtain the information about the baryonic density of the Universe from CMB, we can check the agreement between the data extracted from the CMB and the predictions of nucleosynthesis.

- *Klopot:*

You mentioned that the Planck Satellite was placed on the line between Sun, Earth and the Moon. Is this true ?

- *Riazuelo:*

It is actually on the Earth-Sun line. We want the Sun and the Earth to be in the same direction with respect to the satellite. It is made possible by the fact that if you have a test mass in the gravitational field of two objects (here the Earth and the Sun), you have five points where it can be at rest. These points are called Lagrange points. Three of them are along the Earth-Sun axis: one is beyond Earth, one is between the Earth and Sun, and the last one is behind the Sun. The two other Lagrange points form an equilateral triangle with the Sun and the Earth. From a dynamical point of view, only those two points, called L4 and L5 are stable, while the three others are unstable. However, the instability timescale is about 6 months. This means that by doing some orbit correction maneuvers every month or two months, a satellite can be put in an almost stable trajectory. Now, any satellite that needs to be cooled down at low temperatures because it is searching in the infrared or the microwave domain is going to be sent to L2 as it can stay there for a long time with low fuel expense and can be simultaneously shielded from Earth and Sun since both are in the same direction. Moreover, this L2 point is not very far away, only 1.5 million from Earth. This is really the perfect place because the cruising time is not too long (three months for Planck) and there are not many telemetry issues because it is not too far. The successor to the Hubble telescope, the James Webb telescope, will be sent there too.

- *Klopot:*

On the plot showing thermal stability of the mirror, there are irregular spikes. Could you please comment on this?

- *Riazuelo:*

This was made because of some adjustments due to the cooling system. As they started to loose some Helium 3 in the system, they were oblidged to make some adjustments. Some of the spikes actually monitor solar activity. It is not cosmic rays that we receive from the Sun but highly energetic particles. I could not tell you more about this but the point is that the expected stability was achieved so these small spikes were not an issue in the end. I know that there are also some moving mechanical parts within the cooling system and this induces vibrations that one has to counteract by producing vibrations of opposite phases. Sometimes, it does not work very well so maybe some of the spikes could come from this. In the end, these spikes never became a serious issue.

- *Savitskyi:*

How do you extract the useful data from the CMB maps? Do you perform a cluster fitting or apply a projection on some particular axis?

- *Riazuelo:*

Actually, we are interested in the power spectra, not the map itself. For a cosmological model that is statistically isotropic, the map itself is not of crucial interest. You do not extract the power spectrum from the map, for example. The map can be useful to detect some features that are beyond statistical isotropy. For example, if one is interested in finding some preferred direction effects, or in the study of the non-Gaussianity, you may use the real space map in addition to, say, the full three-point correlation function. Effect such a "cold spot" (of the level of 10^{-4}) are not seen in power spectrum analysis but are seen from study of the map itself. The CMB map can also be useful for studies of the correlation with known foregrounds (like a galactic emission) as well as to check the data processing methods and compare different algorithms.

- *Ferrara:*

Could you please remind the original results of BICEP2 and why they got a high ratio of tensor to scalar perturbations (r=0.2).

- *Riazuelo:*

BICEP2 is ground-based experiment. It can observe the CMB at 100 GHz, but it can't do it at 300 GHz because of Earth's atmosphere. The part of the sky observed by BICEP2 was chosen because it was quite clean from foreground emission. What about the polarization measurement? Actually, the CMB is very weakly polarized, while the foreground can be highly polarized, but how much was not known until very recently. In order to know whether polarization observed at 100 GHz comes from CMB or foregrounds, you need to observe the sky at 300 GHz where emission is dominated by foregrounds. The BICEP2 measurements themselves could not answer this question because they could not observe the sky at 300 GHz. So the BICEP2 people took an unofficial polarization fraction map of the sky extracted from a public Planck presentation in 2013. From this information, the BICEP2 people convinced themselves that the foreground polarization was much smaller than their signal. However, they interpreted the information incorrectly and made a mistake, being confused by the fact that the "fraction", taken from Planck, also were containing the non-polarized (by definition) monopole contribution of the Cosmic Infrared Background (which appears in the denominator of the polarization fraction), which made the ratio so small. Later Planck released the polarization data of the foreground at 300 GHz and, assuming that foreground polarization does not depend on frequency, evaluated the polarization amount of dust at 100 GHz. The comparison with BICEP2 polarization measurement at 100 GHz showed that all the BICEP2 polarization was compatible with a foreground origin, and consequently, there is no evidence for cosmological polarization from BICEP2 measurements.

-. *'t Hooft:*

In the analysis of the Planck data, the points lined exactly behind the densest points of the galaxy. They must have been difficult to analyze in principle. Are those spots bad spots?

- *Riazuelo:*

When we perform the power spectrum analysis, we discard data that are not reliable enough, so that instead of considering the whole sky, we use some masks which hide the Galactic region. The fraction of the map we use to compute the power spectrum with is typically one third of the full map, depending on how conservative we choose to be. The rest of the map is not very reliable for CMB.

- *'t Hooft:*

Did you use the two-point correlation function to guess what is the information for these unreliable parts ?

- *Riazuelo:*

When you compute the 2-point correlation function in order to extract cosmological information, you do not take into account these parts.

- *'t Hooft:*

In your analysis, you assume that the temperature is the same at all frequencies. Is there evidence for this? Could there be a frequency dependence for the temperature and the polarization as well?

- *Riazuelo:*

If one assumes that the polarisation of the CMB comes from Compton scattering, then it is a result of classical electrodynamics that the polarisation fraction is independent of the frequency. When you extract the foreground emissions from the data, you have to do some modelling of the way these foregrounds are polarized as a in order to make the component separation. All the crosschecks that we have done indicate that our foreground modelling is consistent with the data. Up to the amount of crosschecks we can do, we think that what we do is reliable. For example, when Planck was launched, we knew that we would have some effects from synchrotron, Bremsstrahlung, thermal dust, point source, CIB, etc. However we did not know that we would have a foreground emission arising from spinning dust. This spinning dust is a discovery from the Planck experiment because we could not solve our component separation algorithms without including it. In a sense, the component separation technique can check its own consistency. Moreover, some foreground components are compatible with the cosmology we extract from the data. For example, among the foregrounds we extract, there is the Cosmic Infrared Background, which is the integrated thermal emission of dust from distant galaxies.

These galaxies are responsible for part of the lensing that we observe in CMB maps. Therefore, we expect to see a spatial correlation between the lensing map extracted from CMB and CIB intensity map. This is something that we observe and this a very strong indication that our data processing pipeline is reliable.

LECTURER: A. RIAZUELO

Planck Satellite: Latest Data and Proof of Inflation

Scientific Secretaries V. Prochazka, N. Gulitskiy

DISCUSSION II

- *Meyer:*

In a lecture performed by professor Ferreira, we discussed about the weak lensing effect and there has to be or there should be a systematic error in the Planck's data because it contradicts or it is not comparable with general relativity. Are you aware of anything like that or have you a different explanation, could you make a statement about that?

- *Riazuelo*:

Well, the statement which I can do about that is that there are a few tensions between Planck's and external data and the leading tension is the one between Planck and the weak lensing that was mentioned this morning. So, since the Planck data seem so precise, it is tempting to say that we are going to give good marks or bad marks to the rest of the world, and conclude that if some data are not compatible with Planck cosmology, then these data are wrong. But this is a very conservative point of view in the sense that maybe it is some new unknown physics or maybe it is a new unknown systematic, connecting the Planck's data and the weak lensing data. The point is that there is a space mission named EUCLID that should have an exclusively precise weak lensing data. So at that stage if there is a more significant disagreement between various data sets after EUCLID then we would have a great reason to worry about that but at that moment I would say it is too premature. For example you may remember some unrelated but similar stories that occurred years ago with CMB and non Gaussianity. For years people studied the COBE satellite data and did not find any non Gaussianity. So they were happy because from the point of view of inflation you expect not to have any important amount of non Gaussianity. Then, in 1998 there was a paper which made a different analysis claiming a 7-sigma detection of non Gaussianity. There was a lot of hype about that, because if true, it was an important result, but then it was realized that there was unaccounted systematic in the COBE data that was responsible for this non Gaussianity. So there are many situations where some tensions between model and data actually help to find some overlooked systematic. So that is the reason why people don't get excited early about tensions because they have to know that some extra processing may be need to be done.

Campbell:

What do you think about the tension between Planck's weak lensing result and general relativity, whereby is it caused?

- *Riazuelo*:

I have no idea. What I know is that if you want to have very precise weak lensing data, for example from the EUCLID spacecraft, you face an extremely complicated problem because you need to have exquisitely precise small scale data about structure formation. And for this you need to take into account a lot of astrophysical processes in galaxies, e.g., radiative transfer, etc. So there are many potential sources of errors in any modeling of the small scale lensing, which could make you believe that you have found some deviations from general relativity. But it is just because you did not take into account some more mundane astrophysical effects about physics of galaxies. And people who try to reproduce galaxies formation from numerical simulations know that there are some things that they fail to reproduce at small scales. So we don't have enough trust in our numerical simulations before being able to say that we are sure that there is a discrepancy between the data until observations. The modeling therefore may be not as accurate as we hope it already is.

- *'t Hooft:*

There is of course a big danger in this argument because all scientists are interested in propagating new data and to make people believe in the big bang theory of everything and so on. So at some point you must say "I'm sure I think this data is in agreement with theory and I'm sure it is confirmed with theory."

- *Riazuelo*:

Making the most precise data processing is a difficult task, this is just what I'm saying. Two years ago, we knew that the Planck data had some problems because they did not perfectly agree with the WMAP data whereas they should have, but there was no obvious inconsistency within the Planck data. Everything can seem to be ok but there might be something that you overlooked.

- *Ferrara:*

It is interesting that you can produce exactly the same inflationary potential as you wrote from supersymmetry – namely the D-term inflation. With that there is no extra scalar, there is just one scalar field with exactly the same potential as you have. So in supersymmetry you can reproduce exactly the same potential, which is favored by data as you say.

- *Riazuelo*:

It is favored by data because it has very low gravitational wave content so that is why it is not really constrained. Maybe the potential you were talking about is somewhere here.

- *Ferrara*:

No, it is the potential you just showed.

- *Riazuelo*:

The Higgs inflation potential was historically introduced in a completely different context.

- *Ferrrara:*

I know, but I am talking about the Starobinsky model. The potential which comes from the Starobinsky model is actually the potential you showed there.

- *Riazuelo*:

I agree. The people from this paper (arXiv:1303.3787) have done extensive review of everything that was considered in the literature. They did not give any opinion on whether these models were well motivated or not. The large field inflation was introduced historically because it allowed for some exact solution of the background inflationary field. Only much later it was shown that these potentials were justified in the supersymmetric framework. I fully agree that the reason why you are proposing something happens to be less strong that what it happens to be afterwards. At the moment there is a non-negligible fraction of models that are already excluded by the data. And there would more of them that could be excluded if we had better constraints on the r parameter – the maximum allowed amplitude of the gravitational waves. So we are getting better and better at ruling out these inflationary models independently of their underlying theoretical justification. There is in principle hope that one could get extremely strong constraints on the amount of gravitational waves. There are a lot of models where the amount of gravitational waves you expect is larger than the lowest bound from observations. In the distant future – 30 years from now, we might be able to kill large amount of models and only then you might want to discuss which is the best justified one among the ones that have survived. Presently people do not care much about the theoretical justifications because for example there are models with almost identical potential but completely different theoretical justifications. From the point of view of observational data, these models are equally likely. So in my opinion it is a bit premature to use such theoretical arguments.

- *Ferrara:*

I think it is a little bit too naive to present the potential in this way. I believe you got it from the review "Encyclopaedia Inflationaris". That is already bit old. Nowadays you have one-parameter family of models called alpha-attractors, put forward by Linde, Carlos, myself and others where you can go between different inflationary potentials, raise the ratio of tensor/scalar perturbation, etc., by varying one parameter. So I think that this list is I would say incomplete and not only, it is also a random way of choosing a potential.

- *Riazuelo*:

There are so many things in literature that it is hard to present them in a unified manner. At least these authors present more than 60-70 potentials in a unified manner which is already a good thing. But maybe it could have been done in another way. The point is that they have developed a set of tools which allows you to compare a large class of models to any existing set of data. They were the first to do that. Usually, when people compare different models they choose only a small subset of all this, without trying to get a more global overview. But deciding what is the best thing to do is probably a matter of taste.

- *Gulitskiy:*

I would ask about the frame 19 of your presentation with the anomalous scaling in the gravitation. I also didn't understand how it was calculated so I would ask you to explain this moment or give some links to the papers in which this problem is considered.

- *Riazuelo*:

Ok. This review (arXiv:1303.3787) gives you all the references for each of the models that you have hear. So if you look at this paper you can find around 10 references with the ways to give the potential. So the best is to look at this paper and at the references. It is a wide review of the existing literature, which talks about what has been published. The only correction that they do to the existing literature is in the technical derivation of the spectral of cosmological transformations which was not correctly done in some of the papers. But the choice of the potential is taken as it was in the existing literature and they refer to the historical papers about this or that and it would be simply to understand how the potential was originally built or motivated. So it is better that first you look at this review and then at the papers which studied this or that specific model of inflation.

- *Sola*:

I just want to comment that in this list of inflationary models, most contain scalar fields to make the inflation work. In case of the original Starobinsky it is the most economic recipe. The scalar field that is used in the potential that you have shown is not a new field that you put in by hand. It is just the reshuffling of the same degrees of freedom of the gravitational potential of the R^2. You move from the Jordan to the Einstein frame and play with conformal transformations to display the same structure in terms of the potential. It is the work of gravitation. And moreover it is perfectly consistent with the data. In contrast with the possibility that this phi field is the Higgs field. Which of course could work too, but then one would need to make sure that this scalar is equal to the one found in LHC. But in the Starobinsky case you just get it for free so it is completely economical recipe.

- *Riazuelo*:

Yes. In the classification of this inflationary encyclopedia Higgs-inflation is the only 0-parameter model that they extract from literature exactly for this reason. The only unknown comes from the reheating temperature, the amount of time that elapses between the moment when you produce the CMB fluctuation and the moment when inflation ends. In a way, it is the most predictive scenario. And it has more direct connection with real world than the others so you might find it appealing. Now, the point is that these values of r predicted by this model, which are around 10^{-2}, is something that in principle you can extract with ambitious next generation observations. The current limit is 10^{-1} and the best that you can achieve with these parameters is 10^{-4}. You cannot go below because the gravitational wave signal below this threshold will be drowned into some astrophyscical effects that are not really exciting. People have designed some experiments which, depending on how they work, could reach a sensitivity to this parameter of 10^{-3} or $3*10^{-4}$.

- *Sola:*

Excuse me what is this parameter? I cannot read.

- *Riazuelo*:

This is the r-parameter, i.e., the gravitational waves to density waves power spectrum ratio at some scale. So this Higgs inflation in the next decade should be shown to be either compatible or incompatible with very precise data. So it is really a testable scenario, which is a nice thing. There are many inflationary models that have been proposed which have such a low amount of gravitational waves already known so that we will never be able to constrain them beyond some bound.

Neutrino Physics in China

L. WEN

The lecture notes are not included in this volume

LECTURER: L. WEN

Neutrino Physics in China

Scientific Secretaries: G. Benato, M. Centis Vignali

DISCUSSION

- *Savitsky:*

Regarding the Daya Bay results, today you presented the curve of energy response as a function of positron energy. I'm wondering why is the ratio of the reconstructed energy to the true energy becomes constant after the two MeV threshold and exceeds to the level of 2% instead of approaching the true energy.

- *Wen:*

The non-linearity comes from different reasons. The quenching effect is more significant in the low energy region, particularly below 1 MeV. Another effect is the Cherenkov light: in liquid scintillator the Cherenkov light has a larger fraction in the short wavelength and it can be absorbed and re-emitted again, and this is energy dependent, so you see the different effect. The last uncertainty is the non-linearity of the electronic chain. When we put these effects together we need to build a model to describe the non-linearity. The 2% effect at high energy is, because we anchor our energy scale using a cobalt-60 source, so the value 1 is reached at 2.5 MeV, which means above 2.5 MeV you see this effect reaching a plateau at a value 2% higher. To extract this curve we both use the gamma sources and the gamma spectrum, and the non-linearity is smaller at higher energy.

- *Sawitsky:*

Why is this ratio difference not covered by the uncertainty band? Why is the reconstructed energy 2% away from the true energy?

- *Spiro:*

I am surprised that you try to calibrate the positron response with the gamma rays, because the gamma rays will induce Compton effect, and you will have an electron spectrum which is not representative of mono-energetic positrons. So how do you deconvolute this?

- *Wen:*

We use Monte Carlo to get the distribution of the primary electrons and use it to do the convolution with the electron response. Going back to the other question, the 2% effect just depends on how you define the reconstructed energy. In our case we reconstruct the energy from the visible light, which has a peculiar energy-dependence: first we correct for the non-uniformity, and then we get this reconstructed energy, which is the energy before the non-linearity correction. After we correct for non-linearity, the reconstructed energy is much closer to the true energy.

- *Benato:*

Is the ^{12}B spectrum used in the calibration obtained from a source or from the experimental background?

- *Wen:*

The spectrum is originated using data from cosmic muons, this provide a large statistical sample. After background subtraction we obtain a clean spectrum.

- *Benato:*

I asked because comparing this spectrum with the data used for the energy calibration using gamma sources, I see that the threshold being used is around 4 MeV, while the gamma data have points starting at about 0.5 MeV. Is this due to the presence of other background contributions at lower energy?

- *Wen:*

Yes, since the energy of the natural radioactivity extends to 3.5 MeV, I used a cut at higher energy to get rid of these events.

- *Centis Vignali:*

In the estimated rate of solar neutrino events in the Juno detector you expect between 10 and 1000 events per day. Why do you have such a big uncertainty on this value?

- *Wen:*

Different channels must be considered for the solar neutrinos, and some of these channels are not easily seen using liquid scintillator. So here I gave just a rough estimate of the rate. We are particularly interested in the neutrinos from ^{7}Be and ^{8}B, which have different rates.

- *Centis Vignali:*

How do you reconstruct the direction of the neutrinos interacting in a liquid scintillator detector? And, which is the limiting factor in the pointing resolution?

- *Wen:*

You mean pointing to a supernova?

- *Centis Vignali:*

Yes.

- *Wen:*

We use inverse beta decay to detect supernova neutrinos. The neutron generated in this process moves in the same direction as the incoming neutrino. By looking at the position where the positron is emitted and the one where the capture of the neutron occurs, it is possible to estimate the direction from which came the neutrino. The thermalization of the neutron reduces the pointing accuracy. The position of the supernova can be estimated combining the information of several events. Most recently we have also been studying the possibility to use the Cherenkov light emitted by the positron to determine its direction. This also helps the reconstruction of the neutrino's direction.

«ETTORE MAJORANA» FOUNDATION AND CENTRE FOR SCIENTIFIC CULTURE
TO PAY A PERMANENT TRIBUTE TO ARCHIMEDES AND GALILEO GALILEI, FOUNDERS OF MODERN SCIENCE AND TO ENRICO FERMI, THE "ITALIAN NAVIGATOR", FATHER OF THE WEAK FORCES

INTERNATIONAL SCHOOL OF SUBNUCLEAR PHYSICS
THE FUTURE OF OUR PHYSICS INCLUDING NEW FRONTIERS
53rd Course – ERICE-SICILY: 24 JUNE – 3 JULY 2015

NEW TALENTS CELEBRATION

ANTONINO ZICHICHI
University of Bologna and INFN, Italy
CERN, Geneva, Switzerland
Enrico Fermi Centre, Rome, Italy
Pontifical Academy of Sciences, Vatican City
World Federation of Scientists, Beijing, Geneva, Moscow, New York
Ettore Majorana Foundation and Centre for Scientific Culture, Erice, Italy

Thursday, 2 July 2015
Paul A.M. Dirac Lecture Hall – Patrick M.S. Blackett Institute

NEW TALENTS CELEBRATION

A. Zichichi

Let us start the **Celebration of our New Talents** by calling your attention on what has been done – one week before our School activity started its 53rd Course devoted to "*the Future of our Physics, including new Frontiers*" – in Bologna.

The oldest University in the world has been able to devote ***Three Days*** to **Science**, **Technology** and **Culture** with 30.000 people participating into 60 events in order to focus their interest in these three great achievements of the human intellectual activities.

You are coming from 31 Countries.

If what we have been able to do in Bologna would be done in your Countries, our Science would have top-priority, as shown in Figure 1, where the Logical Structure of the so-called "New Manhattan Project" is illustrated.

Let me give you a very brief synthesis of the special volume published for this great event, whose purpose is the implementation of the "New Manhattan Project" for the future of our planet.

Those of You interested can ask our Centre a copy of the Book in order to have You in our group of physicists who want to be engaged in this new venture.

The red-circles in Figure 1 indicate the projects which are being studied; some of them have effectively started.

Figure 1

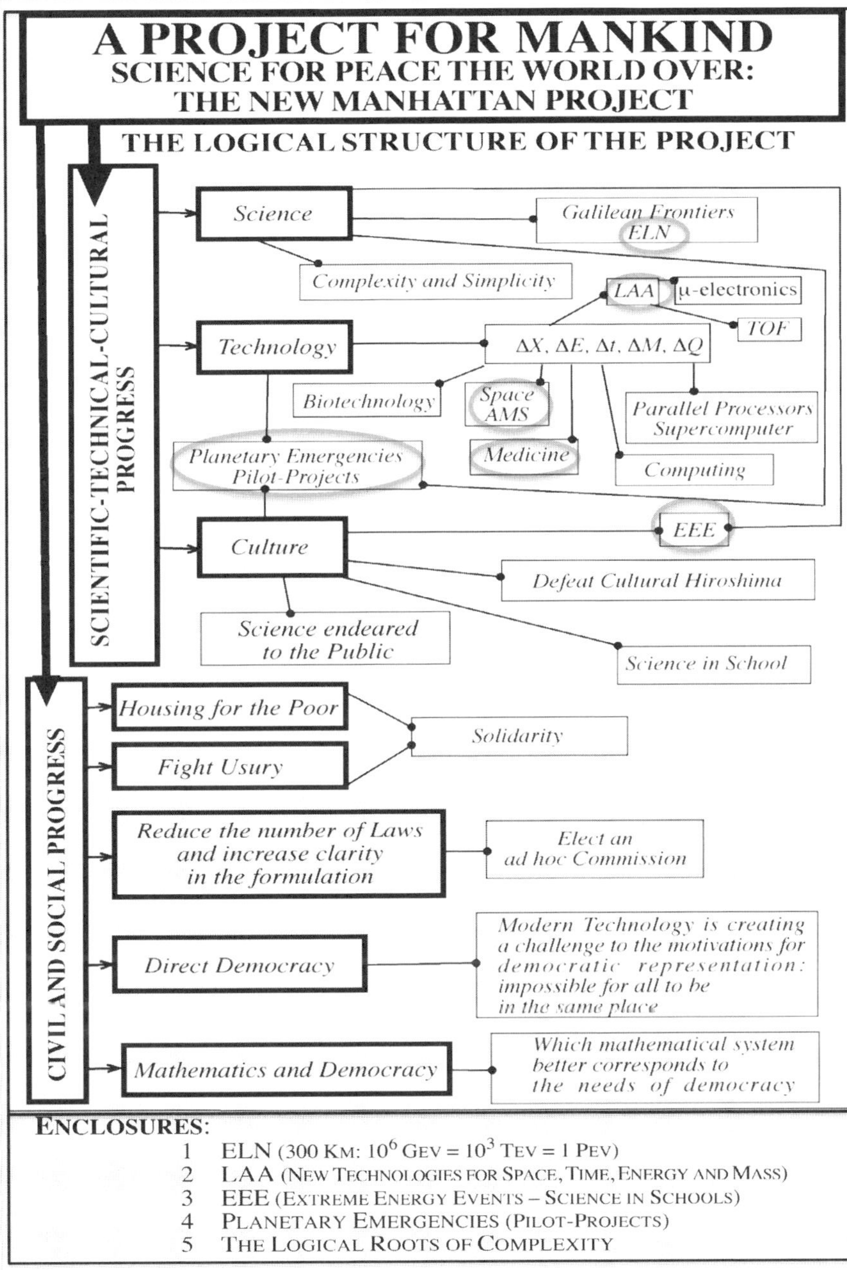

ENCLOSURES:

1 ELN (300 KM: 10^6 GEV = 10^3 TEV = 1 PEV)
2 LAA (NEW TECHNOLOGIES FOR SPACE, TIME, ENERGY AND MASS)
3 EEE (EXTREME ENERGY EVENTS – SCIENCE IN SCHOOLS)
4 PLANETARY EMERGENCIES (PILOT-PROJECTS)
5 THE LOGICAL ROOTS OF COMPLEXITY

This book, whose first page is reproduced in Figure 2, is a contribution to the *THREE DAYS* that the University of Bologna has dedicated to Science, Technology and Culture, so that they enter into the Heart of Everybody.

Figure 2

We are in the **Cultural Hiroshima**, as Fermi feared in the middle of the last Century, when (1945) the world was in full Political Hiroshima.

The *Three Days* of Bologna have two objectives: to defeat the Cultural Hiroshima and to solve the problems of Planetary Emergencies, proposing the implementation of a project (the New Manhattan Project) so that the great achievements of Science come soon in every day life.

In Figure 3 the Political and Cultural Hiroshima are briefly described.

Figure 3

⊙ **POLITICAL HIROSHIMA (1945)**
The existence of Science is totally forgotten (for decisions at very high political level) as if it were not **thanks** to Science that the **new** war **Technologies**, million times more powerful, had been implemented.

⊙ **CULTURAL HIROSHIMA (NOW)**
The existence of Science is totally forgotten for decisions (at any political level) and in every day life as if it is not **thanks** to Science that **new Technologies** have been invented: technologies which allow to win life-threatening illness, technologies that liberate all of us from rough work and that allow humanity to have a standard of living never before achieved.

THE NEW MANHATTAN PROJECT

Science for Peace the World Over

The New Manhattan Project ≡ A Project for Mankind
would be pure utopia if he did not have roots in what has been
actually achieved, starting from the University of Bologna,
in the fight against the Planetary Emergencies:
hundred pilot-projects implemented in 50 Countries.
The scientific-technological results obtained demonstrate
that the problems of the **72 Planetary Emergencies** can be solved.

We live in a culture that ignores the **72 Planetary Emergencies**.
In **1985** the two most powerful Heads of State (USA, USSR)
Reagan and Gorbachev,
said in Geneva that
the enemy number one of Peace in the world
are the secret Laboratories; they declared the will to open them,
as proposed by the scientists signatories of the
Erice Statement, for a Science without secrets and without borders.
Unfortunately, even today, the secret Laboratories are closed and
the only solution is to fight the secret in Science at the origins,
with the implementation of the "**New Manhattan Project**".

The Culture of our time, called modern,
ignores the great achievements of Science
and brings us back to before **1985**,
when the **Future of all Nations** was in danger because of the
60 thousand H-bombs
(each a million tons of TNT equivalent)
accumulated in the arsenals of the two superpowers.
Today, the danger lies in the existence of the **72 Planetary Emergencies**
that threaten the **Future of all Nations** North-South-East-West.

Those who speak of Science
without having ever been able to make
a discovery or an invention
are responsible for the
Cultural Hiroshima in which we live and that
the *Three Days* of Bologna
want to bring the focus of all people in the world.

How?
Denouncing the existence of as **many as 72 Planetary Emergencies**
and proposing the implementation of a
"**New Manhattan Project**",
so that the great achievements of Science produce
– quickly – technological inventions
to use in everyday life.

Not letting tens of
thousands of years
as it happened with the **Wheel** and the **Fire,**
or one hundred years
as it happened with the discovery of the electron.
But within a few years as shown by the **Manhattan Project**.

Enrico Fermi feared, already half a Century ago, that
the Political Hiroshima would have been followed by
the **Cultural Hiroshima**
in which we are immersed.

The only way to rid
the **Future of our Planet**
from the nightmare of
the **72 Planetary Emergencies**
is the realization of the "**New Manhattan Project**".

ONE OF THE FIRST SESSIONS DEVOTED TO THE ERICE STATEMENT AND TO THE FUTURE OF OUR PLANET.

(From left) Luigi Dadda, Pierre A. Piroué, Enrico Bignami, Yuval Ne'eman, Richard L. Garwin, John C. Eccles, Eugene P. Wigner, A.Z., Edward Teller, Paul A.M. Dirac, George Charpak.

Without the activities of the Ettore Majorana Centre neither the Erice Statement nor the Erice Geist for a Science without secrets and without borders would have been possible to imagine.

DATA ON ACTIVITIES OF THE ETTORE MAJORANA FOUNDATION AND CENTRE FOR SCIENTIFIC CULTURE SINCE 1963

126 SCHOOLS,
1.738 COURSES,
120.680 PARTICIPANTS(*)
COMING FROM **932** UNIVERSITIES
AND LABORATORIES OF **140** NATIONS.

(*) **134** Nobel Laureates (85 of them were awarded the Nobel Prize after their participation to the Ettore Majorana Schools and 49 were already Nobel when they started to take part in the Centre activities).

Prof. Antonino Zichichi
Presidente
Ettore Majorana Foundation
and Centre for Scientific Culture
Via Guarnotta, 26
91016 ERICE (TP)

Il Presidente della Repubblica

Caro Nino,

Ti ringrazio molto per la tua gentile attenzione.

Ti prego di ringraziare la tua comunità scientifica, particolarmente per il ricordo di [illegible]. Con amicizia

Giorgio Napolitano

1988

To celebrate the
25th Anniversary of
the "Ettore Majorana" International Centre for Scientific Culture.
(From left: A.Z., Sergio Mattarella and the Mayor of Erice).

From left: Minko Balkanski, Abdus Salam, Kay M.B. Siegbahn, the Minister for Relations with Parliament Sergio Mattarella, A.Z., Tsung-Dao Lee, Herwig F. Schopper and Qian Jiadong.

From left: Enzo Boschi, Minko Balkanski, Abdus Salam, Kay M.B. Siegbahn, the Minister for Relations with Parliament Sergio Mattarella, A.Z., Tsung-Dao Lee, Herwig F. Schopper, Qian Jiadong, Arthur S. Wightman, Günther Harbeke and Arthur I. Miller.

THE FORGOTTEN ½ Mc PUZZLE WHERE FROM QED STARTED

Dear new Talents, as you know among the four interactions so far discovered (gravitational, weak, electromagnetic, and strong) the electromagnetic interaction is the one we know best.

The reason I decided recall the $(^1/_2)$ Mc puzzle is twofold:

first, its solution has played a great role in establishing the validity of QED;

second, it is a very instructive example of what can happen in physics: a wrong experimental result and an incorrect theoretical calculation.

Nowadays nobody talks about this puzzle, probably because its solution was not a breakdown of QED but of the puzzle itself. On the other hand, when a puzzle is afflicting a certain branch of physics we never know the answer "a priori": think of the "ϑ-τ" puzzle in weak interactions. This is why the solution of a puzzle should merit the same interest in spite of breakdowns. Finally, it is worth pointing out that only in 1960 a really satisfactory solution of the intricate puzzle has been obtained.

THE ORIGIN OF QED AND THE ½ MEGACYCLE PUZZLE

The origin of QED is the discovery by Blackett and Occhialini of the (e^+e^-) production in cosmic rays. This discovery produced the first virtual effect in physics the vacuum polarization. Very little was known of QED. There were some contradictory optical evidence that the two levels of the hydrogen atom

$$2S_{1/2}, \quad 2P_{1/2}$$

were not degenerate as the validity of the Coulomb law indicated. The only theoretical prediction concerning the occurrence of a shift between these levels was recognized by **Uehling**, **Heisenberg**, and **Weisskopf**.

The theoretical prediction of Weisskopf was to – 27.08 Mc. Lamb told me that his experiment was designed in order to detect this extremely small value of the shift.

The sign and magnitude of the first experimentally measured shift, called Lamb–shift was:

$$+ 1{,}000 \pm 100 \text{ Mc},$$

in complete disagreement with the expected value, and this is the starting-point of modern QED, from 1932 with Blackett and 1947 with Lamb.

The ½ Mc puzzle begins at the end of this development, when a complete theoretical understanding of the Lamb-Rutherford shift had been achieved. The theory was then predicting

$$+ 1{,}057.12 \pm 0.15 \text{ Mc/sec},$$

while Lamb and Rutherford measured

$$+ 1{,}057.77 \pm 0.10 \text{ Mc/sec}.$$

As the theoretical uncertainty represents upper and lower limits, the discrepancy of ~ 0.5 Mc/sec was beyond any doubt: this is the ½ Mc puzzle. QED has been faced with this puzzle for many years. We will see how subtle was the solution.

Let us recall briefly the theory of the $2S_{1/2} - 2P_{1/2}$ level shift in hydrogen-like atoms. Imagine an electron rotating around a point-like nucleus. In so far as we neglect the interactions of the electron with its own electromagnetic field, the energy levels $2S_{1/2}$, $2P_{1/2}$ are perfectly degenerate. In terms of Feynman diagrams this situation would be described by Fig. 1, where the full line represents an electron, the wavy line a photon, and the cross a nucleus. Notice that in this diagram the electron must be considered in a bound state in the initial and final states. Figure 1 represents an electron interacting via an intermediate photon with a nucleus. This diagram, as we have said, gives complete degeneracy between the two levels

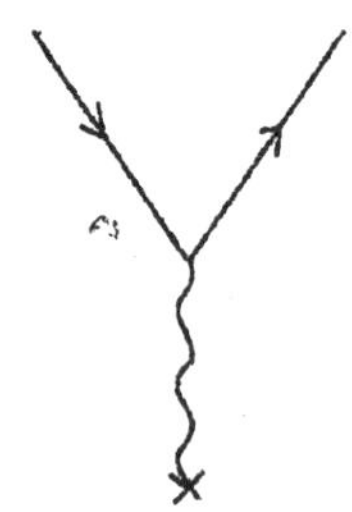

Fig. 1

$$2S_{1/2} - 2P_{1/2} ,$$

but Figure 1 does not represent physics. A real electron interacts with its own electromagnetic field: it emits and absorbs virtual photons with a probability per unit time which depends on the coupling constant.

Let us consider what virtual processes an electron can have while interacting with a nucleus.

1st – The second-order radiative shift

The electron emits a virtual photon, interacts with the nucleus, then it reabsorbs the virtual photon (see Fig. 2).

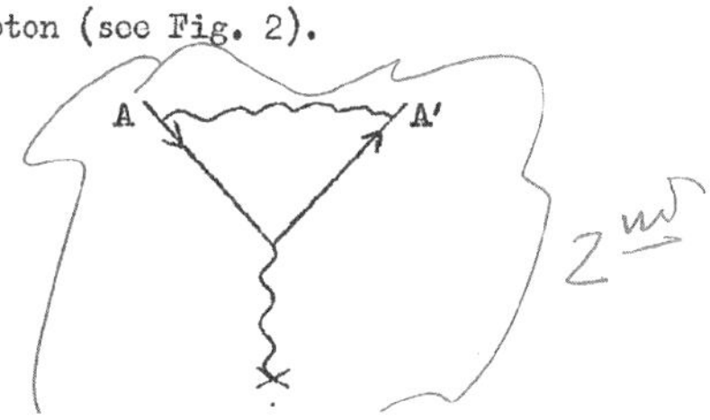

Fig. 2

Feynmann diagram for the radiative correction to the Lamb shift.

The physical fact that an electron can emit and reabsorb a virtual photon produces a shift between the energy levels $2S_{1/2} - 2P_{1/2}$ which amounts to

$$+ 1,077.55 \text{ Mc/sec}$$

For reasons which will appear clearer in the following, it is useful to separate from this total effect that part which is due to the fact that the electron has an anomalous magnetic moment

$$\frac{\alpha}{2\pi} .$$

In other words, suppose that we did not know anything about self-interactions of the electron with its own e.m. field, but we measured experimentally the magnetic moment of the electron to be

$$\mu = \mu_0 \left[1 + \frac{\alpha}{2\pi} \right]$$

where μ_0 is the electron Bohr magneton; we could ask what is the contribution to the Lamb shift coming from the fact that the electron has a magnetic moment which is not exactly the one predicted by the Dirac theory, i.e. μ_0, but contains an extra bit which amounts to $\alpha/2\pi$. The answer would be

$$+ 67.71 \text{ Mc/sec} .$$

More precisely, an effect such as that described by Fig. 2 will turn out to provide one contribution proportional to the invariant square of the four-momentum exchanged with the external potential, as well as one proportional to the components of this energy-momentum via the combination

$$\sigma_{\nu\mu} J_{\nu} \; ; \; \left[\sigma_{\mu\nu} = \frac{1}{2} \left(\gamma_{\mu}\gamma_{\mu} - \gamma_{\nu}\gamma_{\mu} \right) \right] .$$

It is this last contribution which is identified as being due to the presence of an anomalous magnetic moment and which amounts to 67.71 Mc/sec, as mentioned already. .While the first contribution gives

$$1{,}009.84 \text{ Mc/sec}$$

and is usually called the second-order radiative shift.

2nd - The fourth-order radiative shift

This shift is due to the following sequence (see Fig. 3): the electron emits a photon; it emits another photon; it interacts with the nucleus; it reabsorbs the photon; and it reabsorbs the other photon.

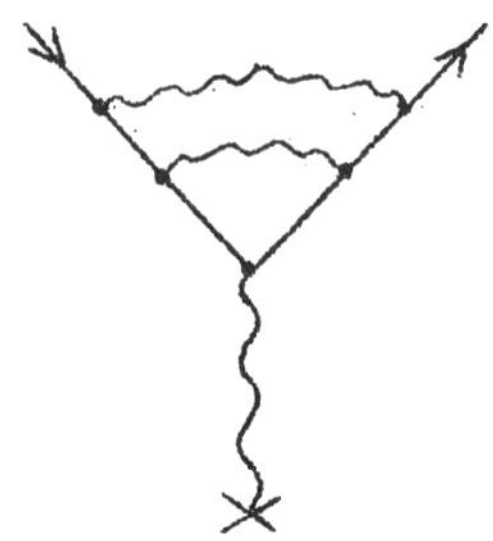

Fig. 3

Typical diagram for the fourth-order radiative shift.

Here, as for the second-order radiative shift, we can keep separate the contribution to the shift coming from the fact that the electron has an anomalous magnetic moment (of order α^2). The fact that the electron has a magnetic moment

$$\mu = \mu_0 \left[1 + \frac{\alpha}{2\pi} - 0.328 \left(\frac{\alpha}{\pi}\right)^2 \right]$$

means that the part $-0.328\ (\alpha/\pi)^2$ will contribute to the shift of fourth order. Its contribution is, in fact,

$$- 0.11 \text{ Mc/sec.}$$

(As we will see later, the fact that this contribution was calculated to be -0.94 Mc/sec instead of -0.11 Mc/sec is at the origin of the puzzle.) What remains of the fourth-order radiative shift is

$$+ 0.24 \pm 0.13 \text{ Mc/sec}.$$

The theoretical uncertainty of ± 0.13 Mc/sec is due to the complication of some terms which are only estimated by upper and lower bounds. This uncertainty is, in fact, the main contribution to the over-all theoretical uncertainty in the knowledge of the shift.

3rd – Second-order vacuum polarization

As we have mentioned at the beginning, this is the first effect which was theoretically put forward by Weisskopf and all other theoretical fellows to predict a shift between the

$$2S_{1/2}, 2P_{1/2}$$

energy levels. It is due to the fact that the intermediate photon (i.e. the photon which is exchanged between the electron and the nucleus) can create a virtual electron-positron pair (see Figure 4) as experimentally discovered by Blackett.

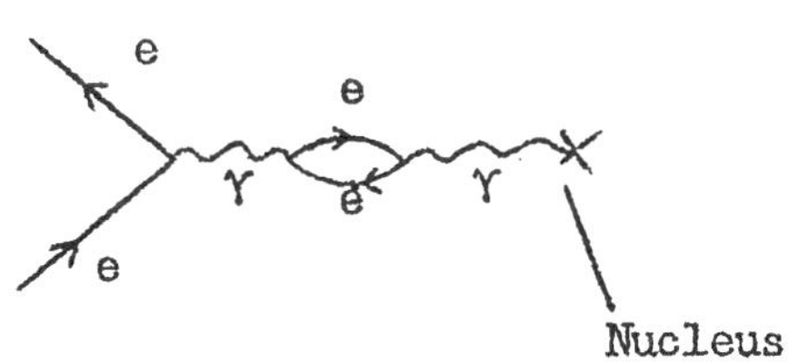

Figure 4

This pair annihilates into a photon, and the photon is then absorbed by the nucleus. The contribution to the shift due this virtual process amounts to

$$-27.08 \text{ Mc / sec.}$$

All other effects are reported in Table 1 which is derived from Petermann and Feynman.

Table 1

Lamb-Rutherford shift : $2S_{1/2} - 2P_{1/2}$

		Effect No.	H Mc/sec	D Mc/sec	He^+ Mc/sec	Typical diagram
1st order in the external potential	2nd order radiative shift	1a	1009.84	1010.63	13168.3	
		1b	67.71	67.77	1084.7	
	4th order radiative shift	2a	0.24 ± 0.13	0.24 ± 0.13	3.9 ± 2	
		2b	− 0.11 (−0.94)	− 0.11	− 1.7	
	2nd order vacuum polarization	3	− 27.08	− 27.11	− 433.9	
	4th order vacuum polarization	4	− 0.24	− 0.24	− 3.8	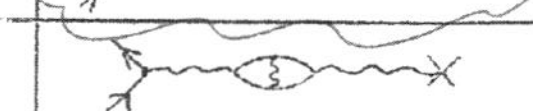
2nd order in e.p.	2nd order radiative shift	5	7.14	7.13	228.4	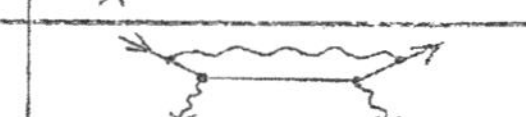
3rd order in e.p.	2nd order radiative shift	6	− 0.25 ± 0.07	− 0.25 ± 0.07	− 9.5 ± 4	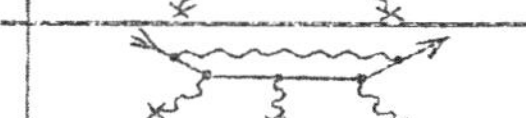
finite nucleus mass		7	0.36	0.18	2.5	$\frac{m_N}{m_e} \neq \infty$
finite nucleus size		8	0.12 ± 0.02	0.73 ± 0.02	7.1 ± 1.0	
		Theoretical	1057.73 ± 0.22	1058.97 ± 0.22	14046.0 ± 7.0	
		Experimental	1057.77 ± 0.10	1059.00 ± 0.10	14040.2 ± 4.5	

In the Figure all the effects for hydrogen, deuterium, and ionized helium are reported. The last column contains a typical diagram corresponding to each effect.

As we see from the table, the theoretical prediction for the shift in hydrogen is

$$1{,}057.73 \pm 0.22 \text{ Mc/sec}$$

and the experimental result gives

$$1{,}057.77 \pm 0.10 \text{ Mc/sec} .$$

The theoretical prediction agrees so well with the experimental result after the solution of the $\left(^1/_2 \text{ Mc}\right)$ puzzle. In fact in **1960**, the theoretical prediction was

$$1{,}057.95 \pm 0.15 \text{ Mc/sec} .$$

This was because the term No. 6, i.e. the second order radiative shifts in third order external potential, had not yet been calculated. At the time of the puzzle (**up to 1957**) the theoretical value for the shift was

$$1{,}057.12 \pm 0.15 \text{ Mc/sec} .$$

This was because the contribution to the second order radiative shift (term 2 b) coming from the fourth order anomalous electron magnetic moment was calculated to give a contribution of

– 0.94 Mc / sec instead of – 0.11 Mc / sec.

On the other hand, the reason why this effect was estimated to be

– 0.94 Mc / sec instead of – 0.11 Mc / sec

lies in the fact that the anomalous magnetic moment of the electron was experimentally and theoretically incorrect; the amusing fact being the coincidence between a **wrong theoretical** calculation with a **wrong experimental** measurement.

At that time there was **no direct experiment of the (g–2) type for the electron**, and in order to know the electron magnetic moment three experiments were necessary.

One was measuring the proton magnetic moment (μp) in Bohr magnetons (μ_0), μp / μ_0. The other two experiments were measuring (μe / μp) using two different experimental methods.

The theoretical value consisted of two terms, one of the order α, the other of order α^2. The first term, calculated by Schwinger[7)] and Luttinger[8)] amounted to

$$\frac{\alpha}{2\pi},$$

the other, calculated by Karplus and Kroll, amounted to

$$-2.973\left(\frac{\alpha}{\pi}\right)^2.$$

This value was used to calculate the shift in the $2S_{1/2} - 2P_{1/2}$ energy levels, with the result that the term 2b) had to be -0.94 Mc/sec instead of the correct -0.11 Mc/sec. Correct

Theory result

Wrong

In 1957, Franken and Liebes[14] repeated the measurement of μ_0/μ_p and found

$$\frac{\mu_0}{\mu_{p(oil)}} = 657.462 \longrightarrow \text{(New)}$$

with a limit error stated to be 0.008. This means that this result was not in agreement with the previous one. On the other hand, this result combined with $\mu_e/\mu_{p(oil)}$ now gave

$$\mu_e = (1.001166 \pm 0.000012)\, \mu_0^e$$

$$1.001145 \quad \text{Th}$$

which disagrees with the theoretical one. The observation that this value of the electron magnetic moment would bring the theoretical prediction of the Lamb shift in the direction desired in order to reduce the discrepancy represents the starting point towards the solution of the $\frac{1}{2}$ Mc puzzle. The credit goes to Kàllen and Petermann and, independently, to Sommerfeld. Encouraged by this possibility, Petermann and Sommerfeld repeated the fourth order calculations of the electron magnetic moment and found that the correct value was *)

$$- 0.328 \left(\frac{\alpha}{\pi}\right)^2 .$$

When this value is introduced in the Lamb shift calculation the term 2b) becomes

$$- 0.11 \text{ Mc/sec}$$

and the total shift is now theoretically predicted to be

for hydrogen	1,057.95 ± 0.15	←	Theor.
	1,057.77 ± 0.10	Exp.	
for deuterium	1,059.21 ± 0.16	←	Theor.
	1,059.00 ± 0.10	Exp.	
for He^+	14,055.8 ± 3.0	←	Theor.
	14,043.0 ± 3.0	Exp.	

We had gained too much. It was, in fact, pointed out by Petermann that both for H and D the theoretical values exceeded the experimental ones by about 0.2 Mc/sec; while for He^+ the excess was of ~ 13 Mc/sec. It was disturbing to notice that the discrepancy between theory and experiment had been reduced, changed sign but remained of the same sign in all three cases: this was, in fact, considered a continuation of the puzzle. Petermann remarked that if we assumed that the excess was due to the third order in the external potential radiative shift (Z^6 terms), by noticing that for H and D the measured excess was 0.2 megacycles it should have been expected for the He^+ to be Z^6 times

*) We recall that the last experiment on the electron (g-2), performed by Crane and Wilkinson [Bull.Am.Phys.Soc. 7, 470 (1962)] gives the following value:

$$\frac{g-2}{2} = \left\{\frac{\alpha}{2\pi} - \left(\frac{\alpha}{\pi}\right)^2 \left[0.327 \pm 0.005\right]\right\}$$

to be compared with

$$\frac{g-2}{2} = \left\{\frac{\alpha}{2\pi} - \left(\frac{\alpha}{\pi}\right)^2 0.328\right\} .$$

larger, that is 64 times 0.2 Mc / sec $\simeq$ 13 Mc / sec, which was in fact the existing excess.

Those interested in a detailed description of what has been reported see:

THE ½ Mc PUZZLE AND THE PRESENT EXPERIMENTAL STATUS OF QUANTUM ELECTRODYNAMICS(*)

by

A. Zichichi
CERN, Geneva, Switzerland

(*) Lecture given at the International School of Physics "Enrico Fermi", Varenna, August 1962, and at the International School of Physics "Ettore Majorana", Erice, June 1963.

Last year (1961) the third order external potential radiative shift was calculated by Layzer (see Table 1) and was found to be in perfect agreement with the predictions **made by Petermann**. It is only one year that theoretical predictions for the **Lamb shift** really agree within errors with the experimental values. QED, born experimentally in **1932** with the experimental discovery of the simultaneous production of (e^+e^-) pair giving rise to the virtual effect theoretically studied by Weisskopf, Heisenberg and Uehling becomes after **28 years a theory without puzzles**. The agreement between theory and experiment being $\sim \pm 2 \times 10^{-4}$. The most significant consequence of this agreement is the validity of the Coulomb law. **As mentioned by**

Feynman (Solvey Report) if we suppose that the Coulomb law is altered in such a way that the potential from a charge is given by

$$\frac{1}{\gamma(1+\epsilon)}$$

in the range of "ϒ" of the order of 10^{-8} cm, then the Lamb experiment tells us that ε is less than $\mathbf{10^{-10}}$. In fact, the **total energy level** of the hydrogen atom amounts to **~ $\mathbf{10^9}$ Mc / sec**. The Lamb experiment verifies the degeneracy of the two levels

$$2S_{1/2} - 2P_{1/2}$$

up to a few parts in $\mathbf{10^{10}}$, because the spitting of the levels is accounted for by the radiative corrections and no effects larger than ~ 2×10^{-10} can be left out to violate the degeneracy of the two levels. The validity of the Coulomb law for distances of order 10^{-8} cm is established with **the fantastic accuracy of a few parts in $\mathbf{10^{10}}$**.

DIPLOMAS

Thirty-two Diplomas were open for competition among the participants. They have been awarded as follows:

- Richard ARNOWITT Diploma to:
  Vasilisa NIKIFOROVA
  INR, Moscow, Russian Federation

- John S. BELL Diploma to:
  Shion CHEN
  The University of Tokyo, Japan

- Patrick M.S. BLACKETT Diploma to:
  Alexandros ANASTASIOU
  Imperial College London, UK

- Nicola CABIBBO Diploma to:
  Maria Vittoria GARGIULO
  University of Salerno, Italy

- James CHADWICK Diploma to:
  Julie HOGAN
  FermiLab, Batavia, IL, USA

- Sidney COLEMAN Diploma to:
  Max ZOLLER
  Karlsruher Institute für Tecnologie, Karlsruhe, Germany

- Richard H. DALITZ Diploma to:
  Maximiliano PUCCIO
  University of Torino, Italy

- Amos DE-SHALIT Diploma to:
  Matteo CENTIS VIGNALI
  Institute for Experimental Physics
  Hamburg University, Hamburg, Germany

- Paul A.M. DIRAC Diploma to:
  Rob KNOOPS
  CERN-TH, Geneva, Switzerland

- Bruno FERRETTI Diploma to:
  Loredana BELLANTUONO
  University of Bari, Italy

- Richard P. FEYNMAN Diploma to:
  Anders KVELLESTAD
  Department of Physics, University of Oslo, Norway

- Vladimir N. GRIBOV Diploma to:
  Alisa KATANAEVA
  Saint-Petersburg State University, Saint-Petersburg, Russian Federation

- Robert HOFSTADTER Diploma to:
  Anton SOKOLOV
  Moscow State University, Moscow, Russian Federation

- Gunnar KÄLLEN Diploma to:
  Hugo PRAGER
  School of Physics and Astronomy
  University of Southampton, Highfield, UK

- Seymour J. LINDENBAUM Diploma to:
  Liron BARAK
  CERN, Geneva, Switzerland

- Yuval NE'EMAN Diploma to:
  Christian Lynge Holm CHRISTENSEN
  Albert Einstein Center, Institute for Theoretical Physics
  University of Bern, Switzerland

- Giuseppe P.S. OCCHIALINI Diploma to:
  Ke LI
  School of Physics, Shandong University and Institute of High Energy Physics, Beijing, China

- Oreste PICCIONI Diploma to:
  Ann-Kathrin PERREVOORT
  Heidelberg Physikalisches Institut
  Ruprecht-Karls-Universitat, Heidelberg, Germany

- Bruno PONTECORVO Diploma to:
  Arnd BEHRING
  DESY, Zeuthen, Germany

- Giampietro PUPPI Diploma to:
  Samuel ABREU
  The University of Edinburgh, Scotland, UK

- Isidor I. RABI Diploma to:
  Taras ZAGOSKIN
  V.N. Karazin Kharkiv National University, Kharkiv, Ukraine

- Giulio RACAH Diploma to:
  Bianca Letizia CERCHIAI
  Department of Mathematics, University of Milano, Italy

- Bruno ROSSI Diploma to:
  Mykola SAVITSKYI
  DESY-CMS, Hamburg, Germany

- Giorgio SALVINI Diploma to:
  Giovanni BENATO
  University of Zurich, Switzerland

- Julian S. SCHWINGER Diploma to:
  Yaroslav KLOPOT
  JINR, Dubna, Russia

- Victor F. WEISSKOPF Diploma to:
  Lukas VARNHORST
  Bergische Universität Wuppertal, Germany

- Eugene P. WIGNER Diploma to:
  Violetta SAGUN
  Bogolyubov Institute for Theoretical Physics, Kyiv, Ukraine

- Arthur S. WIGHTMAN Diploma to:
  Emanuele MESSINA
  University of Catania, Italy

- Bjorn H. WIIK Diploma to:
  Alexander VERKHEEV
  Joint Institute for Nuclear Research
  Dubna, Moscow Region, Russian Federation

- Kenneth G. WILSON Diploma to:
  Alexander LEITHES
  Queen Mary University of London, UK

- Chien Shiung WU Diploma to:
  Sarah CAMPBELL
  Columbia University
  Brookhaven National Laboratory, Upton, NY, USA

- Bruno ZUMINO Diploma to:
  Arnaud BAGUET
  Université de Lyon
  CNRS École Normale Supérieure de Lyon, France

NEW TALENTS AWARDS

The Awards for the best 'New Talent – Speakers' have been attributed to:

- **Mykola SAVITSKYI**
  DESY-CMS, Hamburg, Germany
  EXPERIMENTAL PHYSICS

- **Manuel MEYER**
  Institute for Theoretical Physics, University of Bern, Switzerland
  THEORETICAL PHYSICS

- Award for the BEST QUESTION to:
  Andràs LÀSZLO
  MTA Wigner Research Centre for Physics, Budapest, Hungary

- Award for the BEST SCIENTIFIC SECRETARY to:
  Shion CHEN
  The University of Tokyo, Japan

- Award for the BEST STUDENT to:
  Sarah CAMPBELL
  Columbia University, Brookhaven National Laboratory, Upton, NY, USA

PARTICIPANTS

Samuel ABREU *France/Portugal*	The University of Edinburgh, Scotland, UK
Malte ALBRECHT *Germany*	Ruhr-Universität Bochum, Germany
Alexandros ANASTASIOU *Cyprus*	Imperial College London, UK
Dustin ANDERSON *USA*	CERN, Geneva, Switzerland CALTECH, Pasadena, CA, USA
Evgeny ANDRONOV *Russia*	Saint-Petersburg State University, Russian Federation
Adam ARSLANALIIEV *Ukraine*	V.N. Karazin Kharkiv National University Kharkiv, Ukraine
Nikita ASTRAKHANTSEV *Russia*	MIPT and ITEP, Moscow, Russian Federation
Ibrahim ATAYEV *Azerbaijan*	Physics Institute of National Academy of Sciences Baku State University, Baku, Azerbaijan
Yegor AUSHEV *Ukraine*	National Technical University of Ukraine Kyiv Polytechnic Institute, Kyiv, Ukraine
Aleksandr AVDOSHKIN *Russia*	MIPT and ITEP, Moscow, Russian Federation
Arnaud BAGUET *France*	Université de Lyon CNRS - École Normale Supérieure de Lyon, France
Rinaldo BALDINI FERROLI *Italy*	INFN-LNF, Frascati, Italy
Liron BARAK *Israel*	CERN, Geneva, Switzerland
Arnd BEHRING *Germany*	DESY, Zeuthen, Germany
Loredana BELLANTUONO *Italy*	University of Bari, Italy
Pavel A. BELOV *Russia*	Saint-Petersburg State University, Russian Federation
Giovanni BENATO *Italy*	University of Zurich, Switzerland

Giorgio BENEDEK *Italy*	Milano-Bicocca University, Italy
Sergio BERTOLUCCI *Italy*	CERN, Geneva, Switzerland
Alessandro BETTINI *Italy*	INFN & Padoa University, Italy and Canfranc Underground Lab, Spain
Liudmila BISHLER *Russia*	Moscow State University, Russian Federation
Natalia BLIANKINSHTEIN *Russia*	MIPT, Moscow, Russian Federation
Rodrigo BUFALO *Brazil*	Department of Physics, University of Helsinki, Finland
Sarah CAMPBELL *USA*	Columbia University Brookhaven National Laboratory, Upton, NY, USA
Cosimo CANTINI *Italy*	CERN, Geneva, Switzerland
Matteo CENTIS VIGNALI *Italy*	Institute for Experimental Physics Hamburg University, Hamburg, Germany
Bianca Letizia CERCHIAI *Italy*	Department of Mathematics, University of Milano, Italy
Shion CHEN *Japan*	The University of Tokyo, Japan
Christian Lynge Holm CHRISTENSEN *Denmark*	Albert Einstein Center, Institute for Theoretical Physics University of Bern, Switzerland
Luisa CIFARELLI *Italy*	INFN & University of Bologna, Italy
Cesareo A. DOMINGUEZ *South Africa*	University of Cape Town, South Africa
Francesca DORDEI *Italy*	Physikalisches Institut, Heidelberg University, Germany
Vitalii DROHAN *Ukraine*	Kyiv State University and Bogolyubov Institute for Theoretical Physics, Kyiv, Ukraine
Tatiana DROZHZHOVA *Russia*	Saint-Petersburg State University, Russian Federation

Michael J. DUFF *UK*	Imperial College London, UK
Olga DUNIN-BARKOVSKAYA *Russia*	Sternberg Astronomical Institute Moscow State University, Russian Federation
Petr DUNIN-BARKOWSKI *Russia*	ITEP, Moscow, Russian Federation
Gia DVALI *Georgia, USA*	New York University, NY, USA
Sergio FERRARA *Italy*	CERN, Geneva, Switzerland
Pedro G. FERREIRA *Portugal*	University of Oxford, UK
Fernando FERRONI *Italy*	INFN, Rome, Italy
Dario FRANCIA *Italy*	Scuola Normale Superiore, Pisa, Italy
David FRENKLAKH *Russia*	MIPT and ITEP, Moscow, Russian Federation
Harald FRITZSCH *Germany*	Ludwig Maximilians Universitaet, Muenchen, Germany
Maria Vittoria GARGIULO *Italy*	University of Salerno, Italy
Paolo GIUBELLINO *Italy*	University of Turin, Italy
Ivan GNESI *Italy*	University of Turin, Italy and Centro Studi e Ricerche Enrico Fermi, Rome, Italy
Dmitrii GRAD *Russia*	Saint-Petersburg State University, Russian Federation
Maura GRAZIANI *Italy*	INFN and University of Perugia, Italy
Barbara GUERZONI *Italy*	University of Bologna, Italy
Nikolay GULITSKIY *Russia*	Saint-Petersburg State University, Russian Federation
Dieter HAIDT	EPJC–DESY,

Germany	Hamburg, Germany
Francis HALZEN *Belgium/USA*	University of Wisconsin-Madison, Madison, WI, USA
Gergö HAMAR *Hungary*	Wigner Research Centre for Physics, Budapest, Hungary
Zafrul HASAN *India*	Dept. of Applied Physics, ZH College of Engin. and Technology Aligarh Muslim University, Aligarh, India
Luis HERNANDEZ *Mexico*	Department of Physics, University of Cape Town Rondebosch, South Africa
Julie HOGAN *USA*	Fermilab, Batavia, IL, USA
Narmin HUSEYNOVA *Azerbaijan*	Institute for Physical Problems of Baku State University Baku, Azerbaijan
Evgenii IEVLEV *Russia*	Saint-Petersburg State University, Russian Federation
Gianluca INGUGLIA *Italy*	DESY, Hamburg, Germany
Iryna IVANCHENKO *Ukraine*	Bogoliyubov Institute for Theoretical Physics Kyiv, Ukraine
Mikhail IVANOV *Russia*	EPFL SB ITP LPPC, Lausanne, Switzerland
Inna IVANOVA *Russia*	Faculty of Mathematics of HSE Moscow, Russian Federation
Oleksii IVANYTSKYI *Ukraine*	Bogolyubov Institute for Theoretical Physics National Academy of Sciences, Kyiv, Ukraine
Peter JENNI *Switzerland*	CERN, Geneva, Switzerland
Samuel David JONES *UK*	School of Physics and Astronomy University of Southampton, Highfield, UK
Polina KAKIN *Russia*	Saint-Petersburg State University, Russian Federation
Daniil KALINOV *Russia*	Faculty of Mathematics of HSE Moscow, Russian Federation
Alisa KATANAEVA *Russia*	Saint-Petersburg State University, Russian Federation

Manjit KAUR *India*	Panjab University, Chandigarh, India
Anisa KHATUN *India*	Aligarh Muslin University, Aligarh (U.P.), India
Yaroslav KLOPOT *Ukraine*	JINR, Dubna, Russia
Rob KNOOPS *Belgium*	CERN-TH, Geneva, Switzerland
Kateryna KORNIEIEVA *Ukraine*	Joint Institute for Nuclear Research Dubna, Russian Federation
Christian P. KORTHALS-ALTES *The Netherlands*	CNRS-Luminy, Marseille, France
Mariia KOSTENKO *Russia*	Physical Faculty, Saint-Petersburg State University Saint Petersburg, Russian Federation
Vladimir KOVALENKO *Russia*	Saint-Petersburg State University, Russian Federation
Nikolay KOZYREV *Russia*	Bogolyubov Institute for Theoretical Physics Dubna, Moscow Region, Russian Federation
David KRALJIC *Slovenia*	Oxford University, UK
Daniel KROFF *Brazil/Portugal*	Instituto de Fisica, Universidade do Rio de Janeiro, Brazil
Anders KVELLESTAD *Norway*	Department of Physics, University of Oslo, Norway
Greg LANDSBERG *USA*	Brown University, Providence, RI, USA CERN, Geneva, Switzerland
Dag Toppe LARSEN *Norway*	Jagiellonian University, Kraków, Poland
Rasmus Normann LARSEN *Denmark*	Department of Physics and Astronomy Stony Brook University, Stony Brook, NY, USA
Andràs LÀSZLO *Hungary*	MTA Wigner Research Centre for Physics Budapest, Hungary
Nikita LEBEDEV *Russia*	Saint-Petersburg State University, Russian Federation

Leonid LEDENTSOV
Russia
Sternberg Astronomical Institute
Moscow State University, Moscow, Russian Federation

Alexander LEITHES
UK
Queen Mary University of London, UK

Ke LI
China
School of Physics, Shandong University
and Institute of High Energy Physics, Beijing, China

Sándor LÖKÖS
Hungary
Eötvös Loránd University, Budapest, Hungary

Árpád László LUKÁCS
Hungary
School of Technology
Aristotle University of Thessaloniki, Greece

Sergii MAMEDOV
Ukraine
G.V. Kurdyumov Institute for Metal Physics
NAS of Ukraine, Kyiv, Ukraine

Shahin MAMMADOV
Azerbaijan
Institute for Physical Problems of Baku State Univ.,
Baku, Azerbaijan

Alessio MARRANI
Italy
Department Physics and Astronomy G. Galilei
University of Padova, Italy

Nazarii MEDIUKH
Ukraine
Taras Shevchenko National University of Kyiv, Ukraine

Jack MEDLEY
UK
Higgs Centre for Theoretical Physics
The University of Edinburgh, UK

Emanuele MESSINA
Italy
University of Catania, Italy

Manuel MEYER
Switzerland
Institute for Theoretical Physics
University of Bern, Switzerland

Peter MINKOWSKI
Switzerland
University of Bern, Switzerland

Marco MIRRA
Italy
University of Naple, Italy

Andrey MOROZOV
Russia
ITEP, Moscow, Russian Federation

Tetiana MOSKALETS
Ukraine
V.N. Karazin Kharkiv National University
Kharkiv, Ukraine

Slava MUKHANOV
Russia
Max Planck Institute, Munich, Germany

Dmitrii NEVEROV
Saint-Petersburg State University, Russian Federation

Russia

Vasilisa NIKIFOROVA Institute for Nuclear Research
Russia Russian Academy of Sciences (INR RAS)
Moscow, Russian Federation

Francesco NOFERINI University of Bologna, Italy
Italy

Tamsin NOONEY Queen Mary University of London, UK
UK

Oleg Olegovich NOVIKOV Saint-Petersburg State University, Russian Federation
Russia

Dermot Oisin O'BRIEN School of Physics and Astronomy
UK University of Southampton, Highfield, UK

Alexandra OSTROVSKAYA Math Faculty of HSE, Moscow, Russian Federation
Russia

Patrick OWEN Imperial College London, UK
UK

Luca PARISI University of Salerno, Italy
Italy

Kirill PAVLENKO MIPT and ITEP, Moscow, Russian Federation
Russia

Maxim PERFILOV Moscow State University, Russian Federation
Russia

Ann-Kathrin PERREVOORT Heidelberg Physikalisches Institut
Germany Ruprecht-Karls-Universitat, Heidelberg, Germany

Vladimir A. PETROV IHEP - Division of Theoret. Physics
Russia Protvino, Russian Federation

Daniil PONOMARENKO National Research Nuclear University MEPhI
Russia Moscow, Russian Federation

Fedor POPOV ITEP, Moscow, Russian Federation
Russia

Stanislav POSLAVSKY Division of Theoretical Physics
Russia Protvino, Moscow Region, Russian Federation

Vladimir PROCHAZKA The University of Edinburgh, UK
Slovak

Maximiliano PUCCIO University of Torino, Italy
Italy

Anna RADOVSKAYA *Russia*	P.N. Levedev Physical Institute of RAS Moscow, Russian Federation
Stefano RAGAZZI *Italy*	Gran Sasso National Laboratory, L'Aquila, Italy
Anastasiia REBESH *Ukraine*	Bogolyubov Institute for Theoretical Physics of the NAS Kyiv, Ukraine
Alain RIAZUELO *France*	Institut d'Astrophysique de Paris, France
Lucio ROSSI *Italy*	CERN, Geneva, Switzerland
Dmytro RUBANOV *Ukraine*	V.N. Karazin Kharkiv National University Kharkiv, Ukraine
Igor RUDENOK *Ukraine*	Taras Shevchenko National University of Kyiv, Ukraine
Augusto SAGNOTTI *Italy*	Scuola Normale Superiore, Pisa, Italy
Violetta SAGUN *Ukraine*	Bogolyubov Institute for Theoretical Physics Kyiv, Ukraine
Adel SALAKH *Ukraine*	Kyiv State University and Bogolyubov Institute for Theoretical Physics Kyiv, Ukraine
Mykola SAVITSKYI *Ukraine*	Desy-CMS, Hamburg, Germany
Stefan SCHAEL *Germany*	Physikalisches Institut B, RWTH, Aachen, Germany
Elizaveta SEMENOVA *Russia*	Saint-Petersburg State University, Russian Federation
Alla SEMENOVA *Russia*	National Research Center Kurchatov Saint-Petersburg Nuclear Physics (PNPI) Gatchina, Leningrad District, Russian Federation
Andrey SHKERIN *Russia*	EPFL SB ITP LPPC, Lausanne, Switzerland
Katherin SHTEJER DIAZ *Cuba*	Department of Physics, University of Torino, Italy
Rosa SIMONIELLO *Italy*	CERN, Geneva, Switzerland

Anton SOKOLOV *Russia*	Moscow State University, Moscow, Russian Federation
Joan SOLÀ *Spain*	University of Barcelona, Spain
Nikita SOPENKO *Russia*	ITEP, Moscow, Russian Federation
Michel SPIRO *France*	IN2P3, CNRS, Paris, France
Stephen STOPYRA *UK*	Imperial College London, UK
Michael J. TANNENBAUM USA	Brookhaven National Laboratory, Upton, NY, USA
Oleg TELEGIN *Russia*	Math Faculty of HSE, Moscow, Russian Federation
Daniele TERESI *Italy*	School of Physics and Astronomy University of Manchester, UK
Gerardus 't HOOFT *The Netherlands*	Utrecht University, The Netherlands
Anna TOKAREVA *Russia*	Institute for Nuclear Research of RAS Moscow, Russian Federation
Lukas VARNHORST *Germany*	Bergische Universität Wuppertal, Germany
Alexander VERKHEEV *Russia*	Joint Institute for Nuclear Research Dubna, Moscow Region, Russian Federation
Ilya VILKOVISKIY *Kazhakhstan*	MIPT, Moscow, Russian Federation
Dmitry VLASENKO *Russia*	Physics Faculty, Southern Federal University Rostov-on-Don, Russian Federation
Weiping WANG *China*	University of Science and Technology of China Hefei, Anhui, China
Simon WEHLE *Germany*	Belle & Belle II, DESY, Hamburg, Germany
Lingjian WEN *China*	Institute of High Energy Physics, CAS, Beijing, China
Horst WENNINGER	CERN, Geneva, Switzerland

Germany

Christoph WIEGAND *Germany*	Karlsruher Institut fur Technologie, TTP Karlsruhe, Germany
Guy WILKINSON *UK*	Oxford University, UK
Victoria YANKELEVICH *Russia*	South Federal University Rostov-on-Don, Russian Federation
Gizem YAVUZ *Turkey*	Istanbul University, Beyazit/Fatih-Istanbul, Turkey
Ivan YELETSKIKH *Ukraine*	Joint Institute for Nuclear Research Dubna, Russian Federation
Muhammad YOUNIS *Pakistan*	Department of Physics COMSATS Institute of Information Technology Islamabad, Pakistan
Taras ZAGOSKIN *Ukraine*	V.N. Karazin Kharkiv National University Kharkiv, Ukraine
Vaclav ZATLOUKAL *Czech*	Czech Technical University in Prague Praha, Czech Republic
Yegor ZENKEVICH *Russia*	Moscow State University Moscow, Russian Federation
Sofia ZHIDKOVA *Russia*	Moscow State University, Moscow, Russian Federation
Antonino ZICHICHI Italy	CERN, Geneva, Switzerland; and INFN & University of Bologna, Italy
Radek ZLEBCIK *Czech*	Faculty of Mathematics and Physics Charles University at Prague, Czech Republic
Michele ZOCCALI *Italy*	Cambridge University, DAMTP, Cambridge, UK
Max ZOLLER *Germany*	Karlsruher Institute für Tecnologie, Karlsruhe, Germany